MW01672623

Instructor's Resource Guide for

CALCULUS: AN APPLIED APPROACH

FIFTH EDITION

Larson / Edwards

Bruce H. Edwards
Department of Mathematics
University of Florida

Houghton Mifflin Company Boston New York

Sponsoring Editor: Jack Shira
Managing Editor: Cathy Cantin
Senior Associate Editor: Maureen Ross
Associate Editor: Michael J. Richards
Assistant Editor: Carolyn Johnson
Supervising Editor: Karen Carter
Project Editor: Patty Bergin
Editorial Assistant: Christine E. Lee
Art Supervisor: Gary Crespo
Marketing Manager: Ros Kane
Marketing Assistant: Erin Dionne
Design: Henry Rachlin
Composition and Art: Meridian Creative Group

Printed in the U.S.A.

ISBN: 0-395-93342-0

23456789-EB-03 02 01 00 99

PREFACE

This instructor's guide is a supplement to *Calculus: An Applied Approach,* Fifth Edition, *Brief Calculus: An Applied Approach,* Fifth Edition, and *Brief Calculus: An Applied Worktext,* by Ron Larson and Bruce H. Edwards. All references to chapters, sections, theorems, definitions and examples apply to these texts.

Part I of the guide contains teaching strategies. In Notes to the New Teacher and the Chapter Summaries, I have made comments and suggestions based on my own teaching experiences.

Part II of the guide contains solutions to even-numbered text exercises.

Part III of the guide contains suggested solutions for the Special Features in the text: Group Discussions, Technology, Discovery, Warm Ups, and Post-Graduation Exam Questions.

I would appreciate very much hearing about any errors or other suggestions for improvement.

I would like to thank the staff at Larson Texts, Inc. for preparing this manuscript for publication.

Bruce H. Edwards
Department of Mathematics
University of Florida

CONTENTS

PART I

TEACHING STRATEGIES

In these notes you will find general suggestions to guide the new instructor in preparing for and conducting his or her class. For specific suggestions regarding teaching strategies, assignments, and topics that are traditionally difficult for students, refer to the chapter comments in this guide.

NOTES TO THE NEW TEACHER

The key to presenting a good lecture is to prepare for class. First of all, read the text carefully. Then make notes (to be used in class) so you won't neglect to cover anything. Work out the examples that you intend to use in detail. You might want to use the even-numbered problems as examples and assign the odd-numbered problems.

Work out the assignments that you give. This will help you anticipate some of the students' difficulties. It will also give you some idea of how long an assignment takes so you can make adjustments on future assignments.

You may find it worthwhile to have your students write up and turn in several carefully chosen problems each week. This will give you a chance to look over the notation the student is using as well as his or her approach to the problems. If you insist on correct notation and an organized format, the students will be forced to write better mathematics.

There are chapter tests available in this guide. You may choose to use these or make up your own. Whichever you do, grade the tests immediately and fairly, and return them to the students, the sooner the better. Students like to know what their grades are throughout the semester, so establish a grading policy, put it in writing, and stick to it.

There are several other policies that you need to formulate before the semester begins: make-up exams for tests missed, class attendance, disturbance in class, and cheating. One way to discourage make-up exams is by stating that any test missed will be made up during the last class meeting of the semester. Encourage regular class attendance. Do not put up with class disturbances. Most students will respond to a quiet comment about the disturbance; however, occasionally you may have to ask a student to leave. Cheating occurs everywhere. You should look for it and proctor your exams carefully. Review your school's policy regarding cheating with your students, then enforce that policy if necessary.

It is often helpful to the new teacher to find someone in the department who cares a lot about teaching and who has had experience with the course you are teaching. This person can guide you as to the pace of the class and length of assignments.

If English is not your native language, you may find the following suggestions helpful.

- At the beginning of each class (or at the first class of the week), outline the material you plan to cover in the class (or in the week of classes).
- When you use a math term for the first time, write it on the board.
- Write on the board all key terms that you use in your lectures.
- When you write definitions, theorems, and examples on the board, always label them as such.

Encourage class participation by taking questions during your lectures. You'll be able to judge student comprehension and adjust your lecture accordingly. When students come to you for help, you may want to discuss good study habits, such as those outlined below.

Above all, remember that you are in charge of the class, not your students. Good luck!

Suggested Study Habits for Students

- As you are reading the text or preparing homework assignments, make a note of anything you don't understand and ask about it in class.
- Take notes during class. After class, review your notes and rewrite them if necessary to fill in any missing details. If you don't understand a concept, question your instructor about it during the next class or during his or her office hours.
- As you start each chapter of the text, read the chapter overview to begin to develop an understanding of what you will learn in the chapter.
- Read the assigned material in each section before it is covered in class. After attending class, reread the assigned section to make sure that you understand the material.
- Remember that learning calculus is a step-by-step process. Always keep up with and complete your assignments because you must understand each topic in order to learn the next.
- When you are working problems for homework assignments, show every step in your solution. Then, if you make an error, it will be easier for you or your instructor to find it.
- Study with another student or small group of students, especially when preparing for a test.
- On the day of the test, allow yourself plenty of time to get to the testing location.
- Review your tests carefully, making sure you understand where and how you made errors.
- If you are having a problem, be sure to take advantage of your school's tutoring services for extra help.

For additional teaching strategies, consult the following sources.

Case, Bettye Ann. 1988. *Responses to the Challenge: Keys to Improved Instruction by Teaching Assistants and Part-Time Instructors*. Providence, RI: Mathematical Association of America.

Grabe, Mark, and Grabe, Cindy. 1998. *Learning with Internet Tools: A Primer.* Boston, MA: Houghton Mifflin Company.

Orlich, Donald C., et al. 1998. *Teaching Strategies: A Guide to Better Instruction*, Fifth Edition. Boston, MA: Houghton Mifflin Company.

Ornstein, Allan et al. 1997. *Foundations of Education*, Sixth Edition. Boston, MA: Houghton Mifflin Company.

Ryan, Kevin, and Cooper, James. 1998. *Those Who Can Teach*, Eighth Edition. Boston, MA: Houghton Mifflin Company.

Consult current math journals such as *The Mathematics Teacher, College Mathematics Journal, Computers in Mathematics and Science Teaching,* or *MAA-AMS Notes* for specific teaching strategies in a variety of areas.

Guidelines for Using *Brief Calculus: An Applied Worktext*

Philosophy

Many reform books ask professors to take greater responsibility for students' learning. *Brief Calculus: An Applied Worktext* asks students to take greater responsibility for their own learning, which is the only way they will really learn.

Core Thread

The Core Thread is the recommended path through the worktext. It is designed to provide students with the structure they need to develop strong math study skills—while they are learning the concepts of this course. The goal is to focus students' attention on filling in the most essential blanks and on solving the assigned homework problems to learn the key ideas. With this unique format, students will interact with each example instead of skimming through it and glossing over important details.

The Core Thread consists of Example Steps, Try Its, and Take Ownership boxes with blanks for students to write in their answers. Selected steps in solutions to examples (Example Steps) are omitted, forcing students to provide the next logical step in the problem. After every example, students have the opportunity to solve a related Try It on their own to reinforce the concepts presented in the preceding example. Finally, Take Ownership boxes encourage students to make general comments about a concept or rule in their own words. This enhances understanding and further reinforces the ideas in the section. All Core Thread features appear in the same color so students can easily identify them.

Students are also expected to write in answers to selected problems from end-of-section exercises on the answer sheets provided. Since the answer sheets are two-sided, the student can cut out the strip and hand it in if the professor so chooses without losing any text material or exercises.

As their confidence levels increase, students may go back and read the exposition and/or write their answers in the blanks in selected Technology boxes, Discovery boxes, and Study Tips to enhance their understanding.

Answers to all of the write-ins are provided at the end of the worktext in the form of a double-sided page for each section. Students are expected to tear out this page and check their answers as they work through the Core Thread (and optional additional write-ins) in each section.

Web Component

The worktext also features Web Case Studies and Web Exercises for each chapter. The Web Case Studies show how a real person or company uses mathematics to help make an informed consumer or business decision. The Web Case Study consists of a company profile, background information, and related links on the book's web site, as well as related problems in the worktext, along with room to write in the answers. Open-ended research problems called Web Exercises are identified on each chapter opener in the text. On the web site, guided questions help students do their research. See HYPERLINK http://www.hmco.com for the Web Case Studies, Web Exercises, and additional aids for students and instructors.

Annotated Instructor's Edition

Answers to all of the write-ins and to the selected end-of-section exercises are shown in place in the Annotated Instructor's Edition.

An Additional Note

In addition to the Core Thread and the other interactive features described here, *Brief Calculus: An Applied Worktext* contains all of the content from *Brief Calculus: An Applied Approach*, Fifth Edition.

Technology-Required Exercises

Many exercises in the Applied Calculus Series by Ron Larson and Bruce H. Edwards can be solved using technology. A graphing calculator icon in the texts identifies all exercises in which the student is instructed to use a graphing utility or symbolic computer algebra system for solutions or to check solutions. The solutions of many other exercises may also be facilitated by the use of appropriate technology. For your convenience in planning homework assignments, these exercises are listed below.

Chapter 0 A Precalculus Review

Section	Page	Exercises
0.3	0-18	51, 52
0.5	0-32	49

Chapter 1 Functions, Graphs, and Limits

Section	Page	Exercises
1.1	9	31, 32, 39, 40
1.2	21–23	5–10, 21–34, 43–50, 51–58, 61–65, 70–74
1.3	34–35	35–44, 49–64, 69, 72, 76–82
1.4	45–48	17–22, 47–52, 66, 70–76
1.5	60	49–52, 55, 57–61
1.6	70–71	41, 42, 47, 48, 50, 52, 53
Review	76–79	16, 25, 26, 31–34, 47, 48, 52, 59–64, 105, 106

Chapter 2 Differentiation

Section	Page	Exercises
2.1	92	55–58, 63–65
2.2	104	59, 60
2.3	117–119	9, 10, 33, 36, 37, 44–46
2.4	129–130	35–40, 45–48, 55–58
2.5	138–139	31–40, 61–66, 72
2.6	146	43–44, 54
2.7	153	27–32, 37, 38
2.8	161	23, 25
Review	167–169	33–38, 43, 44, 75, 96

Chapter 3 Applications of the Derivative

Section	Page	Exercises
3.1	180	19–28, 35, 39
3.2	189–190	13–18, 27–30, 31–34, 41–43, 45
3.3	198–200	15–18, 31–42, 47–50, 53, 54, 56–62
3.4	209	36, 37
3.5	217–219	13, 14, 23, 27–32, 39, 40
3.6	229–230	35, 36, 58, 60, 63
3.7	238–239	21–30, 41–44, 51–53
3.8	247	25–28
Review	252–255	9–29, 31–34, 45, 47, 49, 59–64, 71, 89, 90, 93

Chapter 4 Exponential and Logarithmic Functions

Section	Page	Exercises
4.1	267–268	41–44, 57–64
4.2	276	31–35, 40–43
4.3	283–285	19–22, 76, 78–80
4.4	294	67–72, 79, 81
4.5	303	35–38
Review	309–311	25, 26, 65, 66

Chapter 5 Integration and Its Applications

Section	Page	Exercises
5.1	322–324	19–24, 37–42, 72
5.2	331–332	27–32, 47, 48, 55–58
5.3	338–339	25–32, 53, 54
5.4	349–350	35–38, 51–56, 79–82
5.5	358–359	31–34, 49
5.6	365–366	13–16, 25, 26, 31
5.7	373	32–34
Review	378–381	31, 32, 60, 71, 72, 77, 79–82

Chapter 6 Techniques of Integration

Section	Page	Exercises
6.1	391	37–44, 55, 58, 59
6.2	399–401	27–30, 49–54, 65, 66
6.3	410–411	45–53
6.4	421	35–40, 59, 60
6.5	430–431	19–25, 35–38, 40–42
Review	446–449	21–24, 28, 33, 34

Chapter 7 Functions of Several Variables

Section	Page	Exercises
7.2	469	53–56
7.4	489	75
7.7	517–519	5–14, 19–42, 47
7.8	527	47–52
Review	540–543	31–34, 81–86

Chapter 8 Trigonometric Functions

Section	Page	Exercises
8.1	552	9–20
8.2	563–564	33–40, 47–56, 68–70
8.3	573–575	27–32, 47–50, 55–60, 65–69
8.4	584–585	55–60, 63, 67–78
8.5	593–594	39–46, 55, 56, 58, 61, 65–68
8.6	601–603	11–14, 61–64, 70, 71
Review	608–611	5–12, 55, 56, 79, 80, 96

Chapter 9 Probability and Calculus

Section	Page	Exercises
9.1	622–623	20, 39
9.2	629–630	1–14, 29–32, 37
9.3	639–640	7–10, 19–26, 52–55
Review	646	21–24, 49, 50

Chapter 10 Series and Taylor Polynomials

Section	Page	Exercises
10.1	656–658	19–26, 59, 61, 72
10.2	669–670	17–20, 31–40, 49
10.3	678	53, 54
10.4	688	55–58
10.5	697–698	9–16
10.6	706	17–26
Review	712–714	9–12, 63–66, 91–94, 96

Chapter Summaries

CHAPTER SUMMARIES

CHAPTER 0
A Precalculus Review

Section Topics

0.1 **The Real Line and Order**—the real number line; order and intervals on the real line; solving inequalities and applications.

0.2 **Absolute Value and Distance on the Real Line**—concept of absolute value of a real number; distance on the real line; intervals defined by absolute value; midpoint of an interval.

0.3 **Exponents and Radicals**—expressions involving exponents or radicals; domain of an algebraic expression.

0.4 **Factoring Polynomials**—techniques of factoring; factoring polynomials of degree three or more; the rational root theorem.

0.5 **Fractions and Rationalization**—operations with fractions; radical expressions; rationalization techniques.

Chapter Comments

Chapter 0 should be a review chapter for most students, since it covers basic algebraic concepts from high school algebra. It should be covered quickly or skipped entirely; it is an optional chapter. However, there may be students in your class who may benefit from reviewing some of the material in this chapter. Important concepts needed for later work include: the definition of absolute value and absolute value inequalities from Section 0.2; laws of exponents, the definition of fractional exponents, and the type of factoring exhibited in Example 2 on page 0-21 from Section 0.3.

You may wish to discuss the domain of a radical algebraic expression in Section 0.3 in order to make Section 1.4 on domain of functions procede more smoothly. Students should be familiar with the type of factoring done in Section 0.4, but some may be a bit rusty on the Rational Root theorem and/or synthetic division. Usually at least two examples of these are necessary (be sure that the leading coefficient of your polynomial is not always a "1"). These concepts may be needed to solve some exercises in later chapters (for example, finding critical points). In Section 0.5, make sure the students know how to rationalize radical expressions (numerical and algebraic) so that their answers in the future are in the proper form (it is very frustrating for students to get the right answer and then look in the back of the book to see a "different" answer, which automatically makes them think that their answer is wrong!). Point out the rationalization techniques listed in Section 0.5 for fractions containing radicals. You may wish to use the review exercises at the end of Chapter 0 to ascertain whether you actually need to cover this material in class or not.

CHAPTER 1
Functions, Graphs, and Limits

Section Topics

1.1 **The Cartesian Plane and the Distance Formula**—the Cartesian Plane; the distance formula; the midpoint formula.

1.2 **Graphs of Equations**—the graph of an equation; intercepts of a graph; circles; points of intersection; mathematical models.

1.3 **Lines in the Plane and Slope**—linear equations and slope; finding an equation of a line; parallel lines and perpendicular lines.

1.4 **Functions**—definition of function; the graph of a function; function notation; combinations of functions; inverse functions.

1.5 **Limits**—the limit of a function; properties of limits; techniques for evaluation of limits; one-sided limits; unbounded behavior.

1.6 **Continuity**—definition of continuity; continuity on a closed interval; the greatest integer function.

Chapter Comments

Chapter 1 is both a continuation of the review material from Chapter 0 (for e.g. Section 1.3) and an introduction of new concepts vital to the understanding of calculus (Sections 1.5 and 1.6). In Section 1.1 be sure to include the distance formula in your lecture, since this is used to develop the equation of a circle in Section 1.2. It is important that the students are able to take the general form of the equation of a circle and write it in standard form, since this involves completing the square, a tool that calculus students need. Use a simple equation like $y = x^2 - 4x + 3$ to discuss x- and y-intercepts (this equation can be graphed easily on a graphing calculator). By increasing the constant term of this polynomial you can easily get a parabola that has no x-intercepts. Be sure that the students can solve nonlinear systems of two equations in x and y—this will facilitate finding the area between two curves in Chapter 4.

In Section 1.3 require the memorization of the definition of slope, carefully distinguishing between zero slope and undefined slope (i.e. a vertical line). Emphasize point-slope form of the equation of a line, since this form will be needed when writing tangent line equations in Chapter 2. In Section 1.4, make your students memorize your favorite definition of function (they often don't distinguish between equation and function). They should know how to use the vertical line test to see whether a graph represents a function or not. Students need to be familiar with all the topics discussed in Section 1.4, including domain and range, functional notation, composition, and inverses. Looking ahead, a good example to do here is calculating $f(x + \Delta x)$ so that the students get comfortable with the "delta" notation.

The topic of limits is usually difficult for most students at the start. Carefully go over the definition of limit and use as many examples as time permits. Use the graph of the function to explain the behavior. Be sure to cover the properties of limits and operations with limits (legal ways to evaluate limits!). The examples are needed so that students do not get into the habit of just plugging in the number that x is approaching; the fact that the limit of a polynomial can be found by just plugging in the number should also come out in these examples. Indeterminate forms may also be used as examples, where you need to point out that without knowing the algebra it is impossible to evaluate the limit (a nice time to point out that $0/0$ is an indeterminate form). Be sure to cover at least one example where cancelization is used and one where rationalization is used, since these appear in the exercises. Stress the fact that a limit only exists if it is a finite real number (infinity is not a number!); this is discussed via Examples 9 and 10 in Section 1.5. If time permits, talk about limits of the form $0/0$, $0/C$, $C/0$ where $C \neq 0$.

In Section 1.6 make sure your students know how to show a function $f(x)$ is continuous at $x = a$ using the three part definition (I require them to memorize it). A good heuristic definition is that the graph of a continuous function can be drawn without lifting up the pen. A discussion of removable versus nonremovable discontinuities will aid students with finding vertical asymptotes. Don't forget to point out classes of functions that are always continuous (e.g. polynomials).

CHAPTER 2
Differentiation

Section Topics

2.1 **The Derivative and the Slope of a Graph**—tangent lines and the slope of a curve; finding the slope of a curve by the limit process; the derivative of a function; differentiability and continuity.

2.2 **Some Rules for Differentiation**—the constant rule; the power rule; the constant multiple rule; the sum and difference rule.

2.3 **Rates of Change: Velocity and Marginals**—average rate of change; instantaneous rate of change; marginals.

2.4 **The Product and Quotient Rules**—the product rule; the quotient rule.

2.5 **The Chain Rule**—the chain rule; the general power rule; algebraic simplification techniques; summary of differentiation rules.

2.6 **Higher-Order Derivatives**—second-, third-, and higher-order derivatives; acceleration.

2.7 **Implicit Differentiation**—implicit and explicit functions; implicit differentiation.

2.8 **Related Rates**—functions that change with time; applications involving finding the time rate of change of a function.

Chapter Comments

The material presented in Chapter 2 forms the basis for the remainder of differential calculus. Much of it should be committed to memory, beginning with the definition of the derivative found in Section 2.1. Finding the derivative using the four-step process will help your students remember this definition. Your students need to understand, using a simple function like $y = f(x) = x^2$, that the tangent line slope is the limiting value of the secant line slopes as $\Delta x \to 0$. They should be able to write the equation of a tangent line to a curve at a given point; make sure that they do not use the function $f'(x)$ as the value of the slope. They must distinguish between the derivative function and the value of the derivative at a given point.

As you teach the chapter, vary your notations for the derivative, using dy/dx, $f'(x)$, and y' so that the students get comfortable with these different notations. Terminology is also important; instead of always saying "take the derivative" you can say "differentiate". At this time you also could discuss the contributions of Leibniz and Newton to the discovery of calculus.

Section 2.2 introduces some rules for finding derivatives (i.e. "shortcuts") that simplify the process. Section 2.3 considers two important applications of the derivative: velocity and marginals. Be sure that the students equate average rate of change with secant line slope and instantaneous rate of change with tangent line slope.

Two additional rules, the product rule and the quotient rule, are presented in Section 2.4. Be sure that the students learn the rules in words, not just in symbolic form. For example, I use $(FS)' = FS' + SF'$ to help the students with the product rule.

The chain rule will require at least two lectures to cover completely. Students need considerable practice with this and the accompanying algebra. Many students can take the derivative of a function like $f(x) = x^2\sqrt{3 - x^2}$ but cannot simplify it. Insist that they learn to factor and write the answer without negative exponents. Strive to get the answer in the form given in the answer key. This will help them later when they need to set the derivative equal to zero.

Section 2.6 considers higher-order derivatives. Be sure to point out to students that there are practical uses for these, such as acceleration discussed in this section and the approximate integration formulas discussed in Section 6.5.

Implicit differentiation is often a difficult topic for students to grasp. Be sure to have them think of y as a function of x and therefore y^2 is equivalent to $[f(x)]^2$. In this way Section 2.7 can be related to the chain rule covered in Section 2.5.

Section 2.8 is a good application of the chain rule–solving related rate problems. Start out with a familiar formula such as $A = \pi r^2$ and find dA/dt; take care to discuss units and also the significance of a negative sign in an answer. Go through a complete problem step by step; the procedure outlined in the text on page 156 should help. Choose the assignment carefully.

CHAPTER 3
Applications of the Derivative

Section Topics

3.1 **Increasing and Decreasing Functions**—increasing and decreasing functions; critical numbers and their use; applications.

3.2 **Extrema and the First-Derivative Test**—relative extrema; the first-derivative test; absolute extrema; applications of extrema (max/min problems).

3.3 **Concavity and the Second-Derivative Test**—concavity and points of inflection; the second-derivative test; application: law of diminishing returns.

3.4 **Optimization Problems**—optimization problems using the derivative.

3.5 **Business and Economic Applications**—optimization problems in business and economics; price elasticity of demand; summary of business terms and formulas.

3.6 **Asymptotes**—vertical asymptotes and infinite limits; horizontal asymptotes and limits at infinity; applications.

3.7 **Curve Sketching: A Summary**—curve sketching using the first and second derivatives; summary of simple graphs.

3.8 **Differentials and Marginal Analysis**—differentials; marginal analysis; formulas for differentials; error propagation.

Chapter Comments

The first two sections are basic applications of the first derivative: regions of increase/decrease, relative extrema, and the first-derivative test. Be sure that the student can distinguish between a relative extremum and an absolute extremum. The third section gives the basic results concerned with the second derivative: concavity (i.e. the shape of a curve) and inflection points. It also gives the second-derivative test for relative extrema. All of this material is extremely important for a complete understanding of the rest of the chapter. Go over the first three sections slowly and carefully, using a plethora of examples to illustrate the concepts. Students often confuse points on the curve with critical numbers and inflection numbers. I find a simple chart for the first and second derivative (with critical numbers and inflection numbers) is very useful. Frequent review is also a good idea, along with a comparison of the graphs of the function $f(x)$ and its first and second derivatives.

The applied extrema problems in Section 3.4 should go well if the students have a thorough understanding of the first three sections. Stress the difference between a primary equation (the one which is to be maximized/minimized) and a secondary equation. It is also important to have the function to be maximized/minimized in terms of one variable. Follow the guidelines on page 202. Do not hurry through this material. Section 3.5 should require less time, since it is essentially the same as Section 3.4 with an economics flavoring.

Many important ideas are covered in Section 3.6. Point out to the students that vertical asymptotes are non-removable discontinuities, which were discussed in Section 1.6. In order to present horizontal asymptotes, you need first to discuss limits at infinity. Put three examples on the board, like the ones in Example 6 on page 225. Compare the results of taking the limit and finish with the rules for limits at infinity for rational functions found on page 225. From these results discuss horizontal asymptotes. Make sure that the students write the *equations* of vertical and horizontal asymptotes: i.e. $x = 3$ or $y = 8$.

Section 3.7 makes use of the concepts learned in the first three sections (plus previous material) to construct the graphs of various types of functions. It will be helpful if the discussion of a curve (intercepts, symmetry, asymptotes, increasing/decreasing, relative extrema, concavity and points of inflection) would be included in a step-by-step procedure, with the last being the actual sketching of the curve, using all of the calculus information to obtain an accurate sketch. You will need several days to discuss this thoroughly, perhaps concentrating on a different type of function each day (polynomial, rational, radical).

Section 3.8 discusses differentials. Students need to be comfortable with differentials in order to integrate, so do not go too quickly through this material. The basic emphasis is the "tangent line approximation" to a function $y = f(x)$; that is, if delta x is small, then $f(x + \Delta x) = y + \Delta y \approx y + f'(x)\,dx$.

CHAPTER 4
Exponential and Logarithmic Functions

Section Topics

4.1 **Exponential Functions**—exponential functions; natural exponential functions; compound interest and other applications.

4.2 **Derivatives of Exponential Functions**—rules for differentiation of exponential functions; applications of derivatives of exponential functions.

4.3 **Logarithmic Functions**—logarithms to the base e; properties of logarithms; solving exponential and logarithmic equations; applications.

4.4 **Derivatives of Logarithmic Functions**—derivatives of logarithmic functions; using properties of logarithms to simplify differentiation problems.

4.5 **Exponential Growth and Decay**—exponential growth and decay; population growth; continuously compounded interest.

Chapter Comments

Students should be familiar with exponential and logarithmic functions but frequently are not, so present these functions as though they were new ideas. The first two sections concentrate on the exponential functions, with a natural emphasis on the base e. Be sure to use the exponential functions to reiterate previously covered ideas, such as increasing/decreasing functions, relative extrema, and concavity.

Logarithmic functions are introduced in Section 4.3; remind students that the logarithm of a negative number is not defined. Emphasize the "function and inverse" relationship of the exponential and logarithmic functions. You may wish to use the implicit function $x = e^y$ to derive the derivative of the logarithmic function $y = \ln x$. Point out to students, using an example like $y = \ln\sqrt{\dfrac{x^2 + 4x}{x^3 + 3}}$, the advantage of using the properties of logarithms to rewrite the function *before* they differentiate it.

Section 4.5 does not require much time and you can spend as little time on this section as desired. Try to do at least one example of exponential growth and decay (continuously compounded interest and/or radioactive decay) so that the students do understand the diversity of applications of the natural exponential function.

CHAPTER 5
Integration and Its Applications

Section Topics

5.1 **Antiderivatives and Indefinite Integrals**—antiderivatives; integral notation; finding antiderivatives; particular solutions; applications.

5.2 **The General Power Rule**—the general power rule; substitution.

5.3 **Exponential and Logarithmic Integrals**—rule for integration of exponential functions; rule for integrating du/u.

5.4 **Area and the Fundamental Theorem of Calculus**—area; definite integrals and the fundamental theorem of calculus; properties of definite integrals; marginal analysis; average value of a function; integration of even and odd functions.

5.5 **The Area of a Region Bounded by Two Graphs**—area of a region bounded by two curves; consumer and producer surplus; applications.

5.6 **The Definite Integral as the Limit of a Sum**—approximating the definite integral with the midpoint rule; the definite integral as the limit of a sum.

5.7 **Volumes of Solids of Revolution**—the disk method; the washer method; applications.

Chapter Comments

The first section of this chapter stresses correct terminology and notation. Recognizing different ways to express an algebraic expression is important here; make sure the students know that all radical expressions must be changed to exponential form. When you introduce antiderivatives, insist upon having the proper differential when students write an integral problem. Take off points on homework and quizzes for incorrect notation.

Your students should memorize the basic integration rules in Section 5.1. It is probably better that the students learn these in *words* instead of functions (i.e. integral of a sum is sum of integrals, etc.). Be sure to do a number of examples in which it is necessary to do preliminary algebra on the integrand before the rules of integration can be applied. Point out that all integration problems can be checked directly by differentiation of the answer (to produce the integrand).

The general power rule is introduced in Section 5.2. If your students have a good understanding of the chain rule, then this section should not give students a great deal of trouble. Work many examples and allow plenty of time for students to comprehend this material. An example such as $\int x(x^2 + 3)^2\,dx$ (which can be done either by a *u*-substitution or by multiplying out the integrand) may show students that the *u*-substitution technique is OK. Again differentiation of the answer is the final proof that the antiderivative you have found is indeed the correct one. I prefer to always use the technique of substitution, which is also presented in this section. Be sure to cover it, since it will come up again in Chapter 6.

Section 5.3 presents exponential and logarithmic integrals. Students tend to confuse problems such as $\int \frac{2x\,dx}{(x^2 + 3)^2}$ and $\int \frac{2x\,dx}{x^2 + 3}$; be sure to point out the difference. Do a lot of *u*-substitution problems to emphasize this important technique.

Section 5.4 presents the fundamental theorem of calculus. Be sure that your students know this theorem and can use it to evaluate definite integrals. Remind students that they learned how to graph functions in Chapter 3 and you expect them to use this knowledge when doing area problems. Be sure that the students use the right limits for the area problems; do not accept integrals with the upper and lower limits interchanged. It is important to do a few simple examples where the definite integral does **NOT** give the area; remind the students that the definite integral gives the area only when the function being integrated is positive and continuous over the region of integration. The remaining topics in this section (marginal analysis, average value, and integration of even and odd functions) may be covered as time permits. It is important that the students grasp the distinction between an *indefinite* integral (answer is a function) and a *definite* integral (answer is a number).

In covering Section 5.5 be sure to use the representative rectangle (a slice of area) when doing all problems so that the students set up the problems correctly. Remind the students that area (and therefore ΔA) cannot be negative!

Section 5.6 presents the midpoint rule; a good argument for numerical integration techniques is an impossible problem such as $\int_0^1 e^{-x^2}\,dx$. The definition of an integral as the limit of a sum of the form $\lim_{n\to\infty} \sum_{i=0}^{n} f(x_i)\Delta x$ can be discussed as much as time permits. You may also wish to bring in the idea of the upper and lower sums, which always "sandwich in" the value of the integral and make the proper comparison with inscribed and circumscribed rectangles. You may wish also to mention the fact that the two most important calculus tools (derivatives and definite integrals) have definitions involving *limits*. The trapezoidal rule is done in the exercises; it is up to you whether you wish to make use of this or not.

The last section discusses volume problems. Emphasize that drawing a rough sketch of the area to be rotated is enough to set up the volume problems correctly.

CHAPTER 6
Techniques of Integration

Section Topics

6.1 **Integration by Substitution**—review of basic integration formulas; integration by substitution; substitution and definite integrals.

6.2 **Integration by Parts and Present Value**—integration by parts; present value.

6.3 **Partial Fractions and Logistics Growth**—integration of rational functions using partial fractions; solution of logistics differential equations.

6.4 **Integration Tables and Completing the Square**—integration using tables; reduction formulas; completing the square.

6.5 **Numerical Integration**—trapezoidal rule; Simpson's rule; error analysis of these methods.

6.6 **Improper Integrals**—improper integrals with infinite limits of integration; improper integrals with discontinuous integrands.

Chapter Comments

This chapter introduces some more sophisticated techniques of integration, starting with the technique which should already be familiar—integration by substitution. The first section contains a review of all integration covered to this point, as well as a detailed description of the substitution technique. Be sure that the students realize that when a change of variable is made (i.e. a u-substitution), the limits must change also if the students do not wish to change back to the original variable.

When presenting integration by parts, point out to your students that this technique is particularly useful for integrands that are *products* of algebraic and transcendental functions (and in fact the derivation of the formula $\int u\,dv = uv - \int v\,du$ comes directly from the product rule for differentiation). Do a number of examples and take time to go over the exercises from this section.

Partial fractions will take at least two days; one day to go over the necessary algebra and the second day to actually do integration problems. Stress the fact that this technique should only be used on algebraic integrands with the degree of the numerator less than the degree of the denominator.

Section 6.4 looks at integration by tables. Frequently algebra must be done on the integrand in order to get the problem at hand to match a formula in the table. One algebra technique needed is completing the square, previously discussed in Section 1.2 with the standard equation for a circle.

A good introduction to the numerical integration section is an impossible problem such as $\int_0^1 e^{-x^2}\,dx$ or $\int_0^2 \sqrt{1 + x^3}\,dx$. This is a good opportunity to convince students that not all functions have an antiderivative. If you are pressed for time, the derivation of the formula for Simpson's rule can be eliminated, but not the rule itself. I usually stress the idea that the trapezoidal rule is essentially a linear approximation of the integrand over each subinterval while Simpson's rule is a quadratic approximation.

Example 1 of Section 6.6 is a good introduction of improper integrals. Be sure to carefully define the two types of improper integrals: those with discontinuous integrands and those with infinite limits of integration. Show that overlooking the fact that a function has a discontinuity in the region of integration can lead to erroneous results, e.g. $\int_{-1}^{1} \frac{dx}{x^2}$.

CHAPTER 7
Functions of Several Variables

Section Topics

7.1 **The Three-Dimensional Coordinate System**—the three-dimensional coordinate system; distance and midpoint formulas; the equation of a sphere.

7.2 **Surfaces in Space**—equations of planes in space; quadric surfaces.

7.3 **Functions of Several Variables**—functions of several variables; the graph of a function of two variables; applications.

7.4 **Partial Derivatives**—partial derivatives of a function $z = f(x, y)$; interpretation of partial derivatives; partial derivatives of a function of three variables; higher-order partials.

7.5 **Extrema of Functions of Two Variables**—relative extrema of a function $z = f(x, y)$; the first-partials test for relative extrema; the second-partials test for relative extrema; applications of extrema of functions of two variables.

7.6 **Lagrange Multipliers**—Lagrange multipliers with one constraint; Lagrange multipliers with two constraints.

7.7 **Least Squares Regression Analysis**—measuring the accuracy of a mathematical model; least squares regression line; least squares regression quadratic.

7.8 **Double Integrals and Area in the Plane**—double integrals; finding area using double integrals.

7.9 **Applications of Double Integrals**—volume of a solid region; average value of a function over a region.

Chapter Comments

Most of the ideas presented in Section 7.1 are extensions of similar ideas presented in the *xy*-plane, so they should look familiar to your students. However, students often have difficulty visualizing lines and surfaces in 3-D space. So as you present Sections 7.2 and 7.3, use as many visual aids as possible, such as transparencies.

Make sure that the students memorize the definition of $\partial f/\partial x$ and $\partial f/\partial y$ just as they memorized the definition of dy/dx in Chapter 2. You may want to mention that the partial of f with respect to x is just the rate of change of $f(x, y)$ in the x-direction; also mention the fact that this is just the ordinary tangent line to the curve $z = f(x, y)$ restricted to the plane $y = c$.

Section 7.5 examines extrema in 3-D space. The ideas and terminology parallel what was done in the *xy*-plane. Be sure to note that critical points occur where one partial derivative is undefined or where both partials are zero. The idea of a saddle point can be nicely discussed with a curve like $z = y^2 - x^2$. Point out to students that the second-partials test for extrema can fail, just like the second-derivative test for functions of one variable.

In order to do Section 7.6 students need to solve systems of equations. Encourage them to be as creative as possible in this regard. You may wish to mention that in general, only a linear system will have a unique solution. A nonlinear system may have a number of solutions, depending on the degree of the equation.

The method of least squares, discussed in Section 7.7, is an important technique for finding the "best" linear model for a set of data. The quadratic version may be omitted if you are pressed for time.

I have found that the most difficult part of multiple integration for students is setting up the limits. Go slowly and carefully over this every time, emphasizing the vertical element with the *dydx* order and the horizontal element with the *dxdy* order. Point out that the variable of integration cannot appear in either limit of integration and that the outside limits must always be constants. Also stress the fact that the order of integration (Fubini's theorem) does not affect the value of the integral. However, it does often affect the difficulty of the integration. Sketching the region is an integral part of making the decision on the order of the integration.

CHAPTER 8
Trigonometric Functions

Section Topics

8.1 **Radian Measure of Angles**—angles and degree measure; radian measure; triangles.

8.2 **The Trigonometric Functions**—the six trigonometric functions; trigonometric identities; evaluation of trigonometric functions; solving trigonometric equations.

8.3 **Graphs of Trigonometric Functions**—graphs of trigonometric functions; limit of a trigonometric function; applications.

8.4 **Derivatives of Trigonometric Functions**—derivatives of trigonometric functions; relative extrema of trigonometric functions; applications.

8.5 **Integrals of Trigonometric Functions**—integration of trigonometric functions.

8.6 **L'Hôpital's Rule**—indeterminate forms and limits; L'Hôpital's rule.

Chapter Comments

In Section 8.1, students are introduced to radian measure, which is essential for doing calculus with trigonometric functions; stress this point to the students. Be sure that they know the special angles $\frac{\pi}{6}$, $\frac{\pi}{4}$, $\frac{\pi}{3}$, $\frac{\pi}{2}$ and their multiples up to 2π without hesitation. In Section 8.2, it is necessary to memorize the definition of the trigonometric (sometimes called circular) functions and identities given in Section 8.2. It takes some work but the payoff is well worth the effort.

If you are pressed for time, just cover the six basic graphs in Section 8.3 and omit the changes in amplitude and period. In Section 8.4, if you do not have time to prove all of the derivative formulas, present a few in class and assign the rest for homework. The proofs are straightforward and enjoyable for the students. When covering the derivative formulas, you might wish to point out the function-cofunction correspondence thus:

Derivatives of 3 basic functions	Derivatives of their cofunctions
$d/dx(\sin x) = \cos x$	$d/dx(\cos x) = -\sin x$
$d/dx(\tan x) = \sec^2 x$	$d/dx(\cot x) = -\csc^2 x$
$d/dx(\sec x) = \sec x \tan x$	$d/dx(\csc x) = -\csc x \cot x$

The integration of trigonometric functions is not easy for most students. Go slowly through the section and make sure that the students know the six integral formulas given in Section 8.6. Be sure to point out that they can always check their integration by differentiating the answer; if they know the derivative formulas well, there is really nothing new to learn.

The last section covers the classical indeterminate forms $0/0$ and $\frac{\infty}{\infty}$. You may wish to point out that the derivative is actually an indeterminate form $(0/0)$ which in general becomes determinate. A good calculator exercise is to try to evaluate a limit like $\lim_{x\to 0} \frac{\sin x}{x}$, which cannot be done with any kind of algebraic maneuver. Make sure the students know that L'Hôpital's rule can only be applied in an indeterminate situation of the form $0/0$ or $\frac{\infty}{\infty}$. Use a number of examples to enable the students to distinguish between determinate forms $\left(\text{like } \lim_{x\to 0} \frac{\cos x}{x}\right)$ and indeterminate forms.

NB: The following exercises cannot be worked without an understanding of the material from the chapter indicated in parentheses:

Section 8.4: #45–48 (Appendix D)

CHAPTER 9
Probability and Calculus

Section Topics

9.1 **Discrete Probability**—sample space of an experiment; discrete random variables; probability of an event; expected value or mean; variance and standard deviation.

9.2 **Continuous Random Variables**—continuous random variables and probability density functions; applications.

9.3 **Expected Value and Variance**—expected value or mean; variance and standard deviation; median; special probability density functions.

Chapter Comments

Chapter 9 is a brief introduction to probability theory. Section 9.1 discusses topics in probability for a discrete random variable. This section contains terminology and definitions.

Sections 9.2 and 9.3 discuss these same ideas for a continuous random variable. Section 9.3 presents three commonly used continuous density functions, including the standard normal probability density function.

Some of the concepts in this chapter may be difficult for students who have previously never seen counting techniques or probability. Start out slowly with a number of examples of sample spaces for different events, carefully counting the number of ways a certain event can occur.

CHAPTER 10
Series and Taylor Polynomials

Section Topics

10.1 **Sequences**—sequences; limit of a sequence; pattern recognition for sequences; applications.

10.2 **Series and Convergence**—sigma notation; infinite series; the nth term test for divergence; geometric series; applications.

10.3 ***p*-Series and the Ratio Test**—p-series; approximating the sum of a p-series; the ratio test.

10.4 **Power Series and Taylor's Theorem**—power series; radius and interval of convergence for a power series; Taylor and Maclaurin series; a basic list of power series.

10.5 **Taylor Polynomials**—Taylor polynomials and approximation; Taylor's theorem with remainder; applications to probability.

10.6 **Newton's Method**—Newton's method; convergence of Newton's method.

Chapter Comments

The first three sections of this chapter cover sequences and series of constant terms. Since an infinite series, by definition, converges only if its sequence of partial sums converges, it is important that the student have a good understanding of sequences in order to understand series. Section 10.1 presents sequences and the necessary terminology. Section 10.2 introduces series, the divergence test, and geometric series. Be sure to emphasize via examples that knowing that the nth term of a series approaches zero tells you nothing about the convergence or divergence of the series.

Section 10.3 discusses p-series and the ratio test. Be sure to point out that the ratio test fails if $L = 1$. Some students may need help with the algebra involved in simplifying the ratio $\dfrac{a_{n+1}}{a_n}$.

Sections 10.4 and 10.5 cover series with variable terms, which often is difficult for students to grasp. I like to start out with a geometric series like $1 + x + x^2 + \cdot \cdot \cdot = \sum_{n=0}^{\infty} x^n$ since this is already familiar to students and the partial sums are easy to graph. It also points out that such a series (like a function) is only defined on a certain interval (domain). Your students should be able to write a Taylor series for a given function and find the radius of convergence of a power series. It is suggested that they memorize a short list of power series expansions to be able to quickly generate other power series.

The chapter closes with a discussion of Newton's method, a technique for estimating the zeros of a differentiable function. A good example to start with is something like $f(x) = x^3 - 2$, which has only one real zero, and the method will approximate $\sqrt[3]{2}$. As you work through the chapter, some extra motivation is provided by the application problems. Try to integrate at least a few of these into your lecture examples.

NB. The following exercises in the given sections cannot be worked without knowing the material from the chapter indicated in parentheses:

Section 10.2: #54–55 (Chapter 9)

Section 10.5: #29–30 (Chapter 9)

Section 10.5 also contains a subsection entitled "Applications to Probability" that assumes knowledge of terms and concepts from Chapter P.

Guidelines for a Course with Early Trigonometry

The following are guidelines for a course in which trigonometry is introduced earlier and used throughout the text. Chapter and section numbers and titles refer to *Calculus: An Applied Approach*, Fifth Edition by Larson and Edwards. "Prerequisite Topics" indicates the chapters and/or sections that must be covered before introducing the trigonometry topics.

Note that Chapter 8 Trigonometry can be covered any time after Section 6.6. If an earlier integration of trigonometry is required see below.

Trigonometry Topic	Prerequisite Topics
8.1 Radian measure of angles	Cover any time
8.2 The trigonometric functions	1.2 Graphs of Equations and 1.4 Functions
8.3 Graphs of trigonometric functions	1.5 Limits
8.4 Derivatives of trigonometric functions	Chapter 2 Differentiation (derivatives) and Section 3.2 Extrema and the First-Derivative Test (extrema)
8.5 Integrals of trigonometric functions	6.1 Integration by Substitution (except exercises using trapezoidal rule and Simpson's rule require Section 6.5 Numerical Integration)
8.6 L'Hôpital's rule	Chapter 4 Exponential and Logarithmic Functions

PART II

SOLUTIONS TO EVEN-NUMBERED EXERCISES

CHAPTER 0
A Precalculus Review

CHAPTER 0
A Precalculus Review

Section 0.1 The Real Line and Order

2. Since $-3678 = -\frac{3678}{1}$, it is rational.

4. $3\sqrt{2} - 1$ is irrational because $\sqrt{2}$ is irrational.

6. $\frac{22}{7}$ is rational.

8. $0.81778\overline{177}$ is rational since it has a repeating decimal expansion.

10. $2e$ is irrational since e is irrational.

12. $x + 1 < \dfrac{2x}{3}$

$3x + 3 < 2x$

$x + 3 < 0$

$x < -3$

(a) No, if $x = 0$, then x is not less than -3.
(b) No, if $x = 4$, then x is not less than -3.
(c) Yes, if $x = -4$, then x is less than -3.
(d) No, if $x = -3$, then x is not less than -3.

14. $-1 < \dfrac{3 - x}{2} \le 1$

$-2 < 3 - x \le 2$

$-5 < -x \le -1$

$5 > x \ge 1$ or $1 \le x < 5$

(a) No, if $x = 0$, then x is not greater than or equal to 1.
(b) Yes, if $x = \sqrt{5}$, then $1 \le x < 5$.
(c) Yes, if $x = 1$, then $1 \le x < 5$.
(d) No, if $x = 5$, then x is not less than 5.

16. $2x > 3$

$x > \frac{3}{2}$

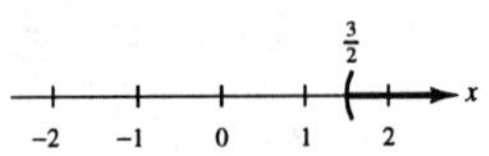

18. $2x + 7 < 3$

$2x < -4$

$x < -2$

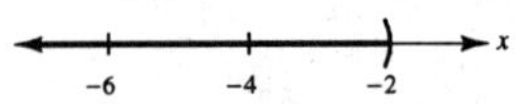

20. $x - 4 \le 2x + 1$

$-x - 4 \le 1$

$-x \le 5$

$x \ge -5$

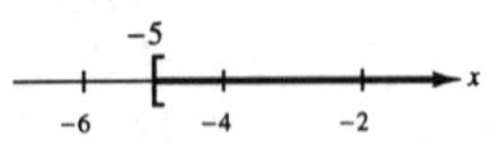

22. $0 \le x + 3 < 5$

$-3 \le x < 2$

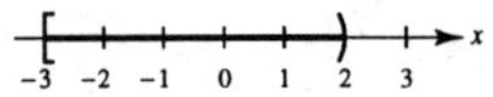

24. $-1 < -\dfrac{x}{3} < 1$

$-3 < -x < 3$

$3 > x > -3$

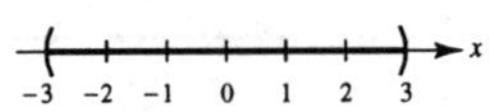

26. $\dfrac{x}{2} - \dfrac{x}{3} > 5$

$6\left(\dfrac{x}{2} - \dfrac{x}{3}\right) > 6(5)$

$3x - 2x > 30$

$x > 30$

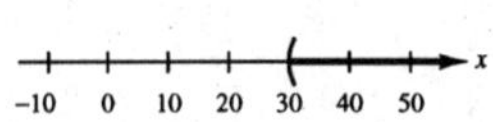

28. $2x^2 + 1 < 9x - 3$

$2x^2 - 9x + 4 < 0$

$(2x - 1)(x - 4) < 0$

Therefore, the solution is $\frac{1}{2} < x < 4$.

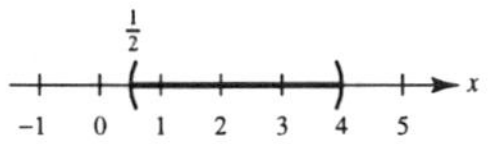

30. Revenue: $R = 2.95x$

Cost: $C = 1.45x + 150$

Profit: $P = R - C$

$= 2.95x - (1.45x + 150) = 1.50x - 150$

Thus, $50 \le 1.50x - 150 \le 200$

$200 \le 1.50x \le 350$

$\frac{400}{3} \le x \le \frac{700}{3}$

$133\frac{1}{3} \le x \le 233\frac{1}{3}$ dozen doughnuts per day.

32. $A = 20$ and $r = 220 - A = 200$.
Let T be the target heart rate. Then

$(0.60)(200) \le T \le (0.90)(200)$

$120 \le T \le 180$

34. Given $a < b$,

(a) $-2a < -2b$ is false.

(b) $a + 2 < b + 2$ is true.

(c) $6a < 6b$ is true.

(d) $\frac{1}{a} < \frac{1}{b}$ is false if $ab > 0$ and true if $ab < 0$.

Section 0.2 Absolute Value and Distance on the Real Line

2. (a) The directed distance from a to b is $-75 - (-126) = 51$.

(b) The directed distance from b to a is $-126 - (-75) = -51$.

(c) The distance between a and b is $|-75 - (-126)| = 51$.

4. (a) The directed distance from a to b is $\frac{112}{75} - \frac{16}{5} = -\frac{128}{75}$.

(b) The directed distance from b to a is $\frac{16}{5} - \frac{112}{75} = \frac{128}{75}$.

(c) The distance between a and b is $\left|\frac{112}{75} - \frac{16}{5}\right| = \frac{128}{75}$.

6. Midpoint $= \dfrac{8.6 + 11.4}{2} = 10$

8. Midpoint $= \dfrac{-4.6 + (-1.3)}{2} = -2.95$

10. $-6 < 2x < 6$

$-3 < x < 3$

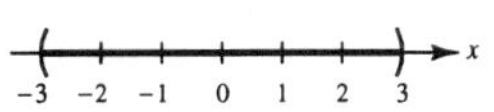

12. $5x < -10$ or $5x > 10$

$x < -2$ $\quad$ $x > 2$

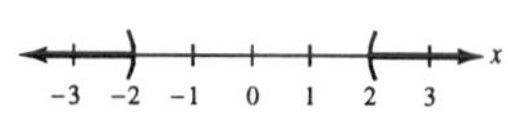

14. $3x + 1 \le -4$ or $3x + 1 \ge 4$

$3x \le -5$ $\quad$ $3x \ge 3$

$x \le -\frac{5}{3}$ $\quad$ $x \ge 1$

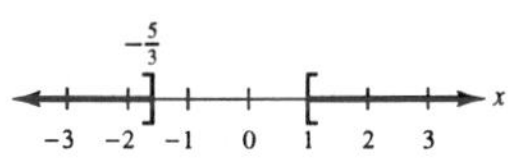

16. $-5 < 2x + 1 < 5$

$-6 < 2x < 4$

$-3 < x < 2$

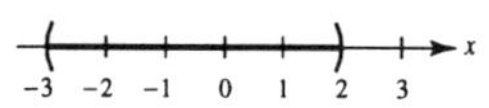

18. $25 - x \le -20$ or $25 - x \ge 20$

$-x \le -45$ $\quad$ $-x \ge -5$

$x \ge 45$ $\quad$ $x \le 5$

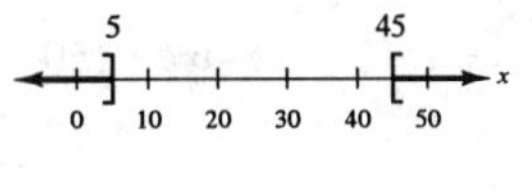

20. $-1 < 1 - \dfrac{2x}{3} < 1$

$-2 < -\dfrac{2x}{3} < 0$

$-6 < -2x < 0$

$3 > x > 0$

$0 < x < 3$

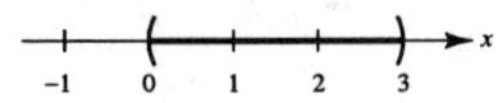

22. $2x - a \le -b$ or $2x - a \ge b$

$2x \le a - b$ $\quad$ $2x \ge a + b$

$x \le \dfrac{a - b}{2}$ $\quad$ $x \ge \dfrac{a + b}{2}$

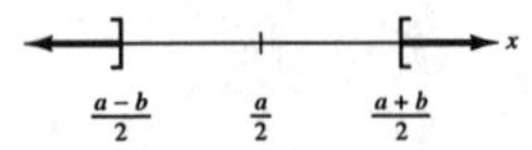

24. $|x| < 3$

26. $|x| \ge 3$

28. $|x + 4| < 3$

30. $|x - 22| < 2$

32. $|x - 3| > 6$

34. $|y - c| < h$

36. $\left|\dfrac{w - 57.5}{7.5}\right| \le 1$

$-1 \le \dfrac{w - 57.5}{7.5} \le 1$

$-7.5 \le w - 57.5 \le 7.5$

$50 \le w \le 65$

38. $|p - 33\frac{1}{8}| \le 2$

40. (a) $|I - 15{,}000| \le 500 \Longrightarrow 14{,}500 \le I \le 15{,}500$

$|I - 15{,}000| \le 0.05(15{,}000)$

$|I - 15{,}000| \le 750 \Longrightarrow 14{,}250 \le I \le 15{,}750$

(b) The given expense is not at variance with the budget restrictions.

42. (a) $|T - 7500| \le 500 \Longrightarrow 7000 \le T \le 8000$

$|T - 7500| \le 0.05(7500)$

$|T - 7500| \le 375 \Longrightarrow 7125 \le T \le 7875$

(b) The given expense is at variance with the budget restrictions.

Section 0.3 Exponents and Radicals

2. $\dfrac{(6)^2}{2} = \dfrac{36}{2} = 18$

4. $7(4)^{-2} = \dfrac{7}{(4)^2} = \dfrac{7}{16}$

6. $3 - 4(3)^{-2} = 3 - \dfrac{4}{(3)^2} = \dfrac{3}{1} - \dfrac{4}{9} = \dfrac{27}{9} - \dfrac{4}{9} = \dfrac{23}{9}$

8. $5(-3)^3 = 5(-27) = -135$

10. $\dfrac{1}{(-4)^{-3}} = \dfrac{1}{1/(-4)^3} = \dfrac{1}{1/(-64)} = (1)\left(\dfrac{-64}{1}\right) = -64$

12. $\sqrt{(1/9)^3} = \left(\sqrt{1/9}\right)^3 = \left(\dfrac{1}{3}\right)^3 = \dfrac{1}{27}$

14. $16^{-3/4} = \dfrac{1}{16^{3/4}} = \dfrac{1}{\left(\sqrt[4]{16}\right)^3} = \dfrac{1}{(2)^3} = \dfrac{1}{8}$

16. $(10^{2/3})^3 = 10^2 = 100$

18. $\dfrac{10{,}000}{1.075^{120}} \approx 1.7021$

20. $\sqrt[6]{325} \approx 2.6221$

22. $(8x^4)(2x^3) = (8)(2)x^{4+3} = 16x^7$

24. $z^{-3}(3z^4) = 3z^{-3+4} = 3z$

26. $(4x^3)^2 = (4)^2x^{(3)(2)} = 16x^6$

28. $\dfrac{r^4}{r^6} = \dfrac{1}{r^{6-4}} = \dfrac{1}{r^2}$

30. $\left(\dfrac{12s^2}{9s}\right)^3 = \left(\dfrac{4s}{3}\right)^3 = \dfrac{64s^3}{27}$

32. $\left(\sqrt[3]{x^2}\right)^3 = (x^{2/3})^3 = x^2$

34. $(2x^2yz^5)^0 = 1$ if $x \neq 0, y \neq 0, z \neq 0.$

36. (a) $\sqrt[3]{\dfrac{16}{27}} = \dfrac{\sqrt[3]{8}\,\sqrt[3]{2}}{\sqrt[3]{27}} = \dfrac{2\sqrt[3]{2}}{3}$

(b) $\sqrt[3]{\dfrac{24}{125}} = \dfrac{\sqrt[3]{8}\,\sqrt[3]{3}}{\sqrt[3]{125}} = \dfrac{2\sqrt[3]{3}}{5}$

38. (a) $\sqrt[4]{(3x^2y^3)^4} = |3x^2y^3| = 3x^2|y|^3$

(b) $\sqrt[3]{54x^7} = \sqrt[3]{27x^6}\,\sqrt[3]{2x} = 3x^2\sqrt[3]{2x}$

40. (a) $\sqrt[5]{96b^6c^3} = \sqrt[5]{32b^5}\,\sqrt[5]{3bc^3} = 2b\sqrt[5]{3bc^3}$

(b) $\sqrt{72(x+1)^4} = \sqrt{36(x+1)^4}\sqrt{2}$

$= 6(x+1)^2\sqrt{2} = 6\sqrt{2}(x+1)^2$

42. $\sqrt{5-2x}$ is defined when $x \leq \frac{5}{2}$.
Therefore, the domain is $(-\infty, \frac{5}{2}]$.

44. $\sqrt[5]{1-x}$ is defined for all real numbers.
Therefore, the domain is $(-\infty, \infty)$.

46. $\dfrac{1}{\sqrt{x+4}}$ is defined when $x > -4$.
Therefore, the domain is $(-4, \infty)$.

48. $\dfrac{\sqrt{x-1}}{x+1}$ is defined when $x \geq 1$.
Therefore, the domain is $[1, \infty)$.

50. $\dfrac{1}{\sqrt{2x+3}}$ is defined when $x > -\dfrac{3}{2}$, and $\sqrt{6-4x}$ is defined when $x \leq \dfrac{3}{2}$.

Therefore, the domain of $\dfrac{1}{\sqrt{2x+3}} + \sqrt{6-4x}$ is $-\dfrac{3}{2} < x \leq \dfrac{3}{2}$ or $\left(-\dfrac{3}{2}, \dfrac{3}{2}\right]$.

52. $A = 7000\left(1 + \dfrac{0.05}{365}\right)^{1000} \approx \$8027.61$

54. $A = P(1+r) + P(1+r)^2 + P(1+r)^3 + \cdots + P(1+r)^n = P(1+r)[1 + (1+r) + (1+r)^2 + \cdots + (1+r)^{n-1}]$

Section 0.4 Factoring Polynomials

2. Since $a = 8$, $b = -2$, and $c = -1$, we have $x = \dfrac{2 \pm \sqrt{4+32}}{16} = \dfrac{2 \pm 6}{16}$.

Thus, $x = \dfrac{2+6}{16} = \dfrac{1}{2}$ or $x = \dfrac{2-6}{16} = -\dfrac{1}{4}$.

4. Since $a = 9$, $b = 12$, and $c = 4$, we have $x = \dfrac{-12 \pm \sqrt{144-144}}{18} = \dfrac{-12 \pm 0}{18} = -\dfrac{2}{3}$.

6. Since $a = 1$, $b = 6$, and $c = -1$, we have $x = \dfrac{-6 \pm \sqrt{36+4}}{2} = \dfrac{-6 \pm 2\sqrt{10}}{2} = -3 \pm \sqrt{10}$.

8. $x^2 + 10x + 25 = (x+5)^2$

10. $9x^2 - 12x + 4 = (3x-2)^2$

12. $2x^2 - x - 1 = (2x + 1)(x - 1)$

14. $x^2 - xy - 2y^2 = (x - 2y)(x + y)$

16. $a^2b^2 - 2abc + c^2 = (ab - c)^2$

18. $x^4 - 16 = (x^2 + 4)(x^2 - 4) = (x^2 + 4)(x + 2)(x - 2)$

20. $y^3 - 64 = y^3 - 4^3 = (y - 4)(y^2 + 4y + 16)$

22. $z^3 + 125 = z^3 + 5^3 = (z + 5)(z^2 - 5z + 25)$

24.
$$\begin{aligned}(x - a)^3 + b^3 &= [(x - a) + b][(x - a)^2 - (x - a)b + b^2] = (x - a + b)[x^2 - 2xa + a^2 - xb + ab + b^2]\\ &= (x - a + b)[x^2 - x(2a + b) + (a^2 + ab + b^2)]\end{aligned}$$

26.
$$\begin{aligned}x^3 - x^2 - x + 1 &= x^2(x - 1) - (x - 1)\\ &= (x - 1)(x^2 - 1)\\ &= (x - 1)(x + 1)(x - 1)\\ &= (x + 1)(x - 1)^2\end{aligned}$$

28.
$$\begin{aligned}x^3 - 5x^2 - 5x + 25 &= x^2(x - 5) - 5(x - 5)\\ &= (x - 5)(x^2 - 5)\end{aligned}$$

30.
$$\begin{aligned}x^3 - 7x^2 - 4x + 28 &= x^2(x - 7) - 4(x - 7)\\ &= (x - 7)(x^2 - 4)\\ &= (x - 7)(x + 2)(x - 2)\end{aligned}$$

32.
$$\begin{aligned}2x^2 - 3x &= 0\\ x(2x - 3) &= 0\\ x &= 0, \tfrac{3}{2}\end{aligned}$$

34.
$$\begin{aligned}x^2 - 25 &= 0\\ (x + 5)(x - 5) &= 0\\ x &= -5, 5\end{aligned}$$

36.
$$\begin{aligned}x^2 - 8 &= 0\\ \left(x + \sqrt{8}\right)\left(x - \sqrt{8}\right) &= 0\\ \left(x + 2\sqrt{2}\right)\left(x - 2\sqrt{2}\right) &= 0\\ x &= \pm 2\sqrt{2}\end{aligned}$$

38.
$$\begin{aligned}(x + 1)^2 - 8 &= 0\\ (x + 1)^2 &= 8\\ x + 1 &= \pm 2\sqrt{2}\\ x &= -1 \pm 2\sqrt{2}\end{aligned}$$

40.
$$\begin{aligned}x^2 + 5x + 6 &= 0\\ (x + 3)(x + 2) &= 0\\ x &= -3, -2\end{aligned}$$

42.
$$\begin{aligned}x^2 + x - 20 &= 0\\ (x + 5)(x - 4) &= 0\\ x &= -5, 4\end{aligned}$$

44.
$$\begin{aligned}x^3 - 216 &= 0\\ x^3 &= 216\\ x &= \sqrt[3]{216} = 6\end{aligned}$$

46.
$$\begin{aligned}x^4 - 625 &= 0\\ x^4 &= 625\\ x &= \pm\sqrt[4]{625} = \pm 5\end{aligned}$$

48.
$$\begin{aligned}2x^3 + x^2 + 6x + 3 &= 0\\ x^2(2x + 1) + 3(2x + 1) &= 0\\ (2x + 1)(x^2 + 3) &= 0\\ x &= -\tfrac{1}{2}\end{aligned}$$

[Note: $x^2 + 3 = 0$ has no real roots.]

50. Since

$$\sqrt{x^2 - 4} = \sqrt{(x + 2)(x - 2)},$$

the roots are $x = \pm 2$. By testing points inside and outside the interval $[-2, 2]$, we find that the expression is defined when $x \le -2$ or $x \ge 2$. Thus, the domain is $(-\infty, -2] \cup [2, \infty)$.

52. Since

$$\sqrt{144 - 9x^2} = \sqrt{9(4 + x)(4 - x)},$$

the roots are $x = \pm 4$. By testing points inside and outside the interval $[-4, 4]$, we find that the expression is defined when $-4 \le x \le 4$. Thus, the domain is $[-4, 4]$.

54. $$\begin{array}{r|rrrr} -1 & 1 & -2 & -1 & 2 \\ & & -1 & 3 & -2 \\ \hline & 1 & -3 & 2 & 0 \end{array}$$

Therefore, the factorization is $x^3 - 2x^2 - x + 2 = (x + 1)(x^2 - 3x + 2) = (x + 1)(x - 1)(x - 2)$.

56. $$\begin{array}{r|rrrrr} 4 & 1 & -16 & 96 & -256 & 256 \\ & & 4 & -48 & 192 & -256 \\ \hline & 1 & -12 & 48 & -64 & 0 \end{array}$$

Therefore, $x^4 - 16x^3 + 96x^2 - 256x + 256 = (x - 4)(x^3 - 12x^2 + 48x - 64)$.

58. Possible rational roots: $\pm 1, \pm 2, \pm 4$

Using synthetic division for $x = 1$, we have the following.

$$\begin{array}{r|rrrr} 1 & 1 & -1 & -4 & 4 \\ & & 1 & 0 & -4 \\ \hline & 1 & 0 & -4 & 0 \end{array}$$

Therefore, we have

$$\begin{aligned} x^3 - x^2 - 4x + 4 &= 0 \\ (x - 1)(x^2 - 4) &= 0 \\ (x - 1)(x + 2)(x - 2) &= 0 \\ x &= 1, \pm 2. \end{aligned}$$

60. Possible rational roots: $\pm 1, \pm 2, \pm 3, \pm 6$

Using synthetic division for $x = 2$, we have the following.

$$\begin{array}{r|rrrr} 2 & 1 & 2 & -5 & -6 \\ & & 2 & 8 & 6 \\ \hline & 1 & 4 & 3 & 0 \end{array}$$

Therefore, we have

$$\begin{aligned} x^3 + 2x^2 - 5x - 6 &= 0 \\ (x - 2)(x^2 + 4x + 3) &= 0 \\ (x - 2)(x + 1)(x + 3) &= 0 \\ x &= 2, -1, -3. \end{aligned}$$

62. Possible rational roots: $\pm 1, \pm 2, \pm 4, \pm\frac{1}{2}, \pm\frac{1}{3}, \pm\frac{1}{6}, \pm\frac{1}{9}, \pm\frac{1}{18}, \pm\frac{2}{3}, \pm\frac{2}{9}, \pm\frac{4}{3}, \pm\frac{4}{9}$

Using synthetic division for $x = \frac{1}{2}$, we have the following.

$$\begin{array}{r|rrrr} \frac{1}{2} & 18 & -9 & -8 & 4 \\ & & 9 & 0 & -4 \\ \hline & 18 & 0 & -8 & 0 \end{array}$$

Therefore, we have

$$\begin{aligned} 18x^3 - 9x^2 - 8x + 4 &= 0 \\ \left(x - \tfrac{1}{2}\right)(18x^2 - 8) &= 0 \\ \left(x - \tfrac{1}{2}\right)(2)(9x^2 - 4) &= 0 \\ (2x - 1)(3x + 2)(3x - 2) &= 0 \\ x &= \tfrac{1}{2}, \pm\tfrac{2}{3}. \end{aligned}$$

64. Possible rational roots: $\pm 1, \pm 3, \pm\frac{1}{2}, \pm\frac{3}{2}, \pm\frac{1}{4}, \pm\frac{3}{4}$

Using synthetic division for $x = -\frac{1}{2}$, we have the following.

$$\begin{array}{r|rrrr} -\frac{1}{2} & 4 & -6 & 2 & 3 \\ & & -2 & 4 & -3 \\ \hline & 4 & -8 & 6 & 0 \end{array}$$

Therefore, we have

$$\begin{aligned} 4x^3 - 6x^2 + 2x + 3 &= 0 \\ \left(x + \tfrac{1}{2}\right)(4x^2 - 8x + 6) &= 0 \\ \left(x + \tfrac{1}{2}\right)(2)(2x^2 - 4x + 3) &= 0 \\ (2x + 1)(2x^2 - 4x + 3) &= 0. \end{aligned}$$

Since $2x^2 - 4x + 3$ has no real solutions, $x = -\frac{1}{2}$ is the only real root.

66. $$\begin{aligned} -200x^2 + 2000x - 3800 &> 1000 \\ 0 &> 200x^2 - 2000x + 4800 \\ 0 &> 200(x^2 - 10x + 24) \\ 0 &> 200(x - 4)(x - 6) \end{aligned}$$

The roots are $x = 4, 6$. By testing points inside and outside the interval $[4, 6]$, we find that the expression is defined when $4 < x < 6$.

Section 0.5 Fractions and Rationalization

2. $\dfrac{2x-1}{x+3}+\dfrac{1-x}{x+3}=\dfrac{x}{x+3}$

4. $\dfrac{5x+10}{2x-1}-\dfrac{2x+10}{2x-1}=\dfrac{5x+10-(2x+10)}{2x-1}=\dfrac{3x}{2x-1}$

6. $\dfrac{5}{x-1}+\dfrac{3}{x}=\dfrac{5x+3(x-1)}{x(x-1)}=\dfrac{8x-3}{x(x-1)}$

8. $\dfrac{x}{x^2+x-2}-\dfrac{1}{x+2}=\dfrac{x}{(x+2)(x-1)}-\dfrac{1}{x+2}=\dfrac{x-(x-1)}{(x+2)(x-1)}=\dfrac{1}{(x+2)(x-1)}$

10. $\dfrac{x}{2-x}+\dfrac{2}{x-2}=\dfrac{-x}{x-2}+\dfrac{2}{x-2}=\dfrac{-x+2}{x-2}=\dfrac{-(x-2)}{x-2}=-1$

12. $\dfrac{Ax+B}{x^2+2}+\dfrac{C}{x-4}=\dfrac{(Ax+B)(x-4)+C(x^2+2)}{(x-4)(x^2+2)}$

$$=\frac{Ax^2-4Ax+Bx-4B+Cx^2+2C}{(x-4)(x^2+2)}=\frac{(A+C)x^2-(4A-B)x-2(2B-C)}{(x-4)(x^2+2)}$$

14. $\dfrac{2}{x+1}+\dfrac{1-x}{x^2-2x+3}=\dfrac{2(x^2-2x+3)+(1-x)(x+1)}{(x+1)(x^2-2x+3)}$

$$=\frac{2x^2-4x+6+x-x^2+1-x}{(x+1)(x^2-2x+3)}=\frac{x^2-4x+7}{(x+1)(x^2-2x+3)}$$

16. $2\sqrt{x}(x-2)+\dfrac{(x-2)^2}{2\sqrt{x}}=\dfrac{2\sqrt{x}(x-2)}{1}\cdot\dfrac{2\sqrt{x}}{2\sqrt{x}}+\dfrac{(x-2)^2}{2\sqrt{x}}$

$$=\frac{4x(x-2)+(x-2)^2}{2\sqrt{x}}=\frac{(x-2)[4x+(x-2)]}{2\sqrt{x}}=\frac{(x-2)(5x-2)}{2\sqrt{x}}$$

18. $-\dfrac{\sqrt{x^2+1}}{x^2}+\dfrac{1}{\sqrt{x^2+1}}=-\dfrac{\sqrt{x^2+1}}{x^2}\cdot\dfrac{\sqrt{x^2+1}}{\sqrt{x^2+1}}+\dfrac{1}{\sqrt{x^2+1}}\cdot\dfrac{x^2}{x^2}$

$$=\frac{-(x^2+1)+x^2}{x^2\sqrt{x^2+1}}=-\frac{1}{x^2\sqrt{x^2+1}}$$

20. $\dfrac{x-1}{x^2+5x+4}+\dfrac{2}{x^2-x-2}+\dfrac{10}{x^2+2x-8}=\dfrac{x-1}{(x+1)(x+4)}+\dfrac{2}{(x+1)(x-2)}+\dfrac{10}{(x-2)(x+4)}$

$$=\frac{(x-1)(x-2)+2(x+4)+10(x+1)}{(x+1)(x-2)(x+4)}$$

$$=\frac{x^2-3x+2+2x+8+10x+10}{(x+1)(x-2)(x+4)}$$

$$=\frac{x^2+9x+20}{(x+1)(x-2)(x+4)}$$

$$=\frac{(x+4)(x+5)}{(x+1)(x-2)(x+4)}$$

$$=\frac{x+5}{(x+1)(x-2)}$$

22. $$\frac{A}{x-5}+\frac{B}{x+5}+\frac{C}{(x+5)^2}=\frac{A(x+5)^2+B(x-5)(x+5)+C(x-5)}{(x-5)(x+5)^2}$$
$$=\frac{A(x^2+10x+25)+B(x^2-25)+Cx-5C}{(x-5)(x+5)^2}$$
$$=\frac{(A+B)x^2+(10A+C)x+5(5A-5B-C)}{(x-5)(x+5)^2}$$

24. $$\left(\sqrt{x^3+1}-\frac{3x^3}{2\sqrt{x^3+1}}\right)\div(x^3+1)=\left(\frac{\sqrt{x^3+1}\left(2\sqrt{x^3+1}\right)-3x^3}{2\sqrt{x^3+1}}\right)\frac{1}{x^3+1}$$
$$=\frac{2(x^3+1)-3x^3}{2(x^3+1)^{1/2}(x^3+1)}=\frac{2-x^3}{2(x^3+1)^{3/2}}$$

26. $$\frac{x(x+1)^{-1/2}-(x+1)^{1/2}}{x^2}=\frac{(x+1)^{-1/2}[x-(x+1)]}{x^2}=\frac{-1}{x^2\sqrt{x+1}}$$

28. $$\frac{\dfrac{2x^2}{3(x^2-1)^{2/3}}-(x^2-1)^{1/3}}{x^2}=\frac{2x^2-(x^2-1)^{1/3}(3)(x^2-1)^{2/3}}{3(x^2-1)^{2/3}}\cdot\frac{1}{x^2}$$
$$=\frac{2x^2-3(x^2-1)}{3x^2(x^2-1)^{2/3}}=\frac{3-x^2}{3x^2(x^2-1)^{2/3}}$$

30. $$\frac{5}{\sqrt{10}}=\frac{5}{\sqrt{10}}\cdot\frac{\sqrt{10}}{\sqrt{10}}=\frac{5\sqrt{10}}{10}=\frac{\sqrt{10}}{2}$$

32. $$\frac{\sqrt{26}}{2}=\frac{\sqrt{26}}{2}\cdot\frac{\sqrt{26}}{\sqrt{26}}=\frac{26}{2\sqrt{26}}=\frac{13}{\sqrt{26}}$$

34. $$\frac{4y}{\sqrt{y+8}}=\frac{4y}{\sqrt{y+8}}\cdot\frac{\sqrt{y+8}}{\sqrt{y+8}}=\frac{4y\sqrt{y+8}}{y+8}$$

36. $$\frac{x\sqrt{x^2+4}}{3}=\frac{x\sqrt{x^2+4}}{3}\cdot\frac{\sqrt{x^2+4}}{\sqrt{x^2+4}}=\frac{x(x^2+4)}{3\sqrt{x^2+4}}$$

38. $$\frac{10(x+2)}{\sqrt{x^2-x-6}}=\frac{10(x+2)}{\sqrt{x^2-x-6}}\cdot\frac{\sqrt{x^2-x-6}}{\sqrt{x^2-x-6}}$$
$$=\frac{10(x+2)\sqrt{x^2-x-6}}{x^2-x-6}=\frac{10(x+2)\sqrt{x^2-x-6}}{(x+2)(x-3)}=\frac{10\sqrt{x^2-x-6}}{x-3}$$

40. $$\frac{13}{6+\sqrt{10}}=\frac{13}{6+\sqrt{10}}\cdot\frac{6-\sqrt{10}}{6-\sqrt{10}}$$
$$=\frac{13\left(6-\sqrt{10}\right)}{36-10}$$
$$=\frac{6-\sqrt{10}}{2}$$

42. $$\frac{x}{\sqrt{2}+\sqrt{3}}=\frac{x}{\sqrt{2}+\sqrt{3}}\cdot\frac{\sqrt{2}-\sqrt{3}}{\sqrt{2}-\sqrt{3}}$$
$$=\frac{x\left(\sqrt{2}-\sqrt{3}\right)}{2-3}$$
$$=x\left(\sqrt{3}-\sqrt{2}\right)$$

44. $$\frac{\sqrt{15}+3}{12}=\frac{\sqrt{15}+3}{12}\cdot\frac{\sqrt{15}-3}{\sqrt{15}-3}$$
$$=\frac{15-9}{12\left(\sqrt{15}-3\right)}$$
$$=\frac{1}{2\left(\sqrt{15}-3\right)}$$

46. $$\frac{x-1}{\sqrt{x}+x}=\frac{x-1}{\sqrt{x}+x}\cdot\frac{\sqrt{x}-x}{\sqrt{x}-x}$$
$$=\frac{(x-1)\left(\sqrt{x}-x\right)}{x-x^2}$$
$$=\frac{(x-1)\left(\sqrt{x}-x\right)}{-x(x-1)}$$
$$=\frac{x-\sqrt{x}}{x}$$

48. $$\frac{10}{\sqrt{x}+\sqrt{x+5}} = \frac{10}{\sqrt{x}+\sqrt{x+5}} \cdot \frac{\sqrt{x}-\sqrt{x+5}}{\sqrt{x}-\sqrt{x+5}}$$
$$= \frac{10\left(\sqrt{x}-\sqrt{x+5}\right)}{x-(x+5)}$$
$$= -2\left(\sqrt{x}-\sqrt{x+5}\right)$$
$$= 2\left(\sqrt{x+5}-\sqrt{x}\right)$$

50. (a) $$C = 6x + \frac{900{,}000}{x}$$
$$= \frac{6x^2 + 900{,}000}{x} = \frac{6(x^2 + 150{,}000)}{x}$$

(b) When $x = 240$,
$$C = 6(240) + \frac{900{,}000}{240} = \$5190.00.$$

CHAPTER 1
Functions, Graphs, and Limits

CHAPTER 1
Functions, Graphs, and Limits

Section 1.1 The Cartesian Plane and the Distance Formula

2. (a) $a = \sqrt{(13 - 1)^2 + (1 - 1)^2} = 12$

$b = \sqrt{(13 - 13)^2 + (6 - 1)^2} = 5$

$c = \sqrt{(13 - 1)^2 + (6 - 1)^2} = 13$

(b) $a^2 + b^2 = (12)^2 + (5)^2 = 169 = c^2$

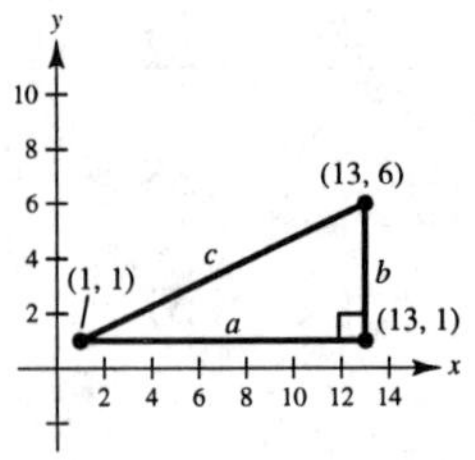

4. (a) $a = \sqrt{(6 - 2)^2 + (-2 + 2)^2} = 4$

$b = \sqrt{(2 - 2)^2 + (5 + 2)^2} = 7$

$c = \sqrt{(2 - 6)^2 + (5 + 2)^2} = \sqrt{65}$

(b) $a^2 + b^2 = (4)^2 + (7)^2 = 65 = c^2$

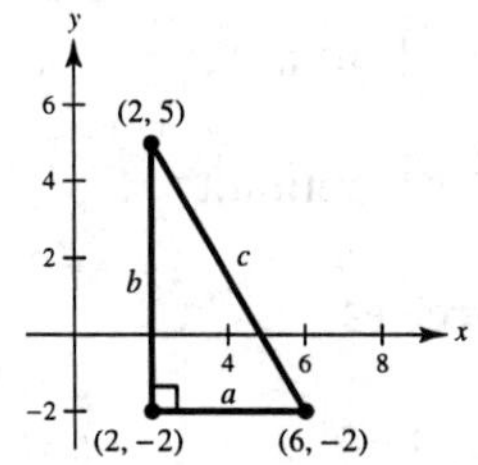

6. (a) See graph.

(b) $d = \sqrt{(-3 - 3)^2 + (2 + 2)^2} = \sqrt{52} = 2\sqrt{13}$

(c) Midpoint $= \left(\frac{-3 + 3}{2}, \frac{2 + (-2)}{2}\right) = (0, 0)$

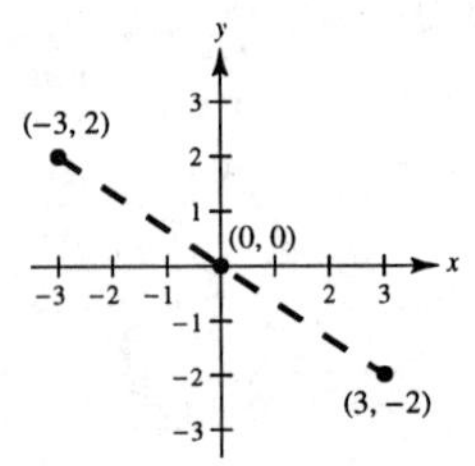

8. (a) See graph.

(b) $d = \sqrt{\left(\frac{5}{6} - \frac{2}{3}\right)^2 + \left(1 + \frac{1}{3}\right)^2} = \sqrt{\frac{1}{36} + \frac{16}{9}} = \frac{\sqrt{65}}{6}$

(c) Midpoint $= \left(\frac{(5/6) + (2/3)}{2}, \frac{1 - (1/3)}{2}\right) = \left(\frac{3}{4}, \frac{1}{3}\right)$

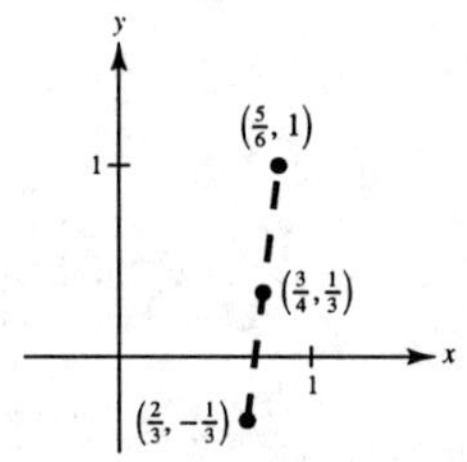

10. (a) See graph.

(b) $d = \sqrt{(-3 - 1)^2 + (7 + 1)^2} = \sqrt{16 + 64} = 4\sqrt{5}$

(c) Midpoint $= \left(\frac{-3 + 1}{2}, \frac{7 - 1}{2}\right) = (-1, 3)$

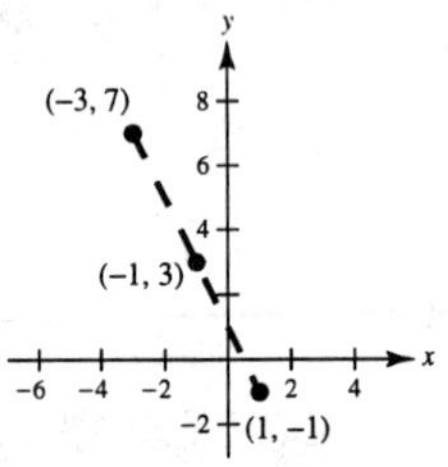

12. (a) See graph.

(b) $d = \sqrt{(-2 - 0)^2 + \left(0 - \sqrt{2}\right)^2} = \sqrt{6}$

(c) Midpoint $= \left(\frac{-2 + 0}{2}, \frac{0 + \sqrt{2}}{2}\right) = \left(-1, \frac{\sqrt{2}}{2}\right)$

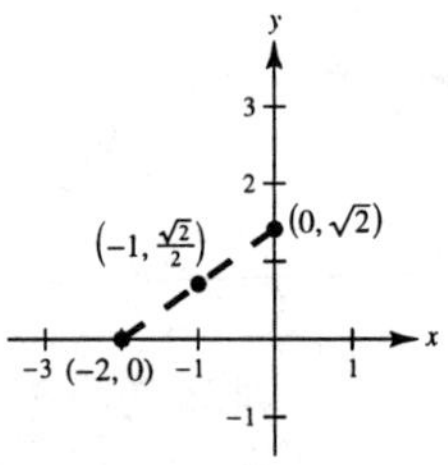

14. $a = \sqrt{(-2-3)^2 + (4-2)^2} = \sqrt{29}$

$b = \sqrt{(3-1)^2 + (2+3)^2} = \sqrt{29}$

$c = \sqrt{(-2-1)^2 + (4+3)^2} = \sqrt{58}$

Since $a = b$ the figure is an isosceles triangle.

[**Note:** It is also a right triangle since $a^2 + b^2 = c^2$.]

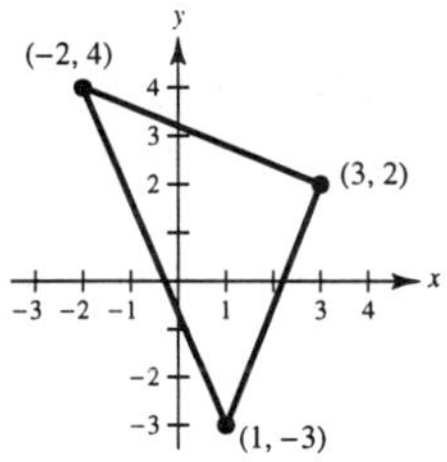

16. $a = \sqrt{(3-0)^2 + (7-1)^2} = 3\sqrt{5}$

$b = \sqrt{(3-4)^2 + (7-4)^2} = \sqrt{10}$

$c = \sqrt{(4-1)^2 + (4+2)^2} = 3\sqrt{5}$

$d = \sqrt{(1-0)^2 + (-2-1)^2} = \sqrt{10}$

Since $a = c$ and $b = d$, the figure is a parallelogram.

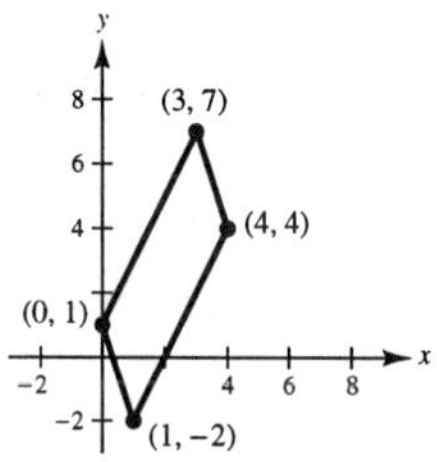

18. $d_1 = \sqrt{(-5-0)^2 + (11-4)^2} = \sqrt{74}$

$d_2 = \sqrt{(0-7)^2 + (4+6)^2} = \sqrt{149}$

$d_3 = \sqrt{(-5-7)^2 + (11+6)^2} = \sqrt{433}$

$d_1 + d_2 \approx 20.80888$

$d_3 \approx 20.80865$

Since $d_1 + d_2 \neq d_3$, the points are not collinear.

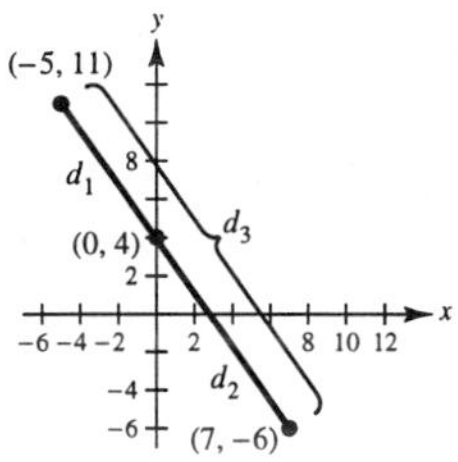

20. $d_1 = \sqrt{(-1-3)^2 + (1-3)^2} = 2\sqrt{5}$

$d_2 = \sqrt{(3-5)^2 + (3-5)^2} = 2\sqrt{2}$

$d_3 = \sqrt{(-1-5)^2 + (1-5)^2} = 2\sqrt{13}$

$d_1 + d_2 \approx 7.30056$

$d_3 \approx 7.21110$

Since $d_1 + d_2 \neq 3$, the points are not collinear.

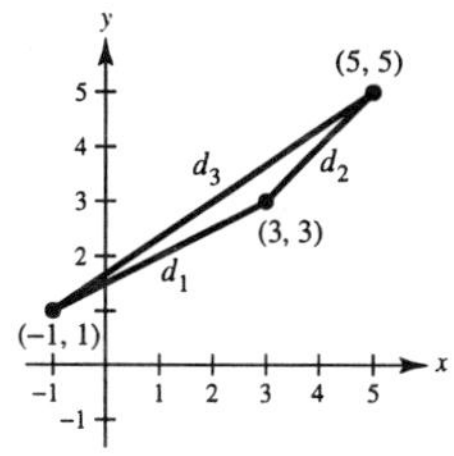

22. $d = \sqrt{(x-2)^2 + (2+1)^2} = 5$

$\sqrt{x^2 - 4x + 13} = 5$

$x^2 - 4x + 13 = 25$

$x^2 - 4x - 12 = 0$

$(x+2)(x-6) = 0$

$x = -2, 6$

24. $d = \sqrt{(5-5)^2 + (y-1)^2} = 8$

$\sqrt{(y-1)^2} = 8$

$(y-1)^2 = 64$

$y - 1 = \pm 8$

$y = 1 \pm 8$

$y = -7, 9$

26. To show that $\left(\frac{2x_1 + x_2}{3}, \frac{2y_1 + y_2}{3}\right)$ is a point of trisection of the line segment joining (x_1, y_1) and (x_2, y_2), we must show that $d_1 = \frac{1}{2}d_2$ and $d_1 + d_2 = d_3$.

$$d_1 = \sqrt{\left(\frac{2x_1 + x_2}{3} - x_1\right)^2 + \left(\frac{2y_1 + y_2}{3} - y_2\right)^2}$$

$$= \sqrt{\left(\frac{x_2 - x_1}{3}\right)^2 + \left(\frac{y_2 - y_1}{3}\right)^2} = \frac{1}{3}\sqrt{(x_2 - x_1)^2 + (y_2 - y_1)^2}$$

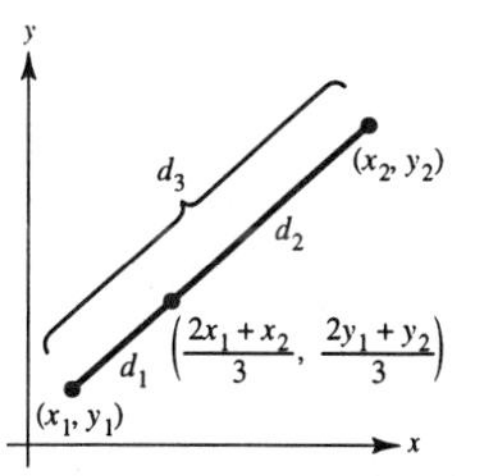

–CONTINUED–

26. —CONTINUED—

$$d_2 = \sqrt{\left(x_2 - \frac{2x_1 + x_2}{3}\right)^2 + \left(y_2 - \frac{2y_1 + y_2}{3}\right)^2}$$

$$= \sqrt{\left(\frac{2x_2 - 2x_1}{3}\right)^2 + \left(\frac{2y_2 - 2y_1}{3}\right)^2} = \frac{2}{3}\sqrt{(x_2 - x_1)^2 + (y_2 - y_1)^2}$$

$$d_3 = \sqrt{(x_2 - x_1)^2 + (y_2 - y_1)^2}$$

Therefore, $d_1 = \frac{1}{2}d_2$ and $d_1 + d_2 = d_3$. The midpoint of the line segment joining $\left(\frac{2x_1 + x_2}{3}, \frac{2y_1 + y_2}{3}\right)$ and (x_2, y_2) is

$$\text{Midpoint} = \left(\frac{\frac{2x_1 + x_2}{3} + x_2}{2}, \frac{\frac{2y_1 + y_2}{3} + y_2}{2}\right) = \left(\frac{x_1 + 2x_2}{3}, \frac{y_1 + 2y_2}{3}\right).$$

28. (a) $\left(\frac{2(1) + 4}{3}, \frac{2(-2) + 1}{3}\right) = (2, -1)$

$\left(\frac{1 + 2(4)}{3}, \frac{-2 + 2(1)}{3}\right) = (3, 0)$

(b) $\left(\frac{2(-2) + 0}{3}, \frac{2(-3) + 0}{3}\right) = \left(-\frac{4}{3}, -2\right)$

$\left(\frac{-2 + 2(0)}{3}, \frac{-3 + 2(0)}{3}\right) = \left(-\frac{2}{3}, -1\right)$

30. $c^2 = 200^2 + 125^2$

$c^2 = 55{,}625$

$c \approx 235.8495$ feet

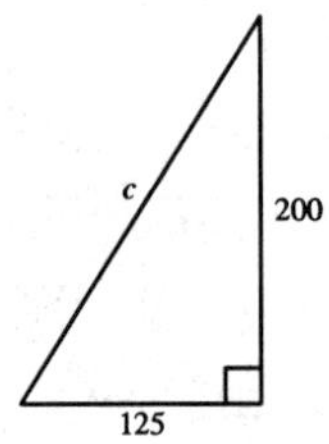

32.

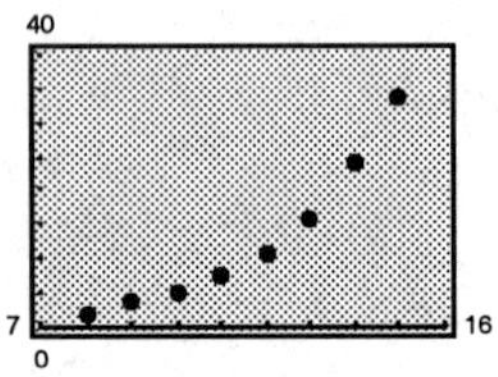

34. (a) $\frac{3850 - 3800}{3800} \approx 0.013 = 1.3\%$ (increase)

(b) $\frac{5600 - 5150}{5150} \approx 0.087 = 8.7\%$ (increase)

36. (a) $\frac{115 - 104}{104} \approx 0.106 = 10.6\%$ (increase)

(b) $\frac{95 - 101}{101} \approx -0.059 = -5.9\%$ (decrease)

38. (a) Revenue midpoint $= \left(\frac{1992 + 1996}{2}, \frac{21{,}970 + 31{,}645}{2}\right) = (1994, 26{,}807.5)$

Profit midpoint $= \left(\frac{1992 + 1996}{2}, \frac{1301.7 + 1865.0}{2}\right) = (1994, 1583.35)$

(b) Actual 1994 revenue: \$28,472 million; Actual 1994 profit: \$1784.0 million

(c) Yes, the increase in revenue per year is $\approx$ \$2419 million.

No, the profit from 1992 to 1994 was \$483 million, and the profit from 1994 to 1996 was \$81 million.

(d) Expenses for 1992: Expenses = Revenue − Profit = 21,970 − 1301.7 = \$20,668.3 million
Expenses for 1994: Expenses = Revenue − Profit = 28,472 − 1784 = \$26,688 million
Expenses for 1996: Expenses = Revenue − Profit = 31,645 − 1865 = \$29,780 million

(e) Answers vary.

40. (a) $(0, 2)$ is translated to $(0 - 3, 2 - 3) = (-3, -1)$

$(1, 3)$ is translated to $(1 - 3, 3 - 3) = (-2, 0)$

$(3, 1)$ is translated to $(3 - 3, 1 - 3) = (0, -2)$

$(2, 0)$ is translated to $(2 - 3, 0 - 3) = (-1, -3)$

(b)

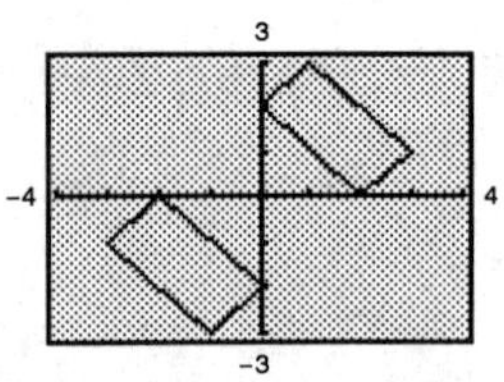

Section 1.2 Graphs of Equations

2. (a) This is a solution point since $(1)^2 + \left(-\sqrt{3}\right)^2 = 4$.

(b) This is not a solution point since $\left(\frac{1}{2}\right)^2 + (-1)^2 = \frac{5}{4} \neq 4$.

(c) This is not a solution point since $\left(\frac{3}{2}\right)^2 + \left(\frac{7}{2}\right)^2 = \frac{29}{2} \neq 4$.

4. (a) This is not a solution point since $(0)^2 - (0)(2) + 4(2) = 8 \neq 3$.

(b) This is a solution point since $(-2)^2 - (-2)\left(-\frac{1}{6}\right) + 4\left(-\frac{1}{6}\right) = 3$.

(c) This is a solution point since $(3)^2 - (3)(-6) + 4(-6) = 3$.

6. The graph of $y = -\frac{1}{2}x + 2$ is a straight line with y-intercept at $(0, 2)$. Thus, it matches (b).

8. The graph of $y = \sqrt{9 - x^2}$ is a semicircle with intercepts $(0, 3)$, $(3, 0)$, and $(-3, 0)$. Thus, it matches (f).

10. The graph of $y = x^3 - x$ has intercepts at $(0, 0)$, $(1, 0)$, and $(-1, 0)$. Thus, it matches (d).

12. The y-intercept is $(0, 3)$. The x-intercepts are $(1, 0)$ and $(3, 0)$.

14. The y-intercept is $(0, 0)$. To find the x-intercepts, let $y = 0$ to obtain

$$0 = x^3 - 4x$$
$$0 = x(x^2 - 4)$$
$$0 = x(x + 2)(x - 2)$$
$$x = 0, \pm 2.$$

Thus, the x-intercepts are $(0, 0)$, $(-2, 0)$ and $(2, 0)$.

16. Since $xy = 4$, we note that $x \neq 0$ and $y \neq 0$. Therefore, there are no intercepts.

18. The y-intercept is $(0, 0)$. To find the x-intercepts, let $y = 0$ to obtain

$$0 = \frac{x^2 + 3x}{(3x + 1)^2}$$
$$0 = x^2 + 3x$$
$$0 = x(x + 3)$$
$$x = 0, -3.$$

Thus, the x-intercepts are $(0, 0)$ and $(-3, 0)$.

20. The y-intercept is $(0, 1)$. There are no x-intercepts since $x^2 + 1 \neq 0$ for any real value of x.

22. The graph of $y = -3x + 2$ is a straight line with slope -3 and $y =$ intercept $(0, 2)$.

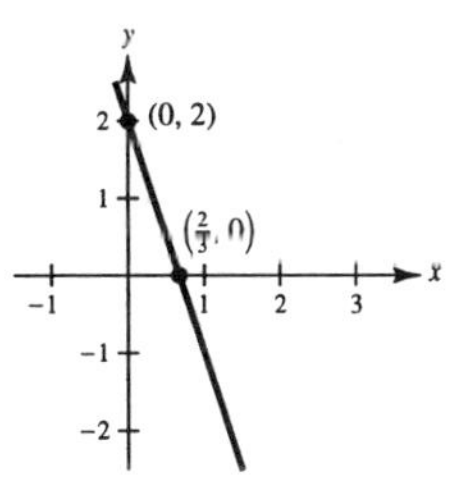

24. The graph of $y = x^2 + 3$ is a parabola with vertex at $(0, 3)$, which is also the only intercept.

x	0	± 1	± 2
y	3	4	7

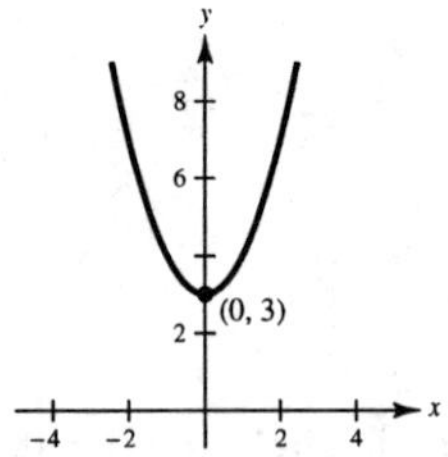

26. Intercepts: $(0, 1)$ and $(1, 0)$

x	0	1	-1	2
y	1	0	2	-7

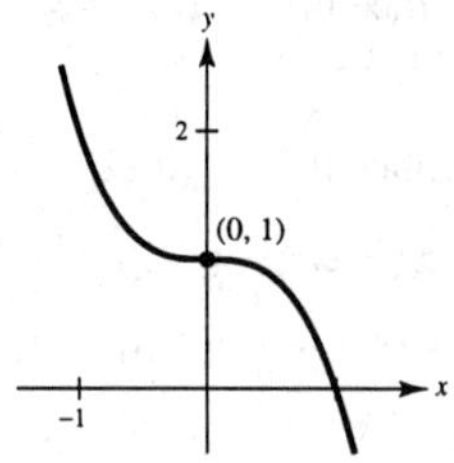

28. The graph of $y = \sqrt{x + 1}$ is a translation of $y = \sqrt{x}$ one unit to the left.

Intercepts: $(-1, 0)$, $(0, 1)$

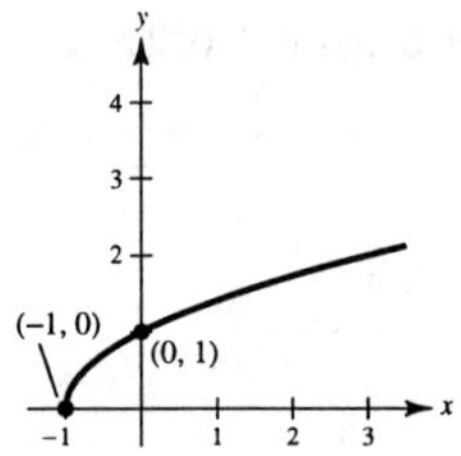

30. Intercepts: $(2, 0)$ and $(0, -2)$

x	2	0	1	3	4
y	0	-2	-1	-1	-2

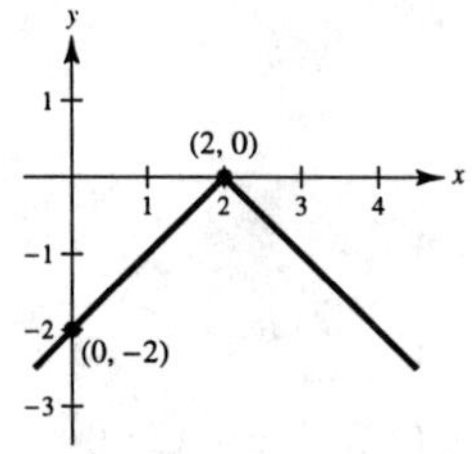

32. Intercept: $(0, 1)$

x	0	± 1	± 2	± 3
y	1	$\frac{1}{2}$	$\frac{1}{5}$	$\frac{1}{10}$

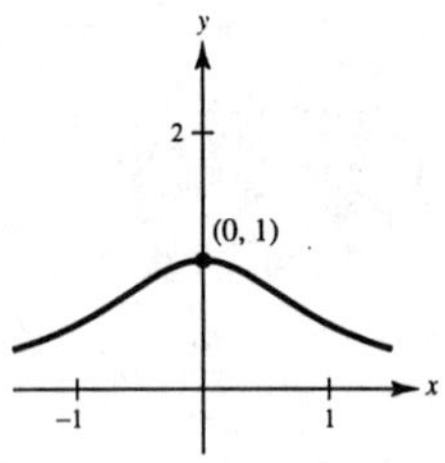

34. Intercepts: $(0, 2)$, $(0, -2)$, $(4, 0)$

x	0	3	4
y	± 2	± 1	0

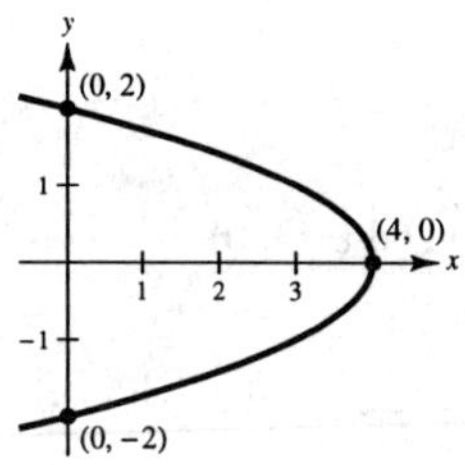

36.
$$\begin{aligned}(x - 0)^2 + (y - 0)^2 &= 5^2\\ x^2 + y^2 &= 25\\ x^2 + y^2 - 25 &= 0\end{aligned}$$

38.
$$\begin{aligned}(x + 4)^2 + (y - 3)^2 &= \left(\tfrac{5}{8}\right)^2\\ x^2 + 8x + 16 + y^2 - 6y + 9 &= \tfrac{25}{64}\\ 64x^2 + 64y^2 + 512x - 384y + 1575 &= 0\end{aligned}$$

40.
$$\begin{aligned}\text{Radius} &= \sqrt{(-1 - 3)^2 + (1 + 2)^2} = 5\\ (x - 3)^2 + (y + 2)^2 &= 5^2\\ x^2 - 6x + 9 + y^2 + 4y + 4 &= 25\\ x^2 + y^2 - 6x + 4y - 12 &= 0\end{aligned}$$

42. Center = midpoint $= \left(\dfrac{-4+4}{2}, \dfrac{-1+1}{2}\right) = (0, 0)$

Radius = distance from the center to an endpoint $= \sqrt{(4-0)^2 + (1-0)^2} = \sqrt{17}$

$$(x-0)^2 + (y-0)^2 = \left(\sqrt{17}\right)^2$$
$$x^2 + y^2 - 17 = 0$$

44. $(x^2 - 2x + 1) + (y^2 + 6y + 9) = 15 + 1 + 9$

$$(x-1)^2 + (y+3)^2 = 25$$

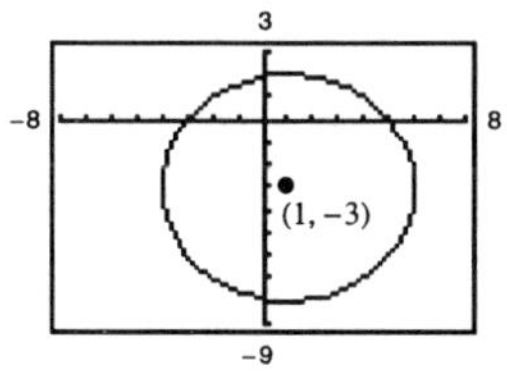

46. $x^2 + y^2 - 2y - \frac{1}{3} = 0$

$$x^2 + (y^2 - 2y + 1) = \tfrac{1}{3} + 1$$
$$x^2 + (y-1)^2 = \tfrac{4}{3}$$

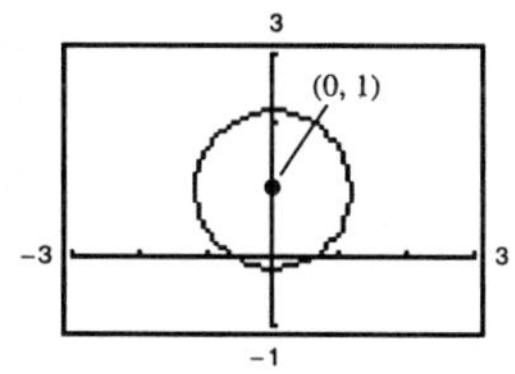

48. $x^2 + y^2 - x + \frac{1}{2}y - \frac{1}{4} = 0$

$$\left(x^2 - x + \tfrac{1}{4}\right) + \left(y^2 + \tfrac{1}{2}y + \tfrac{1}{16}\right) = \tfrac{1}{4} + \tfrac{1}{4} + \tfrac{1}{16}$$
$$\left(x - \tfrac{1}{2}\right)^2 + \left(y + \tfrac{1}{4}\right)^2 = \tfrac{9}{16}$$

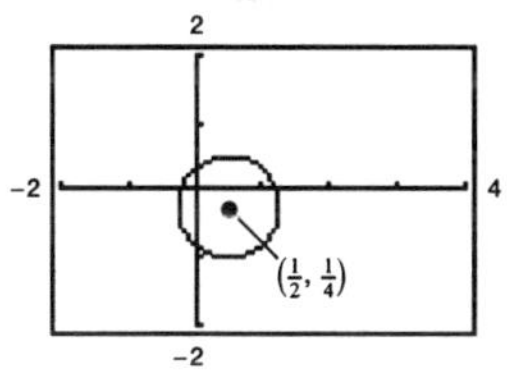

50. $(x^2 - 4x + 4) + (y^2 + 2y + 1) = -3 + 4 + 1$

$$(x-2)^2 + (y+1)^2 = 2$$

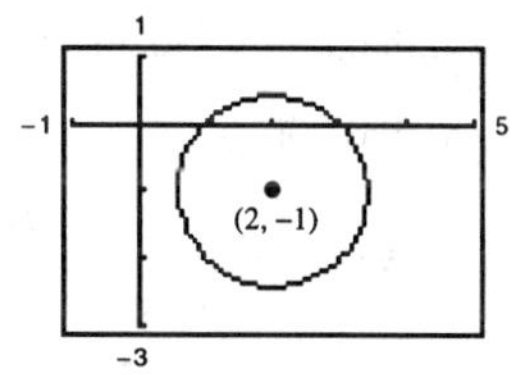

52. The first equation gives $y = 7 - x$. Hence,

$$3x - 2(7 - x) = 5x - 14 = 11$$
$$5x = 25$$
$$x = 5, y = 2$$

54. Solving for y in the second equation yields $y = 2x - 1$ and substituting this value into the first equation gives us the following.

$$x^2 + (2x - 1) = 4$$
$$x^2 + 2x - 5 = 0$$
$$x = -1 \pm \sqrt{6} \text{ by the Quadratic Formula}$$

The corresponding y-values are $y = -3 \pm 2\sqrt{6}$, so the points of intersection are $\left(-1 + \sqrt{6}, -3 + 2\sqrt{6}\right)$ and $\left(-1 - \sqrt{6}, -3 - 2\sqrt{6}\right)$.

56. By equating the y-values for the two equations, we have

$$\sqrt{x} = x$$
$$x = x^2$$
$$0 = x(x-1)$$
$$x = 0, 1.$$

The corresponding y-values are $y = 0, 1$, so the points of intersection are $(0, 0)$ and $(1, 1)$.

58. By equating the y-values for the two equations, we have

$$x^3 - 2x^2 + x - 1 = -x^2 + 3x - 1$$
$$x^3 - x^2 - 2x = 0$$
$$x(x+1)(x-2) = 0$$
$$x = 0, -1, 2.$$

The corresponding y-values are $y = -1, -5$, and 1, so the points of intersection are $(0, -1)$, $(-1, -5)$, and $(2, 1)$.

60. (a) $C_g = 14{,}500 + 1.389\left(\frac{x}{22}\right)$

$= 14{,}500 + \frac{1.389x}{22}$

$C_d = 15{,}450 + 1.389\left(\frac{x}{31}\right)$

$= 15{,}450 + \frac{1.389x}{31}$

(b) $C_g = C_d$

$14{,}500 + \frac{1.389x}{22} = 15{,}450 + \frac{1.389x}{31}$

$\left(\frac{1.389}{22} - \frac{1.389}{31}\right)x = 950$

$0.018329912x \approx 950$

$x \approx 51{,}828$ miles

62. $R = C$

$35x = 6x + 500{,}000$

$29x = 500{,}000$

$x = 500{,}000/29 \approx 17{,}242$ units

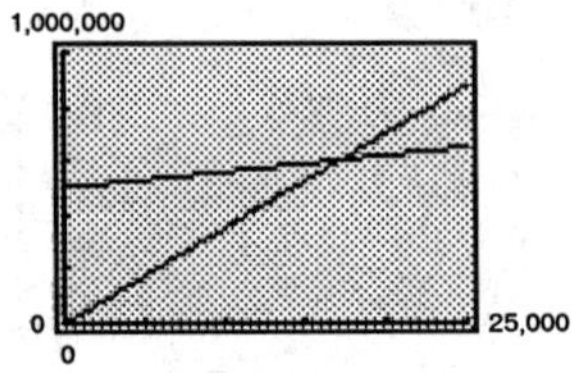

64.

$R = C$

$3.29x = 5.5\sqrt{x} + 10{,}000$

$(3.29x - 10{,}000)^2 = (5.5\sqrt{x})^2$

$10.8241x^2 - 65{,}800x + 100{,}000{,}000 = 30.25x$

$10.8241x^2 - 65{,}830.25x + 100{,}000{,}000 = 0$

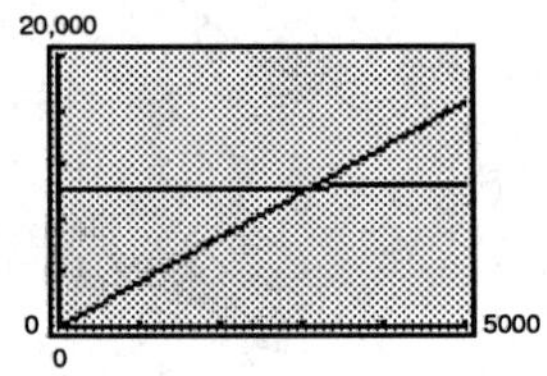

By using the Quadratic Formula, we have $x = \frac{65{,}830.25 \pm \sqrt{3{,}981{,}815.062}}{21.6482}$

$x \approx 3133$ units.

[**Note:** $x = 2949$ units is an extraneous solution.]
You can also solve this problem with a graphing utility by determining the point of intersection of the two equations $y_1 = R = 3.29x$ and $y_2 = C = 5.5\sqrt{x} + 10{,}000$.

66. (a)

Year	1985	1988	1990	1993	1994
Expense	1575	1998	2319	2498	2536
Model	1578	1988	2300	2470	2492

The model is a good fit.

(b) 2000: $y(20) \approx \$5750$ million

68. (a)

Year	1985	1990	1992	1994	1995	1996
Salary	1482.93	1596.20	1716.49	1860.97	1940.99	2025.76

(b) It is probably accurate.

(c) For 2000, $t = 20$ and Salary $\approx \$2408.5$.

70. (a)

Year	1990	1991	1992	1993	1994	1995
Transplants	217.5	1384.4	2531.1	3657.6	4763.9	5850

(b) The actual data is

Year	1990	1991	1992	1993	1994	1995
Transplants	203	1405	2535	3667	4723	5871

The model is a good fit.

(c) For $t = 10$, $y \approx 10{,}978$ transplants.

72. Intercepts: $(0, 6.25), (1.0539, 0), (-10.5896, 0)$

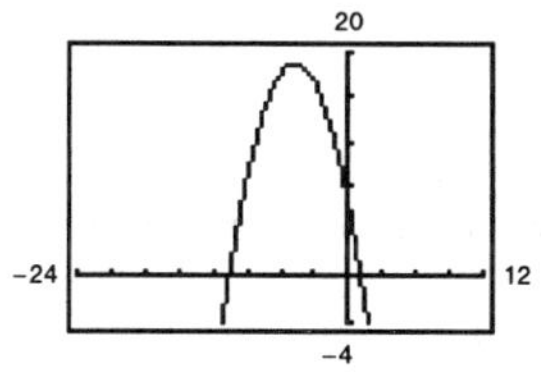

74. Intercept: $(0, 0.4167)$

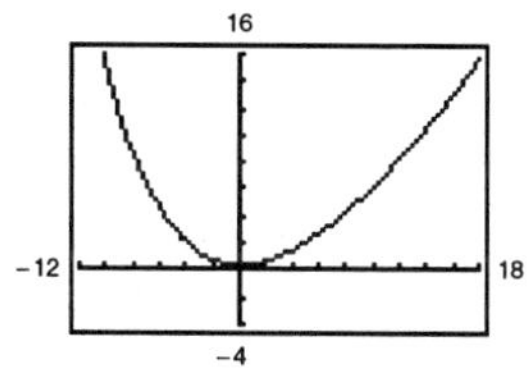

Section 1.3 Lines in the Plane and Slope

2. The slope is 2 since the line rises two units vertically for each unit of horizontal change from left to right.

4. The slope is -1 since the line falls one unit vertically for each unit of horizontal change from left to right.

6. The points are plotted in the accompanying graph and the slope is

$$m = \frac{-3 - 1}{4 + 2} = -\frac{2}{3}.$$

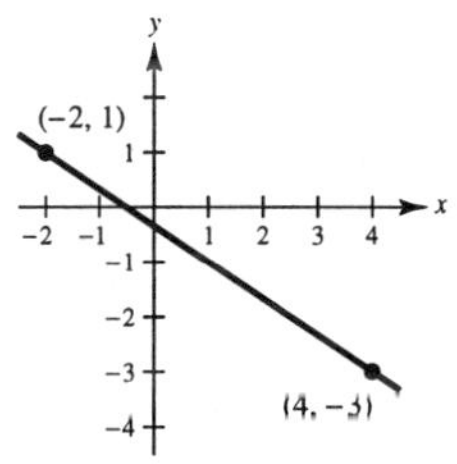

8. The points are plotted in the accompanying graph and the slope is

$$m = \frac{4 + 5}{(5/6) + (3/2)} = \frac{27}{7}.$$

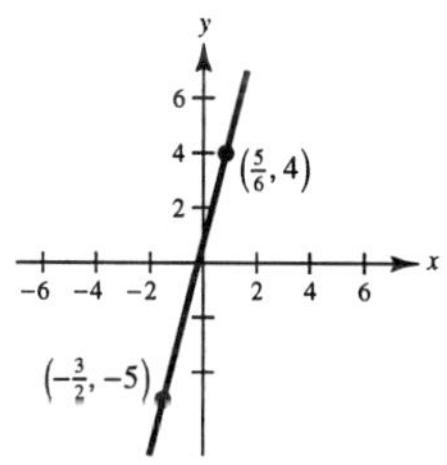

10. The points are plotted in the accompanying graph. The slope is undefined since

$$m = \frac{5 - 1}{2 - 2}.$$

Thus, the line is vertical.

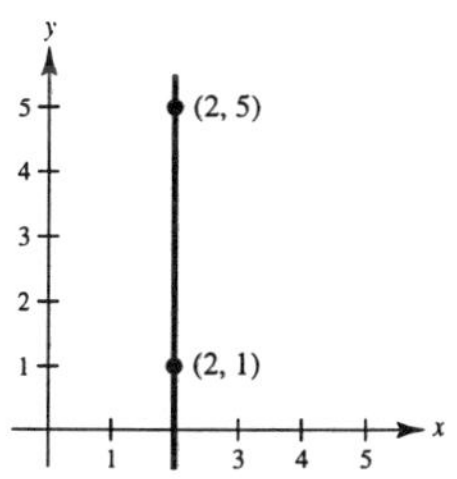

12. The points are plotted in the accompanying graph and the slope is

$$m = \frac{(-1/4) - (3/4)}{(5/4) - (7/8)} = -\frac{8}{3}.$$

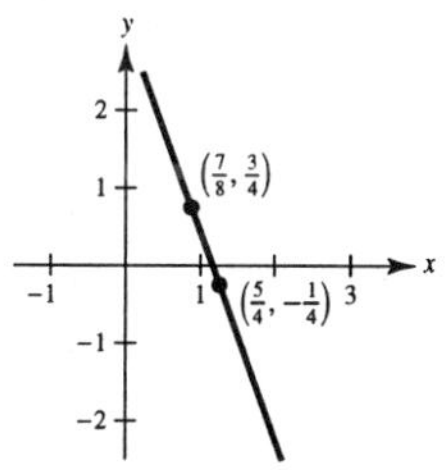

14. The equation of this vertical line is $x = -3$. Therefore, three additional points are $(-3, 0)$, $(-3, 1)$, and $(-3, 2)$.

16. The equation of this line is

$$y + 2 = 2(x + 2)$$
$$y = 2x + 2.$$

Therefore, three additional points are $(-1, 0)$, $(0, 2)$, and $(1, 4)$.

18. The equation of this line is

$$y + 6 = -1(x - 10)$$
$$y = -x + 4.$$

Therefore, three additional points are $(11, -7)$, $(9, -5)$, and $(8, -4)$.

20. The equation of this horizontal line is $y = -1$. Therefore, three additional points are $(0, -1)$, $(1, -1)$, and $(2, -1)$.

22. $2x + y = 40$

$$y = -2x + 40$$

Therefore, the slope is $m = -2$, and the y-intercept is $(0, 40)$.

24. $6x - 5y = 15$

$$y = \frac{6}{5}x - 3$$

Therefore, the slope is $m = \frac{6}{5}$, and the y-intercept is $(0, -3)$.

26. Since the line is horizontal, the slope is $m = 0$, and the y-intercept is $(0, -1)$.

28. The slope of the line is

$$m = \frac{4 - (-4)}{1 - (-3)} = 2.$$

Using the point-slope form, we have

$$y - 4 = 2(x - 1)$$
$$y = 2x + 2$$
$$0 = 2x - y + 2.$$

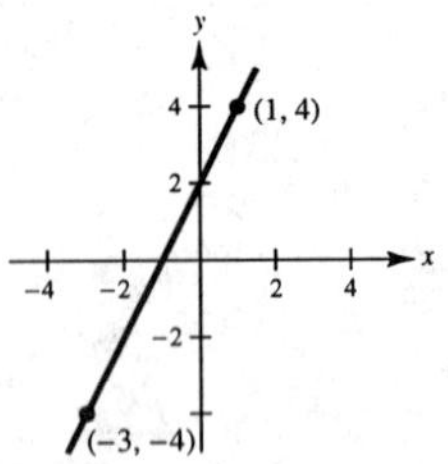

30. The slope of the line is

$$m = \frac{2 - 6}{1 - (-3)} = -1.$$

Using the point-slope form, we have

$$y - 2 = -1(x - 1)$$
$$y = -x + 3$$
$$x + y - 3 = 0.$$

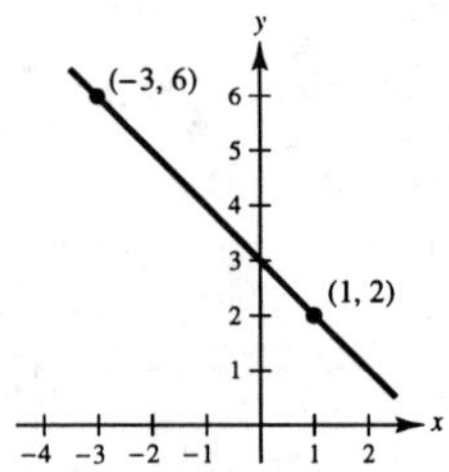

32. The slope of the line is $m = \dfrac{1 - 1}{10 - 6} = 0.$

The line is horizontal and its equation is

$$y = 1$$
$$y - 1 = 0.$$

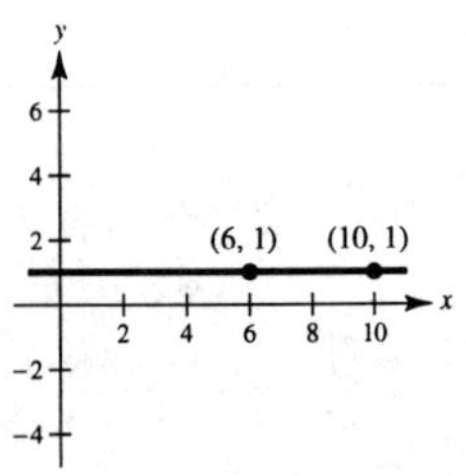

34. The slope of the line is $m = \dfrac{(-1/4) - (3/4)}{(5/4) - (7/8)} = -\dfrac{8}{3}$.

Using the point-slope form, we have

$$y - \frac{3}{4} = -\frac{8}{3}\left(x - \frac{7}{8}\right)$$

$$y - \frac{3}{4} = -\frac{8}{3}x + \frac{7}{3}$$

$$12y - 9 = -32x + 28$$

$$32x + 12y - 37 = 0.$$

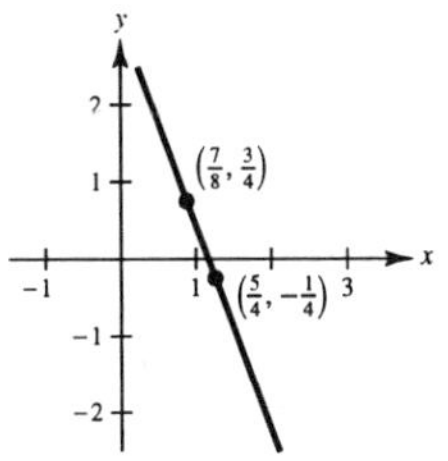

36. Since the slope is undefined, the line is vertical and its equation is $x = -1$.

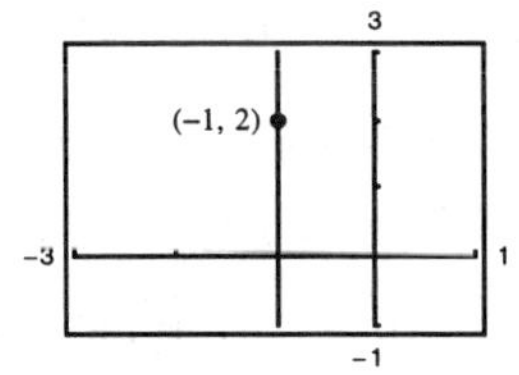

38. Using the point-slope form, we have

$$y + 4 = \tfrac{1}{4}(x + 1)$$

$$4y + 16 = x + 1$$

$$0 = x - 4y - 15.$$

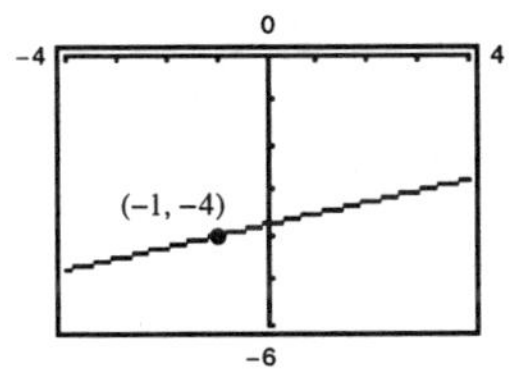

40. Using the point-slope form, we have

$$y - 4 = -\tfrac{3}{5}(x + 2)$$

$$5y - 20 = -3x - 6$$

$$3x + 5y - 14 = 0.$$

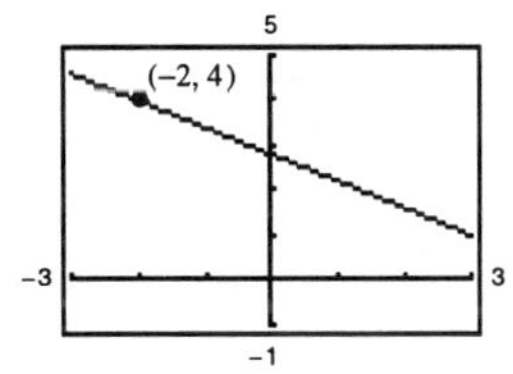

42. Using the point-slope form, we have

$$y + \tfrac{2}{3} = \tfrac{1}{6}(x - 0)$$

$$6y + 4 = x$$

$$0 = x - 6y - 4.$$

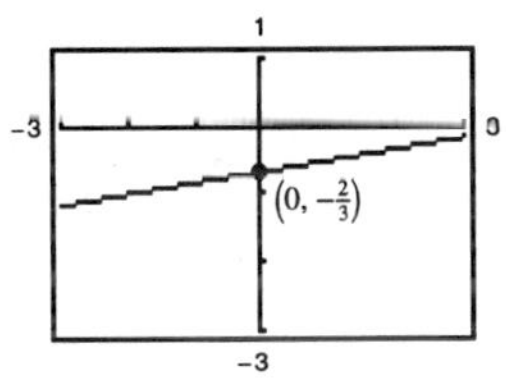

44. Since $m = 0$, the line is horizontal and its equation is $y = 4$.

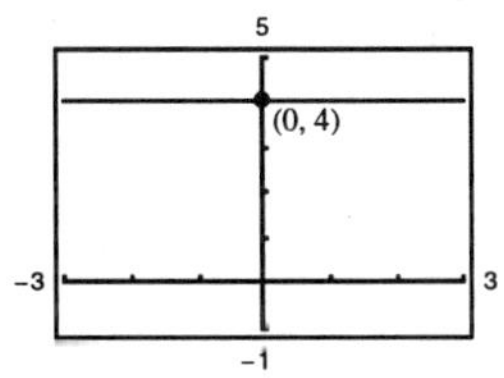

46. The slope of the line joining $(-5, 11)$ and $(0, 4)$ is $\frac{11-4}{-5-0} = \frac{7}{-5} = -\frac{7}{5}$.

The slope of the line joining $(0, 4)$ and $(7, -6)$ is $\frac{4-(-6)}{0-7} = -\frac{10}{7}$.

Since the slopes are different, the points are not collinear.

$$d_1 = \sqrt{(-5-0)^2 + (11-4)^2} = \sqrt{25+49} = \sqrt{74} \approx 8.60233$$
$$d_2 = \sqrt{(7-0)^2 + (-6-4)^2} = \sqrt{49+100} = \sqrt{149} \approx 12.20656$$
$$d_3 = \sqrt{(-5-7)^2 + [11-(-6)]^2} = \sqrt{144+289} = \sqrt{433} \approx 20.80865$$

Since $d_1 + d_2 \neq d_3$, the points are not collinear.

48. Since the line is horizontal, it has a slope of $m = 0$, and its equation is

$$y = 0x + (-5)$$
$$y = -5.$$

50. Given line: $y = 2x - \frac{3}{2}$

(a) Parallel: $m_1 = 2$

$$y - 1 = 2(x - 2)$$
$$0 = 2x - y - 3$$

(b) Perpendicular: $m_2 = -\frac{1}{2}$

$$y - 1 = -\frac{1}{2}(x - 2)$$
$$2y - 2 = -x + 2$$
$$x + 2y - 4 = 0$$

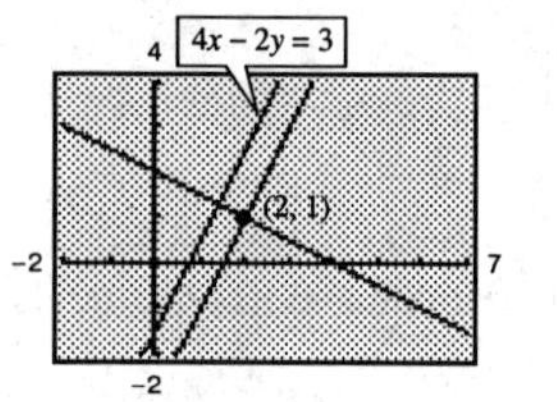

52. Given line: $y = -\frac{5}{3}x$

(a) Parallel: $m_1 = -\frac{5}{3}$

$$y - \frac{3}{4} = -\frac{5}{3}\left(x - \frac{7}{8}\right)$$
$$y - \frac{3}{4} = -\frac{5}{3}x + \frac{35}{24}$$
$$24y - 18 = -40x + 35$$
$$40x + 24y - 53 = 0$$

(b) Perpendicular: $m_2 = \frac{3}{5}$

$$y - \frac{3}{4} = \frac{3}{5}\left(x - \frac{7}{8}\right)$$
$$y - \frac{3}{4} = \frac{3}{5}x - \frac{21}{40}$$
$$40y - 30 = 24x - 21$$
$$0 = 24x - 40y + 9$$

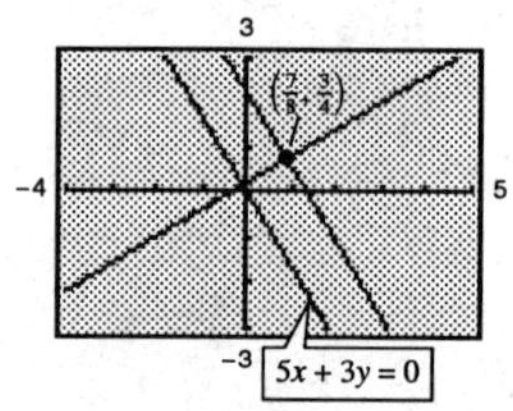

54. Given line: $x = 4$ is vertical.

(a) Parallel: m is undefined at $x = 2$.

(b) Perpendicular: $m = 0, y = 5$

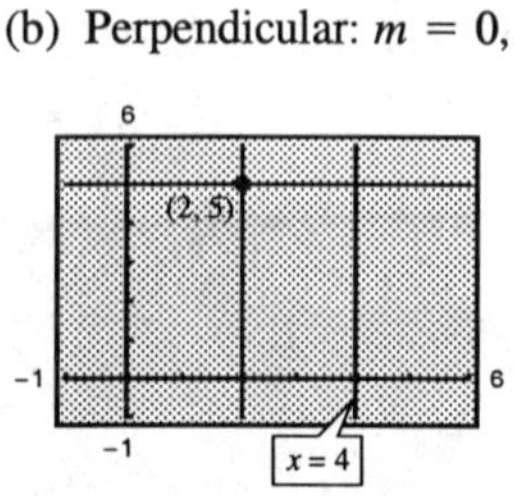

56. Given line: $y = -\frac{1}{4}x + 1$

(a) Parallel: $m_1 = -\frac{1}{4}$

$$y - (-3) = -\frac{1}{4}(x - 12)$$
$$y = -\frac{1}{4}x$$
$$4y + x = 0$$

(b) Perpendicular: $m_2 = 4$

$$y - (-3) = 4(x - 12)$$
$$y = 4x - 51$$
$$y - 4x + 51 = 0$$

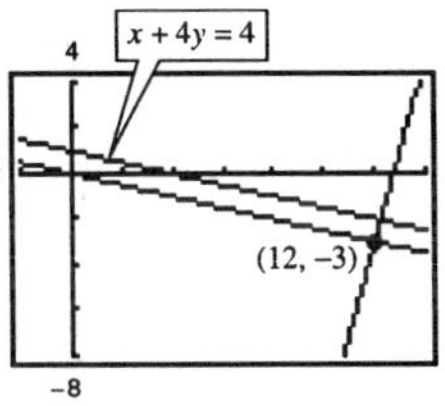

58. $x = 4$ is a vertical line with x-intercept of $(4, 0)$.

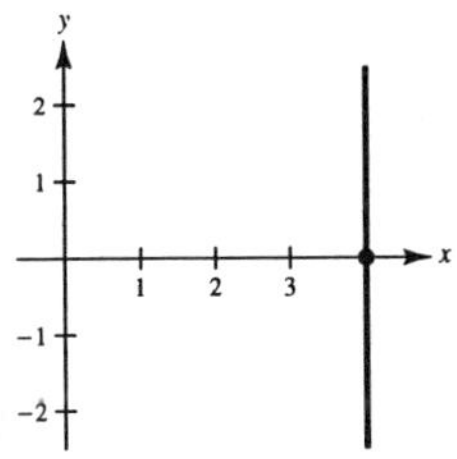

60. $x + 2y + 6 = 0$ has intercepts at $(-6, 0)$ and $(0, -3)$.

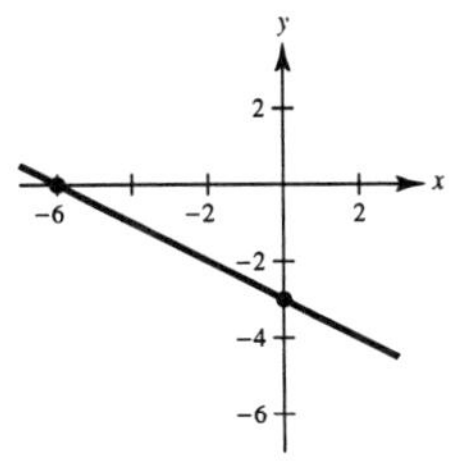

62. $y = 3x + 13$ has intercepts at $(0, 13)$ and $\left(-\frac{13}{3}, 0\right)$.

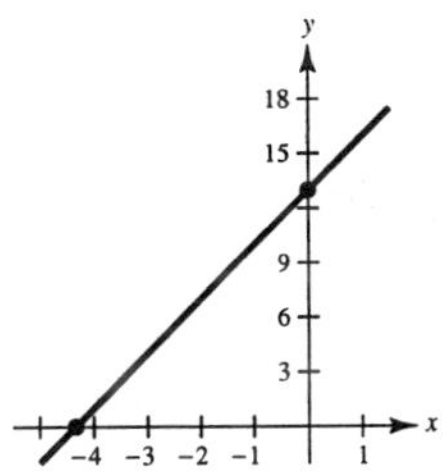

64. $4x + 5y - 20 = 0$ has intercepts at $(5, 0)$ and $(0, 4)$.

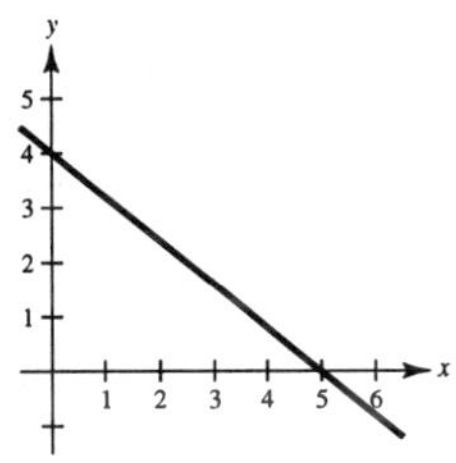

66. $(0, 32), (100, 212)$

$$F - 32 = \frac{212 - 32}{100 - 0}(C - 0)$$
$$F = 1.8C + 32 = \frac{9}{5}C + 32$$

or

$$C = \frac{5}{9}(F - 32)$$

68. $C = 150 + 0.31x$

70. (a) $(0, -45), (100, 115)$

$$C + 45 = \frac{115 + 45}{100 - 0}(A - 0)$$
$$C = \frac{8}{5}A - 45 \text{ or } A = \frac{5}{8}(C + 45)$$

(b) $C = \frac{5}{9}(F - 32) = \frac{8}{5}A - 45$

$$25F - 800 = 72A - 2025$$
$$25F = 72A - 1225$$
$$F = \frac{72}{25}A - 49 \text{ or } A = \frac{25}{72}(F + 49)$$

(c) $C = \frac{8}{5}A - 45 = A$

$$\frac{3}{5}A = 45$$
$$A = 75°$$

(d) $C = \frac{8}{5}(86) - 45 = 92.6°$; $F = \frac{72}{25}(86) - 49 = 198.68°$

(e) $A = \frac{5}{8}(45 + 45) = 56.25°$

72. (a) The equipment depreciates

$$\frac{1025}{5} = \$205 \text{ per year,}$$

so the value is $y = 1025 - 205t$, where $0 \le t \le 5$.

(b) 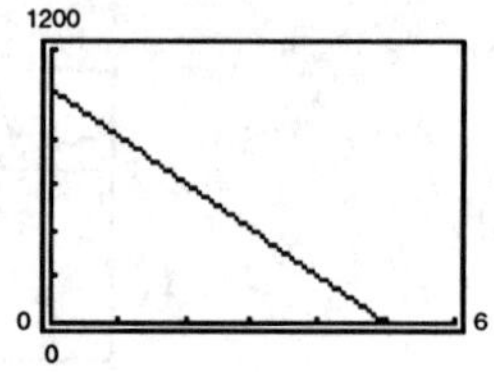

(c) When $t = 3$, the value is \$410.00.

(d) The value is \$600 when $t = 2.07$ years.

74. (a) $C = (5.25 + 9.50)t + 26{,}500 = 14.75t + 26{,}500$

(b) $R = 25t$

(c) $P = R - C = 25t - (14.75t + 26{,}500) = 10.25t - 26{,}500$

(d)
$$\begin{aligned} R &= C \\ 25t &= 14.75t + 26{,}500 \\ 10.25t &= 26{,}500 \\ t &\approx 2585.4 \text{ hours} \end{aligned}$$

76. (a) $W = 2000 + .07S$

(b) $W = 2300 + .05S$

(c) 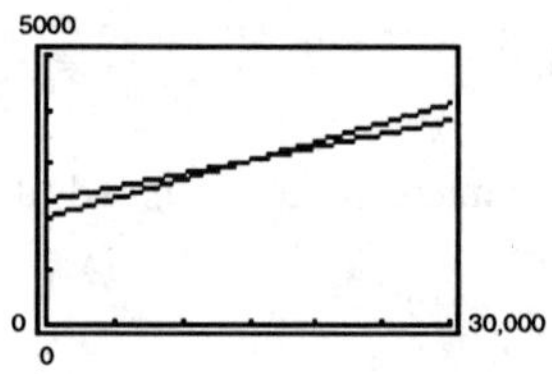

The lines intersect at (15,000, 3050). If you sell \$15,000, then both jobs would yield wages of \$3050.

(d) No. You will make more money (if sales are \$20,000) at your current job ($w = \$3400$) than in the offered job ($w = \$3300$).

78. $C = 30{,}000 + 575x$ where $C \le 100{,}000$.

Thus,
$$\begin{aligned} 30{,}000 + 575x &\le 100{,}000 \\ 575 &\le 70{,}000 \\ x &\le 121.739 \approx 122 \text{ units.} \end{aligned}$$

Therefore, $x \le 121$ units.

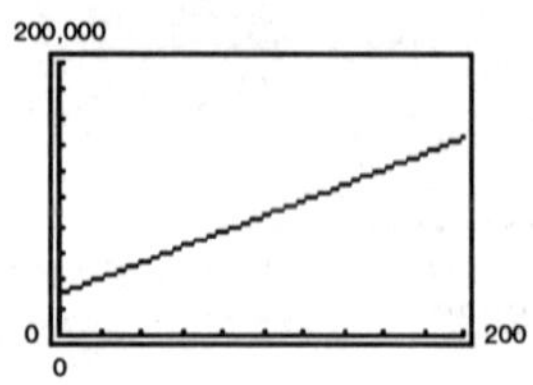

80.
$$\begin{aligned} C = 24{,}900 + 1785x &\le 100{,}000 \\ 1785x &\le 75{,}100 \\ x &\le 42.07 \\ x &\le 42 \text{ units} \end{aligned}$$

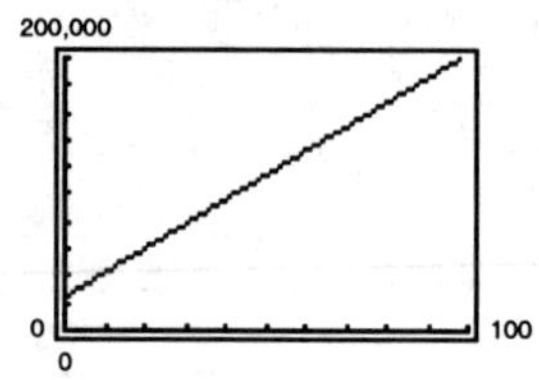

82.
$$\begin{aligned} C = 83{,}620 + 67x &\le 100{,}000 \\ 67x &\le 16{,}380 \\ x &\le 244.48 \\ x &\le 244 \text{ units} \end{aligned}$$

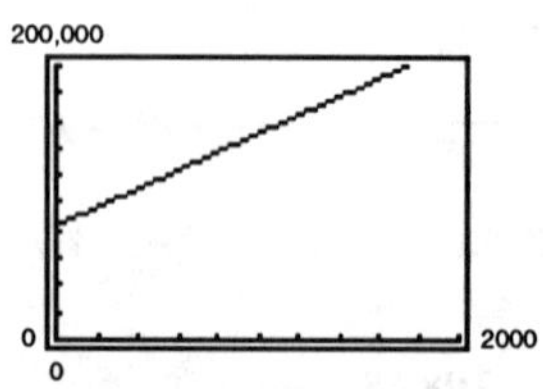

Section 1.4 Functions

2. (a) $f\left(\frac{1}{2}\right) = \left(\frac{1}{2}\right)^2 - 2\left(\frac{1}{2}\right) + 2 = \frac{5}{4}$

(b) $f(-1) = (-1)^2 - 2(-1) + 2 = 5$

(c) $f(c) = c^2 - 2c + 2$

(d) $f(x + \Delta x) = (x + \Delta x)^2 - 2(x + \Delta x) + 2$

$= x^2 + 2x\Delta x + (\Delta x)^2 - 2x - 2\Delta x + 2$

4. (a) $f(2) = |2| + 4 = 6$

(b) $f(-2) = |-2| + 4 = 6$

(c) $f(x^2) = |x^2| + 4 = x^2 + 4$

(d) $f(x + \Delta x) - f(x) = (|x + \Delta x| + 4) - (|x| + 4)$

$= |x + \Delta x| - |x|$

6.
$$\frac{h(2 + \Delta x) - h(2)}{\Delta x} = \frac{(2 + \Delta x)^2 - (2 + \Delta x) + 1 - (4 - 2 + 1)}{\Delta x}$$
$$= \frac{4 + 4\Delta x + (\Delta x)^2 - 2 - \Delta x + 1 - 3}{\Delta x}$$
$$= \frac{\Delta x(3 + \Delta x)}{\Delta x}$$
$$= 3 + \Delta x, \Delta x \neq 0$$

8.
$$\frac{f(x) - f(2)}{x - 2} = \frac{\left(1/\sqrt{x - 1}\right) - 1}{x - 2}$$
$$= \frac{1 - \sqrt{x - 1}}{(x - 2)\sqrt{x - 1}} \cdot \frac{1 + \sqrt{x - 1}}{1 + \sqrt{x - 1}} = \frac{1 - (x - 1)}{(x - 2)\sqrt{x - 1}\left(1 + \sqrt{x - 1}\right)}$$
$$= \frac{2 - x}{(x - 2)\sqrt{x - 1}\left(1 + \sqrt{x - 1}\right)} = \frac{-1}{\sqrt{x - 1}\left(1 + \sqrt{x - 1}\right)}, x \neq 2$$

10. $y = \dfrac{3x + 5}{2}$

y *is* a function of x since there is only one value of y for each x.

12. $y = \pm\sqrt{4 - x}$

y is *not* a function of x since there are two values of y for some x.

14. $(x^2 - 2x + 1) + (y^2 - 4y + 4) = -1 + 1 + 4$

$(x - 1)^2 + (y - 2)^2 = 4$

The graph is a circle; therefore, by the vertical line test, y is *not* a function of x.

16. $y(x^2 + 4) = x^2$

$$y = \frac{x^2}{x^2 + 4}$$

y *is* a function of x since there is only one value of y for each x. [**Note:** It is not a one-to-one function.]

18. Domain: $(-\infty, \infty)$

Range: $(-\infty, \infty)$

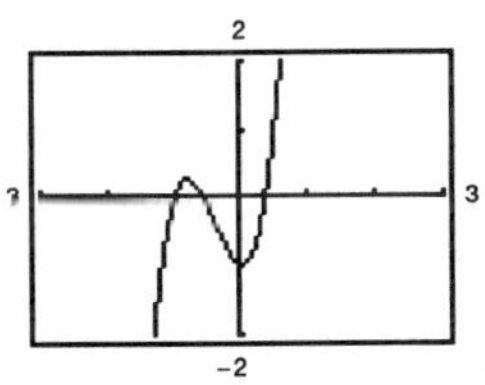

20. Domain: $(0, \infty)$

Range: $(0, \infty)$

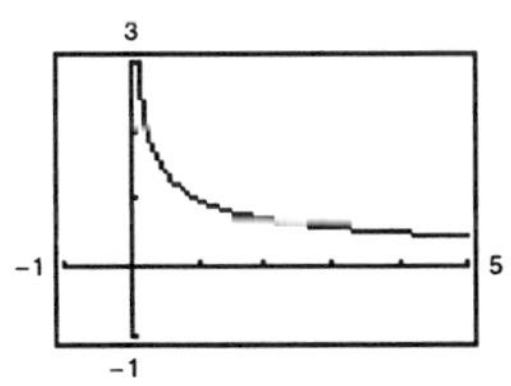

22. Domain: $(-\infty, 1) \cup (1, \infty)$

Range: $(-\infty, -4] \cup [0, \infty)$

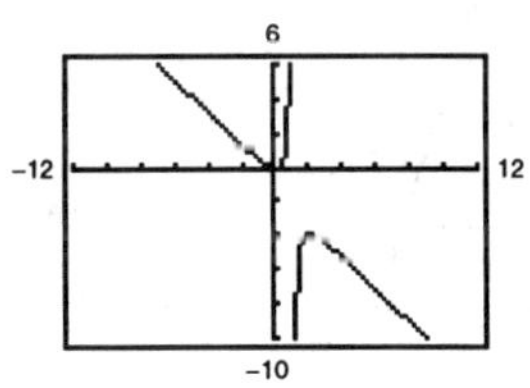

24. Domain: $\left[\frac{3}{2}, \infty\right)$

Range: $[0, \infty)$

26. Domain: $(-\infty, \infty)$

Range: $[0, \infty)$

28. y *is* a function of x.

30. y is *not* a function of x.

32. (a) $f(x) + g(x) = (2x - 5) + 5 = 2x$

(b) $f(x) \cdot g(x) = (2x - 5)(5) = 10x - 25$

(c) $\dfrac{f(x)}{g(x)} = \dfrac{2x - 5}{5}$

(d) $f(g(x)) = f(5) = 2(5) - 5 = 5$

(e) $g(f(x)) = g(2x - 5) = 5$

34. (a) $f(x) + g(x) = \dfrac{x}{x + 1} + x^3 = \dfrac{x^4 + x^3 + x}{x + 1}$

(b) $f(x) \cdot g(x) = \left(\dfrac{x}{x + 1}\right)(x^3) = \dfrac{x^4}{x + 1}$

(c) $\dfrac{f(x)}{g(x)} = \dfrac{\left(\dfrac{x}{x + 1}\right)}{x^3} = \dfrac{1}{x^2(x + 1)}$

(d) $f(g(x)) = f(x^3) = \dfrac{x^3}{x^3 + 1}$

(e) $g(f(x)) = g\left(\dfrac{x}{x + 1}\right) = \left(\dfrac{x}{x + 1}\right)^3 = \dfrac{x^3}{(x + 1)^3}$

36. (a) $f(g(x)) = f\left(\dfrac{1}{x}\right) = \dfrac{1}{1/x} = x, \ x \neq 0$

$g(f(x)) = g\left(\dfrac{1}{x}\right) = \dfrac{1}{1/x} = x, \ x \neq 0$

(b) See accompanying graph.

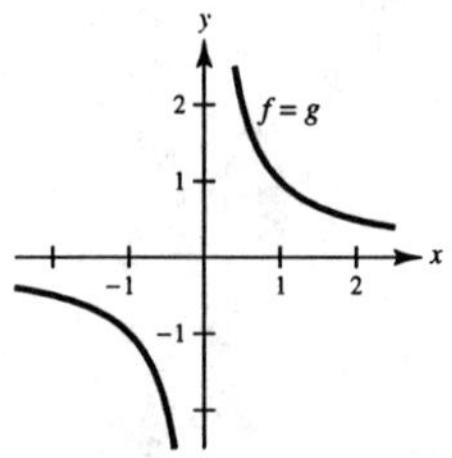

38. (a) $f(g(x)) = f\left(\sqrt[3]{1 - x}\right)$

$= 1 - \left(\sqrt[3]{1 - x}\right)^3 = x$

$g(f(x)) = g(1 - x^3)$

$= \sqrt[3]{1 - (1 - x^3)} = x$

(b) See accompanying graph.

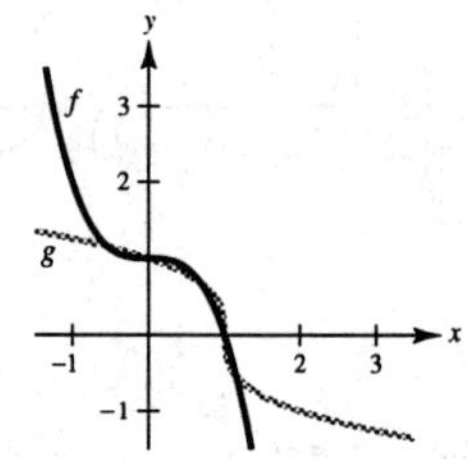

40. $f(x) = 6 - 3x = y$

$x = 6 - 3y$

$y = \dfrac{6 - x}{3}$

$f^{-1}(x) = \dfrac{6 - x}{3}$

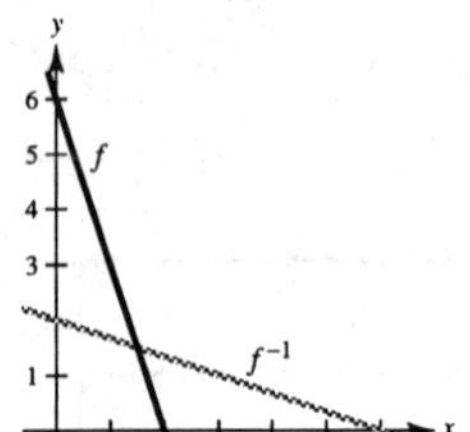

42. $f(x) = x^3 + 1 = y$

$x = y^3 + 1$

$\sqrt[3]{x - 1} = y$

$f^{-1}(x) = \sqrt[3]{x - 1}$

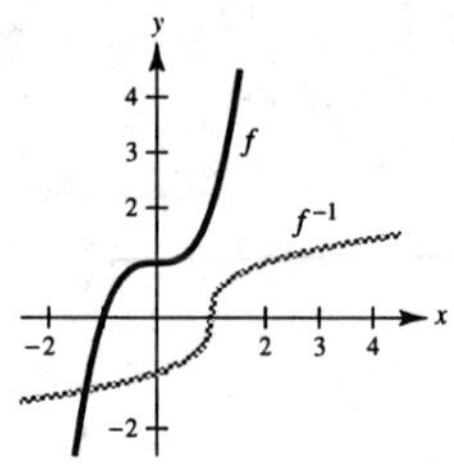

44. $f(x) = \sqrt{x^2 - 4} = y,\ x \ge 2$

$x = \sqrt{y^2 - 4}$

$x^2 + 4 = y^2$

$y = \sqrt{x^2 + 4},\ x \ge 0$

$f^{-1}(x) = \sqrt{x^2 + 4},\ x \ge 0$

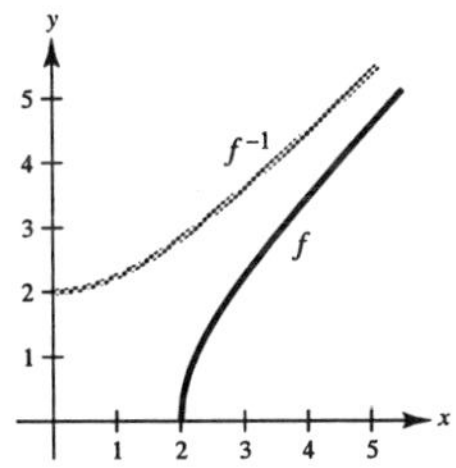

46. $f(x) = x^{3/5} = y$

$x = y^{3/5}$

$x^{5/3} = y$

$f^{-1}(x) = x^{5/3}$

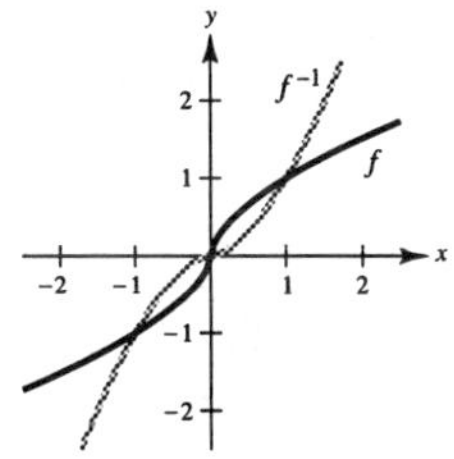

48.

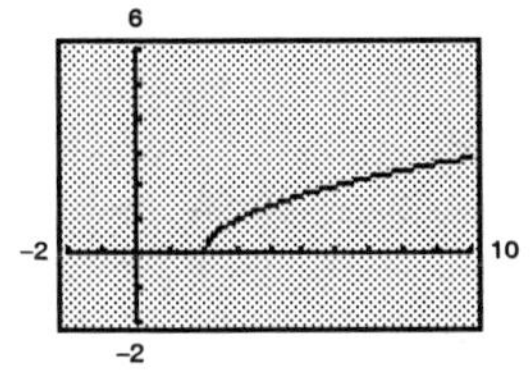

$f(x) = \sqrt{x - 2}$ *is* one-to-one for $x \ge 2$.

$f^{-1}(x) = x^2 + 2$ where $x \ge 0$.

50.

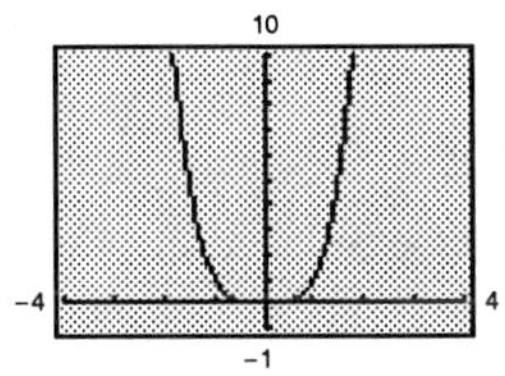

$f(x) = x^4$ is *not* one-to-one since $f(2) = 16 = f(-2)$.

52.

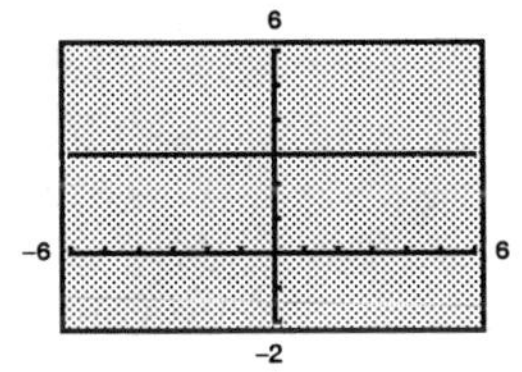

$f(x) = 3$ is *not* one-to-one since $f(0) = 3 = f(1)$.

54. $f(x) = \dfrac{1}{x},\ g(x) = x^2 - 1$

(a) $f(g(2)) = f(2^2 - 1) = f(3) = \dfrac{1}{3}$

(b) $g(f(2)) = g\left(\dfrac{1}{2}\right) = \left(\dfrac{1}{2}\right)^2 - 1 = -\dfrac{3}{4}$

(c) $f\left(g\left(\dfrac{1}{\sqrt{2}}\right)\right) = f\left(\dfrac{1}{2} - 1\right) = f\left(-\dfrac{1}{2}\right) = -2$

(d) $g\left(f\left(\dfrac{1}{\sqrt{2}}\right)\right) = g\left(\sqrt{2}\right) = 2 - 1 = 1$

(e) $f(g(x)) = f(x^2 - 1) = \dfrac{1}{(x^2 - 1)}$

(f) $g(f(x)) = g\left(\dfrac{1}{x}\right) = \left(\dfrac{1}{x}\right)^2 - 1$

56. The data fits the function (a) $f(x) = cx$, with $c = \frac{1}{4}$.

58. The data fits the function (c) $h(x) = c\sqrt{|x|}$, with $c = 3$.

60. (a)

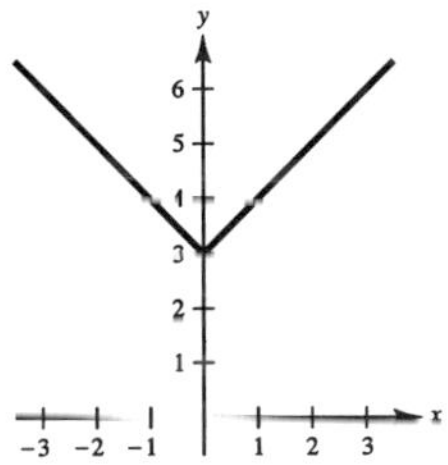

(b)

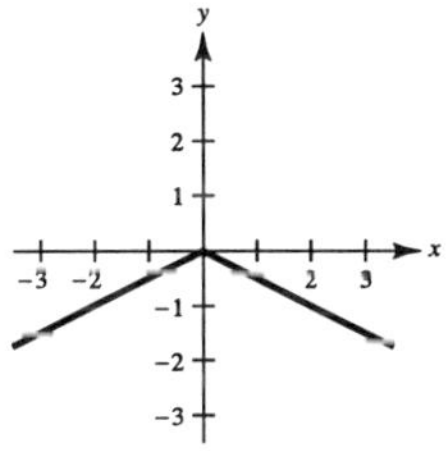

(c)

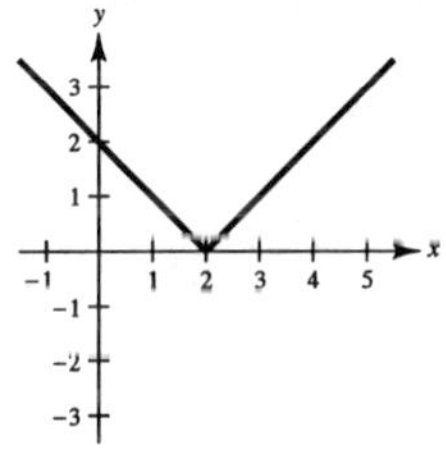

(d)

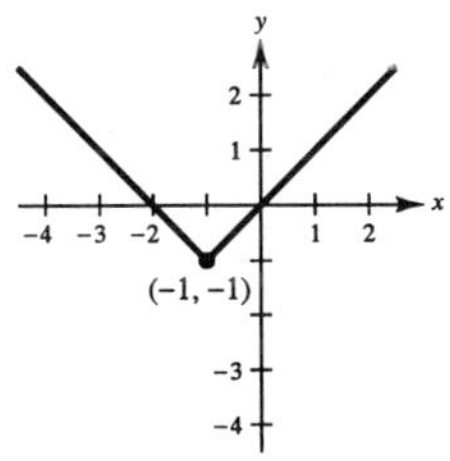

(e)

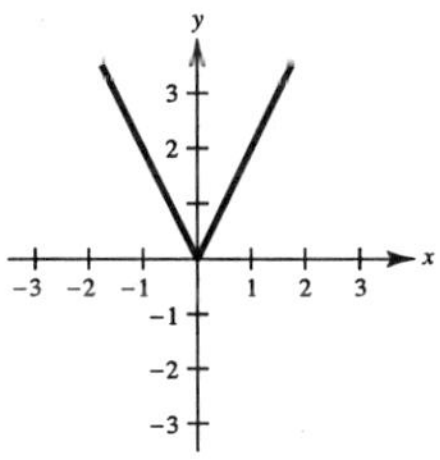

62. Value $= V = 2500x + 750{,}000$

64. (a) $1 + 0.01x = \dfrac{14.75}{p}$

$$x = \frac{(14.75/p) - 1}{0.01} = \frac{14.75 - p}{0.01p} = \frac{100(14.75 - p)}{p} = \frac{1475}{p} - 100$$

(b) $x = \dfrac{100(14.75 - 10)}{10} = 47.5$ units

66. (a) Graphing utility graph

(b) $(25, 14)$ is equilibrium point.

(c) Demand exceeds supply for $x < 25$.

(d) Supply exceeds demand for $x > 25$.

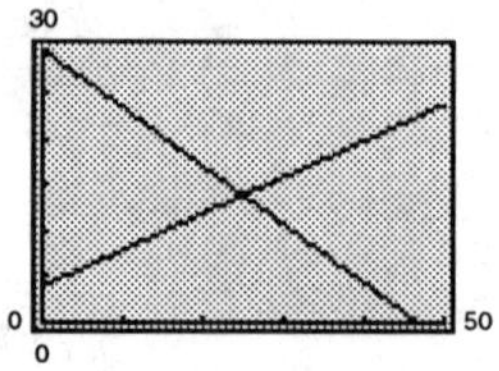

68. (a) Cost $= C = 98{,}000 + 12.30x$

(b) Revenue $= R = 17.98x$

(c) Profit $= R - C = 17.98x - (12.30x + 98{,}000)$

$= 5.68x - 98{,}000$

70. $F(t) = 98 + \dfrac{3}{t + 1}$

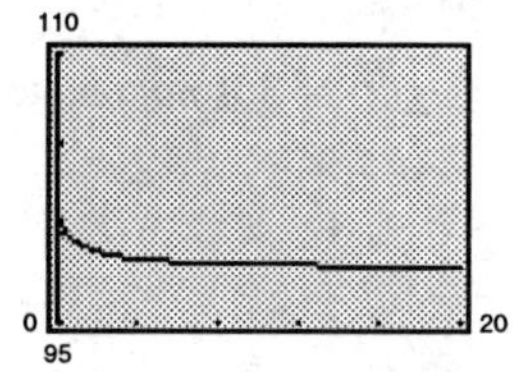

Valid for all $t \geq 0$

72. $f(x) = 2\left(3x^2 - \dfrac{6}{x}\right)$

Zeros: $x = 1.2599$, $(\sqrt[3]{2})$

The function is *not* one-to-one.

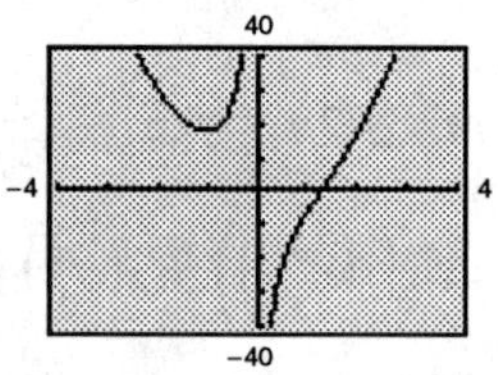

74. $h(x) = 6x^3 - 12x^2 + 4$

Zeros: $x = 3.9129, 0.6289, -0.5418$

The function is *not* one-to-one.

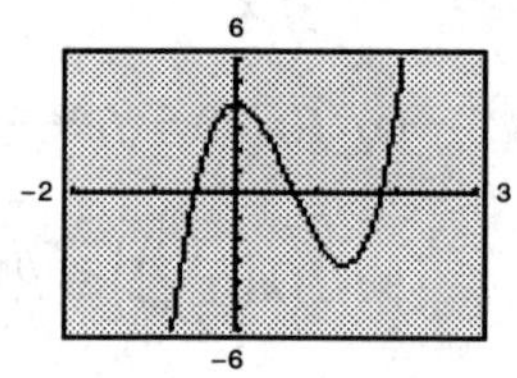

76. $f(x) = \left|\frac{1}{2}x^2 - 4\right|$

Zeros: $x = \pm 2\sqrt{2}$

The function is *not* one-to-one.

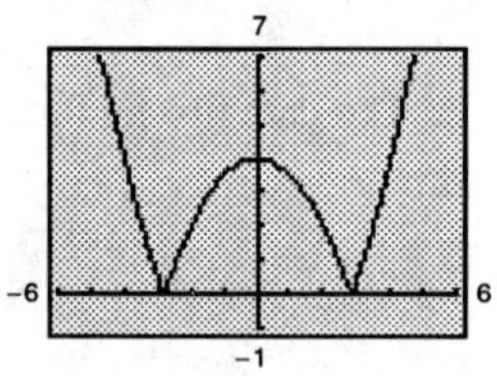

Section 1.5 Limits

2.

x	1.9	1.99	1.999	2	2.001	2.01	2.1
$f(x)$	0.3448	0.3344	0.3334	undefined	0.3332	0.3322	0.3226

$$\lim_{x \to 2} \frac{x - 2}{x^2 - x - 2} = \frac{1}{3}$$

4.

x	1.9	1.99	1.999	2	2.001	2.01	2.1
$f(x)$	72.39	79.20	79.92	undefined	80.08	80.80	88.41

$$\lim_{x \to 2} \frac{x^5 - 32}{x - 2} = 80$$

6.

x	-0.1	-0.01	-0.001	0	0.001	0.01	0.1
$f(x)$	0.3581	0.3540	0.3536	undefined	0.3535	0.3531	0.3492

$$\lim_{x\to 0} \frac{\sqrt{x+2}-\sqrt{2}}{x} = \frac{1}{2\sqrt{2}} \approx 0.354 \text{ (See Exercise 53.)}$$

8.

x	0.5	0.1	0.01	0.001	0
$f(x)$	-0.1	-0.1190	-0.1244	-0.1249	undefined

$$\lim_{x\to 0^+} \frac{[1/(2+x)]-(1/2)}{2x} = -\frac{1}{8} = -0.125$$

10. (a) $\lim_{x\to 1} f(x) = -2$

(b) $\lim_{x\to 3} f(x) = 0$

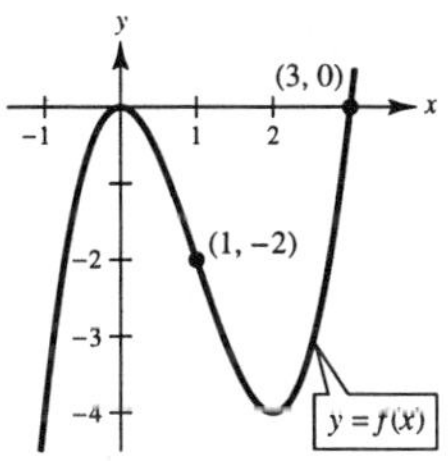

12. (a) $\lim_{x\to -2} h(x) = -5$

(b) $\lim_{x\to 0} h(x) = -3$

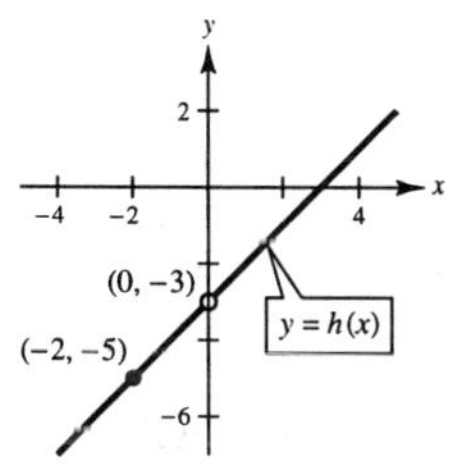

14. (a) $\lim_{x\to c} [f(x) + g(x)] = \lim_{x\to c} f(x) + \lim_{x\to c} g(x) = \frac{3}{2} + \frac{1}{2} = 2$

(b) $\lim_{x\to c} [f(x) \cdot g(x)] = \left[\lim_{x\to c} f(x)\right]\left[\lim_{x\to c} g(x)\right] = \left(\frac{3}{2}\right)\left(\frac{1}{2}\right) = \frac{3}{4}$

(c) $\lim_{x\to c} \frac{f(x)}{g(x)} = \frac{\lim_{x\to c} f(x)}{\lim_{x\to c} g(x)} = \frac{3/2}{1/2} = 3$

16. (a) $\lim_{x\to -2^+} f(x) = -2$

(b) $\lim_{x\to -2^-} f(x) = -2$

(c) $\lim_{x\to -2} f(x) = -2$

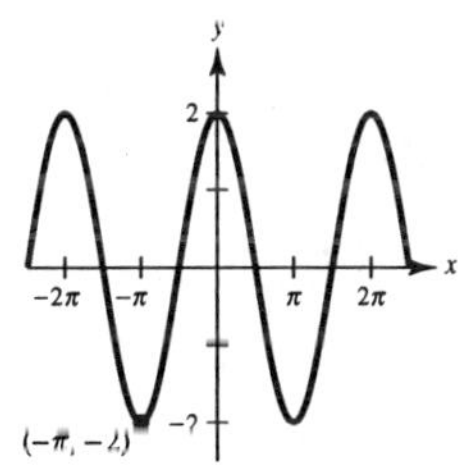

18. (a) $\lim_{x\to -2^+} f(x) = 2$

(b) $\lim_{x\to -2^-} f(x) = 2$

(c) $\lim_{x\to -2} f(x) = 2$

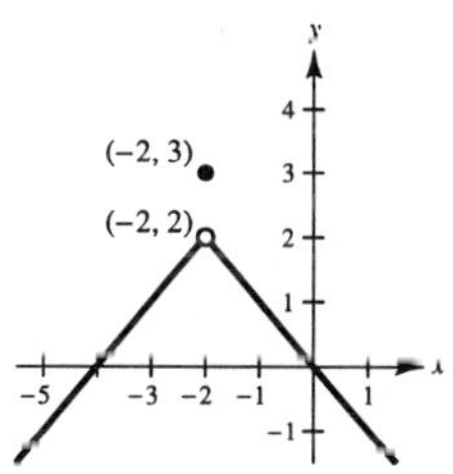

20. (a) $\lim_{x\to -1^+} f(x) = 0$

(b) $\lim_{x\to -1^-} f(x) = 2$

(c) $\lim_{x\to -1} f(x)$ does not exist.

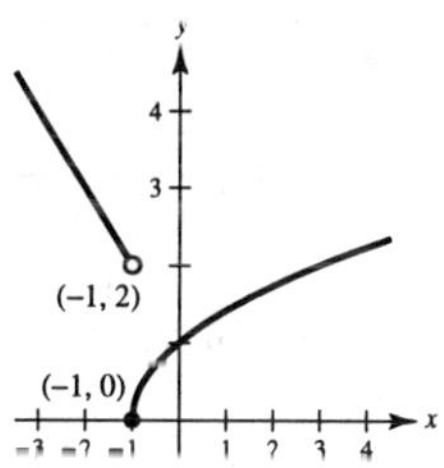

22. $\lim_{x\to 0} 2x - 3 = 2(0) - 3 = -3$

24. $\lim_{x\to 2} (-x^2 + x - 2) = -4 + 2 - 2 = -4$

26. $\lim_{x\to 4} \sqrt[3]{x+4} = \sqrt[3]{4+4} = 2$

28. $\lim_{x\to -2} \frac{3x+1}{2-x} = \frac{3(-2)+1}{2-(-2)} = \frac{-5}{4}$

30. $\lim_{x \to -1} \frac{4x - 5}{3 - x} = \frac{4(-1) - 5}{3 - (-1)} = \frac{-9}{4} = -\frac{9}{4}$

32. $\lim_{x \to -1} \frac{2x^2 - x - 3}{x + 1} = \lim_{x \to -1} \frac{(x + 1)(2x - 3)}{x + 1}$

$= \lim_{x \to -1} (2x - 3) = -5$

34. $\lim_{x \to 2} \frac{2 - x}{x^2 - 4} = \lim_{x \to 2} \frac{-(x - 2)}{(x + 2)(x - 2)}$

$= \lim_{x \to 2} \frac{-1}{x + 2} = -\frac{1}{4}$

36. $\lim_{t \to 1} \frac{t^2 + t - 2}{t^2 - 1} = \lim_{t \to 1} \frac{(t - 1)(t + 2)}{(t + 1)(t - 1)}$

$= \lim_{t \to 1} \frac{t + 2}{t + 1} = \frac{3}{2}$

38. $\lim_{x \to 1} \frac{x^3 - 1}{x - 1} = \lim_{x \to 1} \frac{(x - 1)(x^2 + x + 1)}{x - 1}$

$= \lim_{x \to 1} (x^2 + x + 1) = 3$

40. $\lim_{x \to 2^-} \frac{|x - 2|}{x - 2} = -1$

$\lim_{x \to 2^+} \frac{|x - 2|}{x - 2} = 1$

Therefore, $\lim_{x \to 2} \frac{|x - 2|}{x - 2}$ does not exist.

42. $\lim_{s \to 1^-} f(s) = \lim_{s \to 1^-} s = 1$

$\lim_{s \to 1^+} f(s) = \lim_{s \to 1^+} (1 - s) = 0$

Therefore, $\lim_{s \to 1} f(s)$ does not exist.

44. $\lim_{\Delta x \to 0} \frac{(1 + \Delta x)^3 - 1}{\Delta x} = \lim_{\Delta x \to 0} \frac{1 + 3\Delta x + 3(\Delta x)^2 + (\Delta x)^3 - 1}{\Delta x}$

$= \lim_{\Delta x \to 0} [3 + 3\Delta x + (\Delta x)^2] = 3$

46. $\lim_{\Delta x \to 0} \frac{\sqrt{x + \Delta x} - \sqrt{x}}{\Delta x} = \lim_{\Delta x \to 0} \frac{\sqrt{x + \Delta x} - \sqrt{x}}{\Delta x} \cdot \frac{\sqrt{x + \Delta x} + \sqrt{x}}{\sqrt{x + \Delta x} + \sqrt{x}}$

$= \lim_{\Delta x \to 0} \frac{(x + \Delta x) - x}{\Delta x\left[\sqrt{x + \Delta x} + \sqrt{x}\right]}$

$= \lim_{\Delta x \to 0} \frac{1}{\sqrt{x + \Delta x} + \sqrt{x}}$

$= \frac{1}{2\sqrt{x}}$

48. $\lim_{x \to 4} \frac{\sqrt{x} - 2}{x - 4} = \lim_{x \to 4} \frac{\sqrt{x} - 2}{x - 4} \cdot \frac{\sqrt{x} + 2}{\sqrt{x} + 2}$

$= \lim_{x \to 4} \frac{x - 4}{(x - 4)(\sqrt{x} + 2)}$

$= \lim_{x \to 4} \frac{1}{\sqrt{x} + 2} = \frac{1}{4}$

50. $\lim_{x \to 1^+} \frac{5}{1 - x} = -\infty$

52. $\lim_{x \to 0^-} \frac{x + 1}{x} = -\infty$

54. Because $4 - x^2 \le f(x) \le 4 + x^2$,

$\lim_{x \to 0} (4 - x^2) \le \lim_{x \to 0} f(x) \le \lim_{x \to 0} (4 + x^2)$

$4 \le \lim_{x \to 0} f(x) \le 4.$

Hence, $\lim_{x \to 0} f(x) = 4.$

56. $\lim_{r\to 0.06} A = \lim_{r\to 0.06} 1000\left(1 + \dfrac{r}{4}\right)^{40} = 1814.02$

Yes, the limit exists.

58. $\lim_{x\to 1} \dfrac{x^2 + 6x - 7}{x^3 - x^2 + 2x - 2} = \lim_{x\to 1} \dfrac{(x - 1)(x + 7)}{(x - 1)(x^2 + 2)}$

$= \dfrac{8}{3} \approx 2.667$

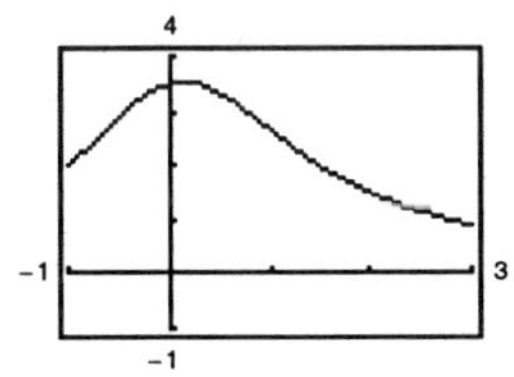

60. $\lim_{x\to -2} \dfrac{4x^3 + 7x^2 + x + 6}{3x^2 - x - 14} = \lim_{x\to -2} \dfrac{(x + 2)(4x^2 - x + 3)}{(x + 2)(3x - 7)}$

$= \dfrac{21}{-13} \approx -1.615$

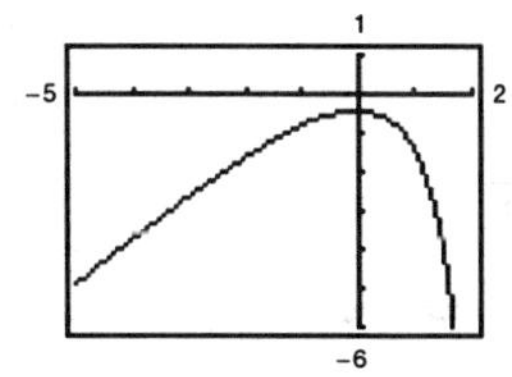

Section 1.6 Continuity

2. The polynomial $f(x) = (x^2 - 1)^3$ is continuous on the entire real line.

4. The rational function $f(x) = \dfrac{1}{4 + x^2}$ is continuous on the entire real line.

6. $f(x) = \dfrac{(x^2 - 1)}{x}$ is continuous on $(-\infty, 0) \cup (0, \infty)$.

8. $f(x) = \dfrac{1}{x^2 - 4}$ is continuous on $(-\infty, -2) \cup (-2, 2) \cup (2, \infty)$.

10. $f(x) = \dfrac{1}{x^2 + 1}$ is continuous on $(-\infty, \infty)$.

12. $f(x) = \dfrac{x}{x^2 - 1}$ is continuous on $(-\infty, -1) \cup (-1, 1) \cup (1, \infty)$.

14. $f(x) = \dfrac{x - 3}{x^2 - 9}$ is continuous on $(-\infty, -3) \cup (-3, 3) \cup (3, \infty)$.

16. $f(x) = \dfrac{x - 1}{x^2 + x - 2} = \dfrac{x - 1}{(x - 1)(x + 2)}$ is continuous on $(-\infty, -2) \cup (-2, 1) \cup (1, \infty)$.

18. $f(x) = \dfrac{[\![x]\!]}{2 + x}$ is continuous on all intervals of the form $(c, c + 1)$ where c is an integer.

20. $\lim_{x\to 2^-} f(x) = \frac{1}{2}(2) + 1 = 2$

$\lim_{x\to 2^+} f(x) = 3 - 2 = 1$

f is continuous on $(-\infty, 2)$ and $(2, \infty)$.

22. $\lim_{x\to 0^-} f(x) = \lim_{x\to 0^-} (|x - 2| + 3) = 5$

$\lim_{x\to 0^+} f(x) = \lim_{x\to 0^+} (x + 5) = 5$

Since $f(0) = 5$, f is continuous on the entire real line.

24. $\lim_{x\to 4^-} \frac{|4-x|}{4-x} = 1$

$\lim_{x\to 4^+} \frac{|4-x|}{4-x} = -1$

$\lim_{x\to 4} \frac{|4-x|}{4-x}$ does not exist.

f is continuous on $(-\infty, 4) \cup (4, \infty)$.

26. $\lim_{x\to c^-} (x - [\![x]\!]) = c - (c-1) = 1$, c is any integer

$\lim_{x\to c^+} (x - [\![x]\!]) = c - c = 0$, c is any integer

f is continuous on all intervals $(c, c+1)$.

28. $h(x) = f(g(x)) = f(x^2+5)$

$= \frac{1}{(x^2+5)-1} = \frac{1}{x^2+4}$

Thus, h is continuous on the entire real line.

30. $f(x) = \frac{5}{x^2+1}$ is continuous on $[-2, 2]$.

[**Note:** f is continuous on the entire real line.]

32. $f(x) = \frac{x}{(x-1)(x-3)}$ has nonremovable discontinuities at $x = 1$ and $x = 3$.

34. $f(x) = \frac{2x^2+x}{x} = \frac{x(2x+1)}{x}$

f has a removable discontinuity at $x = 0$; continuous on $(-\infty, 0) \cup (0, \infty)$.

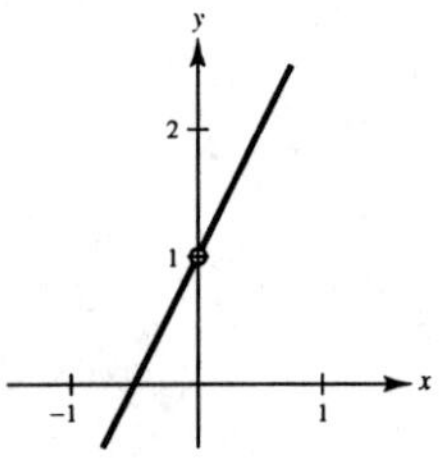

36. $f(x) = \frac{x^4-1}{x^2-1} = \frac{(x^2+1)(x^2-1)}{x^2-1}$

f has removable discontinuities at $x = \pm 1$; continuous on $(-\infty, -1) \cup (-1, 1) \cup (1, \infty)$.

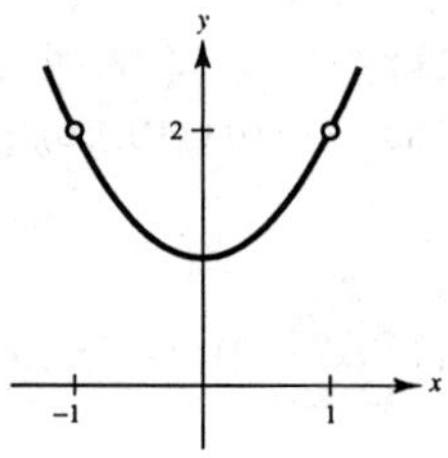

38. $f(x) = x - 2[\![x]\!]$ is continuous on intervals of the form $(c, c+1)$, where c is an integer.

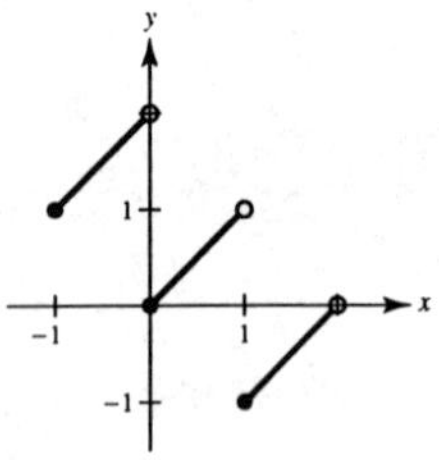

40. $\lim_{x\to -1^-} f(x) = 2$

$\lim_{x\to -1^+} f(x) = -a + b$

$\lim_{x\to 3^-} f(x) = 3a + b$

$\lim_{x\to 3^+} f(x) = -2$

Thus:

$$\begin{aligned} -a + b &= 2 \\ 3a + b &= -2 \\ \hline -4a &= 4 \\ a &= -1 \\ b &= 1 \end{aligned}$$

42. $f(x) = \begin{cases} 2x - 4, & x \le 3 \\ x^2 - 2x, & x > 3 \end{cases}$

f is not continuous at $x = 3$.

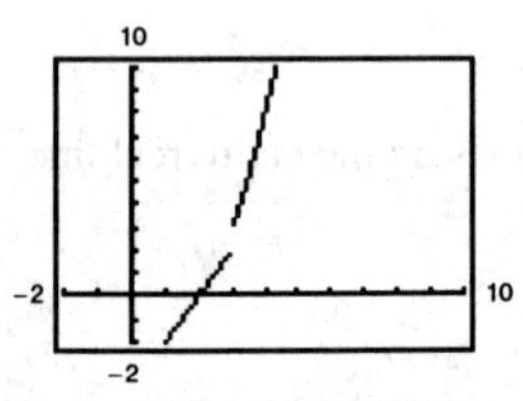

44. $f(x) = x\sqrt{x+3}$ is continuous on $[-3, \infty)$.

46. $f(x) = \dfrac{x+1}{\sqrt{x}}$ is continuous on $(0, \infty)$.

48. $f(x) = \dfrac{x^3 - 8}{x - 2} = \dfrac{(x-2)(x^2+2x+4)}{(x-2)}$

appears to be continuous on $[-4, 4]$. But, it is not continuous at $x = 2$ (removable discontinuity).

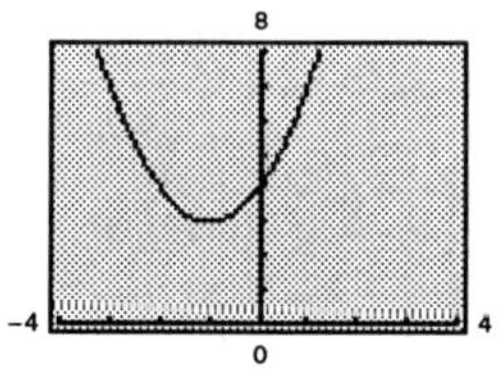

50. $C = \dfrac{2x}{100 - x}$

(a) $[0, 100)$

(b) C is continuous on its domain.

(c) For $x = 75$,

$$C = \frac{2(75)}{100 - 75} = \frac{150}{25} = 6 \text{ million dollars.}$$

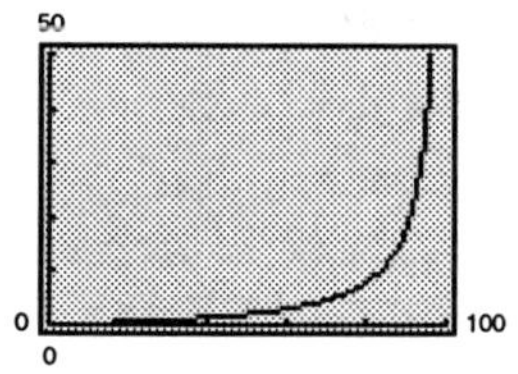

52. (a) Nonremovable discontinuities at $t = 1, 2, 3, 4, 5$

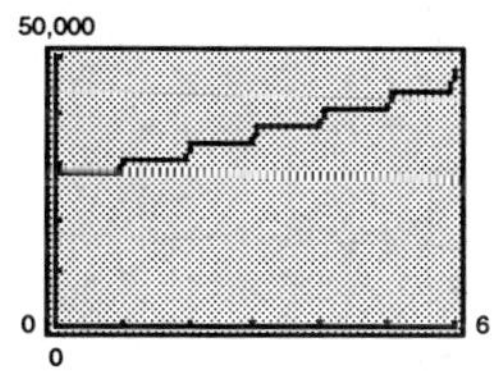

(b) For $t = 5$, $S = \$43{,}850.78$.

54. Yes, a linear model is a continuous function. No, actual revenue would probably not be continuous.

Review Exercises for Chapter 1

2. Population of California (c)

4. IBM Revenues (d)

6. Distance $= \sqrt{(1-4)^2 + (2-3)^2} = \sqrt{25+4} = \sqrt{29}$

8. Distance $= \sqrt{[6-(-3)]^2 + (8-7)^2}$

$= \sqrt{81+1} = \sqrt{82}$

10. Midpoint $= \left(\dfrac{0-4}{2}, \dfrac{0+8}{2}\right) = (-2, 4)$

12. Midpoint $= \left(\dfrac{7-3}{2}, \dfrac{-9+5}{2}\right) = (2, -2)$

14. For 1996, revenue $\approx$ 13 billion

cost $\approx$ 10.5 billion

profit $\approx$ 2.5 billion.

16. Bar graph for data:

Mytilus	107
Gammarus	78
Littorina	65
Nassarius	112
Arbacia	6
Mya	18

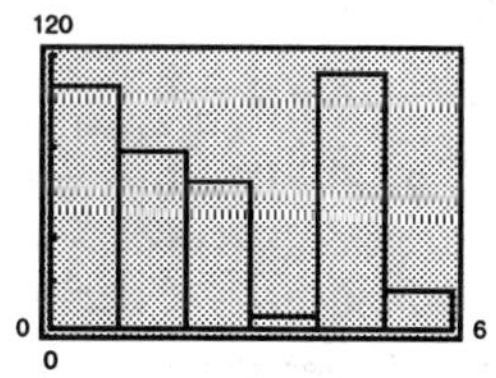

18. $y = 4 - 3x$

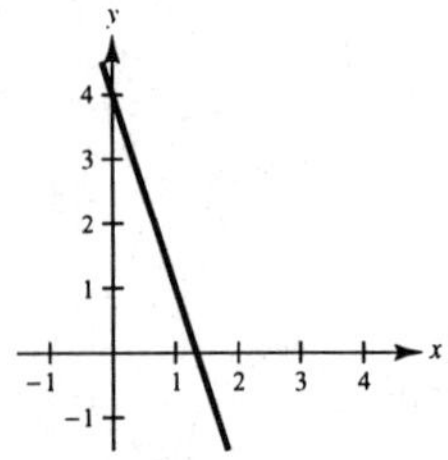

20. $y = 1 - x^2$

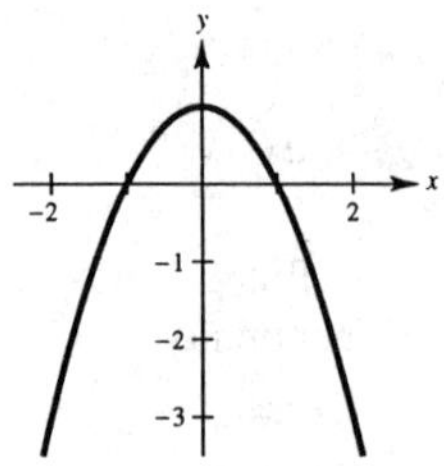

22. $y = |2x - 3|$

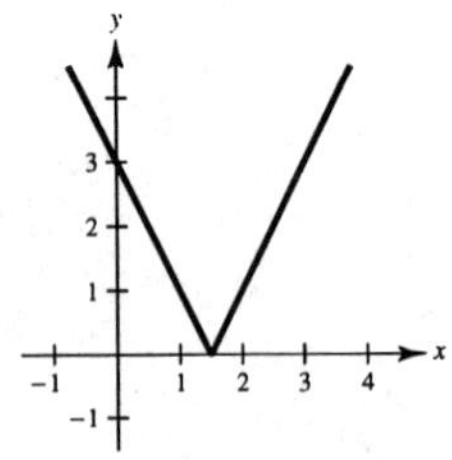

24. $y = \sqrt{2x}$

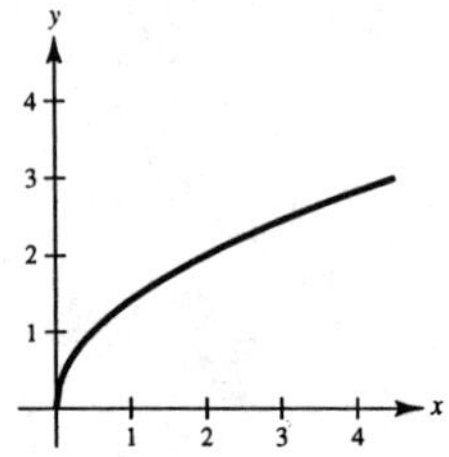

26. $y = (x - 1)^3 + 2(x - 1)^2$

$x = 0 \rightarrow y = (-1)^3 + 2(-1)^2 = 1 \rightarrow (0, 1)$

y-intercept

$y = 0 \rightarrow 0 = (x - 1)^3 + 2(x - 1)^2 = (x - 1)^2(x - 1 + 2)$

$= (x - 1)^2(x + 1) \rightarrow (1, 0), (-1, 0)$

x-intercepts

28.
$$(x - 2)^2 + (y + 1)^2 = r^2$$
$$(-1 - 2)^2 + (7 + 1)^2 = r^2$$
$$9 + 64 = r^2$$
$$73 = r^2$$
$$(x - 2)^2 + (y + 1)^2 = 73$$

30.
$$x^2 + y^2 + 10x + 4y - 7 = 0$$
$$(x^2 + 10x + 25) + (y^2 + 4y + 4) = 7 + 25 + 4$$
$$(x + 5)^2 + (y + 2)^2 = 36$$

Center: $(-5, -2)$
Radius: 6

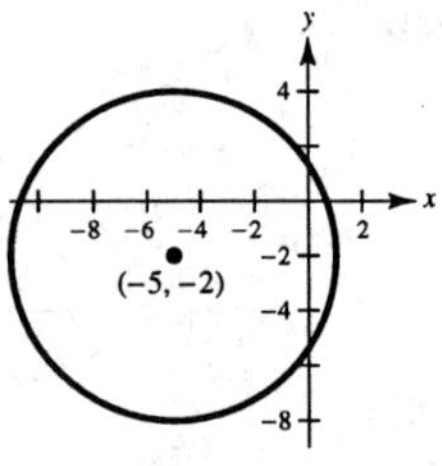

32. $x^2 + y^2 = 5$

$x - y = 1 \rightarrow y = x - 1$

$x^2 + (x - 1)^2 = 5$

$2x^2 - 2x + 1 = 5$

$2x^2 - 2x - 4 = 0$

$2(x - 2)(x + 1) = 0$

$x = -1, 2$

Points of intersection: $(-1, -2), (2, 1)$

34. $y = \sqrt{x}$ $\qquad \sqrt{x} = x$

$y = x \qquad x = x^2$

$x^2 - x = 0$

$x(x - 1) = 0$

Points of intersection: $(0, 0), (1, 1)$

36. (a) $C = 6000 + 6.50x$

$R = 13.90x$

(b) $C = R$

$6000 + 6.5x = 13.9x$

$6000 = 7.4x$

$x \approx 810.81$, or 811 units

38. $-\frac{1}{3}x + \frac{5}{6}y = 1$

$\frac{5}{6}y = \frac{1}{3}x + 1$

$y = \frac{2}{5}x + \frac{6}{5}$

Slope: $m = \frac{2}{5}$

y-intercept: $\left(0, \frac{6}{5}\right)$

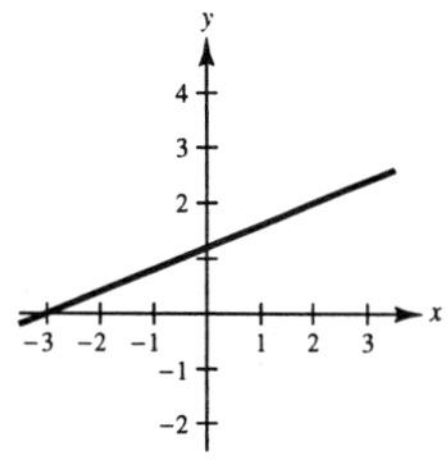

40. $x = -3$

Slope: undefined (vertical line)
No y-intercept

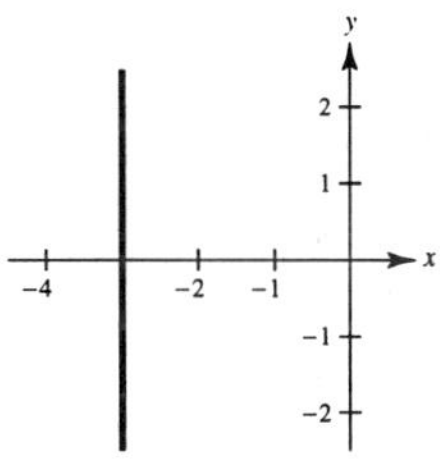

42. $3.2x - 0.8y + 5.6 = 0$

$8y = 32x + 56$

$y = 4x + 7$

Slope: 4
y-intercept: $(0, 7)$

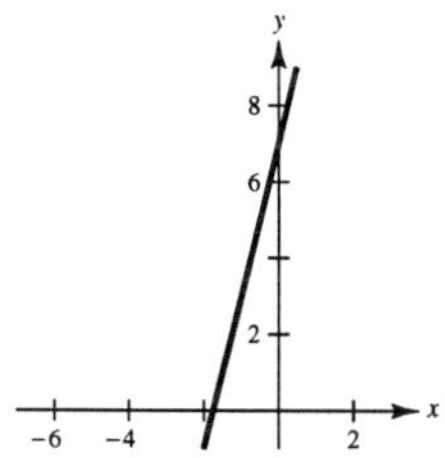

44. Slope $= \dfrac{7 - 5}{-5 - (-1)} = \dfrac{2}{-4} = -\dfrac{1}{2}$

46. Slope $= \dfrac{-3 - (-3)}{-1 - (-11)} = \dfrac{0}{10} = 0$ (horizontal line)

48. $y - (-3) = \frac{1}{2}[x - (-3)]$

$y = \frac{1}{2}x - \frac{3}{2}$

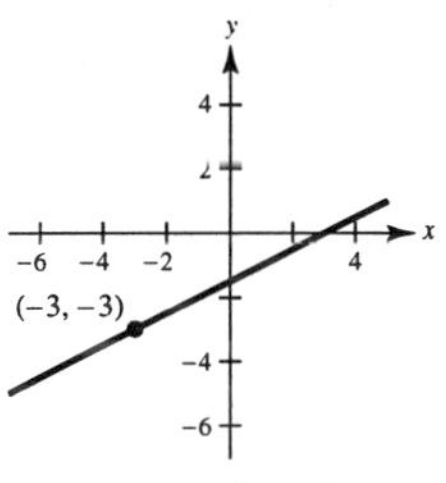

50. (a) Slope $= 0 \rightarrow y = -3$

(b) Slope undefined $\rightarrow x = 1$

(c) $-4x + 5y = -3 \rightarrow y = \frac{4}{5}x - \frac{3}{5}$

$y - (-3) = \frac{4}{5}(x - 1)$

$y = \frac{4}{5}x - \frac{19}{5}$

(d) $5x - 2y = 3y = \frac{5}{2}x - \frac{3}{2}$

Slope of perpendicular is $-\frac{2}{5}$.

$y - (-3) = -\frac{2}{5}(x - 1)$

$y = -\frac{2}{5}x - \frac{13}{5}$

52. $(0, 117{,}000), (9, 0)$

$$m = \frac{117{,}000}{-9} = -13{,}000$$

(a) $v = -13{,}000(t - 9) = -13{,}000t + 117{,}000$

(b) Graphing utility (at the right)

(c) $v(4) = 65{,}000$

(d) $v = 84{,}000$ when $t \approx 2.54$ years

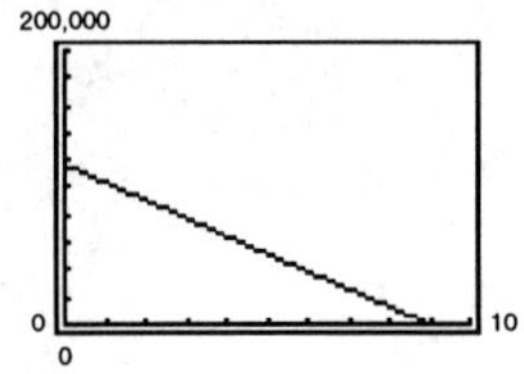

54. $x^2 + y^2 = 4$

No

56. $y = |x + 4|$

Yes

58. $f(x) = x^2 + 4x + 3$

(a) $f(0) = 0^2 + 4(0) + 3 = 3$

(b) $f(x - 1) = (x - 1)^2 + 4(x - 1) + 3 = x^2 + 2x$

(c) $f(x + \Delta x) - f(x) = (x + \Delta x)^2 + 4(x + \Delta x) + 3 - (x^2 + 4x + 3)$

$= 2x\Delta x + (\Delta x)^2 + 4\Delta x$

60. $f(x) = 2$

Domain: $(-\infty, \infty)$
Range: $\{2\}$

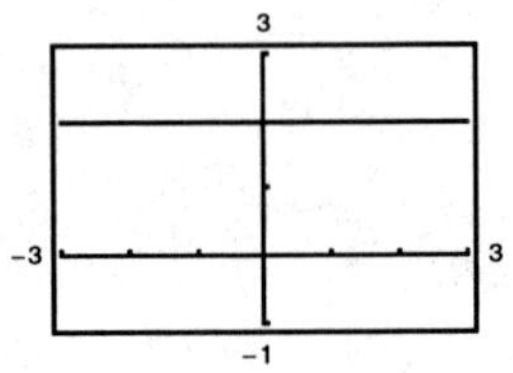

62. $f(x) = \dfrac{x - 3}{x^2 + x - 12} = \dfrac{(x - 3)}{(x - 3)(x + 4)}$

$= \dfrac{1}{x + 4}, \; x \neq 3$

Domain: $(-\infty, -4) \cup (-4, 3) \cup (3, \infty)$
Range: $(-\infty, 0) \cup \left(0, \frac{1}{7}\right) \cup \left(\frac{1}{7}, \infty\right)$

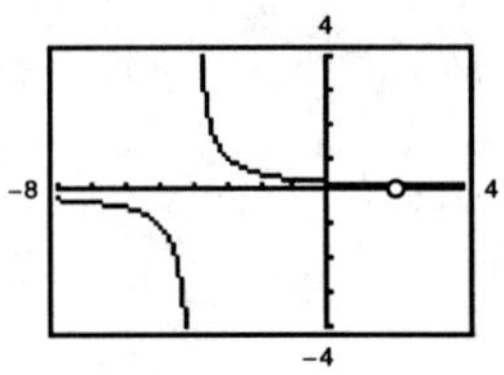

64. $f(x) = \dfrac{-12}{13}x - \dfrac{7}{8}$

Domain: $(-\infty, \infty)$
Range: $(-\infty, \infty)$

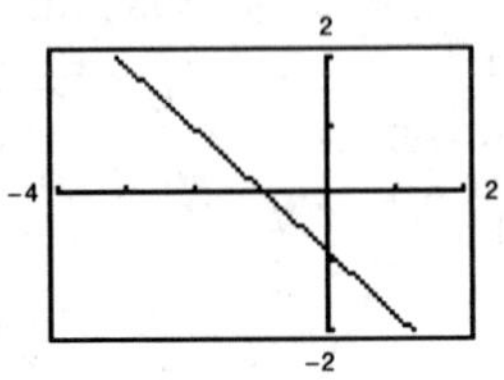

66. (a) $f(x) + g(x) = 2x - 3 + \sqrt{x + 1}$

(b) $f(x) - g(x) = 2x - 3 - \sqrt{x + 1}$

(c) $f(x)g(x) = (2x - 3)\sqrt{x + 1}$

(d) $\dfrac{f(x)}{g(x)} = \dfrac{2x - 3}{\sqrt{x + 1}}$

(e) $f(g(x)) = f\left(\sqrt{x + 1}\right) = 2\sqrt{x + 1} - 3$

(f) $g(f(x)) = g(2x - 3) = \sqrt{(2x - 3) + 1} = \sqrt{2x - 2}$

68. $f(x) = |x + 1|$ does not have an inverse by the horizontal line test.

70. $f(x) = x^3 - 1$ has an inverse by the horizontal line test.

$$\begin{aligned} y &= x^3 - 1 \\ x &= y^3 - 1 \\ x + 1 &= y^3 \\ y &= \sqrt[3]{x+1} \\ f^{-1}(x) &= \sqrt[3]{x+1} \end{aligned}$$

72. $\lim_{x \to 2} (2x + 9) = 2(2) + 9 = 13$

74. $\lim_{x \to 2} \dfrac{5x - 3}{2x + 9} = \dfrac{5(2) - 3}{2(2) + 9} = \dfrac{7}{13}$

76. $\lim_{t \to 1} \dfrac{t + 1}{t - 2} = \dfrac{1 + 1}{1 - 2} = -2$

78. $\lim_{t \to 2^-} \dfrac{t + 1}{t - 2} = -\infty$

$\lim_{t \to 2^+} \dfrac{t + 1}{t - 2} = \infty$

$\lim_{t \to 2} \dfrac{t + 1}{t - 2}$ does not exist.

80. $\lim_{x \to 3^-} \dfrac{x^2 - 9}{x - 3} = \lim_{x \to 3^-} (x + 3) = 6$

82. $\lim_{x \to 1/2} \dfrac{2x - 1}{6x - 3} = \lim_{x \to 1/2} \dfrac{1}{3} = \dfrac{1}{3}$

84. $$\lim_{s \to 0} \frac{\left(1/\sqrt{1+s}\right) - 1}{s} = \lim_{s \to 0} \frac{1 - \sqrt{1+s}}{s\sqrt{1+s}} \cdot \frac{1 + \sqrt{1+s}}{1 + \sqrt{1+s}}$$
$$= \lim_{s \to 0} \frac{1 - (1 + s)}{s\sqrt{1+s}\left(1 + \sqrt{1+s}\right)} = \lim_{s \to 0} \frac{-1}{\sqrt{1+s}\left(1 + \sqrt{1+s}\right)} = -\frac{1}{2}$$

86. $$\lim_{\Delta x \to 0} \frac{[1 - (x + \Delta x)^2] - (1 - x^2)}{\Delta x} = \lim_{\Delta x \to 0} \frac{1 - x^2 - 2x\Delta x - (\Delta x)^2 - 1 + x^2}{\Delta x}$$
$$= \lim_{\Delta x \to 0} \frac{\Delta x(-2x - \Delta x)}{\Delta x} = \lim_{\Delta x \to 0} (-2x - \Delta x) = -2x$$

88.

x	1.1	1.01	1.001	1.0001
$f(x)$	-0.3228	-0.3322	-0.3332	-0.3333

$\lim_{x \to 1^+} \dfrac{1 - \sqrt[3]{x}}{x - 1} = -\dfrac{1}{3}$

90. The statement $\lim_{x \to 0} x^3 = 0$ is true.

92. The statement $\lim_{x \to 0} \sqrt[3]{x} = 0$ is true.

94. The statement $\lim_{x \to 3} f(x) = 1$ is true since

$\lim_{x \to 3^-} f(x) = \lim_{x \to 3^-} (x - 2) = 1$

$\lim_{x \to 3^+} f(x) = \lim_{x \to 3^+} (-x^2 + 8x - 14) = 1.$

96. $f(x) = \dfrac{x + 2}{x}$ is continuous on the intervals $(-\infty, 0) \cup (0, \infty)$.

98. $f(x) = \dfrac{x + 1}{2x + 2}$ is continuous on the intervals $(-\infty, -1) \cup (-1, \infty)$.

100. $f(x) = \dfrac{3x^2 - x - 2}{x - 1}$ is continuous on the intervals $(-\infty, 1) \cup (1, \infty)$.

102. $f(x)$ is continuous on $(-\infty, \infty)$.

104. $\lim_{x \to 1^-} f(x) = \lim_{x \to 1^-} (x + 1) = 2$

$\lim_{x \to 1^+} f(x) = \lim_{x \to 1^+} (2x + a) = 2 + a$

Thus, $2 = 2 + a$ and $a = 0$.

106. (a) Graphing utility graph of $A = (1/4)[\![x/24]\!]$ not continuous at $x = 24n$, n a positive integer. Graph as shown below.

(b) If $x = 1500$, $A = \$15.50$.

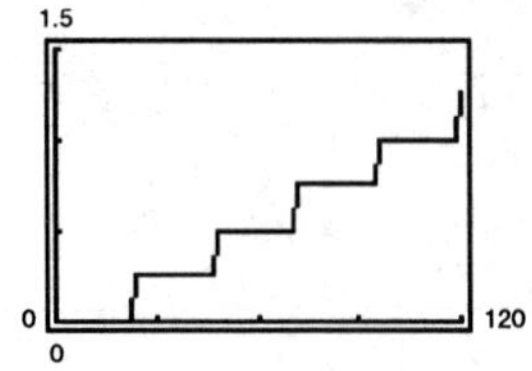

CHAPTER 2
Differentiation

CHAPTER 2
Differentiation

Section 2.1 The Derivative and the Slope of a Graph

2. The tangent line at (x_1, y_1) has a negative slope. The tangent line at (x_2, y_2) has a positive slope.

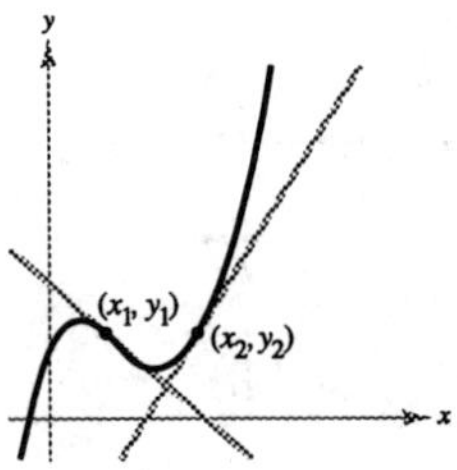

4. The tangent line at (x_1, y_1) has zero slope. The tangent line at (x_2, y_2) has a positive slope.

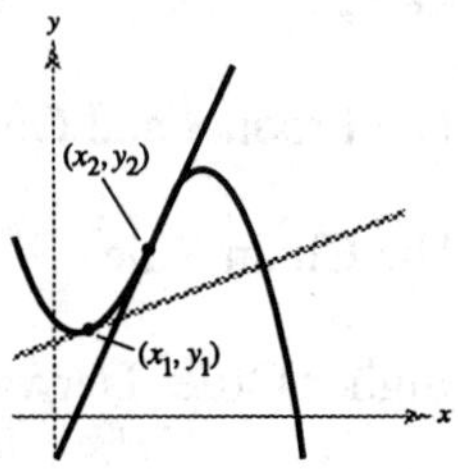

6. The slope is $m = \frac{4}{3}$.

8. The slope is $m = \frac{1}{4}$.

10. The slope is $m = -3$.

12. In 1992 ($t = 2$), the slope is about zero.

In 1995 ($t = 5$), the slope is about 1700.

14. $f(x + \Delta x) = -4$

$$f(x + \Delta x) - f(x) = -4 - (-4) = 0$$

$$\frac{f(x + \Delta x) - f(x)}{\Delta x} = \frac{0}{\Delta x} = 0$$

$$\lim_{\Delta x \to 0} \frac{f(x + \Delta x) - f(x)}{\Delta x} = 0$$

16. $f(x + \Delta x) = \frac{1}{2}(x + \Delta x) = \frac{1}{2}x + \frac{1}{2}\Delta x$

$$f(x + \Delta x) - f(x) = \frac{1}{2}\Delta x$$

$$\frac{f(x + \Delta x) - f(x)}{\Delta x} = \frac{1}{2}$$

$$\lim_{\Delta x \to 0} \frac{f(x + \Delta x) - f(x)}{\Delta x} = \frac{1}{2}$$

18. $f(x + \Delta x) = 1 - (x + \Delta x)^2$

$$= 1 - x^2 - 2x\Delta x - (\Delta x)^2$$

$$f(x + \Delta x) - f(x) = -2x\Delta x - (\Delta x)^2$$

$$\frac{f(x + \Delta x) - f(x)}{\Delta x} = -2x - \Delta x$$

$$\lim_{\Delta x \to 0} \frac{f(x + \Delta x) - f(x)}{\Delta x} = -2x$$

20. $f(x + \Delta x) = (x + \Delta x) - (x + \Delta x)^2 = x + \Delta x - x^2 - 2x\Delta x - (\Delta x)^2$

$$f(x + \Delta x) - f(x) = \Delta x - 2x\Delta x - (\Delta x)^2$$

$$\frac{f(x + \Delta x) - f(x)}{\Delta x} = 1 - 2x - \Delta x$$

$$\lim_{\Delta x \to 0} \frac{f(x + \Delta x) - f(x)}{\Delta x} = 1 - 2x$$

22. $f(x + \Delta x) = \sqrt{2(x + \Delta x)}$

$$f(x + \Delta x) - f(x) = \sqrt{2(x + \Delta x)} - \sqrt{2x}$$

$$= \frac{\sqrt{2x + 2\Delta x} - \sqrt{2x}}{1} \cdot \frac{\sqrt{2x + 2\Delta x} + \sqrt{2x}}{\sqrt{2x + 2\Delta x} + \sqrt{2x}}$$

$$= \frac{2\Delta x}{\sqrt{2x + 2\Delta x} + \sqrt{2x}}$$

$$\frac{f(x + \Delta x) - f(x)}{\Delta x} = \frac{2}{\sqrt{2x + 2\Delta x} + \sqrt{2x}}$$

$$\lim_{\Delta x \to 0} \frac{f(x + \Delta x) - f(x)}{\Delta x} = \frac{1}{\sqrt{2x}}$$

24. $f(t + \Delta t) = (t + \Delta t)^3 + (t + \Delta t)^2 = t^3 + 3t^2\Delta t + 3t(\Delta t)^2 + (\Delta t)^3 + t^2 + 2t(\Delta t) + (\Delta t)^2$

$$f(t + \Delta t) - f(t) = 3t^2\Delta t + 3t(\Delta t)^2 + (\Delta t)^3 + 2t(\Delta t) + (\Delta t)^2$$

$$\frac{f(t + \Delta t) - f(t)}{\Delta t} = 3t^2 + 3t(\Delta t) + (\Delta t)^2 + 2t + \Delta t$$

$$\lim_{\Delta t \to 0} \frac{f(t + \Delta t) - f(t)}{\Delta t} = 3t^2 + 2t$$

26. $g(s + \Delta s) = \dfrac{1}{s + \Delta s - 1}$

$$g(s + \Delta s) - g(s) = \frac{1}{s + \Delta s - 1} - \frac{1}{s - 1}$$

$$= \frac{(s - 1) - (s + \Delta s - 1)}{(s + \Delta s - 1)(s - 1)}$$

$$= -\frac{\Delta s}{(s + \Delta s - 1)(s - 1)}$$

$$\frac{g(s + \Delta s) - g(s)}{\Delta s} = -\frac{1}{(s + \Delta s - 1)(s - 1)}$$

$$\lim_{\Delta s \to 0} \frac{g(s + \Delta s) - g(s)}{\Delta s} = -\frac{1}{(s - 1)^2}$$

28. $f(x + \Delta x) = 2(x + \Delta x) + 4$

$$f(x + \Delta x) - f(x) = 2x + 2\Delta x + 4 - (2x + 4) = 2\Delta x$$

$$\frac{f(x + \Delta x) - f(x)}{\Delta x} = 2$$

$$\lim_{\Delta x \to 0} \frac{f(x + \Delta x) - f(x)}{\Delta x} = 2$$

At $(1, 6)$, the slope of the tangent line is $m = 2$.

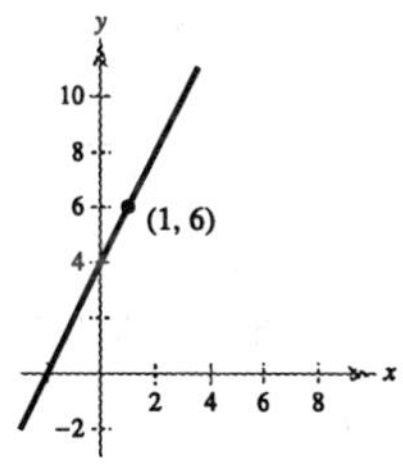

30. $f(x + \Delta x) = 6$

$$f(x + \Delta x) - f(x) = 6 - 6 = 0$$

$$\frac{f(x + \Delta x) - f(x)}{\Delta x} = \frac{0}{\Delta x} = 0$$

$$\lim_{\Delta x \to 0} \frac{f(x + \Delta x) - f(x)}{\Delta x} = 0$$

At $(-2, 6)$, the slope of the tangent line is $m = 0$.

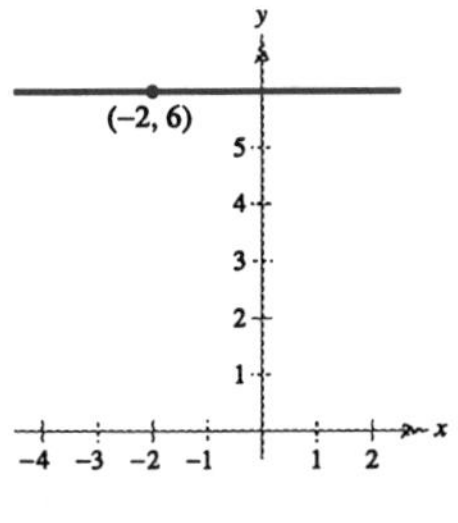

32. $f(x + \Delta x) = (x + \Delta x)^2 + 2(x + \Delta x) + 1$

$$= x^2 + 2x\Delta x + (\Delta x)^2 + 2x + 2\Delta x + 1$$

$$f(x + \Delta x) - f(x) = 2x\Delta x + (\Delta x)^2 + 2\Delta x$$

$$\frac{f(x + \Delta x) - f(x)}{\Delta x} = 2x + \Delta x + 2$$

$$\lim_{\Delta x \to 0} \frac{f(x + \Delta x) - f(x)}{\Delta x} = 2x + 2$$

At $(-3, 4)$, the slope of the tangent line is $m = 2(-3) + 2 = -4$.

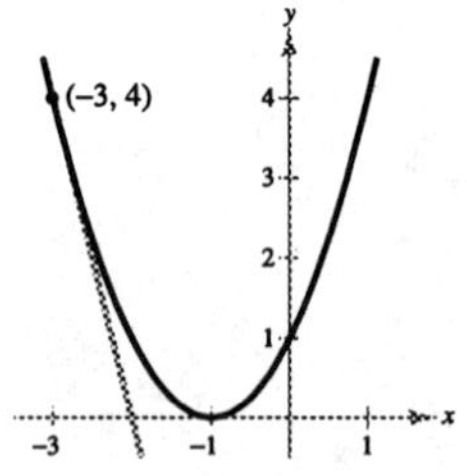

34. $f(x + \Delta x) = (x + \Delta x)^3 + 3$

$$= x^3 + 3x^2\Delta x + 3x(\Delta x)^2 + (\Delta x)^3 + 3$$

$$f(x + \Delta x) - f(x) = 3x^2\Delta x + 3x(\Delta x)^2 + (\Delta x)^3$$

$$\frac{f(x + \Delta x) - f(x)}{\Delta x} = 3x^2 + 3x\Delta x + (\Delta x)^2$$

$$\lim_{\Delta x \to 0} \frac{f(x + \Delta x) - f(x)}{\Delta x} = 3x^2$$

At $(-2, -5)$, the slope of the tangent line is $m = 3(-2)^2 = 12$.

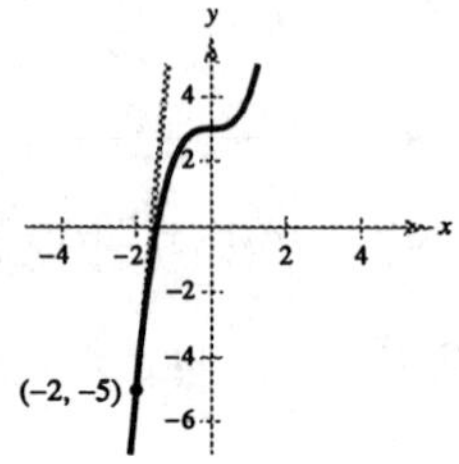

36. $f(x + \Delta x) = \sqrt{2(x + \Delta x) - 2}$

$$f(x + \Delta x) - f(x) = \sqrt{2x + 2\Delta x - 2} - \sqrt{2x - 2}$$

$$= \frac{\sqrt{2x + 2\Delta x - 2} - \sqrt{2x - 2}}{1} \cdot \frac{\sqrt{2x + 2\Delta x - 2} + \sqrt{2x - 2}}{\sqrt{2x + 2\Delta x - 2} + \sqrt{2x - 2}}$$

$$= \frac{2\Delta x}{\sqrt{2x + 2\Delta x - 2} + \sqrt{2x - 2}}$$

$$\frac{f(x + \Delta x) - f(x)}{\Delta x} = \frac{2}{\sqrt{2x + 2\Delta x - 2} + \sqrt{2x - 2}}$$

$$\lim_{\Delta x \to 0} \frac{f(x + \Delta x) - f(x)}{\Delta x} = \frac{2}{2\sqrt{2x - 2}} = \frac{1}{\sqrt{2x - 2}}$$

At $(9, 4)$, the slope of the tangent line is $m = \dfrac{1}{\sqrt{2(9) - 2}} = \dfrac{1}{4}$.

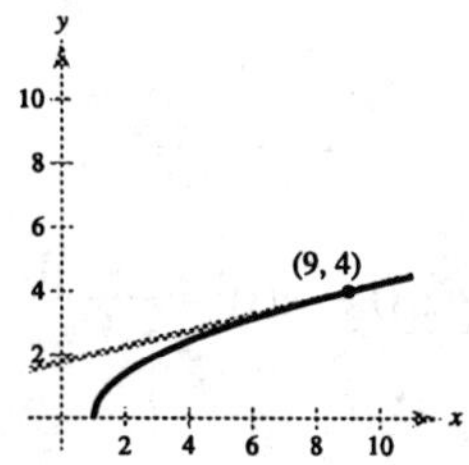

38. $f(x + \Delta x) = -(x + \Delta x)^2 = -x^2 - 2x\Delta x - (\Delta x)^2$

$$f(x + \Delta x) - f(x) = -2x\Delta x - (\Delta x)^2$$

$$\frac{f(x + \Delta x) - f(x)}{\Delta x} = -2x - \Delta x$$

$$\lim_{\Delta x \to 0} \frac{f(x + \Delta x) - f(x)}{\Delta x} = -2x$$

At the point $(-1, -1)$, the slope of the tangent line is $m = -2(-1) = 2$. The equation of the tangent line is

$$y - (-1) = 2[x - (-1)]$$

$$y = 2x + 1.$$

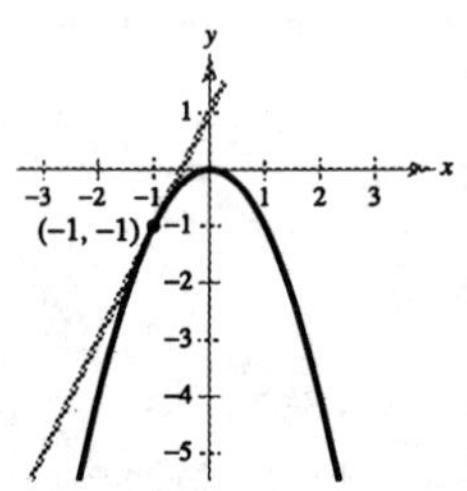

40. $f(x + \Delta x) = 2(x + \Delta x)^2 - 1$

$$= 2x^2 + 4x\Delta x + 2(\Delta x)^2 - 1$$

$$f(x + \Delta x) - f(x) = 4x\Delta x + 2(\Delta x)^2$$

$$\frac{f(x + \Delta x) - f(x)}{\Delta x} = 4x + 2\Delta x$$

$$\lim_{\Delta x \to 0} \frac{f(x + \Delta x) - f(x)}{\Delta x} = 4x$$

At the point $(0, -1)$, the slope of the tangent line is $m = 4(0) = 0$. The equation of the horizontal tangent line is $y = -1$.

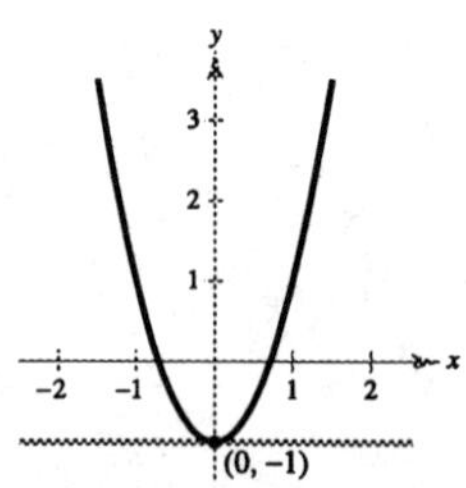

42. $f(x + \Delta x) = \dfrac{1}{x + \Delta x}$

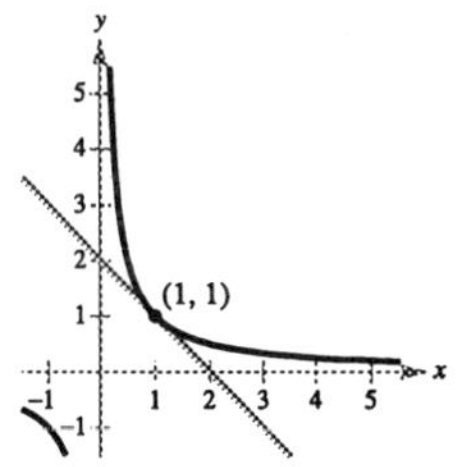

$$f(x + \Delta x) - f(x) = \frac{1}{x + \Delta x} - \frac{1}{x} = \frac{x - (x + \Delta x)}{(x + \Delta x)(x)} = -\frac{\Delta x}{x(x + \Delta x)}$$

$$\frac{f(x + \Delta x) - f(x)}{\Delta x} = -\frac{1}{x(x + \Delta x)}$$

$$\lim_{\Delta x \to 0} \frac{f(x + \Delta x) - f(x)}{\Delta x} = -\frac{1}{x^2}$$

At the point $(1, 1)$, the slope of the tangent line is $m = -\dfrac{1}{(1)^2} = -1$. The equation of the tangent line is

$$y - 1 = -1(x - 1)$$
$$y - 1 = -x + 1$$
$$x + y - 2 = 0.$$

44. $f(x + \Delta x) = (x + \Delta x)^2 + 1 = x^2 + 2x\Delta x + (\Delta x)^2 + 1$

$$f(x + \Delta x) - f(x) = 2x\Delta x + (\Delta x)^2$$

$$\frac{f(x + \Delta x) - f(x)}{\Delta x} = 2x + \Delta x$$

$$\lim_{\Delta x \to 0} \frac{f(x + \Delta x) - f(x)}{\Delta x} = 2x \quad \text{(Slope of tangent line)}$$

Since the slope of the given line is -2, we have $2x = -2$ and $x = -1$. Therefore, at the point $(-1, 2)$, the tangent line parallel to $2x + y = 0$ is

$$y - 2 = -2[x - (-1)]$$
$$y = -2x.$$

46. $f(x + \Delta x) = \dfrac{1}{\sqrt{x + \Delta x}}$

$$f(x + \Delta x) - f(x) = \frac{1}{\sqrt{x + \Delta x}} - \frac{1}{\sqrt{x}} = \frac{\sqrt{x} - \sqrt{x + \Delta x}}{\sqrt{x}\sqrt{x + \Delta x}} \cdot \frac{\sqrt{x} + \sqrt{x + \Delta x}}{\sqrt{x} + \sqrt{x + \Delta x}}$$

$$= \frac{x - (x + \Delta x)}{\sqrt{x}\sqrt{x + \Delta x}\left(\sqrt{x} + \sqrt{x + \Delta x}\right)} = -\frac{\Delta x}{\sqrt{x}\sqrt{x + \Delta x}\left(\sqrt{x} + \sqrt{x + \Delta x}\right)}$$

$$\frac{f(x + \Delta x) - f(x)}{\Delta x} = -\frac{1}{\sqrt{x}\sqrt{x + \Delta x}\left(\sqrt{x} + \sqrt{x + \Delta x}\right)}$$

$$\lim_{\Delta x \to 0} \frac{f(x + \Delta x) - f(x)}{\Delta x} = -\frac{1}{\sqrt{x}\sqrt{x}\left(\sqrt{x} + \sqrt{x}\right)} = -\frac{1}{2x\sqrt{x}}$$

The slope of the tangent line is

$$f'(x) = -\frac{1}{2x\sqrt{x}}.$$

Since the slope of the given line is $-1/2$, we have

$$-\frac{1}{2x\sqrt{x}} = -\frac{1}{2} \text{ and } x = 1.$$

Therefore, at the point $(1, 1)$ the tangent line is parallel to $x + 2y - 6 = 0$. The equation of the tangent line is

$$y - 1 = -\frac{1}{2}(x - 1)$$
$$2y - 2 = -x + 1$$
$$x + 2y - 3 = 0.$$

48. y is differentiable everywhere except the two cusps at $x = \pm 3$.

50. y is not differentiable at $x = 0$. At $(0, 0)$ the graph has a cusp.

52. y is differentiable everywhere except $x = \pm 2$, where f is not defined.

54. Since $x = 1$ is a nonremovable discontinuity, y is differentiable everywhere except at $x = 1$.

56.

x	-2	$-\frac{3}{2}$	-1	$-\frac{1}{2}$	0	$\frac{1}{2}$	1	$\frac{3}{2}$	2
$f(x)$	2	1.125	0.5	0.125	0	0.125	0.5	1.125	2
$f'(x)$	-2	-1.5	-1	-0.5	0	0.5	1	1.5	2

Analytically, the slope of $f(x) = \frac{1}{2}x^2$ is

$$m = \lim_{\Delta x \to 0} \frac{f(x + \Delta x) - f(x)}{\Delta x} = \lim_{\Delta x \to 0} \frac{\frac{1}{2}(x + \Delta x)^2 - \frac{1}{2}x^2}{\Delta x} = \lim_{\Delta x \to 0} \frac{x\Delta x + \frac{1}{2}(\Delta x)^2}{\Delta x} = \lim_{\Delta x \to 0} \left(x + \frac{1}{2}\Delta x\right) = x.$$

58.

x	-2	$-\frac{3}{2}$	-1	$-\frac{1}{2}$	0	$\frac{1}{2}$	1	$\frac{3}{2}$	2
$f(x)$	-6	-3.375	-1.5	-0.375	0	-0.375	-1.5	-3.375	-6
$f'(x)$	6	4.5	3	1.5	0	-1.5	-3	-4.5	-6

$f(x) = -\frac{3}{2}x^2 \qquad f'(x) = -3x$

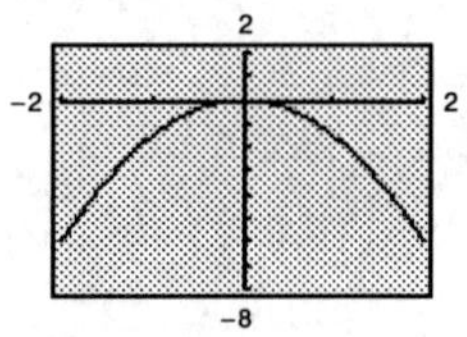

60. False. $f(x) = |x|$ is continuous, but not differentiable at $x = 0$.

62. True

64. $$f'(x) = \lim_{\Delta x \to 0} \frac{f(x + \Delta x) - f(x)}{\Delta x} = \lim_{\Delta x \to 0} \frac{2 + 6(x + \Delta x) - (x + \Delta x)^2 - (2 + 6x - x^2)}{\Delta x}$$

$$= \lim_{\Delta x \to 0} \frac{6\Delta x - 2x\Delta x - (\Delta x)^2}{\Delta x} = \lim_{\Delta x \to 0} (6 - 2x - \Delta x) = 6 - 2x$$

Graphs of $f(x) = 2 + 6x - x^2$ and $f'(x) = 6 - 2x$ (see figure)

The x-intercept of the derivative indicates a point of horizontal tangency for f.

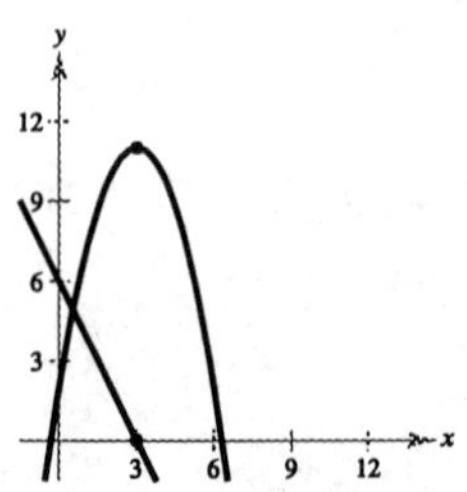

Section 2.2 Some Rules for Differentiation

2. (a) $y = x^{3/2}$

$y' = \frac{3}{2}x^{1/2} = \frac{3\sqrt{x}}{2}$

At $(1, 1)$, $y' = \frac{3}{2}$.

(b) $y = x^3$

$y' = 3x^2$

At $(1, 1)$, $y' = 3$.

4. (a) $y = x^{-1/2}$

$y' = -\frac{1}{2}x^{-3/2} = -\frac{1}{2x\sqrt{x}}$

At $(1, 1)$, $y' = -\frac{1}{2}$.

(b) $y = x^{-2}$

$y' = -2x^{-3} = -\frac{2}{x^3}$

At $(1, 1)$, $y' = -2$.

6. $f'(x) = 0$

8. $g'(x) = 3$

10. $y' = 2t + 2$

12. $y' = 3x^2$

14. $y' = 6x^2 - 2x + 3$

16. $h'(x) = \frac{5}{2}x^{3/2}$

18. $g(x) = 4x^{1/3} + 2$

$g'(x) = 4\left(\frac{1}{3}x^{-2/3}\right) = \frac{4}{3x^{2/3}}$

20. $s'(t) = 4(-1t^{-2}) = -\frac{4}{t^2}$

	Function	*Rewrite*	*Differentiate*	*Simplify*
22.	$y = \frac{2}{3x^2}$	$y = \frac{2}{3}x^{-2}$	$y' = -\frac{4}{3}x^{-3}$	$y' = -\frac{4}{3x^3}$
24.	$y = \frac{\pi}{(3x)^2}$	$y = \frac{\pi}{9}x^{-2}$	$y' = -\frac{2\pi}{9}x^{-3}$	$y' = -\frac{2\pi}{9x^3}$
26.	$y = \frac{4}{x^{-3}}$	$y = 4x^3$	$y' = 12x^2$	$y' = 12x^2$

28. $y = -\frac{1}{2} + \frac{7}{5}x^3$

$y' = \frac{21}{5}x^2$

At $(0, -1/2)$, $y' = 0$.

30. $y = 3x\left(x^2 - \frac{2}{x}\right) = 3x^3 - 6$

$y' = 9x^2$

At $(2, 18)$, $y' = 9(2)^2 = 36$.

32. $f(x) = 3(5 - x)^2 = 75 - 30x + 3x^2$

$f'(x) = -30 + 6x$

At $(5, 0)$, $f'(5) = 0$.

34. $f(x) = x^2 - 3x - 3x^{-2} + 5x^{-3}$

$f'(x) = 2x - 3 + 6x^{-3} - 15x^{-4}$

$= 2x - 3 + \frac{6}{x^3} - \frac{15}{x^4}$

36. $f(x) = x^2 + 4x + \frac{1}{x} = x^2 + 4x + x^{-1}$

$f'(x) = 2x + 4 - x^{-2} = 2x + 4 - \frac{1}{x^2}$

38. $f(x) = \frac{2x^2 - 3x + 1}{x} = 2x - 3 + x^{-1}$

$f'(x) = 2 - x^{-2} = 2 - \frac{1}{x^2} = \frac{2x^2 - 1}{x^2}$

40. $f(x) = (x^2 + 2x)(x + 1) = x^3 + 3x^2 + 2x$

$f'(x) = 3x^2 + 6x + 2$

42. $f(x) = x^{1/3} - 1$

$f'(x) = \frac{1}{3}x^{-2/3} = \frac{1}{3x^{2/3}}$

44. $y' = 3x^2 + 1$

At $(-1, -2)$, the slope is $m = 4$.

The equation of the tangent line is

$$y - (-2) = 4[x - (-1)]$$
$$y + 2 = 4x + 4$$
$$0 = 4x - y + 2.$$

46. $f(x) = \frac{1}{\sqrt[3]{x^2}} = x^{-2/3}$

$f'(x) = -\frac{2}{3}x^{-5/3} = -\frac{2}{3x^{5/3}}$

At $(-1, 1)$, the slope is $\frac{2}{3}$.

$y - 1 = \frac{2}{3}(x + 1) \rightarrow y = \frac{2}{3}x + \frac{5}{3}$

48. $y' = x + 5 = 0$ when $x = -5$.

The function has a horizontal tangent line at the point $\left(-5, -\frac{25}{2}\right)$.

50. $y' = 2x = 0$ when $x = 0$.

The function has a horizontal tangent line at the point $(0, 1)$.

52. (a)

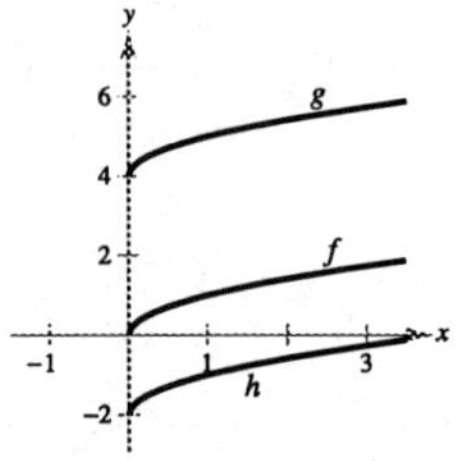

(c)

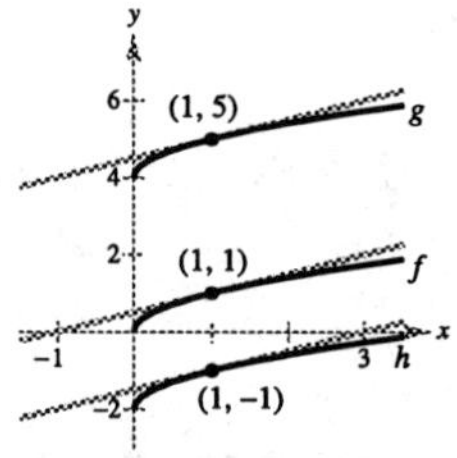

(b) $f'(x) = g'(x) = h'(x) = \frac{1}{2}x^{-1/2} = \frac{1}{2\sqrt{x}}$

$f'(1) = g'(1) = h'(1) = \frac{1}{2}$

54. $R(t) = 58.16t^3 - 449.37t^2 + 1311.9t + 2184.49$

$R'(t) = 174.48t^2 - 898.74t + 1311.9$

(a) For 1991, $t = 1$ and $R'(1) = 587.64$

For 1995, $t = 5$ and $R'(5) = 1180.2$

(b) The answers are very close.

(c) The units are millions of dollars per year per year.

56. $C = 7.75x + 500$

$C' = 7.75$, which equals the variable cost.

58. (a) Men and women seem to have more migraines between 30 and 40 years old. Males have more than females. Wealthy people have fewer migraines than poor people.

(b) The derivatives are positive up to approximately 37 years old. The number of migraines increases up to about 37 years old, then decreases. The units of the derivative are number of migraines per year.

60.

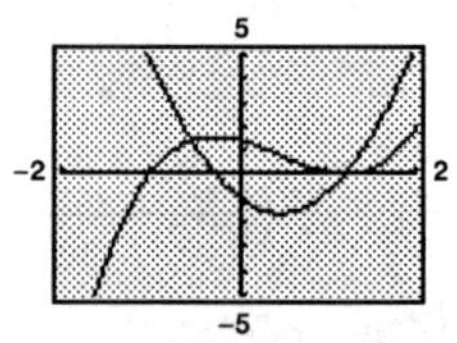

f has horizontal tangents $(y' = 0)$ at $x = 1.2$ and $x = -0.267$.

Section 2.3 Rates of Change: Velocity and Marginals

2. (a) $\dfrac{350-50}{1985-1970} = \dfrac{300}{15} = 20$ billion/year (imports)

(b) $\dfrac{220-50}{1985-1970} = \dfrac{170}{15} \approx 11$ billion/year (exports)

(c) $\dfrac{750-350}{1995-1985} = \dfrac{400}{10} = 40$ billion/year (imports)

(d) $\dfrac{590-210}{1995-1985} = \dfrac{380}{10} = 38$ billion/year (exports)

4. $h(x) = 1 - x^2, [0, 1], h'(x) = -2x$

Average rate of change: $\dfrac{\Delta y}{\Delta x} = \dfrac{h(1)-h(0)}{1-0} = \dfrac{0-1}{1-0} = -1$

Instantaneous rates of change: $h'(0) = 0, h'(1) = -2$

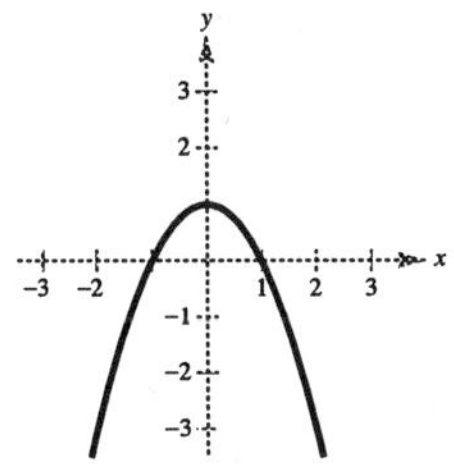

6. $f'(x) = 2x - 6$

Average rate of change:

$$\frac{\Delta y}{\Delta x} = \frac{f(3)-f(-1)}{3-(-1)} = \frac{-10-6}{4} = -4$$

Instantaneous rates of change: $f'(-1) = -8,\ f'(3) = 0$

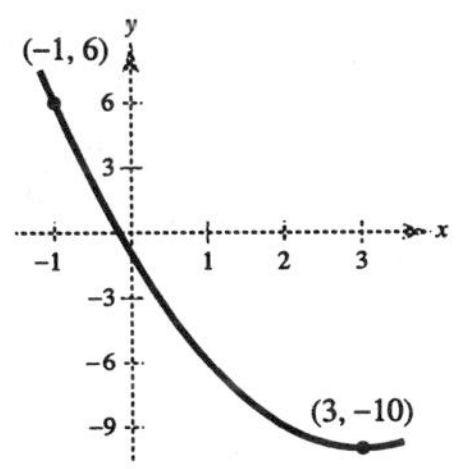

8. $f'(x) = -\dfrac{1}{2}x^{-3/2} = -\dfrac{1}{2x\sqrt{x}}$

Average rate of change:

$$\frac{\Delta y}{\Delta x} = \frac{f(4)-f(1)}{4-1} = \frac{(1/2)-1}{3} = -\frac{1}{6}$$

Instantaneous rates of change: $f'(1) = -\dfrac{1}{2},\ f'(4) = -\dfrac{1}{16}$

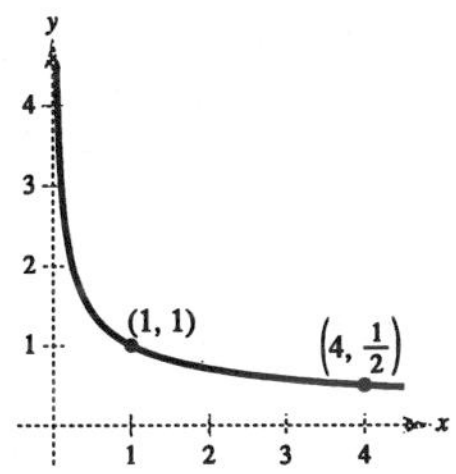

10. $g'(x) = 3x^2$

Average rate of change:

$$\frac{\Delta y}{\Delta x} = \frac{g(1)-g(-1)}{1-(-1)} = \frac{0-(-2)}{2} = 1$$

Instantaneous rates of change: $g'(-1) = 3,\ g'(1) = 3$

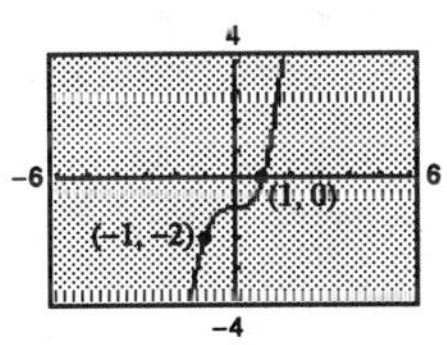

12. (a) $H'(v) = 33\left[10\left(\dfrac{1}{2}v^{-1/2}\right) - 1\right] = 33\left[\dfrac{5}{\sqrt{v}} - 1\right]$

(b) $H'(2) = 33\left[\dfrac{5}{\sqrt{2}} - 1\right] \approx 83.673 \div 3600 \approx 0.023$ kilocalories/meters^3

$H'(5) = 33\left[\dfrac{5}{\sqrt{5}} - 1\right] \approx 40.790 \div 3600 \approx 0.11$ kilocalories/meters^3

14. $s = -16t^2 + 200t$

$v = s' = -32t + 200$

(a) $v(1) = -32(1) + 200 = 168$ ft/sec

(b) $v = s' = -32t + 200 = 0$

$$32t = 200$$

$$t = 6.25$$

The speed is decreasing on the interval (0, 6.25).

(c) The speed is increasing on the interval (6.25, 12.5).

16. First leg: 0.75 km in 20 seconds

Second leg: 0.75 km in 25 seconds

(a) $\dfrac{0.75}{20} = 0.0375$ km/sec $= 37.5$ m/sec

(b) $\dfrac{1.50}{20 + 25} = 0.3\overline{3}$ km/sec $= 33.3\overline{3}$ m/sec

18. $\dfrac{dC}{dx} = 7200$

20. $\dfrac{dC}{dx} = 100\left[0 + 3\left(\dfrac{1}{2}x^{-1/2}\right)\right] = \dfrac{150}{\sqrt{x}}$

22. $\dfrac{dR}{dx} = 30 - 2x$

24. $\dfrac{dR}{dx} = 50\left[20 - \dfrac{3}{2}x^{1/2}\right] = 1000 - 75\sqrt{x}$

26. $\dfrac{dP}{dx} = -0.5x + 2000$

28. $\dfrac{dP}{dx} = -1.5x^2 + 60x - 164.25$

30. $C = 3.6\sqrt{x} + 500$

(a) $C(10) - C(9) \approx 0.584$

(b) $C'(x) = 1.8/\sqrt{x}$

$C'(9) = 0.6$

(c) The answers are very close.

32. $P = 22t^2 + 52t + 10{,}000$

(a) $P(0) = 10{,}000$

$P(10) = 12{,}720$

$P(20) = 19{,}840$

$P(25) = 25{,}050$

The population is growing quadratically.

(b) $\dfrac{dP}{dt} = 44t + 52$

(c) $P'(0) = 52$

$P'(10) = 492$

$P'(20) = 932$

$P'(25) = 1152$

The rate of growth is increasing.

34. $\dfrac{dP}{dx} = 2048\left(\dfrac{1}{2}x^{-1/2}\right) - \dfrac{1}{8}(-2x^{-3})$

$= \dfrac{1024}{\sqrt{x}} + \dfrac{1}{4x^3}$

(a) When $x = 150$, $\dfrac{dP}{dx} \approx 83.61$.

(b) When $x = 175$, $\dfrac{dP}{dx} \approx 77.41$.

(c) When $x = 200$, $\dfrac{dP}{dx} \approx 72.41$.

(d) When $x = 225$, $\dfrac{dP}{dx} \approx 68.27$.

(e) When $x = 250$, $\dfrac{dP}{dx} \approx 64.76$.

(f) When $x = 275$, $\dfrac{dP}{dx} \approx 61.75$.

36. (a) $TR = -10Q^2 + 160Q$

(b) $(TR)' = MR = -20Q + 160$

(c)

Q	0	2	4	6	8	10
Model	160	120	80	40	0	−40
Table	0	130	90	50	10	−30

38. (36,000, 6), (30,000, 7)

$$\text{Slope} = \frac{7-6}{30{,}000 - 36{,}000} = \frac{1}{-6000}$$

$$p - 6 = \frac{-1}{6000}(x - 36{,}000)$$

$$p = \frac{-1}{6000}x + 12 \qquad \text{(demand function)}$$

(a) $P = R - C = xp - (0.20x + 85{,}000)$

$$= x\left(\frac{-1}{6000}x + 12\right) - 0.2x - 85{,}000$$

$$= \frac{-x^2}{6000} + 11.8x - 85{,}000$$

(b)

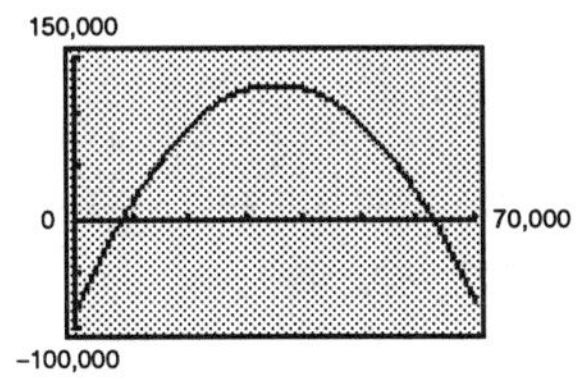

The slope of P is positive at $x = 18{,}000$.

(c) $P'(x) = \dfrac{-x}{3000} + 11.8$

$P'(18{,}000) = 5.8$ dollars

40. $p = \dfrac{50}{\sqrt{x}}, \quad 1 \le x < 8000, \quad C = 0.5x + 500$

$$P = R - C = xp - C = x\left(\frac{50}{\sqrt{x}}\right) - (0.5x + 500)$$

$$= 50\sqrt{x} - 0.5x - 500$$

$$P'(x) = \frac{25}{\sqrt{x}} - 0.5$$

(a) $P'(900) = 0.333$

(b) $P'(1600) = 0.125$

(c) $P'(2500) = 0.0$

(d) $P'(3600) = -0.0833$

$P'(2500) = 0$ indicates that $x = 2500$ is the optimal value of x. Hence, $p = 50/\sqrt{x} = 50/\sqrt{2500} = \$1.00$.

42. (a) $\dfrac{525 - 0}{1900 - 1790} \approx 4.77$ mi/year (b) $\dfrac{285}{1990 - 1900} \approx 3.12$ mi/year (c) $\dfrac{810}{1990 - 1790} \approx 4.05$ mi/year

44. $f(x) = \dfrac{4}{x} = 4x^{-1}, f'(x) = -4x^{-2} = \dfrac{-4}{x^2}$

(a)

x	$\frac{1}{8}$	$\frac{1}{4}$	$\frac{1}{2}$	1	2	3	4	5
$f(x)$	32	16	8	4	2	1.33	1	0.8
$f'(x)$	−256	−64	−16	−4	−1	−0.44	−0.25	−0.16

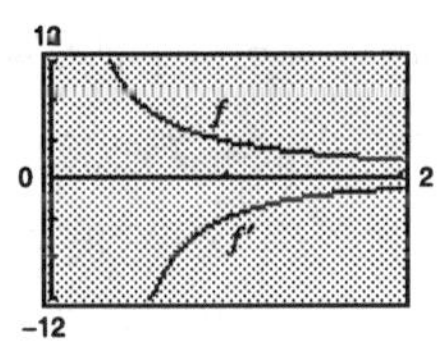

(b) $\left[\frac{1}{4}, \frac{1}{8}\right]$: average rate of change $= \dfrac{16 - 32}{(1/4) - (1/8)} = -128$

$\left[\frac{1}{2}, \frac{1}{4}\right]$: average rate of change $= -32$

$\left[1, \frac{1}{2}\right]$: average rate of change $= -8$

$[2, 1]$: average rate of change $= -2$

$[3, 2]$: average rate of change $= -0.67$

$[4, 3]$: average rate of change $= -0.33$

$[5, 4]$: average rate of change $= -0.2$

46. $f(x) = x^4 - 12x^3 + 52x^2 - 96x + 64$

$f'(x) = 4x^3 - 36x^2 + 104x - 96$

$f'(x) = 0$ when $x = 2, 3, 4$.

Therefore, f has horizontal tangents at $(2, 0)$, $(3, 1)$, and $(4, 0)$.

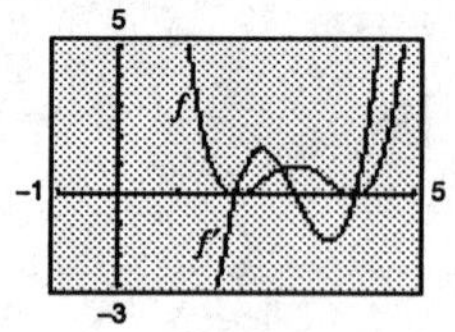

Section 2.4 The Product and Quotient Rules

2. $f'(x) = (x^2 + 1)(2) + (2x + 5)(2x)$

$= 6x^2 + 10x + 2$

$f'(-1) = -2$

4. $f'(x) = \frac{1}{7}(-12x) = -\frac{12x}{7}$

$f'(1) = -\frac{12}{7}$

6. $g'(x) = (x^2 - 2x + 1)(3x^2) + (x^3 - 1)(2x - 2)$

$= 3x^4 - 6x^3 + 3x^2 + 2x^4 - 2x^3 - 2x + 2$

$= 5x^4 - 8x^3 + 3x^2 - 2x + 2$

$g'(1) = 0$

8. $h'(x) = \frac{(x + 3)(2x) - (x^2)(1)}{(x + 3)^2}$

$= \frac{2x^2 + 6x - x^2}{(x + 3)^2} = \frac{x^2 + 6x}{(x + 3)^2}$

$h'(-1) = -\frac{5}{4}$

10. $f'(x) = \frac{(x - 1)(1) - (x + 1)(1)}{(x - 1)^2} = -\frac{2}{(x - 1)^2}$

$f'(2) = -2$

	Function	*Rewrite*	*Differentiate*	*Simplify*
12.	$y = \frac{4x^{3/2}}{x}$	$y = 4x^{1/2}$	$y' = 2x^{-1/2}$	$y' = \frac{2}{\sqrt{x}}$
14.	$y = \frac{4}{5x^2}$	$y = \frac{4}{5}x^{-2}$	$y' = -\frac{8}{5}x^{-3}$	$y' = -\frac{8}{5x^3}$
16.	$y = \frac{x^2 - 4}{x + 2}$	$y = x - 2, x \neq -2$	$y' = 1, x \neq -2$	$y' = 1, x \neq -2$
18.	$y = \frac{3x^2 - 4x}{6x}$	$y = \frac{1}{2}x - \frac{2}{3}, x \neq 0$	$y' = \frac{1}{2}, x \neq 0$	$y' = \frac{1}{2}, x \neq 0$

20. $h(t) = (t^5 - 1)(4t^2 - 7t - 3)$

$h'(t) = (t^5 - 1)(8t - 7) + (5t^4)(4t^2 - 7t - 3)$

$= 8t^6 - 7t^5 - 8t + 7 + 20t^6 - 35t^5 - 15t^4$

$= 28t^6 - 42t^5 - 15t^4 - 8t + 7$

22. $h(p) = (p^3 - 2)^2$

$h(p) = p^6 - 4p^3 + 4$

$h'(p) = 6p^5 - 12p^2 = 6p^2(p^3 - 2)$

24. $f(x) = x^{1/3}(x + 1) = x^{4/3} + x^{1/3}$

$f'(x) = \frac{4}{3}x^{1/3} + \frac{1}{3}x^{-2/3}$

$= \frac{4x^{1/3}}{3} + \frac{1}{3x^{2/3}} = \frac{4x + 1}{3x^{2/3}}$

26. $f(x) = \frac{x^3 + 3x + 2}{x^2 - 1}$

$f'(x) = \frac{(x^2 - 1)(3x^2 + 3) - (x^3 + 3x + 2)(2x)}{(x^2 - 1)^2}$

$= \frac{3x^4 - 3 - 2x^4 - 6x^2 - 4x}{(x^2 - 1)^2}$

$= \frac{x^4 - 6x^2 - 4x - 3}{(x^2 - 1)^2}$

28. $f(x) = (x^5 - 3x)(1/x^2)$

$$f(x) = (x^5 - 3x)\left(\frac{1}{x^2}\right) = x^3 - \frac{3}{x}$$

$$f'(x) = 3x^2 + \frac{3}{x^2} = \frac{3x^4 + 3}{x^2} = \frac{3(x^4 + 1)}{x^2}$$

30. $h(t) = \dfrac{t + 2}{t^2 + 5t + 6}$

$$h'(t) = \frac{(t^2 + 5t + 6)(1) - (t + 2)(2t + 5)}{(t^2 + 5t + 6)^2}$$

$$= \frac{-t^2 - 4t - 4}{(t + 2)^2(t + 3)^2}$$

$$= -\frac{(t + 2)^2}{(t + 2)^2(t + 3)^2}$$

$$= \frac{-1}{(t + 3)^2}, t \neq -2$$

Equivalently, note that

$$h(t) = \frac{t + 2}{(t + 2)(t + 3)} = \frac{1}{t + 3}, t \neq -2$$

32. $f(x) = \dfrac{x + 1}{\sqrt{x}}$

$$f(x) = \frac{x + 1}{\sqrt{x}} = x^{1/2} + x^{-1/2}$$

$$f'(x) = \frac{1}{2}x^{-1/2} - \frac{1}{2}x^{-3/2}$$

$$= \frac{1}{2}\left(\frac{1}{x^{1/2}} - \frac{1}{x^{3/2}}\right)$$

$$= \frac{1}{2}\left(\frac{x - 1}{x^{3/2}}\right)$$

$$= \frac{x - 1}{2x^{3/2}}$$

34. $f(x) = (3x^3 + 4x)(x - 5)(x + 1)$

$$f(x) = (3x^3 + 4x)(x - 5)(x + 1) = (3x^3 + 4x)(x^2 - 4x - 5)$$

$$f'(x) = (3x^3 + 4x)(2x - 4) + (x^2 - 4x - 5)(9x^2 + 4)$$

$$= (6x^4 - 12x^3 + 8x^2 - 16x) + (9x^4 - 36x^3 - 41x^2 - 16x - 20)$$

$$= 15x^4 - 48x^3 - 33x^2 - 32x - 20$$

36. $h(x) = (x^2 - 1)^2, \quad (-2, 9)$

$$h'(x) = 2(x^2 - 1)(2x) = 4x(x^2 - 1)$$

$$h'(-2) = 4(-2)(4 - 1) = -24$$

$$y - 9 = -24(x + 2)$$

$$y = -24x - 39$$

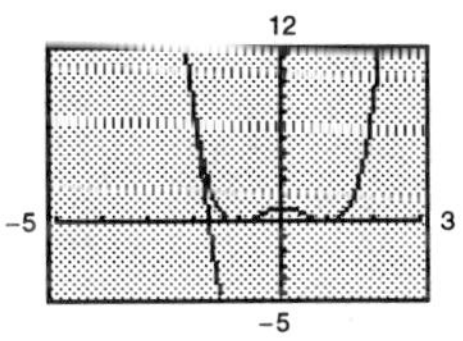

38. $f(x) = \dfrac{\sqrt{x}}{x + 1}, \quad \left(1, \frac{1}{2}\right)$

$$f'(x) = \frac{(x + 1)\frac{1}{2}x^{-1/2} - x^{1/2}}{(x + 1)^2}$$

$$= \frac{\frac{1}{2}x^{-1/2}(x + 1 - 2x)}{(x + 1)^2}$$

$$= \frac{1 - x}{2\sqrt{x}(x + 1)^2}$$

$$y - \frac{1}{2} = 0(x - 1)$$

$$y = \frac{1}{2}$$

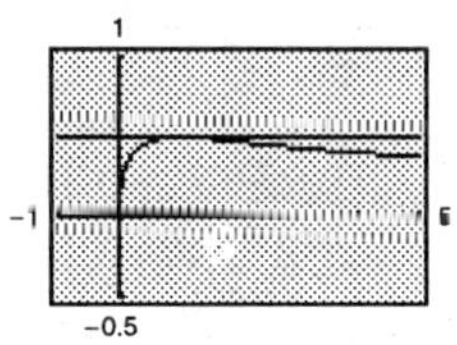

40. $f(x) = (x^2 - x)(x^2 + 1)(x^2 + x + 1),\ (1, 0)$

$f(x) = x^6 + x^4 - x^3 - x$

$f'(x) = 6x^5 + 4x^3 - 3x^2 - 1$

$f'(1) = 6$

$y - 0 = 6(x - 1)$

$y = 6x - 6$

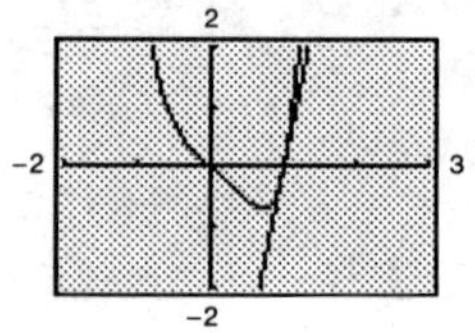

42. $f'(x) = \dfrac{(x^2 + 1)(2x) - (x^2)(2x)}{(x^2 + 1)^2} = \dfrac{2x}{(x^2 + 1)^2}$

$f'(x) = 0$ when $2x = 0$, which implies that $x = 0$. Thus, the horizontal tangent line occurs at $(0, 0)$.

44. $f'(x) = \dfrac{(x^2 + 1)(4x^3) - (x^4 + 3)(2x)}{(x^2 + 1)^2} = \dfrac{2x(x^2 + 3)(x^2 - 1)}{(x^2 + 1)^2}$

$f'(x) = 0$ when $x = 0$ and $x = \pm 1$. Thus, the horizontal tangent line occurs at the points $(0, 3)$, $(1, 2)$, and $(-1, 2)$.

46. $f(x) = x^2(x + 1) = x^3 + x^2$

$f'(x) = 3x^2 + 2x = x(3x + 2)$

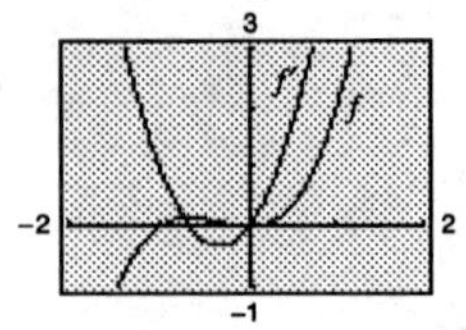

48. $f(x) = x^2(x + 1)(x - 1) = x^4 - x^2$

$f'(x) = 4x^3 - 2x = 2x(2x^2 - 1)$

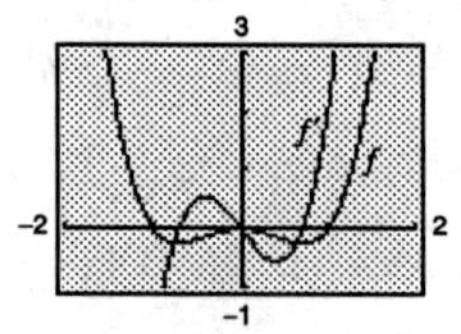

50. $$\frac{dx}{dp} = 0 - 1 - \frac{(p + 1)(2) - (2p)(1)}{(p + 1)^2}$$

$$= -1 - \frac{2}{(p + 1)^2}$$

$$= \frac{-(p + 1)^2 - 2}{(p + 1)^2}$$

$$= \frac{-p^2 - 2p - 3}{(p + 1)^2}$$

When $p = 3$, $\dfrac{dx}{dp} = \dfrac{-9 - 6 - 3}{16} = -\dfrac{9}{8}$.

52. The initial temperature is $T = 10\left(\dfrac{75}{10}\right) = 75$ deg.

$$\frac{dT}{dt} = 10\left[\frac{(t^2 + 4t + 10)(8t + 16) - (4t^2 + 16t + 75)(2t + 4)}{(t^2 + 4t + 10)^2}\right] = 10\left[\frac{-70t - 140}{(t^2 + 4t + 10)^2}\right] = \frac{-700(t + 2)}{(t^2 + 4t + 10)^2}$$

(a) When $t = 1$, $\dfrac{dT}{dt} = \dfrac{-700(3)}{(15)^2} = -\dfrac{28}{3} \approx -9.33$ deg/hr.

(b) When $t = 3$, $\dfrac{dT}{dt} = \dfrac{-700(5)}{(31)^2} = -\dfrac{3500}{961} \approx -3.64$ deg/hr.

(c) When $t = 5$, $\dfrac{dT}{dt} = \dfrac{-700(7)}{(55)^2} = -\dfrac{196}{121} \approx -1.62$ deg/hr.

(d) When $t = 10$, $\dfrac{dT}{dt} = \dfrac{-700(12)}{(150)^2} = -\dfrac{28}{75} \approx -0.37$ deg/hr.

54. $\frac{dP}{dt} = \frac{50(t+2)(1) - (t+1750)(50)}{[50(t+2)]^2} = \frac{50[(t+2) - (t+1750)]}{2500(t+2)^2} = \frac{-1748}{50(t+2)^2} = \frac{-874}{25(t+2)^2}$

(a) When $t = 1$, $\frac{dP}{dt} = \frac{-874}{225} \approx -3.88$ percent/day.

(b) When $t = 10$, $\frac{dP}{dt} = \frac{-874}{3600} = -\frac{437}{1800} \approx -0.24$ percent/day.

56. (a) $P = ax^2 + bx + c$

When $x = 10$, $P = 50$: $50 = 100a + 10b + c$.

When $x = 12$, $P = 60$: $60 = 144a + 12b + c$.

When $x = 14$, $P = 65$: $65 = 196a + 14b + c$.

Solving this system, we have

$a = -\frac{5}{8}$, $b = \frac{75}{4}$, and $c = -75$.

Thus, $P = -\frac{5}{8}x^2 + \frac{75}{4}x - 75$.

(b) 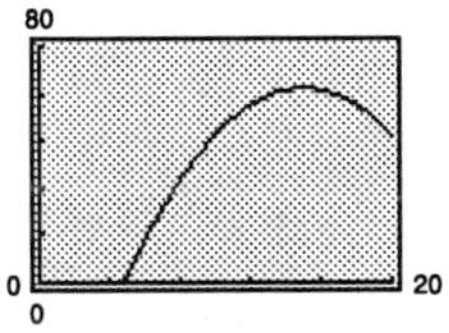

(c) Marginal profit: $P' = -\frac{5}{4}x + \frac{75}{4} = 0 \Rightarrow x = 15$

This is the maximum point on the graph of P.

58. $C = x^3 - 15x^2 + 87x - 73, \quad 4 \le x \le 9$

Marginal cost: $\frac{dC}{dx} = 3x^2 - 30x + 87$

Average cost: $\frac{C}{x} = x^2 - 15x + 87 - \frac{73}{x}$

(a) 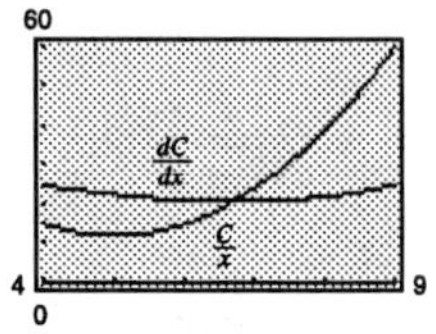

(b) Point of intersection:

$$3x^2 - 30x + 87 = x^2 - 15x + 87 - \frac{73}{x}$$

$$2x^2 - 15x + \frac{73}{x} = 0$$

$$2x^3 - 15x^2 + 73 = 0$$

$$x \approx 6.683$$

When $x = 6.683$, $\frac{C}{x} = \frac{dC}{dx} \approx 20.50$.

Thus, the point of intersection is (6.683, 20.50).

60. $M(t) = \frac{300t}{t^2 + 1} + 8$

(a) $M'(t) = \frac{(t^2+1)(300) - 300t(2t)}{(t^2+1)^2} = \frac{300(1-t^2)}{(t^2+1)^2}$

(b) $M(3) = 98$

$M'(3) = -24$

(c) $M(24) \approx 20.48$

$M'(24) = -0.52$

62. Answers will vary.

Section 2.5 The Chain Rule

	$y = f(g(x))$	$u = g(x)$	$y = f(u)$
2.	$y = (x^2 - 2x + 3)^3$	$u = x^2 - 2x + 3$	$y = u^3$
4.	$y = (x^2 + 1)^{4/3}$	$u = x^2 + 1$	$y = u^{4/3}$
6.	$y = \sqrt{9 - x^2}$	$u = 9 - x^2$	$y = u^{1/2}$
8.	$y = (x + 1)^{-1/2}$	$u = x + 1$	$y = u^{-1/2}$

10. $f(x) = \dfrac{2x}{1 - x^3}$ Quotient Rule (d)

12. $f(x) = \sqrt[3]{x^2} = x^{2/3}$ Simple Power Rule (a)

14. $y' = 4(3x^2 + 1)^3(6x) = 24x(3x^2 + 1)^3$

16. $h'(t) = 4(1 - t^2)^3(-2t) = -8t(1 - t^2)^3$

18. $f'(x) = 3(4x - x^2)^2(4 - 2x) = 6(2 - x)(4x - x^2)^2$

20. $f'(t) = \dfrac{2}{3}(9t + 2)^{-1/3}(9) = 6(9t + 2)^{-1/3} = \dfrac{6}{\sqrt[3]{9t + 2}}$

22. $g(x) = \sqrt{2x + 3} = (2x + 3)^{1/2}$

$$g'(x) = \frac{1}{2}(2x + 3)^{-1/2}(2) = \frac{1}{\sqrt{2x + 3}}$$

24. $y = \sqrt[3]{3x^3 + 4x} = (3x^3 + 4x)^{1/3}$

$$y' = \frac{1}{3}(3x^3 + 4x)^{-2/3}(9x^2 + 4) = \frac{9x^2 + 4}{3(3x^3 + 4x)^{2/3}}$$

26. $y = 2(4 - x^2)^{1/2}$

$$y' = 2\left(\frac{1}{2}\right)(4 - x^2)^{-1/2}(-2x) = \frac{-2x}{\sqrt{4 - x^2}}$$

28. $f(x) = (25 + x^2)^{-1/2}$

$$f'(x) = -\frac{1}{2}(25 + x^2)^{-3/2}(2x) = \frac{-x}{(25 + x^2)^{3/2}}$$

30. $f'(x) = -\dfrac{5}{2}(4 - 3x)^{-7/2}(-3) = \dfrac{15}{2(4 - 3x)^{7/2}}$

32. $g(x) = 3(9x - 4)^4, \quad (2, 115{,}248)$

$$g'(x) = 12(9x - 4)^3(9)$$

$$g'(2) = 12(14)^3(9) = 296{,}352$$

$$y - 115{,}248 = 296{,}352(x - 2)$$

$$y = 296{,}352x - 477{,}456$$

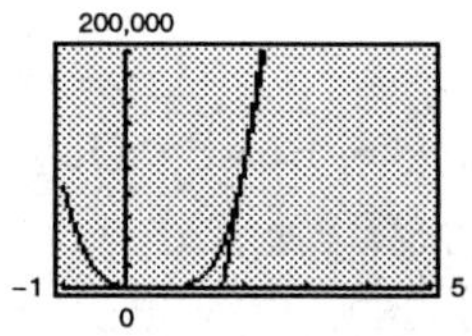

34. $f(x) = x\sqrt{x^2 + 5} = x(x^2 + 5)^{1/2}$

Point: $(2, f(2)) = (2, 6)$

$$f'(x) = x\left[\frac{1}{2}(x^2 + 5)^{-1/2}(2x)\right] + (x^2 + 5)^{1/2}(1)$$

$$= x^2(x^2 + 5)^{-1/2} + (x^2 + 5)^{1/2}$$

$$= (x^2 + 5)^{-1/2}[x^2 + (x^2 + 5)]$$

$$= \frac{2x^2 + 5}{\sqrt{x^2 + 5}}$$

When $x = 2$, the slope is $f'(2) = \frac{13}{3}$ and the equation of the tangent line is

$$y - 6 = \frac{13}{3}(x - 2)$$

$$3y - 18 = 13x - 26$$

$$0 = 13x - 3y - 8$$

$$y = \frac{13}{3}x - \frac{8}{3}.$$

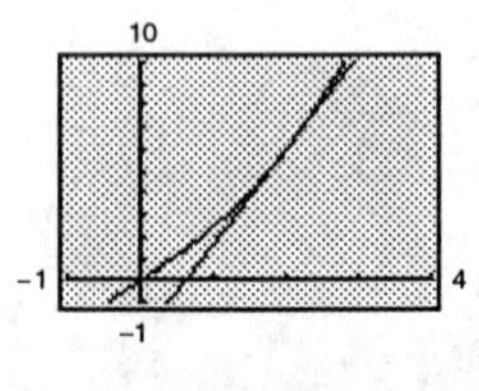

36. $g(x) = (4 - 3x^2)^{-2/3}$

$$g(2) = (-8)^{-2/3} = \frac{1}{4}$$

$$g'(x) = -\frac{2}{3}(4 - 3x^2)^{-5/3}(-6x) = \frac{4x}{(4 - 3x^2)^{5/3}}$$

$$g'(2) = \frac{4(2)}{(-8)^{5/3}} = \frac{8}{-32} = -\frac{1}{4}$$

$$y - \frac{1}{4} = -\frac{1}{4}(x - 2)$$

$$y = -\frac{1}{4}x + \frac{3}{4}$$

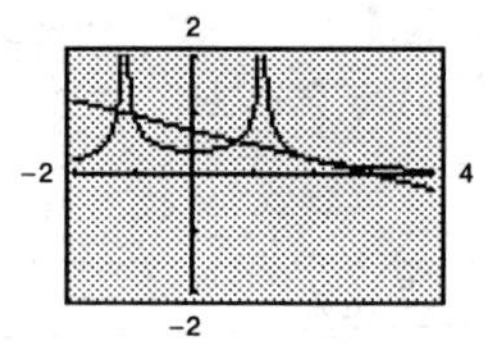

38. $f(x) = \sqrt{\dfrac{2x}{x+2}}$

$f'(x) = \dfrac{\sqrt{2}}{2\sqrt{x}(x+1)^{3/2}}$

f' is never 0.

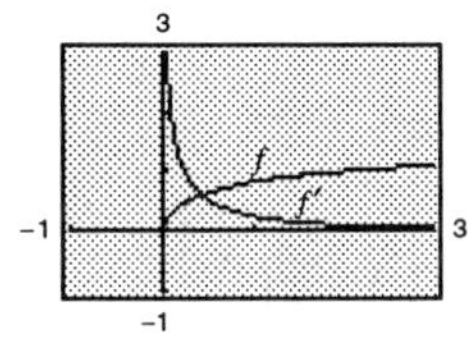

40. $f(x) = \sqrt{x}(2 - x^2)$

$f'(x) = \dfrac{2 - 5x^2}{2\sqrt{x}}$

f has a horizontal tangent when $f' = 0$.

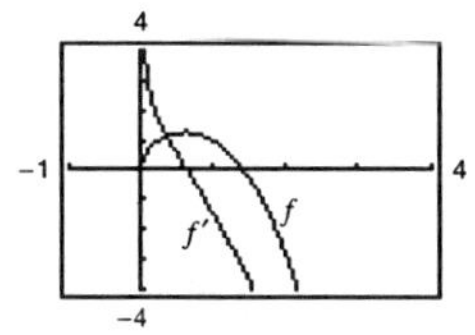

42. $s(t) = \dfrac{1}{t^2 + 3t - 1} = (t^2 + 3t - 1)^{-1}$

$$s'(t) = -1(t^2 + 3t - 1)^{-2}(2t + 3)$$
$$= -\frac{2t + 3}{(t^2 + 3t - 1)^2}$$

44. $f(x) = \dfrac{3}{x^3 - 4} = 3(x^3 - 4)^{-1}$

$$f'(x) = -3(x^3 - 4)^{-2}(3x^2) = \frac{-9x^2}{(x^3 - 4)^2}$$

46. $y = \dfrac{1}{\sqrt{x+2}} = (x + 2)^{-1/2}$

$$y' = -\frac{1}{2}(x + 2)^{-3/2} = \frac{-1}{2(x + 2)^{3/2}}$$

48. $g(x) = \dfrac{3}{\sqrt[3]{x^3 - 1}} = 3(x^3 - 1)^{-1/3}$

$$g'(x) = 3\left(-\frac{1}{3}\right)(x^3 - 1)^{-4/3}(3x^2) = \frac{-3x^2}{(x^3 - 1)^{4/3}}$$

50. $y = t\sqrt{t+1} = t(t + 1)^{1/2}$

$$y' = t\left[\frac{1}{2}(t + 1)^{-1/2}(1)\right] + (t + 1)^{1/2}(1)$$
$$= \frac{1}{2}t(t + 1)^{-1/2} + (t + 1)^{1/2}$$
$$= \frac{1}{2}(t + 1)^{-1/2}[t + 2(t + 1)]$$
$$= \frac{1}{2}(t + 1)^{-1/2}(3t + 2)$$
$$= \frac{3t + 2}{2\sqrt{t + 1}}$$

52. $f(x) = x^3(x - 4)^2$

$$= x^3(x^2 - 8x + 16)$$
$$= x^5 - 8x^4 + 16x^3$$
$$f'(x) = 5x^4 - 32x^3 + 48x^2$$
$$= x^2(5x^2 - 32x + 48)$$
$$= x^2(5x - 12)(x - 4)$$

54. $y = \sqrt{x}(x - 2)^2 = x^{1/2}(x - 2)^2$

$$y' = x^{1/2}[2(x - 2)^1(1)] + (x - 2)^2\left(\frac{1}{2}x^{-1/2}\right)$$
$$= 2\sqrt{x}(x - 2) + \frac{(x - 2)^2}{2\sqrt{x}}$$
$$= \frac{4x(x - 2) + (x - 2)^2}{2\sqrt{x}}$$
$$= \frac{(x - 2)[4x + (x - 2)]}{2\sqrt{x}}$$
$$= \frac{(x - 2)(5x - 2)}{2\sqrt{x}}$$

56. $g(t) = \dfrac{3t^2}{\sqrt{t^2 + 2t - 1}}$

$$g'(t) = \frac{(t^2 + 2t - 1)^{1/2}(6t) - 3t^2\left(\frac{1}{2}\right)(t^2 + 2t - 1)^{-1/2}(2t + 2)}{t^2 + 2t - 1}$$
$$= \frac{(t^2 + 2t - 1)^{-1/2}[(t^2 + 2t - 1)(6t) - 3t^2(t + 1)]}{(t^2 + 2t - 1)}$$
$$= \frac{6t^3 + 12t^2 - 6t - 3t^3 - 3t^2}{(t^2 + 2t - 1)^{3/2}}$$
$$= \frac{3t^3 + 9t^2 - 6t}{(t^2 + 2t - 1)^{3/2}}$$
$$= \frac{3t(t^2 + 3t - 2)}{(t^2 + 2t - 1)^{3/2}}$$

58. $g(x) = \sqrt{x-1} + \sqrt{x+1} = (x-1)^{1/2} + (x+1)^{1/2}$

$$g'(x) = \frac{1}{2}(x-1)^{-1/2}(1) + \frac{1}{2}(x+1)^{-1/2}(1)$$

$$= \frac{1}{2\sqrt{x-1}} + \frac{1}{2\sqrt{x+1}}$$

60. $y = \left(\frac{4x^2}{3-x}\right)^3 = 64\left[\frac{x^6}{(3-x)^3}\right]$

$$y' = 64\left[\frac{(3-x)^3 6x^5 - x^6(3)(3-x)^2(-1)}{(3-x)^6}\right]$$

$$= 64\left\{\frac{x^5[6(3-x) + 3x]}{(3-x)^4}\right\} = 192\left[\frac{x^5(6-x)}{(3-x)^4}\right]$$

62. $s(x) = (x^2 - 3x + 4)^{-1/2}$, $\left(3, \frac{1}{2}\right)$

$$s'(x) = -\frac{1}{2}(x^2 - 3x + 4)^{-3/2}(2x - 3)$$

$$= \frac{3 - 2x}{2(x^2 - 3x + 4)^{3/2}}$$

$$s'(3) = \frac{3-6}{2(4)^{3/2}} = -\frac{3}{16}$$

$$y - \frac{1}{2} = -\frac{3}{16}(x - 3)$$

$$y = -\frac{3}{16}x + \frac{17}{16}$$

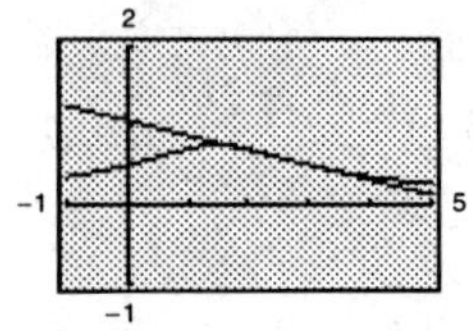

64. $y = \sqrt{\frac{2x}{x+1}} = \left(\frac{2x}{x+1}\right)^{1/2}$

$$y' = \frac{1}{2}\left(\frac{2x}{x+1}\right)^{-1/2}\left[\frac{(x+1)(2) - (2x)(1)}{(x+1)^2}\right]$$

$$= \frac{1}{2}\left(\frac{x+1}{2x}\right)^{1/2} \cdot \frac{2}{(x+1)^2}$$

$$= \frac{(x+1)^{1/2}}{(2x)^{1/2}} \cdot \frac{1}{(x+1)^2}$$

$$= \frac{1}{\sqrt{2x}(x+1)^{3/2}}$$

$$y'(1) = \frac{1}{4}$$

$$y - 1 = \frac{1}{4}(x - 1)$$

$$y = \frac{1}{4}x + \frac{3}{4}$$

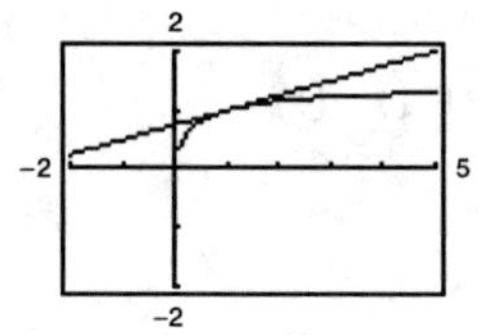

66. $y = \frac{x}{\sqrt{25 + x^2}} = x(25 + x^2)^{-1/2}$

$$y' = x\left[-\frac{1}{2}(25 + x^2)^{-3/2}(2x)\right] + (25 + x^2)^{-1/2}(1)$$

$$= -x^2(25 + x^2)^{-3/2} + (25 + x^2)^{-1/2}$$

$$= (25 + x^2)^{-3/2}[-x^2 + (25 + x^2)]$$

$$= \frac{25}{(25 + x^2)^{3/2}}$$

$$y'(0) = \frac{1}{5}$$

$$y - 0 = \frac{1}{5}(x - 0)$$

$$y = \frac{1}{5}x$$

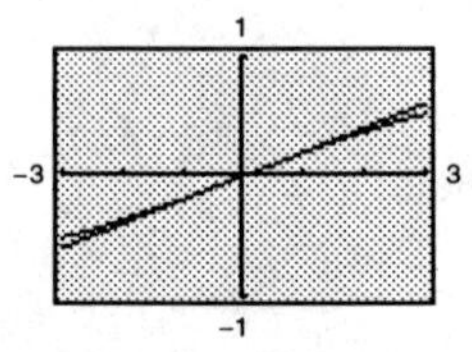

68. $P = 0.25(0.5n^2 + 5n + 25)^{1/2}$

$$P'(n) = \frac{1}{4}\left(\frac{1}{2}\right)(0.5n^2 + 5n + 25)^{-1/2}(n + 5)$$

$$P'(12{,}000) = \frac{12{,}000 + 5}{8[0.5(12{,}000)^2 + 5(12{,}000) + 25]^{1/2}}$$

$$\approx 0.177$$

70. (a) $V = \dfrac{k}{\sqrt{t+1}}$

When $t = 0$, $V = 10{,}000$.

$$10{,}000 = \frac{k}{\sqrt{0+1}} \Rightarrow k = 10{,}000$$

$$V = \frac{10{,}000}{\sqrt{t+1}}$$

(b) $V = 10{,}000(t+1)^{-1/2}$

$$\frac{dV}{dt} = -5000(t+1)^{-3/2}(1) = -\frac{5000}{(t+1)^{3/2}}$$

When $t = 1$,

$$\frac{dV}{dt} = -\frac{5000}{(2)^{3/2}}$$

$$= -\frac{2500}{\sqrt{2}} \approx -\$1767.77.$$

(c) When $t = 3$, $\dfrac{dV}{dt} = -\dfrac{5000}{(4)^{3/2}} = -\$625.00.$

72. $r = \sqrt{1.01t^4 - 6.72t^3 + 4.85t^2 + 2.4t + 330.26}$

(a) $\dfrac{dr}{dt} = \dfrac{4.04t^3 - 20.16t^2 + 9.70t + 2.4}{2\sqrt{1.01t^4 - 6.72t^3 + 4.85t^2 + 2.4t + 330.26}}$

(Chain Rule and Power Rule)

(b)

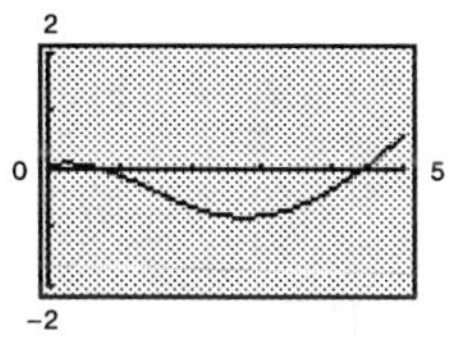

(c) $r(t)$ was changing the most when $t = 3.15$ (min.) and $t = 5$ (max.).

(d) $r'(t) = 0$ when $t = 0.75$.

Section 2.6 Higher-Order Derivatives

2. $f'(x) = 3$

$f''(x) = 0$

4. $f'(x) = 6x + 4$

$f''(x) = 6$

6. $g(t) = t^{-1/3}$

$g'(t) = -\dfrac{1}{3}t^{-4/3}$

$g''(t) = \dfrac{4}{9}t^{-7/3} = \dfrac{4}{9t^{7/3}}$

8. $f(x) = 4(x^2 - 1)^2 = 4(x^4 - 2x^2 + 1)$

$f'(x) = 4(4x^3 - 4x)$

$f''(x) = 4(12x^2 - 4) = 16(3x^2 - 1)$

10. $f(x) = x\sqrt[3]{x} = x^{4/3}$

$f'(x) = \dfrac{4}{3}x^{1/3}$

$f''(x) = \dfrac{4}{9}x^{-2/3} = \dfrac{4}{9x^{2/3}}$

12. $g(t) = -\dfrac{4}{(t+2)^2} = -4(t+2)^{-2}$

$g'(t) = 8(t+2)^{-3}$

$g''(t) = -24(t+2)^{-4} = -\dfrac{24}{(t+2)^4}$

14. $h(s) = s^3(s^2 - 2s + 1) = s^5 - 2s^4 + s^3$

$h'(s) = 5s^4 - 8s^3 + 3s^2$

$h''(s) = 20s^3 - 24s^2 + 6s = 2s(10s^2 - 12s + 3)$

16. $f(x) = x^4 - 2x^3$

$f'(x) = 4x^3 - 6x^2$

$f''(x) = 12x^2 - 12x$

$f'''(x) = 24x - 12 = 12(2x - 1)$

18. $f(x) = (x - 1)^2$

$f'(x) = 2(x - 1)$

$f''(x) = 2$

$f'''(x) = 0$

20. $f(x) = \dfrac{1}{x} = x^{-1}$

$f'(x) = -x^{-2}$

$f''(x) = 2x^{-3}$

$f'''(x) = -6x^{-4} = -\dfrac{6}{x^4}$

22. $f(x) = \sqrt{9 - x^2} = (9 - x^2)^{1/2}$

$f'(x) = \dfrac{1}{2}(9 - x^2)^{-1/2}(-2x) = -x(9 - x^2)^{-1/2}$

$f''(x) = -x\left[-\dfrac{1}{2}(9 - x^2)^{-3/2}(-2x)\right] + (9 - x^2)^{-1/2}(-1)$

$= -x^2(9 - x^2)^{-3/2} - (9 - x^2)^{-1/2}$

$= (9 - x^2)^{-3/2}[-x^2 - (9 - x^2)]$

$= \dfrac{-9}{(9 - x^2)^{3/2}}$

$f''(-\sqrt{5}) = -\dfrac{9}{8}$

24. $f(t) = \sqrt{2t + 3} = (2t + 3)^{1/2}$

$f'(t) = \dfrac{1}{2}(2t + 3)^{-1/2}(2) = (2t + 3)^{-1/2}$

$f''(t) = -\dfrac{1}{2}(2t + 3)^{-3/2}(2) = -(2t + 3)^{-3/2}$

$f'''(t) = \dfrac{3}{2}(2t + 3)^{-5/2}(2) = \dfrac{3}{(2t + 3)^{5/2}}$

$f'''\left(\dfrac{1}{2}\right) = \dfrac{3}{32}$

26. $f''(x) = 20x^3 - 36x^2$

$f'''(x) = 60x^2 - 72x = 12x(5x - 6)$

28. $f'''(x) = 2\sqrt{x - 1} = 2(x - 1)^{1/2}$

$f^{(4)}(x) = 2\left(\dfrac{1}{2}\right)(x - 1)^{-1/2}(1) = \dfrac{1}{\sqrt{x - 1}}$

30. $f(x) = x^3 - 2x$

$f'(x) = 3x^2 - 2$

$f''(x) = 6x$

32. $f(x) = 3x^3 - 9x + 1$

$f'(x) = 9x^2 - 9$

$f''(x) = 18x = 0$

$f''(x) = 0$ when $x = 0$.

34. $f(x) = x^4 - 8x^3 + 18x^2 - 16x + 2$

$f'(x) = 4x^3 - 24x^2 + 36x - 16$

$f''(x) = 12x^2 - 48x + 36 = 12(x^2 - 4x + 3)$

$f''(x) = 0$ when $12(x^2 - 4x + 3) = 0$

$12(x - 1)(x - 3) = 0$

$x = 1, 3.$

36. $f(x) = x\sqrt{4 - x^2} = x(4 - x^2)^{1/2}$

$f'(x) = x\left[\dfrac{1}{2}(4 - x^2)^{-1/2}(-2x)\right] + (4 - x^2)^{1/2}(1)$

$= -x^2(4 - x^2)^{-1/2} + (4 - x^2)^{1/2}$

$= (4 - x^2)^{-1/2}[-x^2 + (4 - x^2)]$

$= (4 - 2x^2)(4 - x^2)^{-1/2}$

$f''(x) = (4 - 2x^2)\left[-\dfrac{1}{2}(4 - x^2)^{-3/2}(-2x)\right] + (4 - x^2)^{-1/2}(-4x)$

$= 2x(2 - x^2)(4 - x^2)^{-3/2} - 4x(4 - x^2)^{-1/2}$

$= 2x(4 - x^2)^{-3/2}[(2 - x^2) - 2(4 - x^2)]$

$= 2x(4 - x^2)^{-3/2}(x^2 - 6) = \dfrac{2x(x^2 - 6)}{(4 - x^2)^{3/2}} = 0$

$f''(x) = 0$ when $x = 0$.

[**Note**: $x = \pm\sqrt{6}$ are not in the domain of f.]

38. $f(x) = \dfrac{x}{x^2 + 1}$

$$f'(x) = \frac{(x^2 + 1)(1) - (x)(2x)}{(x^2 + 1)^2}$$

$$= \frac{1 - x^2}{(x^2 + 1)^2}$$

$$= (1 - x^2)(x^2 + 1)^{-2}$$

$$f''(x) = (1 - x^2)[-2(x^2 + 1)^{-3}(2x)] + (x^2 + 1)^{-2}(-2x)$$

$$= -2x(x^2 + 1)^{-3}[2(1 - x^2) + (x^2 + 1)]$$

$$= -2x(x^2 + 1)^{-3}(3 - x^2)$$

$$= \frac{2x(x^2 - 3)}{(x^2 + 1)^3} = 0$$

$f''(x) = 0$ when $x = 0, \pm\sqrt{3}$.

40. (a) $s(t) = -16t^2 + 1250$

(b) $v(t) = s'(t) = -32t$

$a(t) = v'(t) = -32$

(c) $s(t) = 0$ when $16t^2 = 1250$, or $t = \sqrt{78.125} \approx 8.8$ sec.

(d) $v(8.8) \approx -282.8$ ft/sec

42. $s(t) = -8.25t^2 + 66t$

$v(t) = s'(t) = -16.50t + 66$

$a(t) = s''(t) = -16.50$

t	0	1	2	3	4	5
$s(t)$	0	57.75	99	123.75	132	123.75
$v(t)$	66	49.50	33	16.50	0	−16.50
$a(t)$	−16.50	−16.50	−16.50	−16.50	−16.50	−16.50

Once the brakes are applied, the car decelerates at a constant 16.50 ft/sec^2 and requires 132 ft to come to a complete stop.

44. $f(x) = 3x^3 - 9x$

$f'(x) = 9x^2 - 9$

$f''(x) = 18x$

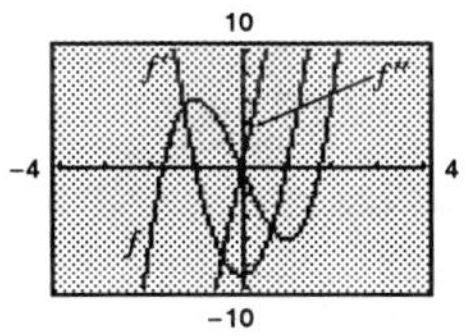

The degrees of the successive derivatives decrease by 1.

46.

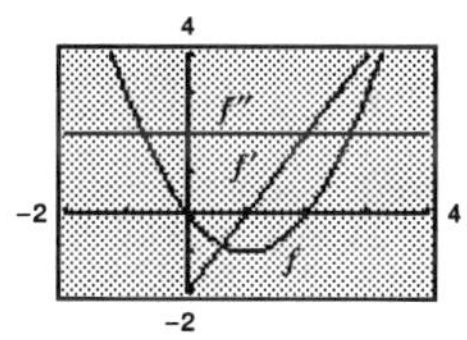

The degrees of the successive derivatives decrease by 1.

48. True. The fifth derivative of a fourth degree polynomial is 0.

50. True. The $(n + 1)^{st}$ derivative of an nth degree polynomial is 0.

52. True. $\dfrac{dC}{dt} = 0$ where $v(t) = C$ is constant.

54. (a) $s(t) = -16t^2 + 48t + 64$

(b) $v(t) = s'(t) = -32t + 48$

$a(t) = v'(t) = -32$

(c) $s(t) = -16t^2 + 48t + 64 = 0 \Longrightarrow$

$-16(t^2 - 3t + 4)$

$= 0 \Longrightarrow (t - 4)(t + 1) = 0 \Longrightarrow$

$t = 4$ sec

(d) $v(t) = -32t + 48 = 0$ when $32t = 48$ or $t = \frac{3}{2}$ sec.

(e) $s\left(\frac{3}{2}\right) = -16\left(\frac{3}{2}\right)^2 + 48\left(\frac{3}{2}\right) + 64 = 100$ feet high

(f)

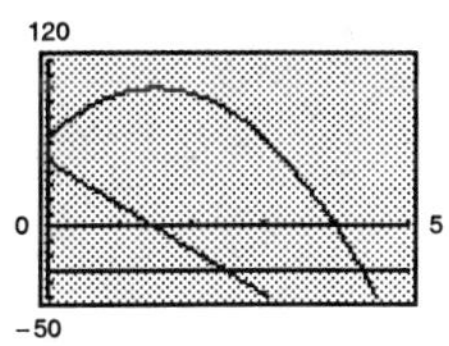

Section 2.7 Implicit Differentiation

2. $\frac{1}{2}x^2 - y = 6x$

$x - y' = 6$

$y' = x - 6$

4. $4x^2y - 3y^{-1} = 0$

$4x^2y' + 8xy + 3y^{-2}y' = 0$

$\left(4x^2 + \frac{3}{y^2}\right)y' = -8xy$

$\frac{4x^2y^2 + 3}{y^2}y' = -8xy$

$y' = \frac{-8xy \cdot y^2}{4x^2y^2 + 3} = \frac{-8xy^3}{4x^2y^2 + 3}$

6. $\frac{xy - y^2}{y - x} = 1$

$xy - y^2 = y - x$

$xy' + y - 2yy' = y' - 1$

$(x - 2y - 1)y' = -y - 1$

$y' = \frac{y + 1}{-x + 2y + 1}$

8. $x^2 - y^2 = 16$

$2x - 2yy' = 0$

$y' = \frac{x}{y}$

At $(4, 0)$, y' is undefined.

10. $x^2 - y^3 = 3$

$2x - 3y^2y' = 0$

$y' = \frac{2x}{3y^2}$

At $(2, 1)$, $y' = \frac{2(2)}{3(1)} = \frac{4}{3}$

12. $x^2y + y^2x = -2$

$x^2y' + 2xy + y^2 + 2yy'x = 0$

$y'(x^2 + 2xy) = -2xy - y^2$

$y' = -\frac{y(2x + y)}{x(x + 2y)}$

At $(2, -1)$, y' is undefined.

14. $x^3 + y^3 = 2xy$

$3x^2 + 3y^2y' = 2xy' + 2y$

$y'(3y^2 - 2x) = 2y - 3x^2$

$y' = \frac{2y - 3x^2}{3y^2 - 2x}$

At $(1, 1)$, $y' = -1$.

16. $\sqrt{xy} = x - 2y$

$\sqrt{x}\sqrt{y} = x - 2y$

$\sqrt{x}\left(\frac{1}{2}y^{-1/2}y'\right) + \sqrt{y}\left(\frac{1}{2}x^{-1/2}\right) = 1 - 2y'$

$\frac{\sqrt{x}}{2\sqrt{y}}y' + 2y' = 1 - \frac{\sqrt{y}}{2\sqrt{x}}$

$$y' = \frac{1 - \frac{\sqrt{y}}{2\sqrt{x}}}{\frac{\sqrt{x}}{2\sqrt{y}} + 2} \cdot \frac{2\sqrt{x}\sqrt{y}}{2\sqrt{x}\sqrt{y}} = \frac{2\sqrt{xy} - y}{x + 4\sqrt{xy}} = \frac{2(x - 2y) - y}{x + 4(x - 2y)} = \frac{2x - 5y}{5x - 8y}$$

At $(4, 1)$, $y' = \frac{1}{4}$.

18.
$$
\begin{aligned}
(x + y)^3 &= x^3 + y^3 \\
3(x + y)^2(1 + y') &= 3x^2 + 3y^2y' \\
3(x + y)^2 + 3(x + y)^2y' &= 3x^2 + 3y^2y' \\
(x + y)^2y' - y^2y' &= x^2 - (x + y)^2 \\
y'[(x + y)^2 - y^2] &= x^2 - (x^2 + 2xy + y^2) \\
y' &= \frac{-(2xy + y^2)}{x^2 + 2xy} = -\frac{y(2x + y)}{x(x + 2y)}
\end{aligned}
$$

At $(-1, 1)$, $y' = -1$.

20. $x^2 + y^2 = 4$

$2x + 2yy' = 0$

$$y' = -\frac{x}{y}$$

At $(0, 2)$, $y' = 0$.

22. $x^2 - y^3 = 0$

$2x - 3y^2y' = 0$

$$y' = \frac{2x}{3y^2}$$

At $(-1, 1)$, $y' = -\frac{2}{3}$.

24. Implicitly: $1 - 2yy' = 0$

$$y' = \frac{1}{2y}$$

Explicitly:
$$
\begin{aligned}
y &= \pm\sqrt{x - 1} \\
&= \pm(x - 1)^{1/2} \\
y' &= \pm\frac{1}{2}(x - 1)^{-1/2}(1) \\
&= \pm\frac{1}{2\sqrt{x - 1}} \\
&= \frac{1}{2\left(\pm\sqrt{x - 1}\right)} \\
&= \frac{1}{2y}
\end{aligned}
$$

At $(2, -1)$, $y' = -\frac{1}{2}$.

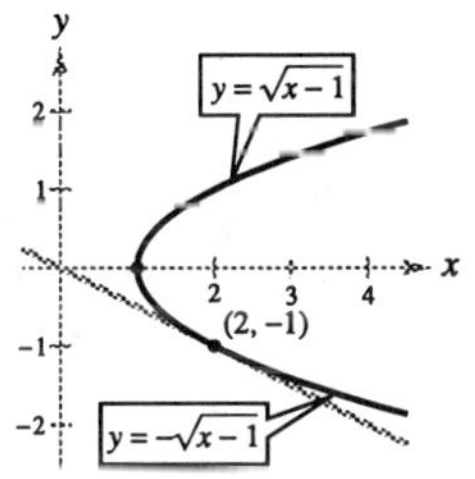

26. Implicitly: $8yy' - 2x = 0$

$$y' = \frac{x}{4y}$$

Explicitly:
$$
\begin{aligned}
y &= \pm\frac{1}{2}\sqrt{x^2 + 7} \\
&= \pm\frac{1}{2}(x^2 + 7)^{1/2} \\
y' &= \pm\frac{1}{4}(x^2 + 7)^{-1/2}(2x) \\
&= \pm\frac{x}{2\sqrt{x^2 + 7}} \\
&= \frac{x}{4\left(\pm\frac{1}{2}\sqrt{x^2 + 7}\right)} \\
&= \frac{x}{4y}
\end{aligned}
$$

At $(3, 2)$, $y' = \frac{3}{8}$.

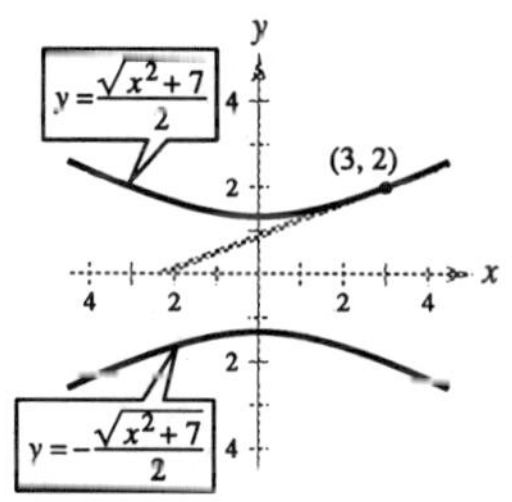

28. $x^2 + y^2 = 9$

$$2x + 2yy' = 0$$

$$y' = -\frac{x}{y}$$

At (0, 3):

$$m = 0$$

$$y - 3 = 0(x - 0)$$

$$y - 3 = 0$$

At $(2, \sqrt{5})$:

$$m = -\frac{2}{\sqrt{5}}$$

$$y - \sqrt{5} = -\frac{2}{\sqrt{5}}(x - 2)$$

$$\sqrt{5}y - 5 = -2x + 4$$

$$2x + \sqrt{5}y - 9 = 0$$

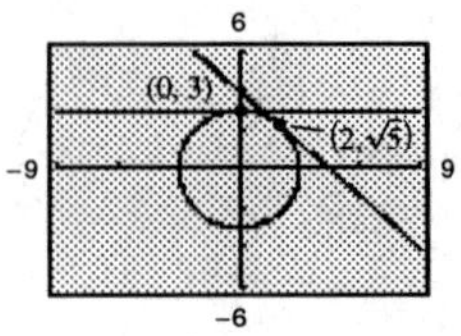

30. $4xy + x^2 = 5$

$$4xy' + 4y + 2x = 0$$

$$y' = -\frac{4y + 2x}{4x}$$

At (1, 1):

$$y' = -\frac{6}{4} = -\frac{3}{2}$$

$$y - 1 = -\frac{3}{2}(x - 1)$$

$$y = -\frac{3}{2}x + \frac{5}{2}$$

At (5, −1):

$$y' = -\frac{-4 + 10}{20} = -\frac{3}{10}$$

$$y + 1 = -\frac{3}{10}(x - 5)$$

$$y = -\frac{3}{10}x + \frac{1}{2}$$

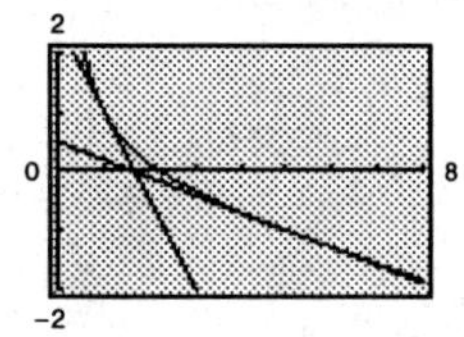

32. $y^2 = \dfrac{x^3}{4 - x}$

$$2yy' = \frac{(4 - x)(3x^2) - (x^3)(-1)}{(4 - x)^2}$$

$$2yy' = \frac{2x^2(6 - x)}{(4 - x)^2}$$

$$y' = \frac{x^2(6 - x)}{y(4 - x)^2}$$

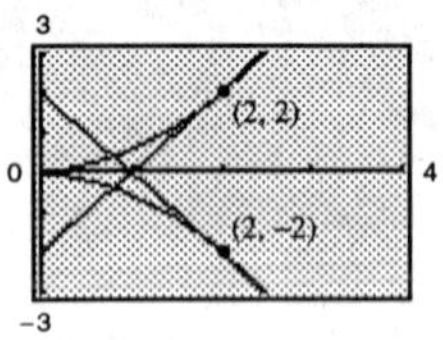

At (2, 2):

$$m = 2$$

$$y - 2 = 2(x - 2)$$

$$y - 2 = 2x - 4$$

$$0 = 2x - y - 2$$

At (2, −2):

$$m = -2$$

$$y + 2 = -2(x - 2)$$

$$y + 2 = -2x + 4$$

$$2x + y - 2 = 0$$

34. $p = \dfrac{725}{0.02x} = \dfrac{72{,}500}{x}$

$$x = \frac{72{,}500}{p} = 72{,}500p^{-1}$$

$$\frac{dx}{dp} = -72{,}500p^{-2} = \frac{-72{,}500}{p^2}$$

36.

$$p = \sqrt{\frac{500 - x}{2x}}, \quad 0 < x \le 500$$

$$p^2 = \frac{500 - x}{2x}$$

$$2xp^2 = 500 - x$$

$$2x(2p) + p^2\left(2\frac{dx}{dp}\right) = -\frac{dx}{dp}$$

$$2p^2\frac{dx}{dp} + \frac{dx}{dp} = -4xp$$

$$\frac{dx}{dp}(2p^2 + 1) = -4xp$$

$$\frac{dx}{dp} = -\frac{4xp}{2p^2 + 1}$$

38. (a)

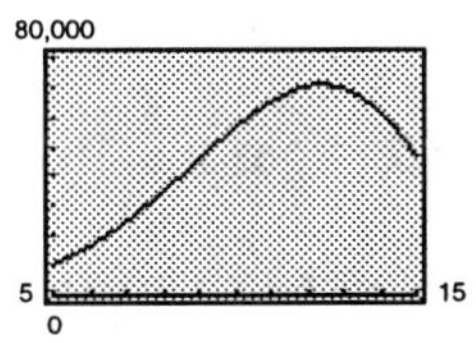

$$\sqrt{79.23y + 1} = -0.088t^4 + 27.03t^2 + 291.14$$

$$y = \frac{[(-0.088t^4 + 27.03t^2 + 291.14)^2 - 1]}{79.23}$$

(b) y' is a maximum at $t = 8.9$, or the end of 1989.

(c)

t	5	6	7	8	9	10	11	12	13	14	15
y	10,495	16,697	24,891	34,805	45,717	56,413	65,230	70,219	69,456	61,556	46,426
y'	5209	7210	9135	10,574	11,047	10,065	7239	2409	−4179	−11,654	−18,257

Section 2.8 Related Rates

2. $y = x^2 - 3x$, $\frac{dy}{dt} = 2x\frac{dx}{dt} - 3\frac{dx}{dt}$, $\frac{dy}{dt} = (2x - 3)\frac{dx}{dt}$, $\frac{dx}{dt} = \left(\frac{1}{2x - 3}\right)\frac{dy}{dt}$

(a) When $x = 3$ and $\frac{dx}{dt} = 2$, $\frac{dy}{dt} = [2(3) - 3](2) = 6$.

(b) When $x = 1$ and $\frac{dy}{dt} = 5$, $\frac{dx}{dt} = \left(\frac{1}{2(1) - 3}\right)(5) = -5$.

4. $x^2 + y^2 = 25$, $2x\frac{dx}{dt} + 2y\frac{dy}{dt} = 0$, $\frac{dy}{dt} = -\frac{x}{y}\frac{dx}{dt}$, $\frac{dx}{dt} = -\frac{y}{x}\frac{dy}{dt}$

(a) When $x = 3$, $y = 4$, and $\frac{dx}{dt} = 8$, $\frac{dy}{dt} = -\frac{3}{4}(8) = -6$.

(b) When $x = 4$, $y = 3$, and $\frac{dy}{dt} = -2$, $\frac{dx}{dt} = -\frac{3}{4}(-2) = \frac{3}{2}$.

6. $V = \frac{4}{3}\pi r^3,\ \frac{dr}{dt} = 2,\ \frac{dV}{dt} = 4\pi r^2\frac{dr}{dt} = 8\pi r^2$

(a) When $r = 6$, $\frac{dV}{dt} = 8\pi(6)^2 = 288\pi$ in^3/min.

(b) When $r = 24$, $\frac{dV}{dt} = 8\pi(24)^2 = 4608\pi$ in^3/min.

8. $V = \frac{4}{3}\pi r^3,\ \frac{dV}{dt} = 4\pi r^2\frac{dr}{dt}$

If dr/dt is constant, dV/dt is *not* constant since it is proportional to the square of r.

10. $V = \frac{1}{3}\pi r^2 h = \frac{1}{3}\pi r^2(3r) = \pi r^3$ [since $h = 3r$]

$\frac{dV}{dt} = 3\pi r^2\frac{dr}{dt} = 6\pi r^2$ [since $dr/dt = 2$]

(a) When $r = 6$, $\frac{dV}{dt} = 6\pi(6)^2 = 216\pi$ in^3/min.

(b) When $r = 24$, $\frac{dV}{dt} = 6\pi(24)^2 = 3456\pi$ in^3/min.

12. $C = 75{,}000 + 1.05x,\ R = 500x - \frac{x^2}{25}$

(a) $\frac{dC}{dt} = 1.05\frac{dx}{dt} = 1.05(250) = 262.5$ dollars/week

(b) $\frac{dR}{dt} = 500\frac{dx}{dt} - \frac{2x}{25}\frac{dx}{dt} = 500(250) - \frac{2(5000)}{25}(250)$

$= 25{,}000$ dollars/week

(c) $P = R - C$

$\frac{dP}{dt} = \frac{dR}{dt} - \frac{dC}{dt} = 25{,}000 - 262.5$

$= 24{,}737.50$ dollars/week

14. $A = 6x^2,\ \frac{dx}{dt} = 3,\ \frac{dA}{dt} = 12x\frac{dx}{dt} = 36x$

(a) When $x = 1$, $\frac{dA}{dt} = 36(1) = 36$ cm^2/sec.

(b) When $x = 10$, $\frac{dA}{dt} = 36(10) = 360$ cm^2/sec.

16. $y = \frac{1}{1 + x^2},\ \frac{dx}{dt} = 2,\ \frac{dy}{dt} = -\frac{2x}{(1 + x^2)^2}\frac{dx}{dt} = -\frac{4x}{(1 + x^2)^2}$

(a) When $x = -2$, $\frac{dx}{dt} = -\frac{4(-2)}{[1 + (-2)^2]^2} = \frac{8}{25}$ cm/min.

(b) When $x = 2$, $\frac{dx}{dt} = -\frac{4(2)}{[1 + (2)^2]^2} = -\frac{8}{25}$ cm/min.

(c) When $x = 0$, $\frac{dx}{dt} = -\frac{4(0)}{[1 + (0)^2]^2} = 0$ cm/min.

(d) When $x = 10$, $\frac{dx}{dt} = -\frac{4(10)}{[1 + (10)^2]^2} = -\frac{40}{10{,}201}$ cm/min.

18. $12^2 + x^2 = y^2$

$\frac{dy}{dt} = -4$

$2x\frac{dx}{dt} = 2y\frac{dy}{dt}$

$\frac{dx}{dt} = \frac{y}{x}\frac{dy}{dt}$

When $y = 13$, $x = 5$ and $\frac{dx}{dt} = \frac{13}{5}(-4) = -10.4$ ft/sec.

As $x \to 0$, $\frac{dx}{dt}$ increases.

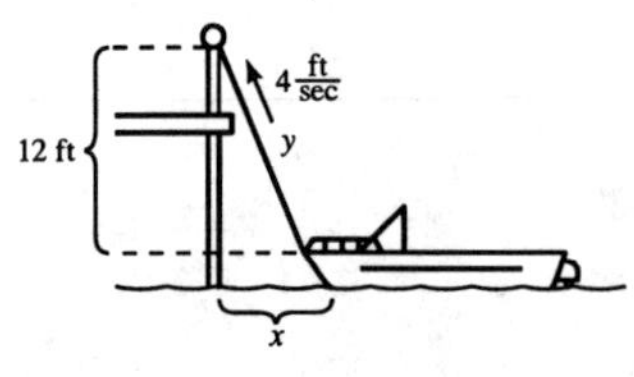

20. $x^2 + 6^2 = s^2$

$2x\frac{dx}{dt} = 2s\frac{ds}{dt}$

$\frac{dx}{dt} = \frac{s}{x}\frac{ds}{dt}$

When $s = 10$, $x = 8$ and $\frac{ds}{dt} = -240$:

$\frac{dx}{dt} = \frac{10}{8}(-240) = -300$ mi/hr.

The **speed** of the plane is 300 mi/hr.

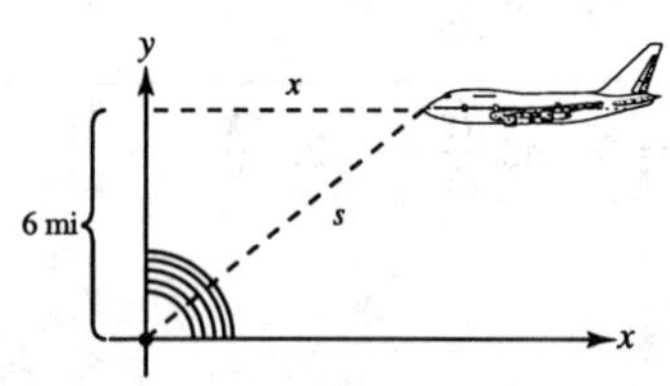

22. $V = \pi r^2 h,\ h = 0.08,\ V = 0.08\pi r^2,\ \dfrac{dV}{dt} = 0.16\pi r\dfrac{dr}{dt}$

When $r = 750$ and $\dfrac{dr}{dt} = \dfrac{1}{2}, \dfrac{dV}{dt} = 0.16\pi(750)\left(\dfrac{1}{2}\right) = 60\pi \approx 188.5$ ft³/min.

24. $P = R - C = xp - C = x(6000 - 0.4x^2) - (2400x + 5200) = -0.4x^3 + 3600x - 5200$

$$\frac{dP}{dt} = -1.2x^2\frac{dx}{dt} + 3600\frac{dx}{dt}$$

$$\frac{dx}{dt} = \frac{1}{3600 - 1.2x^2}\frac{dP}{dt}$$

When $x = 44$ and $\dfrac{dP}{dt} = 6384,\ \dfrac{dx}{dt} = \dfrac{1}{3600 - 1.2(44)^2}(6384) = 5$ units per week.

Review Exercises for Chapter 2

2. Slope $\approx \frac{4}{2} = 2$

4. Slope $\approx \frac{-2}{4} = -\frac{1}{2}$

6. When $t = 2$, slope ≈ 70.

When $t = 4$, slope ≈ 270.

The rate of increase of subscribers is increasing.

8. $f(x) = x^2 + 10, \quad (2, 14)$

$$\begin{aligned} m &= \lim_{\Delta x \to 0} \frac{f(2 + \Delta x) - f(2)}{\Delta x} \\ &= \lim_{\Delta x \to 0} \frac{(2 + \Delta x)^2 + 10 - 14}{\Delta x} \\ &= \lim_{\Delta x \to 0} \frac{(\Delta x)^2 + 4 \cdot \Delta x}{\Delta x} \\ &= \lim_{\Delta x \to 0} (\Delta x + 4) = 4 \end{aligned}$$

10. $f(x) = \dfrac{x + 1}{x}, \quad (1, 2)$

$$\begin{aligned} m &= \lim_{\Delta x \to 0} \frac{\left(\dfrac{1 + \Delta x + 1}{1 + \Delta x}\right) - 2}{\Delta x} \\ &= \lim_{\Delta x \to 0} \frac{2 + \Delta x - 2(1 + \Delta x)}{\Delta x(1 + \Delta x)} \\ &= \lim_{\Delta x \to 0} \frac{-1}{1 + \Delta x} = -1 \end{aligned}$$

12.

$$\begin{aligned} f'(x) &= \lim_{\Delta x \to 0} \frac{f(x + \Delta x) - f(x)}{\Delta x} \\ &= \lim_{\Delta x \to 0} \frac{(x + \Delta x)^2 - 7(x + \Delta x) - 8 - (x^2 - 7x - 8)}{\Delta x} \\ &= \lim_{\Delta x \to 0} \frac{2x \cdot \Delta x + (\Delta x)^2 - 7\Delta x}{\Delta x} \\ &= \lim_{\Delta x \to 0} 2x + \Delta x - 7 = 2x - 7 \end{aligned}$$

14.

$$\begin{aligned} f'(x) &= \lim_{\Delta x \to 0} \frac{\sqrt{x + \Delta x - 5} - \sqrt{x - 5}}{\Delta x} \cdot \frac{\sqrt{x + \Delta x - 5} + \sqrt{x - 5}}{\sqrt{x + \Delta x - 5} + \sqrt{x - 5}} \\ &= \lim_{\Delta x \to 0} \frac{(x + \Delta x - 5) - (x - 5)}{\Delta x\left(\sqrt{x + \Delta x - 5} + \sqrt{x - 5}\right)} \\ &= \lim_{\Delta x \to 0} \frac{1}{\sqrt{x + \Delta x - 5} + \sqrt{x - 5}} = \frac{1}{2\sqrt{x - 5}} \end{aligned}$$

16. $f(x) = -\frac{1}{2}x^2 + 2x$

$f'(x) = -x + 2$

$f'(2) = -2 + 2 = 0$

18. $f(x) = \dfrac{5}{x} = 5x^{-1}$

$f'(x) = -5x^2 = -\dfrac{5}{x^2}$

$f'(1) = \dfrac{-5}{1^2} = -5$

20. $f(x) = -x^2 - 4x - 4$

$f'(x) = -2x - 4$

$f'(-4) = 4$

Tangent line: $y + 4 = 4(x + 4)$

$y = 4x + 12$

22. $y = -|x| + 3$ is not differentiable at $x = 0$.

24. y is not differentiable at $x = -1$.

26. $f'(x) = 0$

28. $g(x) = -x^{-3}$

$g'(x) = 3x^{-4} = \dfrac{3}{x^4}$

30. $h(t) = t^{-1/2}$

$h'(t) = -\dfrac{1}{2}t^{-3/2} = \dfrac{-1}{2t^{3/2}}$

32. $f(x) = 4x^{-4}$

$f'(x) = -16x^{-5} = \dfrac{-16}{x^5}$

34. $h(x) = \dfrac{2}{(3x)^2} = \dfrac{2}{9}x^{-2}, \quad \left(2, \dfrac{1}{18}\right)$

$h'(x) = \dfrac{-4}{9}x^{-3} = \dfrac{-4}{9x^3}$

$h'(2) = -\dfrac{1}{18}$

$y - \dfrac{1}{18} = -\dfrac{1}{18}(x - 2)$

$y = -\dfrac{1}{18}x + \dfrac{1}{6}$

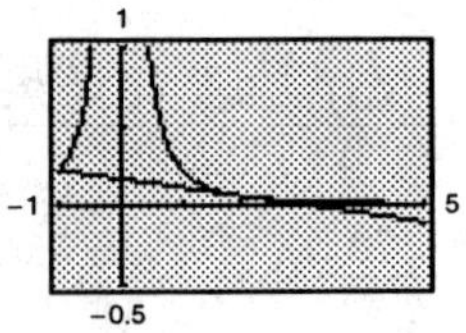

36. $y = x^3 - 5 + \dfrac{3}{x^3} = x^3 - 5 + 3x^{-3}, \quad y(-1) = -9$

$y' = 3x^2 + 3(-3x^{-4}) = 3x^2 - \dfrac{9}{x^4}, \quad y'(-1) = -6$

$y - (-9) = -6[x - (-1)] = -6x - 6$

$y = -6x - 15$

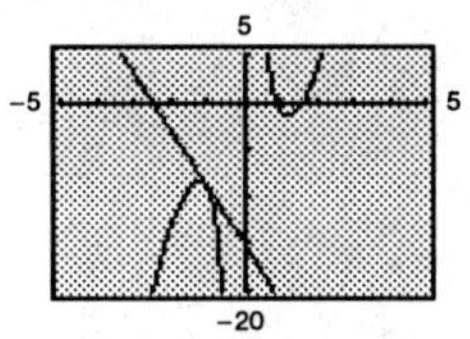

38. $f(x) = 2x^{-3} + 4 - \sqrt{x}, \quad (1, 5)$

$f'(x) = -6x^{-4} - \dfrac{1}{2\sqrt{x}}$

$f'(1) = -6 - \dfrac{1}{2} = -\dfrac{13}{2}$

$y - 5 = -\dfrac{13}{2}(x - 1)$

$y = -\dfrac{13}{2}x + \dfrac{23}{2}$

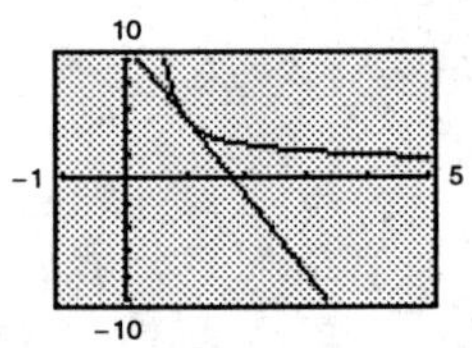

40. $f(x) = x^3 + x, \quad [-2, 2]$

Average rate of change $= \dfrac{f(2) - f(-2)}{2 - (-2)} = \dfrac{10 - (-10)}{4} = 5$

$f'(x) = 3x^2 + 1$

$f'(-2) = 13$

$f'(2) = 13$

42. $S = \dfrac{2163.9 + 666.4t}{1 - 0.18t + 0.01t^2}$

(a) $\dfrac{S(7) - S(3)}{7 - 3} \approx \dfrac{29{,}690 - 7569}{4} \approx 5530$ subscribers/year

(b) $S'(3) \approx 2863$ subscribers/year

$S'(7) \approx 8061$ subscribers/year

(c) The number of subscribers is increasing at a faster rate in 1995 than in 1991.

44. $T = \sqrt{54.89 + 36.11t - 4.63t^2 + 0.19t^3}$

(a)

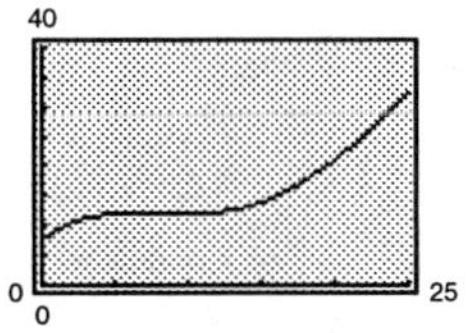

(b) $\dfrac{dT}{dt} = \dfrac{36.11 - 9.26t + 0.57t^2}{2\sqrt{54.89 + 36.11t - 4.63t^2 + 0.19t^3}}$

$T'(0) \approx 2.437$ (1970)

$T'(10) \approx 0.021$ (1980)

$T'(20) \approx 1.870$ (1990)

(c) Yes, $T'(t) > 0$ for $t \geq 0$.

46. $S = 2t^{3/2}, \quad 0 \leq t \leq 8$

$v = 3t^{1/2}$

t	0	2	4	6	8
v	0	4.24	6	7.35	8.49

48. $P = R - C = xp - C$

$= x(1.89 - 0.0083x) - (21 + 0.65x)$

$= -0.0083x^2 + 1.24x - 21$

50. $\dfrac{dC}{dx} = 5.25\left(\dfrac{2}{3}x^{-1/3}\right) = \dfrac{3.5}{\sqrt[3]{x}}$

52. $R = x\left(5 + \dfrac{10}{\sqrt{x}}\right) = 5x + 10x^{1/2}$

$\dfrac{dR}{dx} = 5 + 5x^{-1/2} = 5 + \dfrac{5}{\sqrt{x}} = \dfrac{5(\sqrt{x} + 1)}{\sqrt{x}}$

54. $\dfrac{dP}{dx} = -\dfrac{1}{5}x^2 + 8000x - 120$

56. $y = (3x^2 + 7)(x^2 - 2x)$

$y' = (3x^2 + 7)(2x - 2) + (6x)(x^2 - 2x)$

$= 6x^3 + 14x - 6x^2 - 14 + 6x^3 - 12x^2$

$= 12x^3 - 18x^2 + 14x - 14$

58. $S = \left(4 - \dfrac{1}{t^2}\right)(t^2 - 3t) = 4t^2 - 12t - 1 + 3t^{-1}$

$S' = 8t - 12 - 3t^{-2} = 8t - 12 - \dfrac{3}{t^2}$

60. $f(x) = \dfrac{x^2 + x - 1}{x^2 - 1}$

$f'(x) = \dfrac{(x^2 - 1)(2x + 1) - (x^2 + x - 1)(2x)}{(x^2 - 1)^2}$

$= \dfrac{2x^3 + x^2 - 2x - 1 - 2x^3 - 2x^2 + 2x}{(x^2 - 1)^2}$

$= \dfrac{-x^2 - 1}{(x^2 - 1)^2} = -\dfrac{x^2 + 1}{(x^2 - 1)^2}$

62. $f(x) = \sqrt[3]{x^2 - 1} = (x^2 - 1)^{1/3}$

$f'(x) = \dfrac{1}{3}(x^2 - 1)^{-2/3}(2x) = \dfrac{2x}{3(x^2 - 1)^{2/3}}$

64. $g(x) = (x^6 - 12x^3 + 9)^{1/2}$

$$g'(x) = \frac{1}{2}(x^6 - 12x^3 + 9)^{-1/2}(6x^5 - 36x^2)$$

$$= \frac{3x^5 - 18x^2}{\sqrt{x^6 - 12x^3 + 9}}$$

66. $g(t) = \dfrac{t}{(1-t)^3}$

$$g'(t) = \frac{(1-t)^3(1) - t(3)(1-t)^2(-1)}{(1-t)^6}$$

$$= \frac{(1-t)^3 + 3t(1-t)^2}{(1-t)^6}$$

$$= \frac{(1-t) + 3t}{(1-t)^4} = \frac{2t+1}{(1-t)^4}$$

68. $f(x) = \left(x^2 + \dfrac{1}{x}\right)^5 = (x^2 + x^{-1})^5$

$$f'(x) = 5(x^2 + x^{-1})^4(2x - x^{-2})$$

$$= 5\left(x^2 + \frac{1}{x}\right)^4\left(2x - \frac{1}{x^2}\right)$$

70. $f(x) = [(x-2)(x+4)]^2 = (x^2 + 2x - 8)^2$

$$f'(x) = 2(x^2 + 2x - 8)^1(2x + 2)$$

$$= 2(x-2)(x+4)(2)(x+1)$$

$$= 4(x-2)(x+4)(x+1)$$

72. $f(s) = s^3(s^2 - 1)^{5/2}$

$$f'(s) = s^3\left(\frac{5}{2}\right)(s^2-1)^{3/2}(2s) + 3s^2(s^2-1)^{5/2}$$

$$= s^2(s^2-1)^{3/2}[5s^2 + 3(s^2-1)]$$

$$= s^2(s^2-1)^{3/2}(8s^2 - 3)$$

74. $g(x) = \dfrac{(3x+1)^2}{(x^2+1)^2}$

$$g'(x) = \frac{(x^2+1)^2 2(3x+1)(3) - (3x+1)^2 2(x^2+1)2x}{(x^2+1)^4}$$

$$= \frac{(3x+1)[6(x^2+1) - (3x+1)4x]}{(x^2+1)^3}$$

$$= \frac{(3x+1)(-6x^2 - 4x + 6)}{(x^2+1)^3}$$

76. When $L = 12$,

$$V = \frac{L}{16}(D-4)^2 = \frac{12}{16}(D-4)^2 = \frac{3}{4}(D-4)^2$$

$$\frac{dV}{dD} = \frac{3}{2}(D-4) = (1.5)(D-4).$$

(a) When $D = 8$, $\dfrac{dV}{dD} = (1.5)(8-4) = 6$ board ft/in.

(b) When $D = 16$, $\dfrac{dV}{dD} = (1.5)(16-4) = 18$ board ft/in.

(c) When $D = 24$, $\dfrac{dV}{dD} = (1.5)(24-4) = 30$ board ft/in.

(d) When $D = 36$, $\dfrac{dV}{dD} = (1.5)(36-4) = 48$ board ft/in.

78. $f'(x) = 5x^4 - 6x^2 + 2x$

$$f''(x) = 20x^3 - 12x + 2$$

$$f'''(x) = 60x^2 - 12$$

80. $f(x) = \sqrt{x} = x^{1/2}$

$$f'(x) = \frac{1}{2}x^{-1/2}$$

$$f''(x) = -\frac{1}{4}x^{-3/2}$$

$$f'''(x) = \frac{3}{8}x^{-5/2}$$

$$f^{(4)}(x) = \frac{-15}{16}x^{-7/2} = \frac{-15}{16x^{7/2}}$$

82. $f(x) = x^2 + 3x^{-1}$

$$f'(x) = 2x - 3x^{-2}$$

$$f''(x) = 2 + 6x^{-3} = 2 + \frac{6}{x^3}$$

84. $f'''(x) = 20x^4 - 2x^{-3}$

$f^{(4)}(x) = 80x^3 + 6x^{-4}$

$f^{(5)}(x) = 240x^2 - 24x^{-5}$

$= 240x^2 - \dfrac{24}{x^5}$

86. $s(t) = \dfrac{1}{t^2 + 2t + 1} = (t + 1)^{-2}$

$v(t) = s'(t) = -2(t + 1)^{-3} = -\dfrac{2}{(t + 1)^3}$

$a(t) = s''(t) = 6(t + 1)^{-4} = \dfrac{6}{(t + 1)^4}$

88. $x^2 + 9y^2 - 4x + 3y - 7 = 0$

$2x + 18yy' - 4 + 3y' = 0$

$y'(18y + 3) = 4 - 2x$

$y' = \dfrac{2(2 - x)}{3(6y + 1)}$

90. $y^2 + x^2 - 6y - 2x - 5 = 0$

$2yy' + 2x - 6y' - 2 = 0$

$y'(2y - 6) = 2 - 2x$

$y' = \dfrac{2 - 2x}{2y - 6} = \dfrac{1 - x}{y - 3}$

92. $y^2 - 3 = 2x^{1/2}y^{1/2}$

$2yy' = x^{-1/2}y^{1/2} + x^{1/2}y^{-1/2}y'$

$(2y - x^{1/2}y^{-1/2})y' = x^{-1/2}y^{1/2}$

$y' = \dfrac{x^{-1/2}y^{1/2}}{2y - x^{1/2}y^{-1/2}} = \dfrac{y}{2x^{1/2}y^{3/2} - x}$

At $\left(\frac{1}{8}, 2\right)$,

$y' = \dfrac{2}{2 - \sqrt{2}}$

$y - 2 = \dfrac{2}{2 - \sqrt{2}}\left(x - \dfrac{1}{8}\right)$

94. $y^3 - 2x^2y + 3xy^2 = -1$

$3y^2y' - 2x^2y' - 4xy + 6xyy' + 3y^2 = 0$

$y'(3y^2 - 2x^2 + 6xy) = 4xy - 3y^2$

$y' = \dfrac{4xy - 3y^2}{3y^2 - 2x^2 + 6xy}$

At $(0, -1)$,

$y' = -1$

$y + 1 = -1(x - 0)$

$y = -x - 1$

96. (a) $P = R - C = xp - C = x(211 - 0.002x) - (30x + 1{,}500{,}000) = 181x - 0.002x^2 - 1{,}500{,}000$

(b) $P'(x) = 181 - 0.004x$

$P'(80{,}000) = -139$

(c) 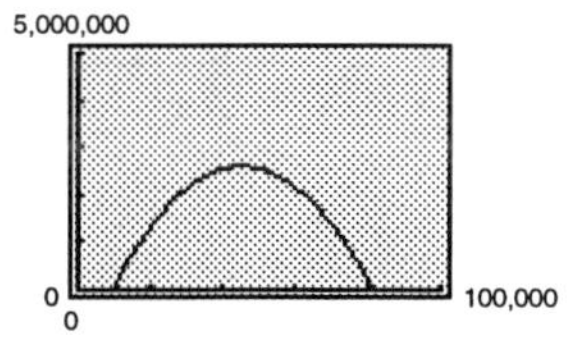

The maximum profit occurs for $x = 45{,}250$, which means $p = 211 - 0.002(45{,}250) = \$120.50$.

CHAPTER 3
Applications of the Derivative

CHAPTER 3
Applications of the Derivative

Section 3.1 Increasing and Decreasing Functions

Solutions to Even-Numbered Exercises

2. $f(x) = x + \dfrac{32}{x^2} = x + 32x^{-2}$

$f'(x) = 1 - 64x^{-3} = \dfrac{x^3 - 64}{x^3}$

At $(2, 10)$, f is decreasing since $f'(2) = -7$.

At $(4, 6)$, f has a critical number since $f'(4) = 0$.

At $\left(8, \frac{17}{2}\right)$, f is increasing since $f'(8) = \frac{7}{8}$.

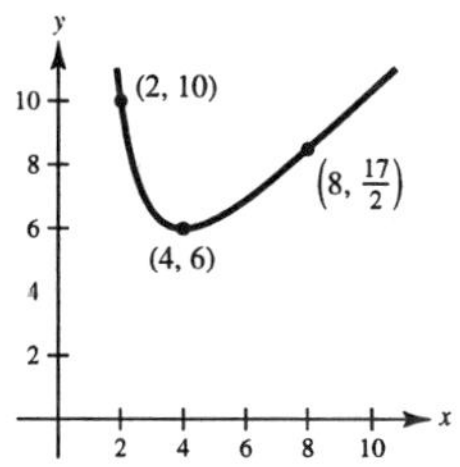

4. $f(x) = -3x\sqrt{x+1} = -3x(x+1)^{1/2}$

$f'(x) = -3x\left[\dfrac{1}{2}(x+1)^{-1/2}(1)\right] + (x+1)^{1/2}(-3)$

$= -\dfrac{3}{2}(x+1)^{-1/2}[x + 2(x+1)]$

$= -\dfrac{3}{2}(x+1)^{-1/2}(3x+2) = -\dfrac{3(3x+2)}{2\sqrt{x+1}}$

At $(-1, 0)$, f has a critical number since $f'(-1)$ is undefined.

At $\left(-2/3, 2\sqrt{3}/3\right)$, f has a critical number since $f'(-2/3) = 0$.

At $(0, 0)$, f is decreasing since $f'(0) = -3$.

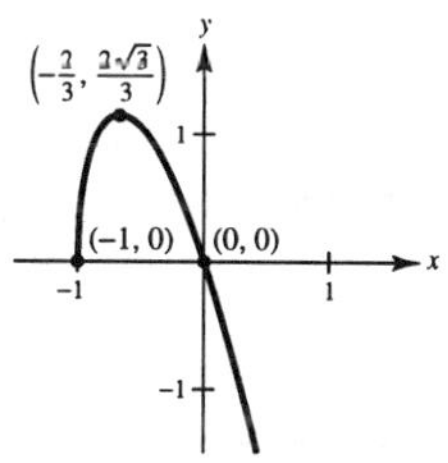

6. $f' = \dfrac{3}{4}x^2 - 3 = \dfrac{3}{4}(x^2 - 4)$

f has critical numbers at $x = \pm 2$. Moreover, f is increasing on $(-\infty, -2)$, $(2, \infty)$ and decreasing on $(-2, 2)$.

8. $f' = \dfrac{(x+1)(2x) - (x^2)(1)}{(x+1)^2} = \dfrac{x^2 + 2x}{(x+1)^2} = \dfrac{x(x+2)}{(x+1)^2}$

f has critical numbers at $x = 0, -2$ and has a discontinuity at $x = -1$. Moreover, f is increasing on $(-\infty, -2)$, $(0, \infty)$ and decreasing on $(-2, -1)$, $(-1, 0)$.

10. $f(x) = 5 - 3x$

$f'(x) = -3$

Since the derivative is negative for all x, the function is decreasing for all x. Thus, there are no critical numbers. f is decreasing on $(-\infty, \infty)$.

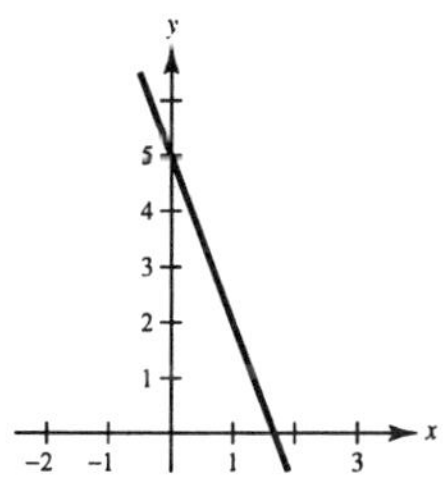

12. $g(x) = (x + 2)^2$

$g'(x) = 2(x + 2) = 0$

Critical number: $x = -2$

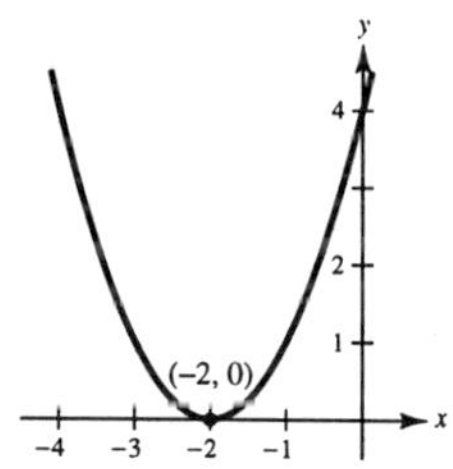

Interval	$-\infty < x < -2$	$-2 < x < \infty$
Sign of g'	$g' < 0$	$g' > 0$
Conclusion	Decreasing	Increasing

14. $y = -(x^2 - 2x) = 2x - x^2$

$y' = 2 - 2x = 0$

Critical number: $x = 1$

Interval	$-\infty < x < 1$	$1 < x < \infty$
Sign of y'	$y' > 0$	$y' < 0$
Conclusion	Increasing	Decreasing

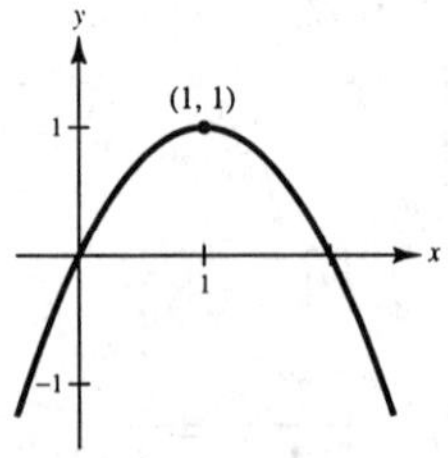

16. $y = (x - 2)^3$

$y' = 3(x - 2)^2 = 0$

Critical number: $x = 2$

Interval	$-\infty < x < 2$	$2 < x < \infty$
Sign of y'	$y' > 0$	$y' > 0$
Conclusion	Increasing	Increasing

y is increasing on $(-\infty, \infty)$.

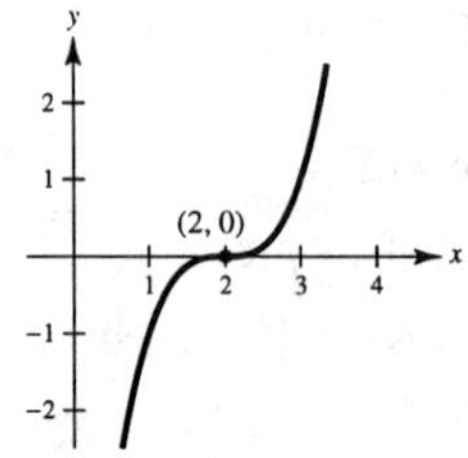

18. $f(x) = \sqrt{4 - x^2}$

$f'(x) = \frac{1}{2}(4 - x^2)^{-1/2}(-2x) = -\frac{x}{\sqrt{4 - x^2}}$

Domain: $-2 \le x \le 2$

Critical numbers: $x = 0, x = \pm 2$

Interval	$-2 < x < 0$	$0 < x < 2$
Sign of f'	$f' > 0$	$f' < 0$
Conclusion	Increasing	Decreasing

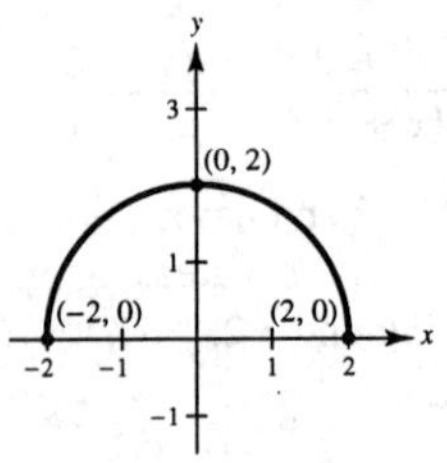

20. $f(x) = x^2 + 8x + 10$

$f'(x) = 2x + 8 = 0$

Critical number: $x = -4$

Interval	$-\infty < x < -4$	$-4 < x < \infty$
Sign of f'	$f' < 0$	$f' > 0$
Conclusion	Decreasing	Increasing

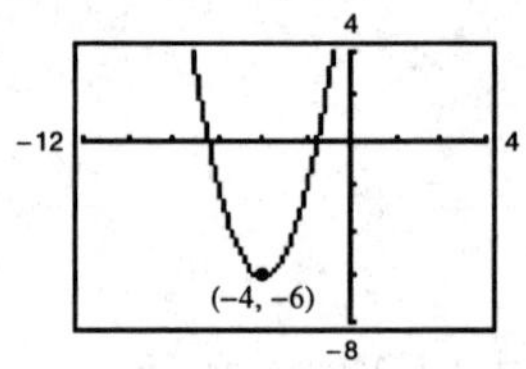

22. $f(x) = (x - 1)^2(x + 2)$

$f'(x) = (x - 1)^2(1) + (x + 2)[(2)(x - 1)(1)] = (x - 1)[(x - 1) + 2(x + 2)] = (x - 1)(3x + 3) = 3(x + 1)(x - 1)$

Critical numbers: $x = \pm 1$

Interval	$-\infty < x < -1$	$-1 < x < 1$	$1 < x < \infty$
Sign of f'	$f' > 0$	$f' < 0$	$f' > 0$
Conclusion	Increasing	Decreasing	Increasing

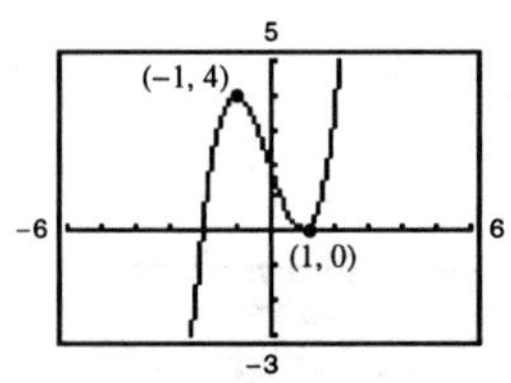

24. $h(x) = \sqrt[3]{x-1} = (x-1)^{1/3}$

$$h'(x) = \frac{1}{3}(x-1)^{-2/3}(1)$$

$$= \frac{1}{3(x-1)^{2/3}}$$

Critical number: $x = 1$

Interval	$-\infty < x < 1$	$1 < x < \infty$
Sign of f'	$f' > 0$	$f' > 0$
Conclusion	Increasing	Increasing

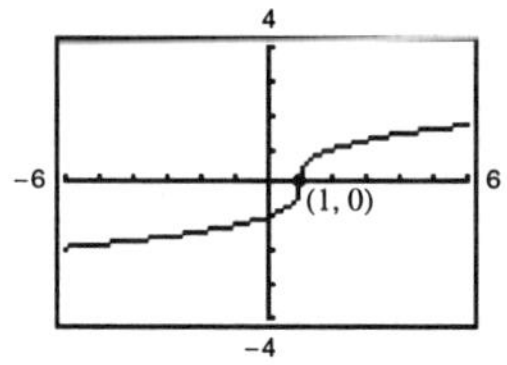

26. $f(x) = x\sqrt{x+1} = x(x+1)^{1/2}$

$$f'(x) = x\left[\frac{1}{2}(x+1)^{-1/2}(1)\right] + (x+1)^{1/2}(1)$$

$$= \frac{1}{2}(x+1)^{-1/2}[x + 2(x+1)]$$

$$= \frac{3x+2}{2\sqrt{x+1}}$$

Domain: $[-1, \infty)$

Critical numbers: $x = -1, -\frac{2}{3}$

Interval	$-1 < x < -\frac{2}{3}$	$-\frac{2}{3} < x < \infty$
Sign of f'	$f' < 0$	$f' > 0$
Conclusion	Decreasing	Increasing

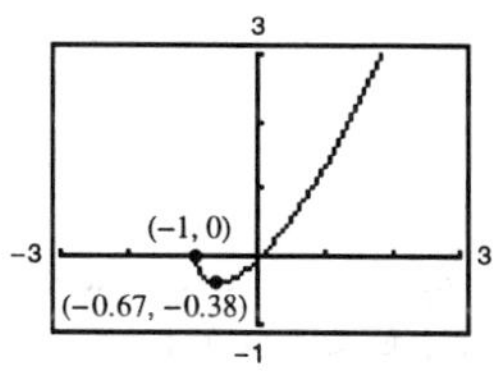

28. $f(x) = y = \dfrac{x^2}{x^2+4}$

$$y' = \frac{(x^2+4)(2x) - (x^2)(2x)}{(x^2+4)^2} = \frac{8x}{(x^2+4)^2}$$

Critical number: $x = 0$

Interval	$-\infty < x < 0$	$0 < x < \infty$
Sign of f'	$f' < 0$	$f' > 0$
Conclusion	Decreasing	Increasing

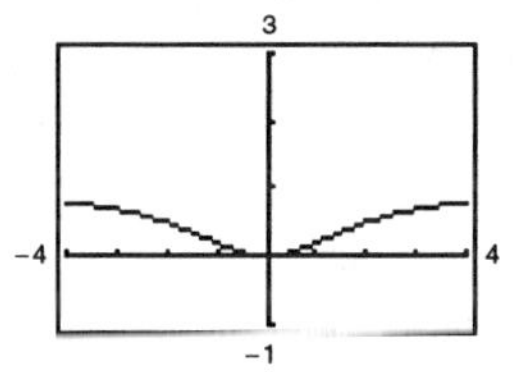

30. $f(x) = \dfrac{x}{x+1}$

$$f'(x) = \frac{(x+1)(1) - (x)(1)}{(x+1)^2} = \frac{1}{(x+1)^2}$$

No critical numbers
Discontinuity: $x = -1$

Interval	$-\infty < x < -1$	$-1 < x < \infty$
Sign of f'	$f' > 0$	$f' > 0$
Conclusion	Increasing	Increasing

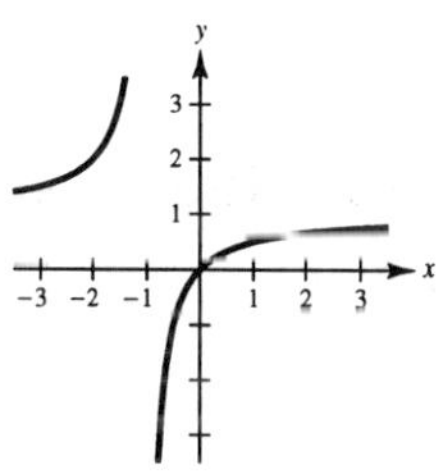

32. $y = f(x) = \begin{cases} 2x + 1, & x \le -1 \\ x^2 - 2, & x > -1 \end{cases}$

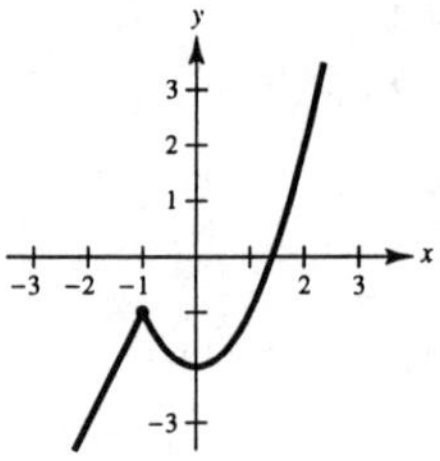

$f'(x) = \begin{cases} 2, & x < -1 \\ 2x, & x > -1 \end{cases}$

$f'(-1)$ is undefined.

Critical numbers: $x = -1, 0$

Interval	$-\infty < x < -1$	$-1 < x < 0$	$0 < x < \infty$
Sign of f'	$f' > 0$	$f' < 0$	$f' > 0$
Conclusion	Increasing	Decreasing	Increasing

34. $y = f(x) = \begin{cases} -x^3 + 1, & x \le 0 \\ -x^2 + 2x, & x > 0 \end{cases}$

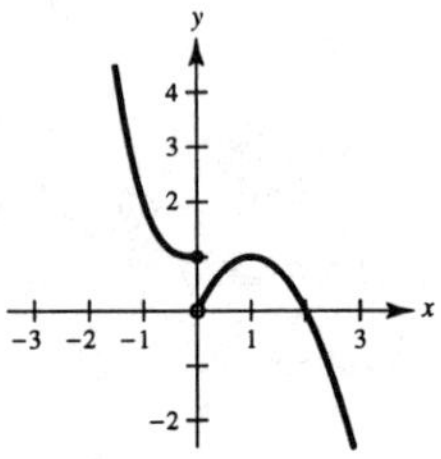

$f'(x) = \begin{cases} -3x^2, & x < 0 \\ -2x + 2, & x > 0 \end{cases}$

$f'(0)$ is undefined.

Critical numbers: $x = 0, 1$

Interval	$-\infty < x < 0$	$0 < x < 1$	$1 < x < \infty$
Sign of f'	$f' < 0$	$f' > 0$	$f' < 0$
Conclusion	Decreasing	Increasing	Decreasing

36. As the temperature increases, the average velocity (the curve's peak) and the spread of velocities increases.

38. $s(t) = -16t^2 + 64t, \quad 0 \le t \le 4$

$s'(t) = -32t + 64 = 0$

The critical number is $t = 2$. Therefore, the ball is moving up on the interval $(0, 2)$ and moving down on $(2, 4)$.

40. $P = 2.36x - \dfrac{x^2}{25{,}000} - 3500 \quad 0 \le x \le 50{,}000$

(a) Increasing on: $(0, 29{,}500)$

Decreasing on: $(29{,}500, 500{,}000)$

(b) You should charge the price that yields sales of $x = 29{,}500$ bags of popcorn.

Section 3.2 Extrema and the First-Derivative Test

2. $f(x) = x^2 + 8x + 10$

$f'(x) = 2x + 8 = 2(x + 4)$

Critical number: $x = -4$

Interval	$(-\infty, -4)$	$(-4, \infty)$
Sign of f'	$-$	$+$
f	Decreasing	Increasing

Relative minimum: $(-4, -6)$

4. $f(x) = (x - 1)^2(x + 2)$

$f'(x) = (x - 1)^2(1) + (x + 2)(2)(x - 1)$

$= (x - 1)[(x - 1) + 2(x + 2)]$

$= (x - 1)(3x + 3)$

$= 3(x - 1)(x + 1)$

Critical numbers: $x = \pm 1$

Interval	$(-\infty, -1)$	$(-1, 1)$	$(1, \infty)$
Sign of f'	$+$	$-$	$+$
f	Increasing	Decreasing	Increasing

Relative maximum: $(-1, 4)$ Relative minimum: $(1, 0)$

6. $g(x) = \frac{1}{5}(x^5 - 5x)$

$g'(x) = \frac{1}{5}(5x^4 - 5) = x^4 - 1$

Critical numbers: $x = \pm 1$

Interval	$(-\infty, -1)$	$(-1, 1)$	$(1, \infty)$
Sign of g'	$+$	$-$	$+$
g	Increasing	Decreasing	Increasing

Relative maximum: $\left(-1, \frac{4}{5}\right)$
Relative minimum: $\left(1, -\frac{4}{5}\right)$

8. $h(x) = (x - 3)^3$

$h'(x) = 3(x - 3)^2$

Critical number: $x = 3$

Interval	$(-\infty, 3)$	$(3, \infty)$
Sign of h'	$+$	$+$
h	Increasing	Increasing

No relative extrema

10. $f(x) = x^4 - 1$

$f'(x) = 4x^3$

Critical number: $x = 0$

Interval	$(-\infty, 0)$	$(0, \infty)$
Sign of f'	$-$	$+$
f	Decreasing	Increasing

Relative minimum: $(0, -1)$

12. $f(x) = x^4 - 32x + 4$

$f'(x) = 4x^3 - 32 = 4(x^3 - 8)$

Critical number: $x = 2$

Interval	$(-\infty, 2)$	$(2, \infty)$
Sign of f'	$-$	$+$
f	Decreasing	Increasing

Relative minimum: $(2, -44)$

14. $f(t) = (t - 1)^{1/3}$

$f'(t) = \frac{1}{3}(t - 1)^{-2/3} = \frac{1}{3\sqrt[3]{(t - 1)^2}}$

Critical number: $t = 1$

Interval	$(-\infty, 1)$	$(1, \infty)$
Sign of f'	$+$	$+$
f	Increasing	Increasing

No relative extrema

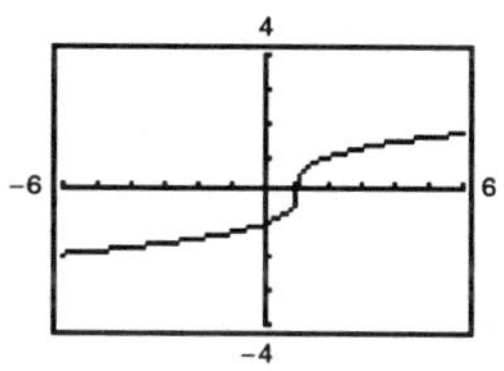

16. $f(x) = x + \frac{1}{x}$

$f'(x) = 1 - \frac{1}{x^2} = 0$

Critical numbers: $x = \pm 1$

Interval	$(-\infty, -1)$	$(-1, 0)$	$(0, 1)$	$(1, \infty)$
Sign of f'	$+$		$-$	$+$
f	Increasing	Decreasing	Decreasing	Increasing

Relative maximum: $(-1, -2)$
Relative minimum: $(1, 2)$

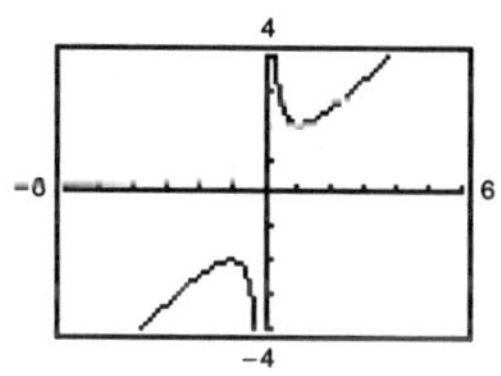

18. $f(x) = \dfrac{4}{x^2 + 1}$

$f'(x) = \dfrac{-8x}{(x^2 + 1)^2} = 0$

Critical number: $x = 0$

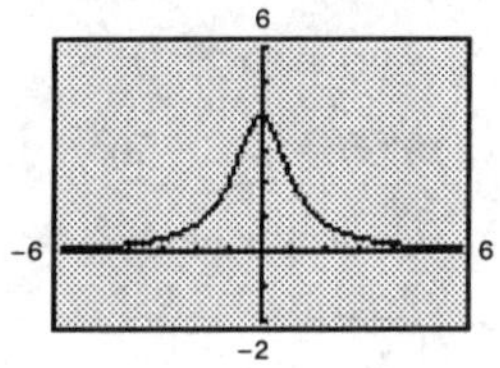

Interval	$(-\infty, 0)$	$(0, \infty)$
Sign of f'	$+$	$-$
f	Increasing	Decreasing

Relative maximum: $(0, 4)$

20. $f(x) = \dfrac{1}{3}(2x + 5), \quad [0, 5]$

$f'(x) = \dfrac{2}{3}$

No critical numbers

x-value	Endpoint $x = 0$	Endpoint $x = 5$
$f(x)$	$\frac{5}{3}$	5
Conclusion	Minimum	Maximum

22. $f(x) = x^2 + 2x - 4, \quad [-1, 1]$

$f'(x) = 2x + 2$

Critical number: $x = -1$ (also an endpoint)

x-value	Endpoint $x = -1$	Endpoint $x = 1$
$f(x)$	-5	-1
Conclusion	Minimum	Maximum

24. $f(x) = 3x^{2/3} - 2x, \quad [-1, 2]$

$f'(x) = 2x^{-1/3} - 2 = \dfrac{2}{x^{1/3}} - 2 = \dfrac{2(1 - x^{1/3})}{x^{1/3}}$

Critical numbers: $x = 0, 1$

x-value	Endpoint $x = -1$	Critical $x = 0$	Critical $x = 1$	Endpoint $x = 2$
$f(x)$	5	0	1	0.762
Conclusion	Maximum	Minimum		

26. $h(t) = \dfrac{t}{t - 2}, \quad [3, 5]$

$h'(t) = \dfrac{-2}{(t - 2)^2}$

No critical numbers

t-value	Endpoint $t = 3$	Endpoint $t = 5$
$h(t)$	3	$\frac{5}{3}$
Conclusion	Maximum	Minimum

28. $g(t) = \dfrac{t^2}{t^2 + 3}, \quad [-1, 1]$

$g'(t) = \dfrac{6t}{(t^2 + 3)^2}$

Critical number: $t = 0$

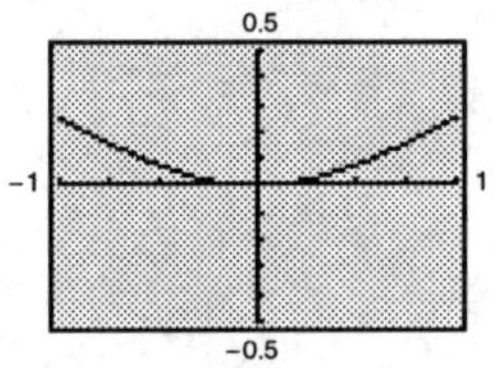

t-value	Endpoint $t = -1$	Critical $t = 0$	Endpoint $t = 1$
$g(t)$	$\frac{1}{4}$	0	$\frac{1}{4}$
Conclusion	Maximum	Minimum	Maximum

30. $g(x) = 4\left(1 + \dfrac{1}{x} + \dfrac{1}{x^2}\right) = 4(1 + x^{-1} + x^{-2}), \quad [-4, 5]$

$g'(x) = 4(-x^{-2} - 2x^{-3}) = 4\left(-\dfrac{1}{x^2} - \dfrac{2}{x^3}\right) = -4\left(\dfrac{x + 2}{x^3}\right)$

Critical number: $x = -2$
Discontinuity: $x = 0$
No maximum

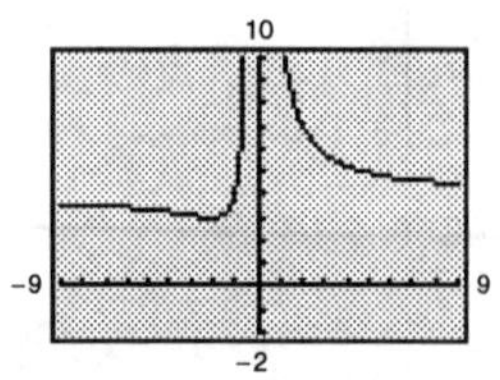

x-value	Endpoint $x = -4$	Critical $x = -2$	Discontinuity $x = 0$	Endpoint $x = 5$
$g(x)$	3.25	3	Undefined	
Conclusion		Minimum		4.96

32. $f(x) = \frac{4}{3}x\sqrt{3-x}, \quad [0, 3]$

Maximum: $\left(2, \frac{8}{3}\right)$
Minima: $(0, 0), (3, 0)$

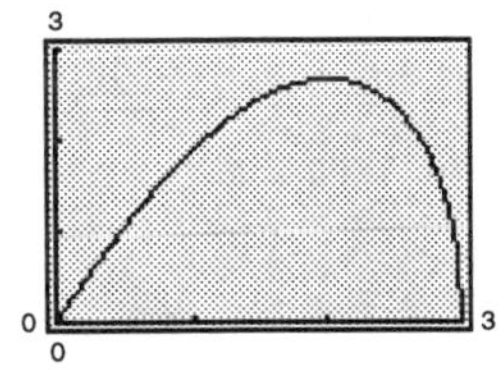

34. $f(x) = 4\sqrt{x} - 2x + 1, \quad [0, 6]$

Maximum: $(1, 3)$
Minimum: $(6, -1.202)$

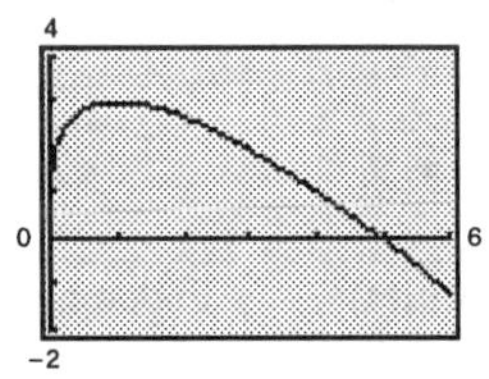

36. $f(x) = \dfrac{8}{x+1}, \quad [0, \infty)$

$$f'(x) = \frac{-8}{(x+1)^2}$$

No critical numbers; f is decreasing on $[0, \infty)$
Maximum: $f(0) = 8$

38. $f(x) = 8 - \dfrac{4x}{x^2+1}, \quad [0, \infty)$

$$f'(x) = -\frac{(x^2+1)(4) - 4x(2x)}{(x^2+1)^2} = \frac{4x^2-4}{(x^2+1)^2} = 0$$

Critical number: $x = 1$
Maximum: $f(0) = 8$
Minimum: $f(1) = 6$

40. $f(x) = \dfrac{1}{x^2+1}, \quad [0, 3]$

$$f'(x) = \frac{-2x}{(x^2+1)^2}$$

$$f''(x) = \frac{2(3x^2-1)}{(x^2+1)^3}$$

$$f'''(x) = \frac{24x(1-x^2)}{(x^2+1)^4}$$

Critical numbers for f'' in $[0, 3]$: $x = 0, x = 1$

x-value	Endpoint $x = 0$	Critical $x = 1$	Endpoint $x = 3$
$\lvert f''(x) \rvert$	2	$\frac{1}{2}$	$\frac{13}{250}$
Conclusion	Maximum		

42. $f(x) = \dfrac{1}{x^2}, \quad [1, 2]$

$$f'(x) = -\frac{2}{x^3}$$

$$f''(x) = \frac{6}{x^4}$$

$$f'''(x) = -\frac{24}{x^5}$$

$$f^{(4)}(x) = \frac{120}{x^6}$$

$$f^{(5)}(x) = -\frac{720}{x^7}$$

No critical numbers of $f^{(4)}$

x-value	Endpoint $x = 1$	Endpoint $x = 2$
$\lvert f^{(4)}(x) \rvert$	120	$\frac{15}{8}$
Conclusion	Maximum	

44. $x = \dfrac{k}{p^3}, \quad p > 1$

$$8 = \frac{k}{(10)^3} \Rightarrow k = 8000$$

$$x = \frac{8000}{p^3} \Rightarrow p = \frac{20}{\sqrt[3]{x}}$$

$C = 4x + 100$

$$P = xp - C = x\left(\frac{20}{\sqrt[3]{x}}\right) - (4x + 100) = 20x^{2/3} - 4x - 100$$

$$\frac{dP}{dx} = \frac{40}{3}x^{-1/3} - 4 = 0 \text{ when } x = \frac{1000}{27} \approx 37 \text{ units.}$$

Since P is increasing on $(1, 1000/27)$ and decreasing on $(1000/27, \infty)$, the profit is maximum when $x = 1000/27 \approx 37$ units and the price is \$6.00.

46. $v = k(R - r)r^2, \quad 0 \le r < R$

$= k(Rr^2 - r^3)$

$\dfrac{dv}{dr} = k(2Rr - 3r^2) = kr(2R - 3r)$

Critical numbers: $r = 0, r = \dfrac{2R}{3}$

r-value	Endpoint $r = 0$	Critical $r = (2R)/3$	Endpoint $r \to R$
v	0	$(4kR^3)/27$	$v \to 0$
Conclusion		Maximum	

The maximum air velocity occurs when $r = 2R/3$.

48. (a) 1956; 3.8 births/woman

(b) 1940–41, 1945–49 most rapidly
1976–91 most slowly

(c) 1944–45, 1949–50 most rapidly
1960–70 most slowly

(d) World War II and Korean Wars

Section 3.3 Concavity and the Second-Derivative Test

2. $f(x) = -x^3 + 3x^2 - 2$

$f'(x) = -3x^2 + 6x$

$f''(x) = -6x + 6$

$f''(x) = 0$ when $x = 1$.

Concave upward on $(-\infty, 1)$
Concave downward on $(1, \infty)$

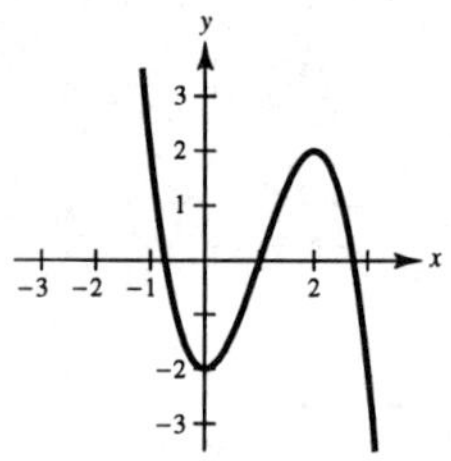

4. $f(x) = \dfrac{x^2 + 4}{4 - x^2}$

$f'(x) = \dfrac{(4 - x^2)2x - (x^2 + 4)(-2x)}{(4 - x^2)^2}$

$= \dfrac{16x}{(4 - x^2)^2}$

$f''(x) = \dfrac{(4 - x^2)^2 16 - 16x \cdot 2(4 - x^2)(-2x)}{(4 - x^2)^4}$

$= \dfrac{-16(4 + 3x^2)}{(4 - x^2)^3}$

f'' is undefined at $x = \pm 2$.
Concave upward on $(-2, 2)$
Concave downward on $(-\infty, -2)$ and $(2, \infty)$

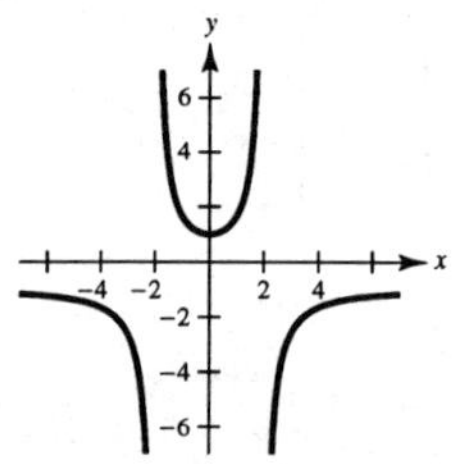

6. $f(x) = x^5 + 5x^4 - 40x^2$

$f'(x) = 5x^4 - 4x^3 - 8x + 1$

$f''(x) = 20x^3 - 12x^2 - 8 = (x - 1)(20x^2 + 8)$

Concave upward on $(1, \infty)$
Concave downward on $(-\infty, 1)$

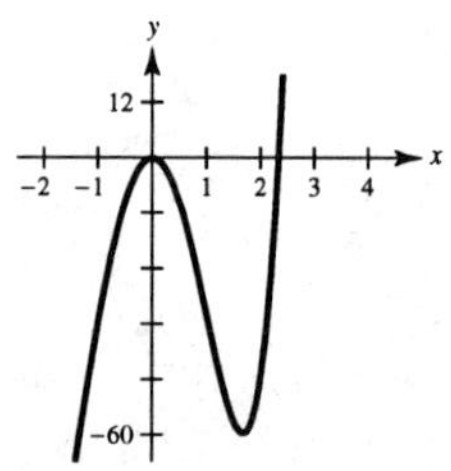

8. $f(x) = (x - 5)^2$

$f'(x) = 2(x - 5) = 0$

Critical number: $x = 5$

$f''(x) = 2$

$f''(5) = 2 > 0$

Thus, $(5, 0)$ is a relative minimum.

10. $f(x) = x^4 - 4x^3 + 3$

$f'(x) = 4x^3 - 12x^2 = 4x^2(x - 3) = 0$

Critical numbers: $x = 0, 3$

$f''(x) = 12x^2 - 24x$

$f''(0) = 0$ Test fails

$f''(3) > 0$

$(3, -25)$ is a relative minimum.
$(0, 3)$ is not a relative extremum.

12. $f(x) = \sqrt{x^2 + 1}$

$$f'(x) = \frac{1}{2}(x^2 + 1)^{-1/2}(2x) = \frac{x}{\sqrt{x^2 + 1}} = 0$$

Critical number: $x = 0$

$$f''(x) = x\left[-\frac{1}{2}(x^2 + 1)^{-3/2}(2x)\right] + (x^2 + 1)^{-1/2}(1)$$

$$= (x^2 + 1)^{-3/2}[-x^2 + (x^2 + 1)]$$

$$= \frac{1}{(x^2 + 1)^{3/2}}$$

$f''(0) = 1 > 0$

Thus, $(0, 1)$ is a relative minimum.

14. $f(x) = \dfrac{x}{x - 1}$

$$f'(x) = \frac{(x - 1)(1) - (x)(1)}{(x - 1)^2} = -\frac{1}{(x - 1)^2}$$

No critical numbers
No relative extrema

16.

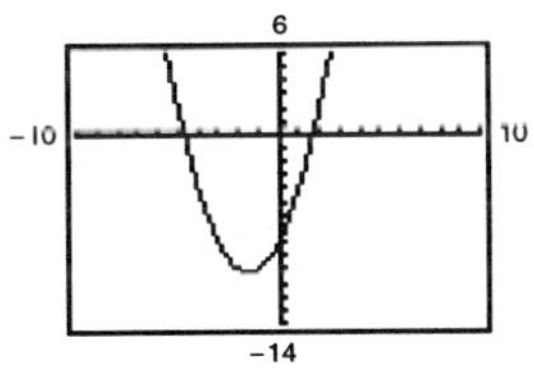

$(-1.5, -10.25)$ is a relative minimum.

18.

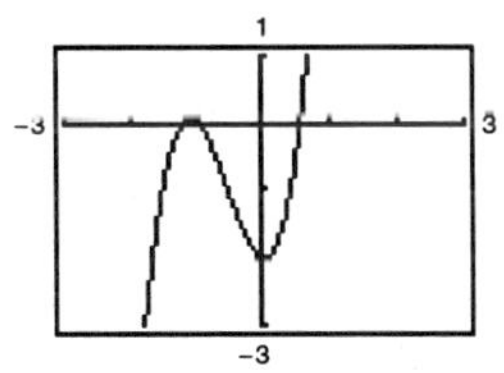

$(-1.11, 0.057)$ is a relative maximum.
$(0, -2)$ is a relative minimum.

20. $f' > 0$, (increasing)
$f'' < 0$, (concave downward)

22. $f' < 0$, (decreasing)
$f'' > 0$, (concave upward)

24. $f(x) = x(6 - x)^2 = x^3 - 12x^2 + 36x$

$f'(x) = 3x^2 - 24x + 36$

$f''(x) = 6x - 24 = 0$ when $x = 4$.

$f''(x) < 0$ on $(-\infty, 4)$

$f''(x) > 0$ on $(4, \infty)$

Thus, $(4, 16)$ is an inflection point.

26. $f(x) = x^4 - 18x^2 + 5$

$f'(x) = 4x^3 - 36x$

$f''(x) = 12x^2 - 36 = 12(x^2 - 3) = 0$ when $x = \pm\sqrt{3}$.

$f''(x) > 0$ on $(-\infty, -\sqrt{3})$

$f''(x) < 0$ on $(-\sqrt{3}, \sqrt{3})$

$f''(x) > 0$ on $(\sqrt{3}, \infty)$

Thus, $(-\sqrt{3}, -40)$ and $(\sqrt{3}, -40)$ are inflection points.

28. $f(x) = -4x^3 - 8x^2 + 32$

$f'(x) = -12x^2 - 16x$

$f''(x) = -24x - 16 = 0 \Rightarrow x = \frac{-2}{3}$

Since f'' changes sign at $x = \frac{-2}{3}$, $\left(\frac{-2}{3}, \frac{800}{27}\right)$ is a point of inflection.

30. $f(t) = (1 - t)(t - 4)(t^2 - 4)$

$f'(t) = -4t^3 + 15t^2 - 20$

$f''(t) = -6t(2t - 5)$

Inflection points: $(0, 16)$, $\left(\frac{5}{2}, \frac{81}{16}\right)$

32. $f(x) = x^3 + 1$

$f'(x) = 3x^2$

Critical number: $x = 0$

$f''(x) = 6x$

$f''(0) = 0$

Since $f'(x) > 0$ when $x \neq 0$ and the concavity changes at $x = 0$, $(0, 1)$ is a point of inflection.
No relative extrema

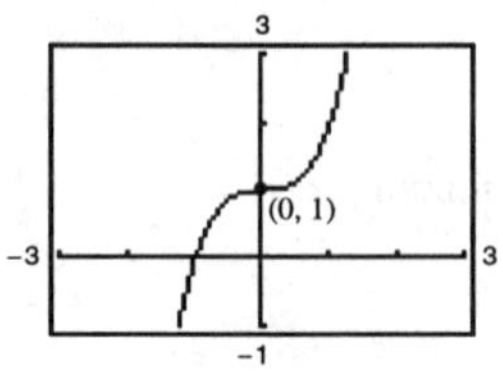

34. $f(x) = x^3 - \frac{3}{2}x^2 - 6x$

$f'(x) = 3x^2 - 3x - 6 = 3(x + 1)(x - 2)$

Critical numbers: $x = -1, x = 2$

$f''(x) = 6x - 3$

$f''(-1) = -9 < 0 \Rightarrow \left(-1, \frac{7}{2}\right)$ relative maximum

$f''(2) = 9 > 0 \Rightarrow (2, -10)$ relative minimum

Point of inflection is $\left(\frac{1}{2}, -\frac{13}{4}\right)$.

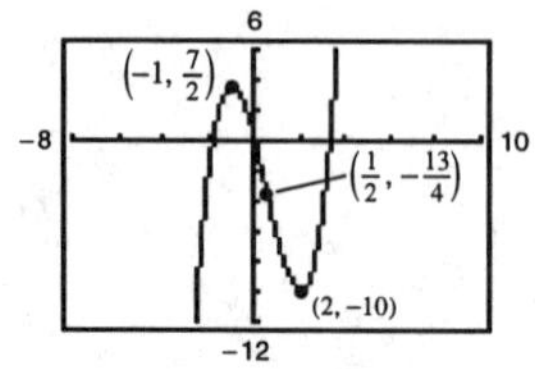

36. $f(x) = 2x^4 - 8x + 3$

$f'(x) = 8x^3 - 8 = 8(x^3 - 1)$

Critical number: $x = 1$

$f''(x) = 24x^2$

$f''(1) = 24 > 0$

Thus, $(1, -3)$ is a relative minimum.

$f''(x) = 24x^2 = 0$ when $x = 0$.

$f''(x) > 0$ on $(-\infty, 0)$ and on $(0, \infty)$

Thus, there are no inflection points.

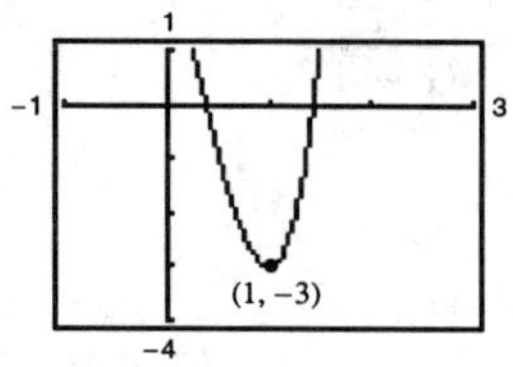

38. $g(x) = (x - 6)(x + 2)^3$

$g'(x) = (x - 6)[3(x + 2)^2] + (x + 2)^3(1)$

$= (x + 2)^2[3(x - 6) + (x + 2)] = (x + 2)^2(4x - 16)$

Critical numbers: $x = -2, 4$

$g''(x) = (x + 2)^2(4) + (4x - 16)[2(x + 2)]$

$= 2(x + 2)[2(x + 2) + (4x - 16)]$

$= 2(x + 2)(6x - 12) = 12(x + 2)(x - 2)$

$g''(-2) = 0$ (Test fails)

$g''(4) = 144 > 0$

Thus, $(4, -432)$ is a relative minimum. By the First-Derivative Test, $(-2, 0)$ is not a relative extremum.

$g''(x) = 12(x + 2)(x - 2) = 0$ when $x = \pm 2$.

$g''(x) > 0$ on $(-\infty, -2)$

$g''(x) < 0$ on $(-2, 2)$

$g''(x) > 0$ on $(2, \infty)$

Thus, $(-2, 0)$ and $(2, -256)$ are points of inflection.

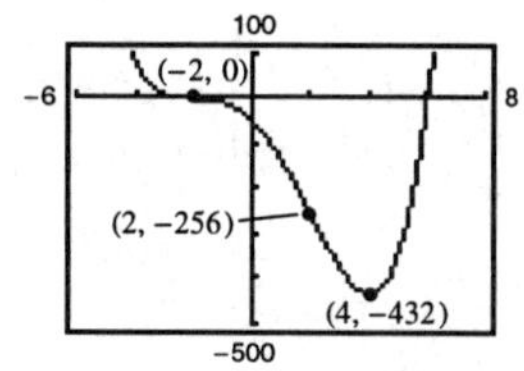

40. $g(x) = \frac{x^3}{16}(x + 4) = \frac{1}{16}(x^4 + 4x^3)$

$g'(x) = \frac{1}{16}(4x^3 + 12x^2) = \frac{1}{4}x^2(x + 3)$

Critical numbers: $x = 0, -3$

$g''(x) = \frac{1}{16}(12x^2 + 24x) = \frac{3}{4}x(x + 2)$

$g''(0) = 0$ (Test fails)

$g''(-3) = \frac{9}{4} > 0$

Thus, $\left(-3, -\frac{27}{16}\right)$ is a relative minimum. By the First-Derivative Test, $(0, 0)$ is not a relative extremum.

$g''(x) = \frac{3}{4}x(x + 2) = 0$ when $x = 0, -2$.

$g''(x) > 0$ on $(-\infty, -2)$

$g''(x) < 0$ on $(-2, 0)$

$g''(x) > 0$ on $(0, \infty)$

Thus, $(-2, -1)$ and $(0, 0)$ are points of inflection.

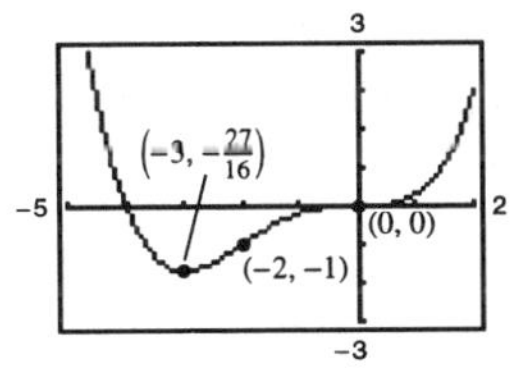

42. $f(x) = x - 4\sqrt{x + 1}$

$f'(x) = 1 - \frac{2}{\sqrt{x + 1}} = 0 \Rightarrow \sqrt{x + 1} = 2 \Rightarrow x = 3$

Domain: $[-1, \infty)$
Critical numbers: $x = -1, 3$

$f''(x) = \frac{1}{(x + 1)^{3/2}}$

$f''(3) = \frac{1}{8} > 0$

Thus, $(3, -5)$ is a relative minimum and $(-1, -1)$ is not a relative extremum. $f''(x) > 0$ on $(-1, \infty)$ and is concave upward. No points of inflection.

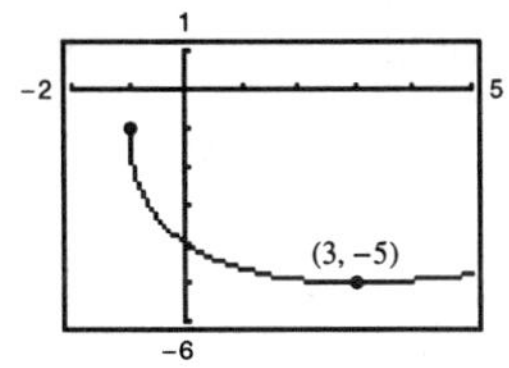

44.

Function	First Derivative	Second Derivative
$f(2) = 0$	$f'(x) > 0, \quad x < 3$	$f''(x) > 0, \quad x \neq 3$
$f(4) = 0$	$f'(3)$ is undefined.	
	$f'(x) < 0, \quad x > 3$	

The function has x-intercepts at $(2, 0)$ and $(4, 0)$. On $(-\infty, 3)$, f is increasing and on $(3, \infty)$, f is decreasing. f has either a relative maximum at $x = 3$ or a discontinuity at $x = 3$. Also, f is concave upward on $(-\infty, 3)$ and $(3, \infty)$.

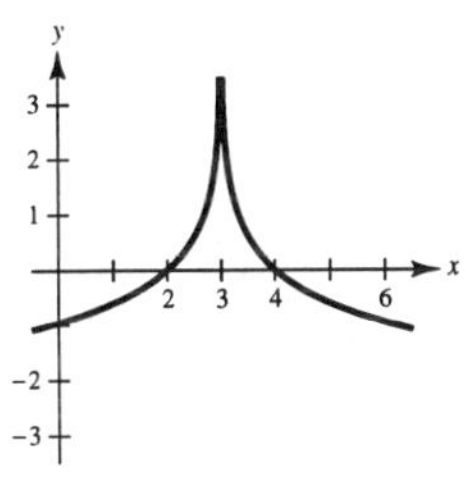

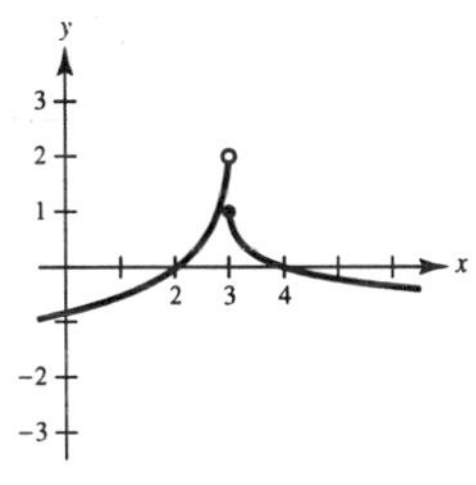

46. (a) f' is positive on $(0, 2)$ where f is increasing.

(b) f' is negative on $(-\infty, 0)$ and $(2, \infty)$ where f is decreasing.

(c) f' is increasing on $(-\infty, 1)$ where f is concave upward.

(d) f' is decreasing on $(1, \infty)$ where f is concave downward.

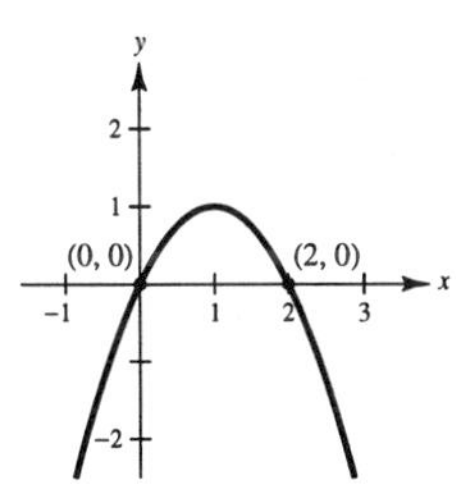

48. $R' = -\frac{4}{9}(3x^2 - 18x)$

$R'' = -\frac{4}{9}(6x - 18) = 0$ when $x = 3$.

$R'' > 0$ on $(0, 3)$

$R'' < 0$ on $(3, 5)$

Since $(3, 36)$ is a point of inflection, it is the point of diminishing returns.

50. $C = 0.002x^3 + 20x + 500$

$\overline{C} = 0.002x^2 + 20 + \dfrac{500}{x}$

$\overline{C}' = 0.004x - \dfrac{500}{x^2} = 0$

$x^3 = \dfrac{500}{0.004}$

$x = 50$ units

52. $N(t) = \dfrac{20t^2}{4 + t^2}, \quad 0 \le t \le 4$

We need to determine when $N'(t)$ is greatest.

$$N'(t) = \frac{(4 + t^2)(40t) - (20t^2)(2t)}{(4 + t^2)^2}$$

$$= 160t(4 + t^2)^{-2}$$

$$N''(t) = 160\{t[-2(4 + t^2)^{-3}(2t)] + (4 + t^2)^{-2}(1)\}$$

$$= 160(4 + t^2)^{-3}[-4t^2 + (4 + t^2)]$$

$$= \frac{160(4 - 3t^2)}{(4 + t^2)^3}$$

$$= 0 \text{ when } t = \pm\frac{2}{\sqrt{3}} = \pm\frac{2\sqrt{3}}{3}.$$

The student is assembling components at the greatest rate when $t = (2\sqrt{3})/3$ or at approximately 8:09 P.M.

54. $S(t) = \dfrac{500{,}000t^2}{36 + t^2}$

$$S'(t) = 500{,}000\left[\frac{(36 + t^2)(2t) - (t^2)(2t)}{(36 + t^2)^2}\right]$$

$$= 36{,}000{,}000t(36 + t^2)^{-2}$$

$$S''(t) = 36{,}000{,}000t[-2(36 + t^2)^{-3}(2t) + (36 + t^2)^{-2}(1)]$$

$$= 36{,}000{,}000(36 + t^2)^{-3}[-4t^2 + (36 + t^2)]$$

$$= \frac{36{,}000{,}000(36 - 3t^2)}{(36 + t^2)^3}$$

$$= 0 \text{ when } t = \pm\sqrt{12} = \pm 2\sqrt{3}.$$

The sales of this new product is increasing at the greatest rate when $t = 2\sqrt{3} \approx 3.464$ years.

56. $f(x) = x(x - 6)^2 = x(x^2 - 12x + 36)$

$= x^3 - 12x^2 + 36x$

$f'(x) = 3x^2 - 24x + 36$

$f''(x) = 6x - 24 = 0$ when $x = 4$.

$f''(x) < 0$ on $(-\infty, 4)$

$f''(x) > 0$ on $(4, \infty)$

Thus, $(4, 16)$ is an inflection point.

To find the relative extrema:

$f'(x) = 3(x^2 - 8x + 12)$

$= 3(x - 2)(x - 6)$

$f'(x) = 0$ when $x = 2, 6$.

$f''(2) < 0$ yields $(2, 32)$ as a relative maximum and $f''(6) > 0$ yields $(6, 0)$ as a relative minimum.

Using the midpoint formula, we have

$$\text{Midpoint} = \left(\frac{2 + 6}{2}, \frac{32 + 0}{2}\right) = (4, 16)$$

which is the same as the point of inflection.

58. $f(x) = -\frac{1}{20}x^5 - \frac{1}{12}x^2 - \frac{1}{3}x + 1, \quad [-2, 2]$

$f'(x) = -\frac{1}{4}x^4 - \frac{1}{6}x - \frac{1}{3}$

$f''(x) = -x^3 - \frac{1}{6}$

Minimum: $(2, -1.6)$
Maximum: $(-2, 2.93)$
Point of inflection: $(-0.55, 1.16)$

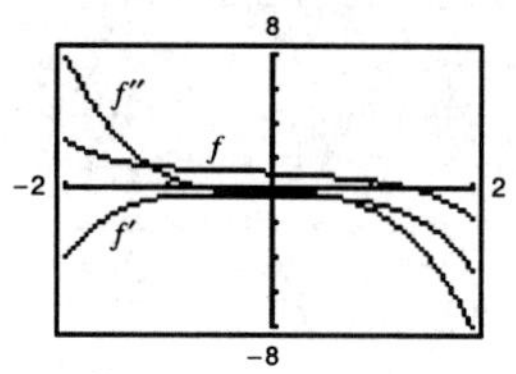

60. $f(x) = \dfrac{x^2}{x^2 + 1}, \quad [-3, 3]$

$$f'(x) = \frac{(x^2 + 1)(2x) - x^2(2x)}{(x^2 + 1)^2} = \frac{2x}{(x^2 + 1)^2}$$

$$f''(x) = \frac{2(1 - 3x^2)}{(x^2 + 1)^3}$$

Relative minimum: $(0, 0)$

Inflection points: $\left(\dfrac{1}{\sqrt{3}}, \dfrac{1}{4}\right), \left(-\dfrac{1}{\sqrt{3}}, \dfrac{1}{4}\right)$

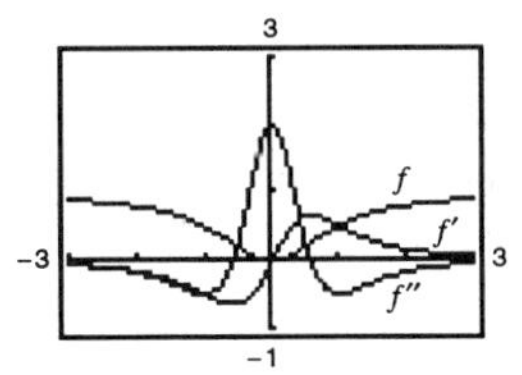

62. $N(t) = -t^3 + 12t^2, \quad 0 \le t \le 12$

(a) $N'(t) = -3t^2 + 24t = 3t(8 - t) = 0 \Rightarrow t = 8$

At $t = 8$, 256 people will be infected.

(b) $N''(t) = -6t + 24 = 0 \Rightarrow t = 4$.

At $t = 4$, the virus is spreading most rapidly.

(c)

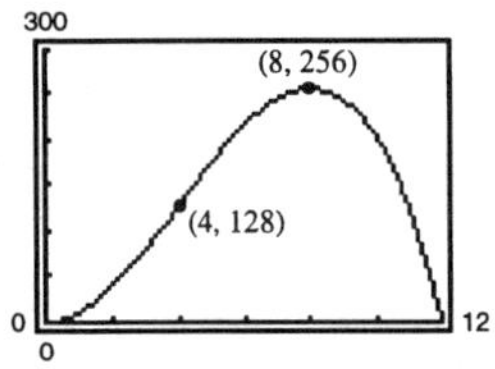

Section 3.4 Optimization Problems

2. Let x be the first number and y be the second number. Then $x + y = S$ and $y = S - x$. Thus, the product of x and y is given by the following.

$$P = xy = x(S - x) = Sx - x^2$$

$$P' = S - 2x$$

$P' = 0$ when $x = S/2$. Since $P''(S/2) = -2 < 0$, the product is maximum when $x = y = S/2$.

4. Let x be the first number and y be the second number. Then $x + 2y = 100$ and $x = 100 - 2y$. Thus, the product of x and y is given by the following.

$$P = xy = (100 - 2y)y = 100y - 2y^2$$

$$P' = 100 - 4y$$

$P' = 0$ when $y = 25$. Since $P''(25) = -4 < 0$, the product is maximum when $x = 100 - 2(25) = 50$ and $y = 25$.

6. Let x be the first number and y be the second number. Then $xy = 192$ and $y = 192/x$. The sum is given by the following.

$$S = x + 3y = x + 3\left(\frac{192}{x}\right) = x + 576x^{-1}$$

$$S' = 1 - 576x^{-2} = \frac{x^2 - 576}{x^2}$$

Critical number: $x = 24$

$$S'' = \frac{1152}{x^3}$$

Since $S''(24) > 0$, the sum is minimum when $x = 24$ and $y = 192/24 = 8$.

8. Let x be the first number and y be the second number. Then $x - y = 50$ and $x = 50 + y$. The product is given by the following.

$$P = xy = (50 + y)y = 50y + y^2$$

$$P' = 50 + 2y$$

$P' = 0$ when $y = -25$. Since $P''(-25) = 2 > 0$, the product is minimum when $x = 50 + (-25) = 25$ and $y = -25$.

10. (a) $A = 2\pi(1)(11) + 2\pi(1)^2 = 24\pi$ square units

$V = \pi(1)^2(11) = 11\pi$ cubic units

(b) $A = 2\pi(2)(4) + 2\pi(2)^2 = 24\pi$ square units

$V = \pi(2)^2(4) = 16\pi$ cubic units

(c) $A = 2\pi(3)(1) + 2\pi(3)^2 = 24\pi$ square units

$V = \pi(3)^2(1) = 9\pi$ cubic units

12. Let x and y be the length and width of the rectangle. Then the area is $xy = 64$ and $y = 64/x$. The perimeter is given by the following.

$$P = 2x + 2y = 2x + 2\left(\frac{64}{x}\right)$$

$$P' = 2 - \frac{128}{x^2} = \frac{2(x^2 - 64)}{x^2}$$

$P' = 0$ when $x = 8$. Since $P'' = 256/x^3$ and $P''(8) > 0$, the perimeter is a minimum when $x = 8$ and $y = 64/8 = 8$ feet.

14. Let x and y be the lengths shown in the figure. Then $xy = 180{,}000$ and $y = 180{,}000/x$. The perimeter is given by the following.

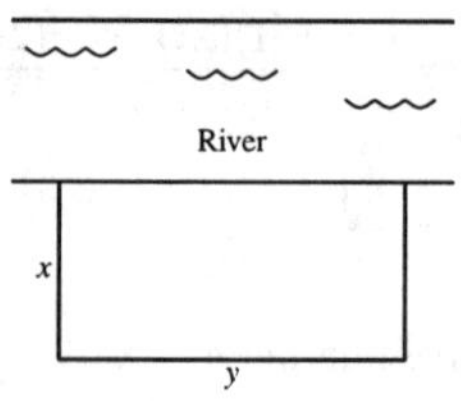

$$P = 2x + y = 2x + \frac{180{,}000}{x}$$

$$P' = 2 - \frac{180{,}000}{x^2} = \frac{2(x^2 - 90{,}000)}{x^2}$$

$P' = 0$ when $x = 300$. Since $P'' = \dfrac{360{,}000}{x^3}$ and $P''(300) > 0$,

P is minimum when $x = 300$ meters and $y = \dfrac{180{,}000}{300} = 600$ meters.

16. Let x be the length of the cut (the height of the box); then the length of the box is $3 - 2x$ and its width is $2 - 2x$. The volume is given by the following.

$$V = x(3 - 2x)(2 - 2x), \quad 0 < x < 1$$

$$= 4x^3 - 10x^2 + 6x$$

$$V' = 12x^2 - 20x + 6$$

$V' = 0$ when $x = (5 \pm \sqrt{7})/6$, but we do not consider $(5 + \sqrt{7})/6$ since it is not in the domain.

Length: $3 - 2\left(\dfrac{5 - \sqrt{7}}{6}\right) \approx 2.215$ feet

Width: $2 - 2\left(\dfrac{5 - \sqrt{7}}{6}\right) \approx 1.215$ feet

Height: $\dfrac{5 - \sqrt{7}}{6} \approx 0.392$ foot

18. Use the values (80, 400) and (81, 368) to determine the linear equation relating the number of trees to the number of apples.

$$y - 400 = \frac{400 - 368}{80 - 81}(x - 80)$$

$$y - 400 = -32(x - 80)$$

$$y = -32x + 2960$$

The number of apples is $N(x) = xy = x(-32x + 2960)$

$$N'(x) = -64x + 2960 = 0$$

$$64x = 2960$$

$$x = 46.25$$

Plant 46 trees, which yields 68,448 apples.

20. Let x and y be the lengths shown in the figure. Then $xy = 30$ and $y = 30/x$. The area of the page is given by the following.

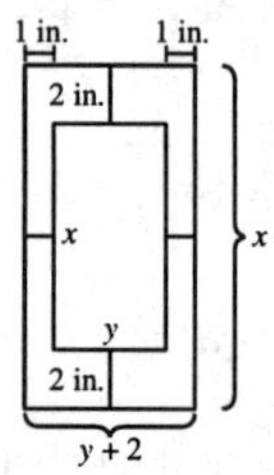

$$A = (x + 4)(y + 2) = (x + 4)\left(\frac{30}{x} + 2\right) = 38 + 2x + \frac{120}{x}$$

$$A' = 2 - \frac{120}{x^2} = \frac{2(x^2 - 60)}{x^2}$$

$A' = 0$ when $x = \sqrt{60} = 2\sqrt{15}$. The area of the page is minimum when dimensions are

$$x + 4 = 2\sqrt{15} + 4 \text{ inches}$$

$$y + 2 = \frac{30}{2\sqrt{15}} + 2 = \sqrt{15} + 2 \text{ inches.}$$

22. The slope of the line connecting $(0, y)$ and $(2, 3)$ is $m = (3 - y)/2$. The slope of the line connecting $(x, 0)$ and $(2, 3)$ is $m = 3/(2 - x)$. By equating these two slopes, we have $(3 - y)/2 = 3/(2 - x) \Rightarrow y = 3 - [6/(2 - x)]$. The area is given by the following.

$$A = \frac{1}{2}bh = \frac{1}{2}x\left(3 - \frac{6}{2 - x}\right) = \frac{3}{2}x - \frac{3x}{2 - x}$$

$$A' = \frac{3}{2} - \frac{6}{(2 - x)^2} = \frac{3[(2 - x)^2 - 4]}{2(2 - x)^2}$$

$A' = 0$ when $x = 0$ or $x = 4$.
The vertices of the triangle occur when $x = 4$ and $y = 6$.
Thus, the vertices are (0, 0), (4, 0), and (0, 6).

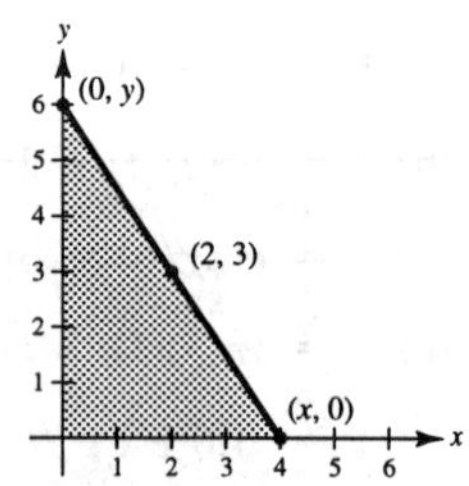

24. The area is given by the following.

$$A = 2xy = 2x\sqrt{r^2 - x^2}$$

$$A' = 2x\left(-\frac{x}{\sqrt{r^2 - x^2}}\right) + 2\sqrt{r^2 - x^2} = \frac{2(r^2 - 2x^2)}{\sqrt{r^2 - x^2}}$$

$A' = 0$ when $x = r/\sqrt{2}$. The length of the rectangle is $2x = \sqrt{2}r$ and the width is $\sqrt{r^2 - x^2} = r/\sqrt{2} = (\sqrt{2}/2)r$.

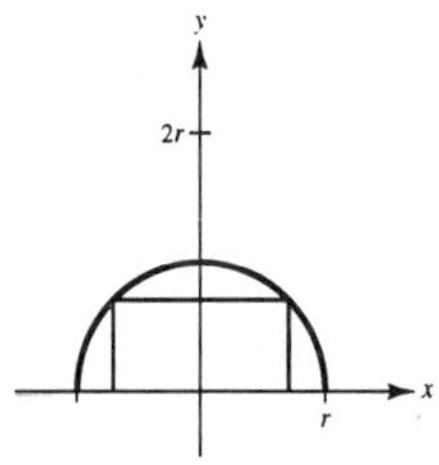

26. The volume is given by the following.

$$V = \pi r^2 h = \pi x^2\left(2\sqrt{r^2 - x^2}\right) = 2\pi x^2\sqrt{r^2 - x^2}$$

$$V' = 2\pi\left[x^2\left(\frac{-x}{\sqrt{r^2 - x^2}}\right) + 2x\sqrt{r^2 - x^2}\right] = 2\pi\left[\frac{2r^2x - 3x^3}{\sqrt{r^2 - x^2}}\right]$$

$V' = 0$ when $x = \sqrt{2/3}\,r$. The maximum volume is

$$V = 2\pi\left(\frac{2}{3}r^2\right)\sqrt{r^2 - \left(\frac{2}{3}\right)r^2} = \frac{4\pi r^3}{3\sqrt{3}}.$$

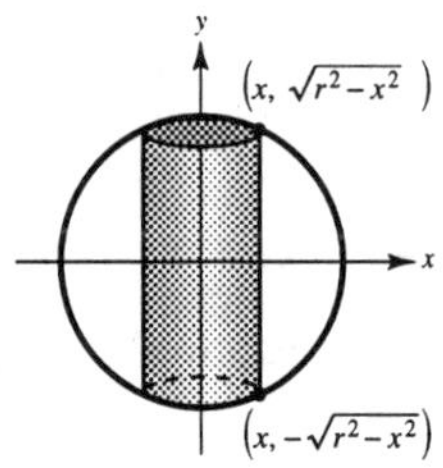

28. The distance between a point (x, y) on the graph and the point $(2, 1/2)$ is

$$d = \sqrt{(x - 2)^2 + \left(y - \frac{1}{2}\right)^2} = \sqrt{(x - 2)^2 + \left(x^2 - \frac{1}{2}\right)^2}$$

and we can minimize d by minimizing its square $L = d^2$.

$$L = (x - 2)^2 + \left(x^2 - \frac{1}{2}\right)^2 = x^4 - 4x + \frac{17}{4}$$

$$L' = 4x^3 - 4$$

$L' = 0$ when $x = 1$ and $y = (1)^2 = 1$. Thus, the point nearest $(2, 1/2)$ is $(1, 1)$.

30. The volume is $\frac{4}{3}\pi r^3 + \pi r^2 h = 12$, thus $h = \frac{12}{\pi r^2} - \frac{4}{3}r$. The surface area is given by the following.

$$A = 4\pi r^2 + 2\pi rh$$

$$= 4\pi r^2 + 2\pi r\left(\frac{12}{\pi r^2} - \frac{4}{3}r\right)$$

$$= 4\pi r^2 + \frac{24}{r} - \frac{8}{3}\pi r^2$$

$$= \frac{4}{3}\pi r^2 + \frac{24}{r}$$

$$A' = \frac{8}{3}\pi r - \frac{24}{r^2} = \frac{8(\pi r^3 - 9)}{3r^2}$$

$A' = 0$ when $r = \sqrt[3]{9/\pi} \approx 1.42$ in.

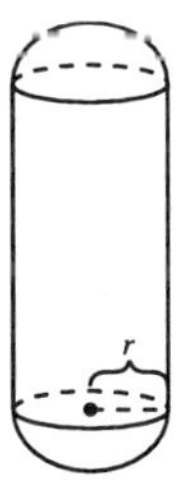

32. The perimeter is given by $3x + 4y = 10$. Thus, $y = \frac{1}{4}(10 - 3x)$. The total area is given by the following.

$$A = \frac{1}{2}bh + y^2 = \frac{1}{2}x\left(\frac{\sqrt{3}x}{2}\right) + \left[\frac{1}{4}(10 - 3x)\right]^2$$

$$= \frac{\sqrt{3}}{4}x^2 + \frac{1}{16}(100 - 60x + 9x^2)$$

$$A' = \frac{\sqrt{3}}{2}x + \frac{1}{16}(-60 + 18x)$$

$$= x\left(\frac{\sqrt{3}}{2} + \frac{9}{8}\right) - \frac{15}{4}$$

$A' = 0$ when $x = \frac{30}{4\sqrt{3} + 9}$ and

$$y = \frac{1}{4}\left[10 - 3\left(\frac{30}{4\sqrt{3} + 9}\right)\right] = \frac{10\sqrt{3}}{4\sqrt{3} + 9}.$$

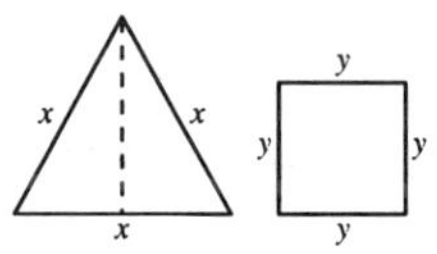

34. Use the points $(1, 4)$, $(2, 3.90)$.

$$y - 4 = \frac{3.90 - 4}{2 - 1}(x - 1)$$

$$y = -0.1x + 4.1 \quad (x = \text{week}, y = \text{dollars/bushel})$$

Now use the points $(1, 120)$, $(2, 124)$.

$$z - 120 = \frac{124 - 120}{2 - 1}(x - 1)$$

$$z = 4x + 116 \quad (x = \text{week}, z = \text{bushels})$$

$$\text{Total value} = f(x) = yz = (-0.1x + 4.1)(4x + 116)$$

$$= -0.4x^2 + 4.8x + 475.6$$

$$f'(x) = -0.8x + 4.8 = 0 \Rightarrow x = 6$$

So, harvest strawberries in the sixth week, which yields $4(6) + 116 = 140$ bushels for a value of $f(6) = \$490$.

36. $\text{Area} = l \cdot w = 2x\sqrt{100 - x^2} = A(x)$

$$A'(x) = \frac{2(100 - 2x^2)}{\sqrt{100 - x^2}}$$

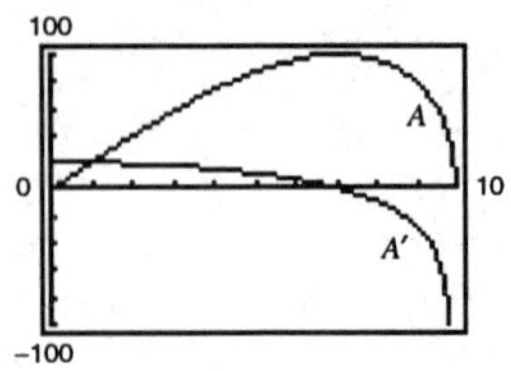

$\text{Length} = l = 10\sqrt{2} \approx 14.14$

$\text{Width} = w = 5\sqrt{2} \approx 7.07$

Section 3.5 Business and Economics Applications

2. $R = 48x^2 - 0.02x^3$

$R' = 96x - 0.06x^2 = x(96 - 0.06x)$

$R' = 0$ when $x = 0$ or $x = 1600$.

Thus, R is maximum when $x = 1600$ units.

4. $R = 30x^{2/3} - 2x$

$$R' = 20x^{-1/3} - 2 = \frac{20}{\sqrt[3]{x}} - 2$$

$R' = 0$ when $x = 1000$.

Thus, R is maximum when $x = 1000$ units.

6. $\overline{C} = 0.001x^2 - 5 + \dfrac{250}{x}$

$$\overline{C}' = 0.002x - \frac{250}{x^2} = \frac{0.002x^3 - 250}{x^2}$$

$\overline{C}' = 0$ when $x = 50$.

Thus, $\overline{C}$ is minimum when $x = 50$ units.

8. $\overline{C} = 0.02x^2 + 55x + \dfrac{1250}{x}$

$$\overline{C}' = 0.04x + 55 - \frac{1250}{x^2} = 0$$

$$\frac{0.04x^3 + 55x^2 - 1250}{x^2} = 0 \Rightarrow x \approx 5 \text{ units}$$

10. $P = xp - C = x(6000 - 0.4x^2) - (2400x + 5200) = -0.4x^3 + 3600x - 5200$

$P' = -1.2x^2 + 3600$

$P' = 0$ when $x = \sqrt{3000} = 10\sqrt{30}$.

Thus, the maximum profit occurs at $x = 10\sqrt{30}$ units and $p = 6000 - 0.4\left(10\sqrt{30}\right)^2 = \$4800$.

12. $P = xp - C = x\left(50 - \frac{\sqrt{x}}{10}\right) - (35x + 500)$

$= 15x - \frac{1}{10}x^{3/2} - 500$

$P' = 15 - \frac{3}{20}x^{1/2}$

$P' = 0$ when $15 = \frac{3}{20}\sqrt{x}$

$100 = \sqrt{x}$

$10{,}000 = x.$

By the First-Derivative Test, P is a maximum when $x = 10{,}000$ units and $p = 50 - \frac{\sqrt{10{,}000}}{10} = \$40.$

14. $C = x^3 - 6x^2 + 13x$

$\overline{C} = x^2 - 6x + 13$

$\overline{C}' = 2x - 6$

$\overline{C}' = 0$ when $x = 3.$

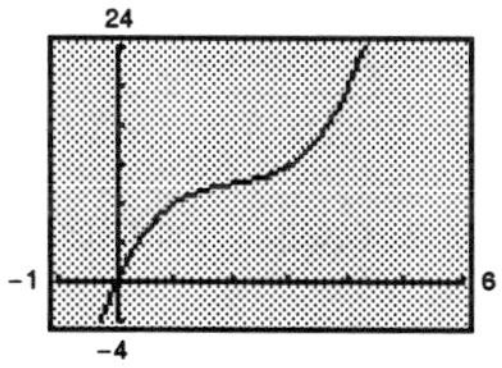

Thus, the average cost is minimum when $x = 3$ units and $\overline{C}(3) = \$4$ per unit.

$C' = \text{marginal cost} = 3x^2 - 12x + 13$

$C'(3) = 4 = \overline{C}(3)$

16. (a) $P = xp - C = x\left(100 - \frac{1}{2}x^2\right) - (50x + 37.5) = -\frac{1}{2}x^3 + 50x - 37.50$

$P' = -\frac{3}{2}x^2 + 50$

$P' = 0$ when $x = \frac{10}{\sqrt{3}}$ units and $p = 100 - \frac{1}{2}\left(\frac{10}{\sqrt{3}}\right)^2 = \frac{250}{3} \approx \$83.33.$

(b) $\overline{C} = 50 + \frac{37.5}{x}$. When $x = \frac{10}{\sqrt{3}}$, the average price is $\overline{C}\left(\frac{10}{\sqrt{3}}\right) = 50 + \frac{37.5}{(10/\sqrt{3})} \approx \$56.50.$

18. $P = -\frac{1}{10}s^3 + 6s^2 + 400$

$P' = -\frac{3}{10}s^2 + 12s = s\left(-\frac{3}{10}s + 12\right)$

$P' = 0$ when $s = 0$ or $s = 40.$

$P'' = -\frac{3}{5}s + 12$

$P''(0) = 12 > 0 \Longrightarrow$ Minimum

$P''(40) = -12 < 0 \Longrightarrow$ Maximum

$P'' = -\frac{3}{5}s + 12 = 0$ when $s = 20.$

The maximum profit occurs when $s = 40$ (or \$40,000) and the point of diminishing returns occurs at $s = 20$ (or \$20,000). The point is (20,000, 2,000,000).

20. Let x = the number of \$30 increases in rent.

Rent $= 540 + 30x$

Number of apartments $= 50 - x$

Cost per apartment $= 36$

Profit = (Rent)(Number of apartments) − (Cost)(Number of apartments)

$= (540 + 30x)(50 - x) - 36(50 - x) = (50 - x)(504 + 30x)$

$P' = (50 - x)(40) + (504 + 30x)(-1) = 996 - 60x$

$P' = 0$ when $x = 16.6$ and the rent is $540 + 30(16.6) = \$1038.$

22. $A = ki^2$

$P = 0.12A - i(A) = ki^2(0.12 - i)$

$P' = ki^2(-1) + (0.12 - i)(2ki) = 0.24ki - 3ki^2 = ki(0.24 - 3i)$

$P' = 0$ when $i = 0$ and $i = 0.08$. The profit is maximum when $i = 8\%$.

24. Let k = cost per mile to run the line overland. Then

$$C = 2k\sqrt{x^2+1} + k(2-x) = k\left[2\sqrt{x^2+1} + (2-x)\right]$$

$$C' = k\left[\frac{2x}{\sqrt{x^2+1}} - 1\right].$$

$$C' = 0 \text{ when } \frac{2x}{\sqrt{x^2+1}} = 1$$

$$2x = \sqrt{x^2+1}$$
$$4x^2 = x^2 + 1$$
$$3x^2 = 1$$
$$x = \sqrt{\frac{1}{3}} = \frac{\sqrt{3}}{3} \text{ mile.}$$

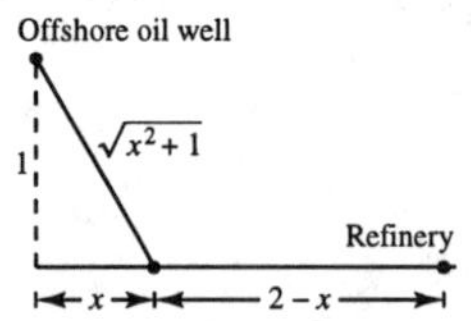

Lay pipe along the coast on land $2 - \sqrt{3}/3$ miles (1.42 miles) from the refinery. Lay $2\sqrt{3}/3$ miles (1.16 miles) in the ocean to the oil well.

26. $C = \left(\frac{v^2 + 360}{720} + 8\right)\left(\frac{110}{v}\right)$

$$= \frac{110}{720}\left(v + \frac{360}{v}\right) + \frac{880}{v}$$

$$C' = \frac{11}{72}\left(1 - \frac{360}{v^2}\right) - \frac{880}{v^2} = \frac{11}{72} - \frac{935}{v^2}$$

$C' = 0$ when $11v^2 = 72.935$

$$v \approx 78.23 \text{ mph.}$$

28. Since $dp/dx = -0.03$, the price elasticity of demand is

$$\eta = \frac{p/x}{dp/dx} = \frac{(5 - 0.03x)/x}{-0.03} = 1 - \frac{5}{0.03x}.$$

When $x = 100$, we have

$$\eta = 1 - \frac{5}{0.03(100)} = -\frac{2}{3}.$$

Since $|\eta(100)| = \frac{2}{3} < 1$, the demand is inelastic.

Elastic: $\left(0, \frac{250}{3}\right)$

Inelastic: $\left(\frac{250}{3}, \frac{500}{3}\right)$

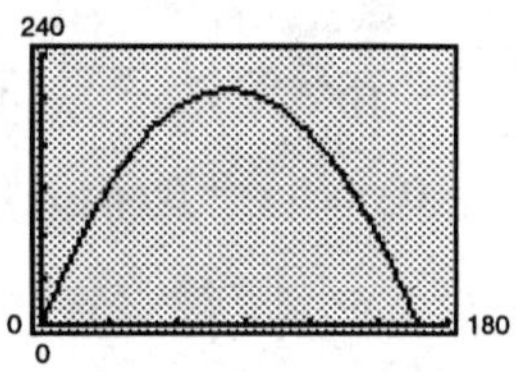

30. Since $\frac{dp}{dx} = -\frac{500}{(x+2)^2}$, the price elasticity of demand is

$$\eta = \frac{p/x}{dp/dx} = \frac{500}{x(x+2)} \cdot \frac{(x+2)^2}{-500} = -\frac{x+2}{x}.$$

When $x = 23$, we have

$$\eta = -\frac{23+2}{23} = -\frac{25}{23}.$$

Since $|\eta(23)| = \frac{25}{23} > 1$, the demand is elastic.

Elastic: $(0, \infty)$

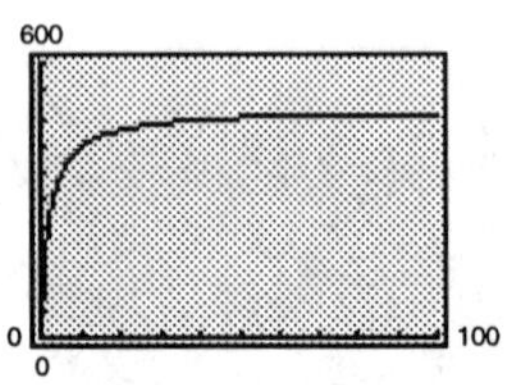

32. Since $\frac{dp}{dx} = -\frac{0.1}{\sqrt{0.2x}}$, the price elasticity of demand is

$$\eta = \frac{p/x}{dp/dx} = \frac{100 - \sqrt{0.2x}}{x} \cdot \frac{\sqrt{0.2x}}{-0.1}$$

$$= 2 - \frac{1000\sqrt{0.2x}}{x}.$$

When $x = 125$, we have

$$\eta = 2 - \frac{1000\sqrt{0.2(125)}}{125} = -38.$$

Since $|\eta(125)| = 38 > 1$, the demand is elastic.

Elastic: $(0, 22{,}342)$

Inelastic: $(22{,}342, 50{,}000)$

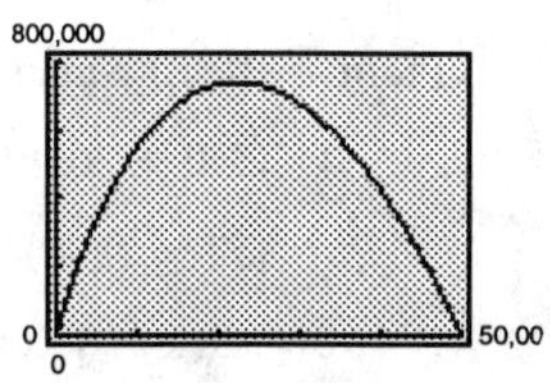

34. (a) $p^3 + x^3 = 9$

$$3p^2\frac{dp}{dx} + 3x^2 = 0$$

$$\frac{dp}{dx} = -\frac{x^2}{p^2}$$

$$\eta = \frac{p/x}{-x^2/p^2} = -\frac{p^3}{x^3}$$

When $x = 2, p = 1$, and $\eta = -\frac{(1)^3}{(2)^3} = -\frac{1}{8}$.

(b) $R = xp = x\sqrt[3]{9 - x^3}$

$$R' = x\left[\frac{1}{3}(9 - x^3)^{-2/3}(-3x^2)\right] + (9 - x^3)^{1/3}(1)$$

$$= (9 - x^3)^{-2/3}[-x^3 + (9 - x^3)] = \frac{9 - 2x^3}{(9 - x^3)^{2/3}}$$

$R' = 0$ when $x = \sqrt[3]{9/2}$ and $p = \sqrt[3]{9/2}$.

(c) $\eta\left(\sqrt[3]{9/2}\right) = -\frac{\left(\sqrt[3]{9/2}\right)^3}{\left(\sqrt[3]{9/2}\right)^3} = -1$

$\left|\eta\left(\sqrt[3]{9/2}\right)\right| = 1$

36. $x = 600 - 50p$

$$p = 12 - \frac{x}{50}$$

$$\frac{dp}{dx} = -\frac{1}{50}$$

When $p = 5, x = 350$, and

$$\eta = \frac{p/x}{dp/dx} = \frac{5/350}{-1/50} = \frac{-5}{7}.$$

Since $|\eta| = \frac{5}{7} < 1$, the demand is inelastic.

38. $x = \frac{a}{p^m}, \quad m > 1$

$$1 = -\frac{am}{p^{m+1}}\frac{dp}{dx}$$

$$\frac{dp}{dx} = -\frac{p^{m+1}}{am}$$

$$\eta = \frac{p/x}{dp/dx} = \frac{p}{x} \cdot \frac{-am}{p^{m+1}}$$

$$= \frac{p}{a/p^m} \cdot \frac{-am}{p^{m+1}} = \frac{p^{m+1}}{a} \cdot \frac{-am}{p^{m+1}} = -m$$

40. $R = 6.59t^4 - 76.46t^3 + 258.22t^2 - 8.71t + 2716.3$

(a) $R' = 26.36t^3 - 229.38t^2 + 516.44t - 8.71$

Graphing R', you see that R' is greatest at $t = 6$ (1996).

(b) Graphing R', you see that R' is least at $t = 0$ (1990).

(c) $R'(6) = 526.01$

$R'(0) = -8.71$

(d)

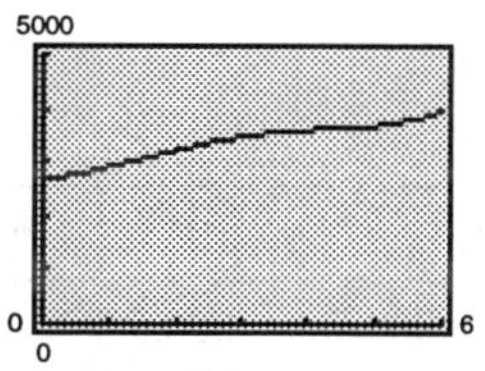

Section 3.6 Asymptotes

2. A horizontal asymptote occurs at $y = 0$ since
$$\lim_{x\to\infty} \frac{4}{(x-2)^3} = 0 \text{ and } \lim_{x\to-\infty} \frac{4}{(x-2)^3} = 0.$$

A vertical asymptote occurs at $x = 2$ since
$$\lim_{x\to2^-} \frac{4}{(x-2)^3} = -\infty \text{ and } \lim_{x\to2^+} \frac{4}{(x-2)^3} = \infty.$$

4. A horizontal asymptote occurs at $y = -1$ since
$$\lim_{x\to\infty} \frac{2+x}{1-x} = -1 \text{ and } \lim_{x\to-\infty} \frac{2+x}{1-x} = -1.$$

A vertical asymptote occurs at $x = 1$ since
$$\lim_{x\to1^-} \frac{2+x}{1-x} = \infty \text{ and } \lim_{x\to1^+} \frac{2+x}{1-x} = -\infty.$$

6. A horizontal asymptote occurs at $y = 0$ since
$$\lim_{x\to\infty} \frac{-4x}{x^2+4} = 0 \text{ and } \lim_{x\to-\infty} \frac{-4x}{x^2+4} = 0.$$

The graph has no vertical asymptotes.

8. A horizontal asymptote occurs at $y = 0$ since
$$\lim_{x\to\infty} \frac{x^2+1}{x^3-8} = 0 \text{ and } \lim_{x\to-\infty} \frac{x^2+1}{x^3-8} = 0.$$

A vertical asymptote occurs at $x = 2$ since
$$\lim_{x\to2^-} \frac{x^2+1}{x^3-8} = -\infty \text{ and } \lim_{x\to2^+} \frac{x^2+1}{x^3-8} = \infty.$$

10. The graph of f has horizontal asymptote at $y = \pm 2$. It has no vertical asymptote and it matches graph (b).

12. The graph of f has a horizontal asymptote at $y = 2$. It has no vertical asymptote and it matches graph (a) since it has a y-intercept at $(0, 2)$.

14. The graph of f has a horizontal asymptote at $y = 2$. It has no vertical asymptote. The y-intercept is $(0, 5)$. Thus, it matches graph (d).

16. $\lim_{x \to -2^-} \frac{1}{x + 2} = -\infty$

18. $\lim_{x \to 1^+} \frac{2 + x}{1 - x} = -\infty$

20. $\lim_{x \to 4} \frac{x^2}{x^2 + 16} = \frac{1}{2}$

22. $\lim_{x \to 0^-} \left(x^2 - \frac{1}{x}\right) = 0 - (-\infty) = \infty$

24. $\lim_{x \to \infty} \frac{5x^3 + 1}{10x^3 - 3x^2 + 7} = \frac{5}{10} = \frac{1}{2}$

26. $\lim_{x \to \infty} \frac{2x^{10} - 1}{10x^{11} - 3} = 0$

28. $\lim_{x \to \infty} \frac{x^3 - 2x^2 + 3x + 1}{x^2 - 3x + 2} = \infty$

30. $\lim_{x \to \infty} \frac{1}{(x + 3)^2} = 0$

32. $\lim_{x \to \infty} \left(\frac{2x^2}{x - 1} + \frac{3x}{x + 1}\right) = \infty + 3 = \infty$

34.

x	10^0	10^1	10^2	10^3	10^4	10^5	10^6
$f(x)$	1	5.132	50.126	500.125	5000.126	50,000.2	500,000

$\lim_{x \to \infty} \left[x^2 - x\sqrt{x(x - 1)}\right] = \infty$

36.

x	-10^6	-10^4	-10^2	10^0	10^2	10^4	10^6
$f(x)$	$-2{,}000{,}000.5$	$-20{,}000.5$	-200.5	1	0.501	0.500	0.500

$\lim_{x \to \infty} \left[x - \sqrt{x(x - 1)}\right] = 0.5,\quad \lim_{x \to -\infty} \left[x - \sqrt{x(x - 1)}\right] = -\infty$

38. Intercepts: $(3, 0), \left(0, \frac{3}{2}\right)$
Horizontal asymptote: $y = 1$
Vertical asymptote: $x = 2$
No relative extrema

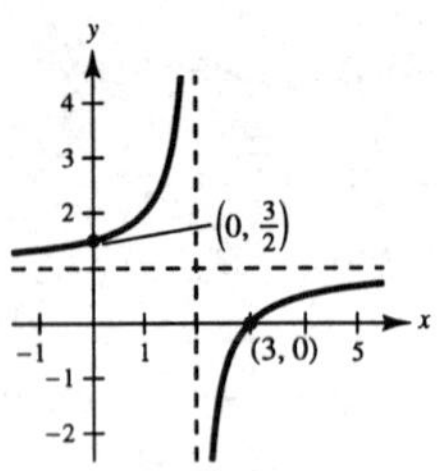

40. Intercept: $(0, 0)$
Horizontal asymptote: $y = 1$
Vertical asymptotes: $x = \pm 3$

$$f'(x) = \frac{-18x}{(x^2 - 9)^2}$$

Relative maximum: $(0, 0)$

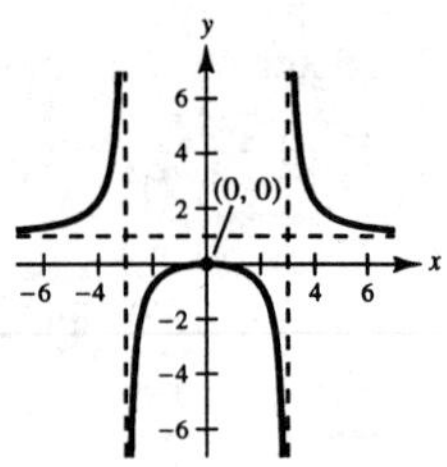

42. Intercept: $(0, 0)$
Horizontal asymptote: $y = 0$
Vertical asymptotes: $x = \pm 2$

$$g'(x) = \frac{-x^2 - 4}{(x^2 - 4)^2}$$

No relative extrema

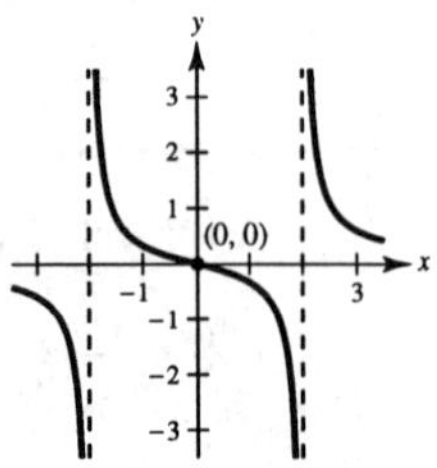

44. $y = \dfrac{4}{x^2}$

No intercepts

Horizontal asymptote: $y = 0$

Vertical asymptote: $x = 0$

$$y' = -\frac{8}{x^3}$$

No relative extrema

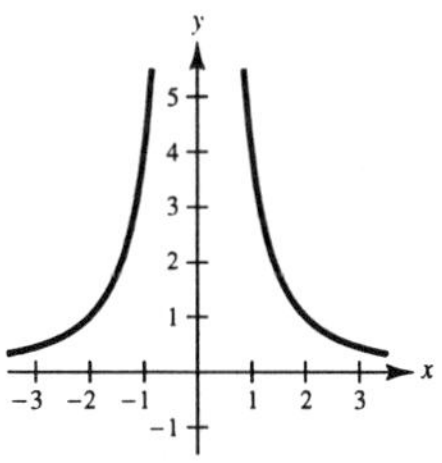

46. Intercept: $(0, 0)$

Horizontal asymptote: $y = 0$

Vertical asymptotes: $x = \pm 1$

$$y' = -\frac{2(x^2 + 1)}{(1 - x^2)^2}$$

No relative extrema

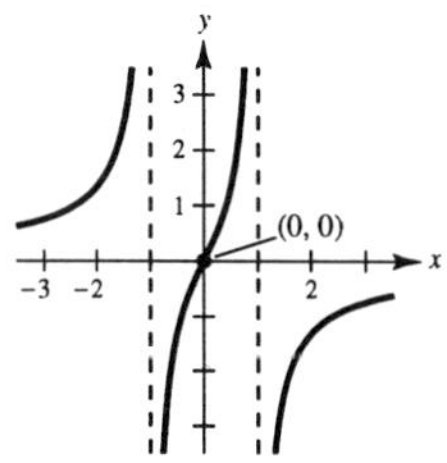

48. Intercept: $(-1, 0)$

Horizontal asymptote: $y = 1$

Vertical asymptote: $x = 0$

$$y' = -\frac{1}{x^2}$$

No relative extrema

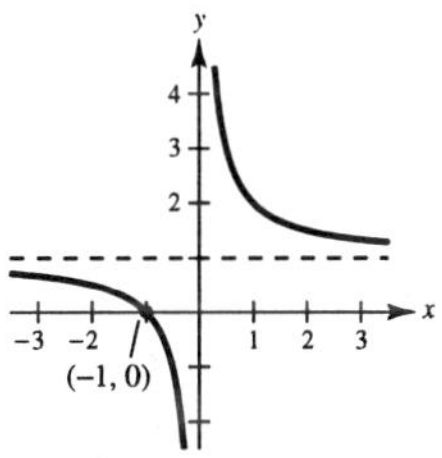

50. $f(x) = \dfrac{x - 2}{(x - 1)(x - 3)}$

Intercepts: $(2, 0), (0, -2/3)$

Horizontal asymptote: $y = 0$

Vertical asymptotes: $x = 1$ and $x = 3$

$$f'(x) = \frac{-(x^2 - 4x + 5)}{(x^2 - 4x + 3)^2}$$

No relative extrema

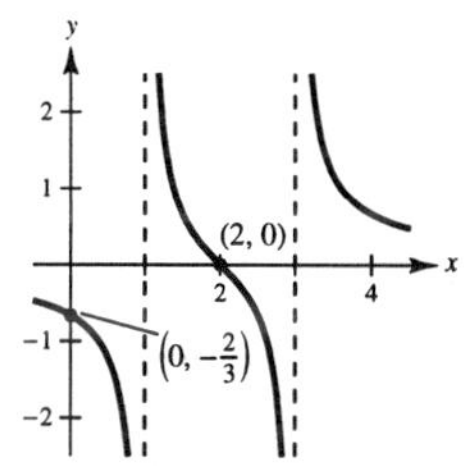

52. $g(x) = \dfrac{x^2 - 9}{x + 3} = \begin{cases} x - 3, & x \neq -3 \\ \text{undefined}, & x = -3 \end{cases}$

Intercept: $(3, 0), (0, -3)$

No asymptotes

No relative extrema

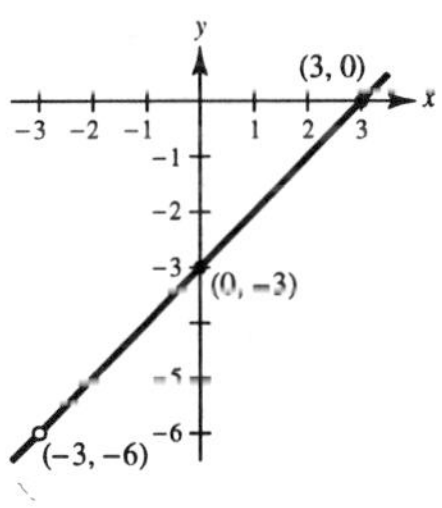

54. Intercept: $(0, 0)$

Horizontal asymptote: $y = 0$

Vertical asymptote: $x = -1$

$$y' = \frac{1 - x}{(x + 1)^3}$$

Relative maximum: $\left(1, \frac{1}{4}\right)$

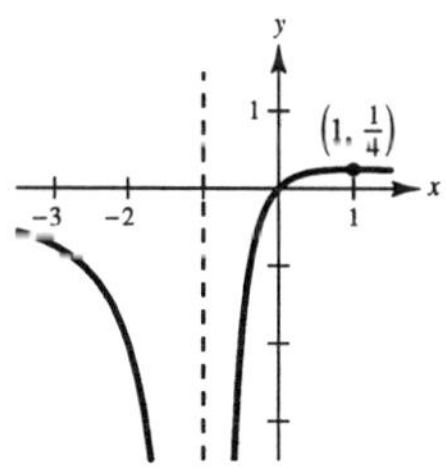

56. $\overline{C} = 0.5 + \dfrac{500}{x}$

$\lim_{x\to\infty} \overline{C} = \lim_{x\to\infty} \left(0.5 + \dfrac{500}{x}\right) = \$0.50$

58. (a) $C(15) \approx \$14{,}117.65$

(b) $C(50) = \$80{,}000$

(c) $C(90) = \$720{,}000$

(d) $\lim_{x\to 100^-} \dfrac{80{,}000}{100 - p} = \infty$

(e)

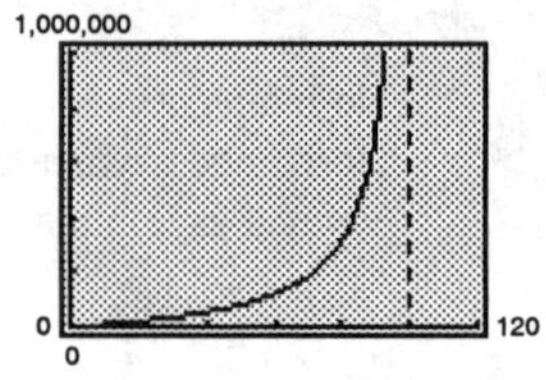

60. (a)

n	1	2	3	4	5	6	7	8	9	10
P	0.50	0.74	0.82	0.86	0.89	0.91	0.92	0.93	0.94	0.95

(b) $\lim_{n\to\infty} \dfrac{0.5 + 0.9(n-1)}{1 + 0.9(n-1)} = 1$

(c)

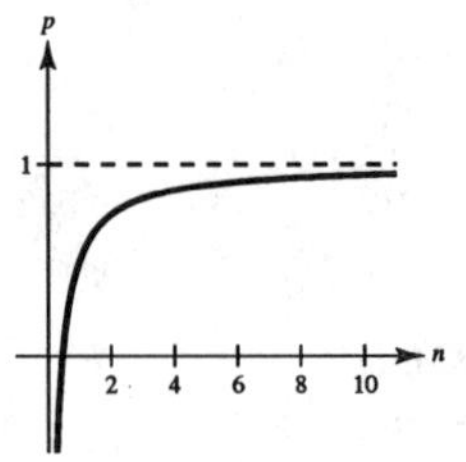

62. (a) $\overline{P} = \dfrac{R - C}{x} = \dfrac{69.9x - (34.5x + 15{,}000)}{x} = \dfrac{35.4x - 15{,}000}{x} = 35.4 - \dfrac{15{,}000}{x}$

(b) $\overline{P}(1000) = \$20.40$

$\overline{P}(10{,}000) = \$33.90$

$\overline{P}(100{,}000) = \$35.25$

(c) $\lim_{x\to\infty} \left(35.4 - \dfrac{15{,}000}{x}\right) = \$35.40$

Section 3.7 Curve Sketching: A Summary

2. $y = 2x^2 - 4x + 1$

$y' = 4x - 4 = 4(x - 1)$

$y'' = 4$

Intercepts: $(0, 1), \left(1 \pm \sqrt{2}/2, 0\right)$
Relative minimum: $(1, -1)$
Concave upward

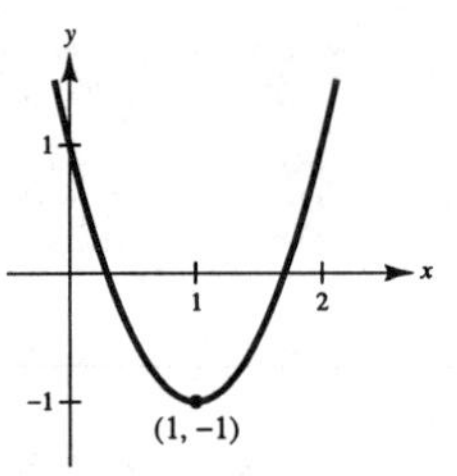

4. $y = -\frac{1}{3}(x^3 - 3x + 2)$

$y' = -\frac{1}{3}(3x^2 - 3) = 1 - x^2$

$y'' = -2x$

Intercepts: $\left(0, -\frac{2}{3}\right), (1, 0), (-2, 0)$
Relative minimum: $\left(-1, -\frac{4}{3}\right)$
Relative maximum: $(1, 0)$
Point of inflection: $\left(0, -\frac{2}{3}\right)$

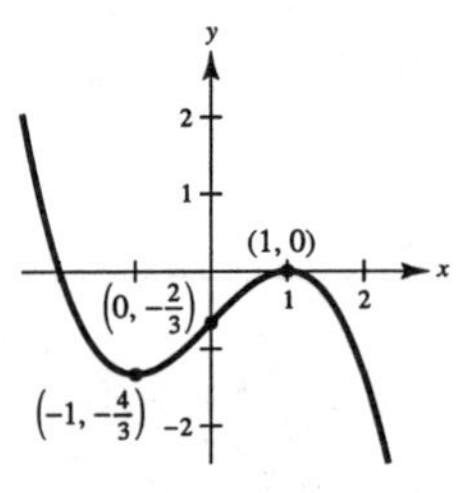

6. $y = x^3 + 3x^2 + 3x + 2$

$y' = 3x^2 + 6x + 3 = 3(x + 1)^2$

$y'' = 6x + 6 = 6(x + 1)$

Intercepts: $(0, 2), (-2, 0)$
No relative extrema
Point of inflection: $(-1, 1)$

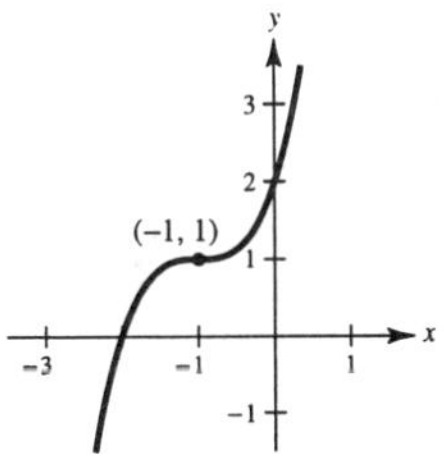

8. $f(x) = \frac{1}{3}(x - 1)^2 + 2$

$f'(x) = \frac{2}{3}(x - 1)$

$f''(x) = \frac{2}{3}$

Intercept: $\left(0, \frac{7}{3}\right)$
Relative minimum: $(1, 2)$
Concave upward

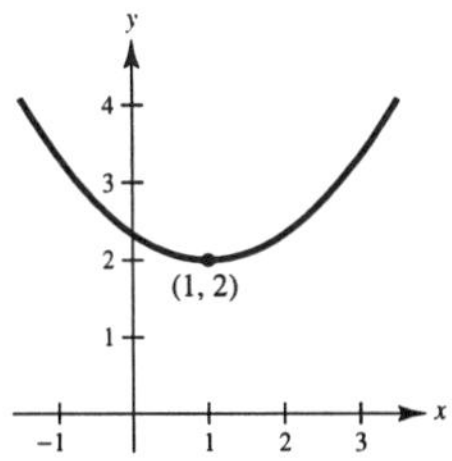

10. $y = 3x^4 - 6x^2 = 3x^2(x^2 - 2)$

$y' = 12x^3 - 12x = 12x(x^2 - 1)$

$y'' = 36x^2 - 12 = 12(3x^2 - 1)$

Intercepts: $(0, 0), \left(\pm\sqrt{2}, 0\right)$
Relative maximum: $(0, 0)$
Relative minima: $(\pm 1, -3)$
Points of inflection: $\left(\pm\sqrt{3}/3, -5/3\right)$

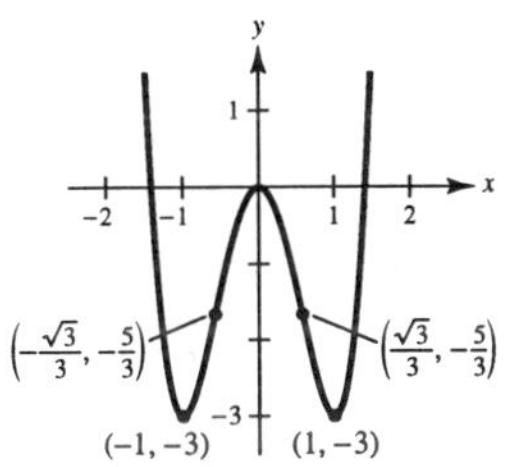

12. $f(x) = -x^3 + 3x^2 + 9x - 2$

$f'(x) = -3x^2 + 6x + 9 = -3(x + 1)(x - 3)$

$f''(x) = -6x + 6 = -6(x - 1)$

Relative maximum: $(3, 25)$
Relative minimum: $(-1, -7)$
Point of inflection: $(1, 9)$

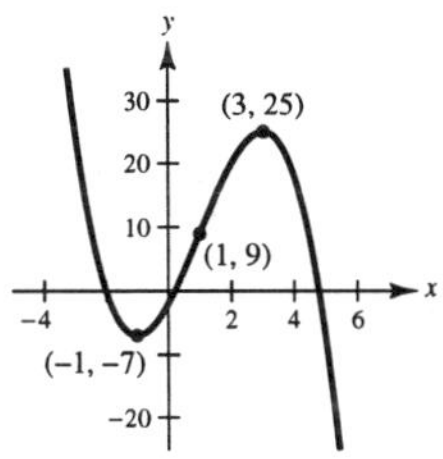

14. $f(x) = x^4 - 4x^3 + 16x - 16$

$f'(x) = 4x^3 - 12x^2 + 16 = 4(x + 1)(x - 2)^2$

$f''(x) = 12x^2 - 24x = 12x(x - 2)$

Relative minimum: $(-1, -27)$
Points of inflection: $(0, -16), (2, 0)$

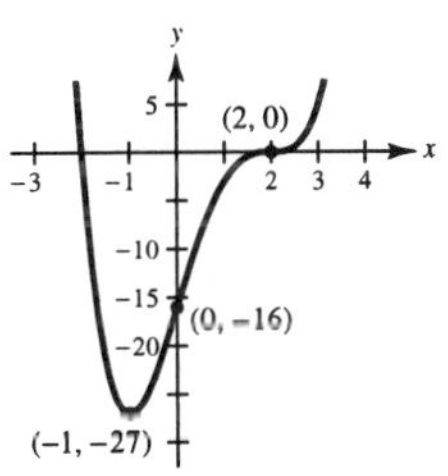

16. $f(x) = x^5 + 1$

$f'(x) = 5x^4$

$f''(x) = 20x^3$

Intercepts: $(0, 1), (-1, 0)$
No relative extrema
Point of inflection: $(0, 1)$

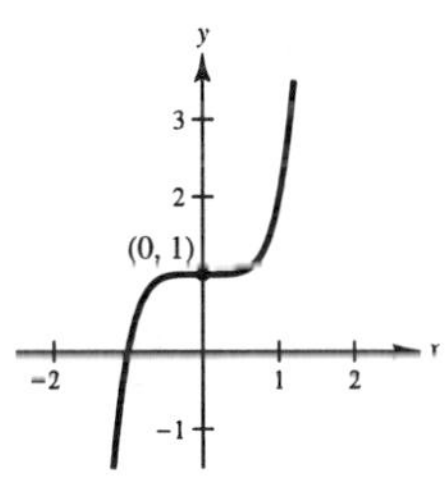

18. $y = (x-1)^5$

$y' = 5(x-1)^4$

$y'' = 20(x-1)^3$

Intercepts: $(0, -1), (1, 0)$
No relative extrema
Point of inflection: $(1, 0)$

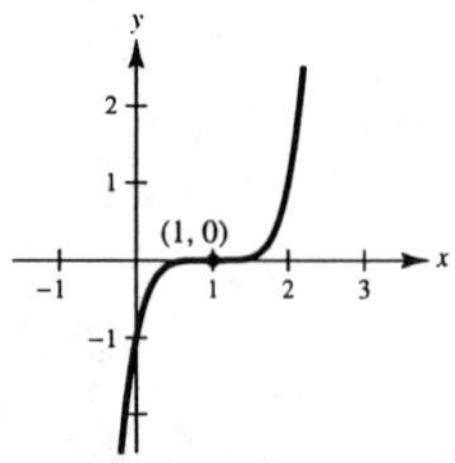

20. $y = |x^2 - 6x + 5|$

$y' = (2x-6)\dfrac{(x^2-6x+5)}{|x^2-6x+5|}$

$y'' = 2\dfrac{(x^2-6x+5)}{|x^2-6x+5|}$

Intercepts: $(0, 5), (1, 0), (5, 0)$
Relative maximum: $(3, 4)$
Relative minima: $(1, 0), (5, 0)$
Points of inflection: $(1, 0), (5, 0)$

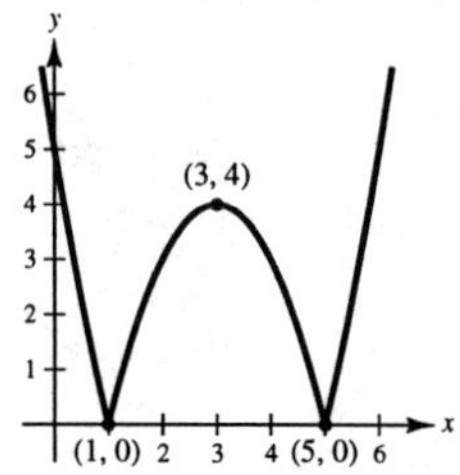

22. $y = \dfrac{x}{x^2+1}$

$y' = \dfrac{1-x^2}{(x^2+1)^2}$

$y'' = \dfrac{2x(x^2-3)}{(x^2+1)^3}$

Intercept: $(0, 0)$
Relative maximum: $(1, 1/2)$
Relative minimum: $(-1, -1/2)$
Points of inflection: $(0, 0), \left(-\sqrt{3}, -\sqrt{3}/4\right), \left(\sqrt{3}, \sqrt{3}/4\right)$

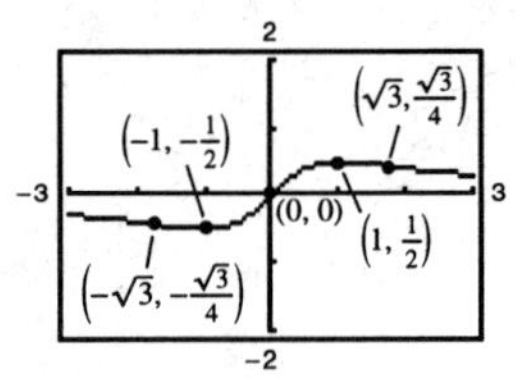

24. $y = 3x^{2/3} - x^2 = x^{2/3}(3 - x^{4/3})$

$y' = 2x^{-1/3} - 2x = \dfrac{2}{\sqrt[3]{x}} - 2x = \dfrac{2(1-x^{4/3})}{\sqrt[3]{x}}$

$y'' = -\dfrac{2}{3}x^{-4/3} - 2 = -\dfrac{2}{3x^{4/3}} - 2$

Intercepts: $(0, 0), \left(\pm\sqrt[4]{27}, 0\right)$
Relative maxima: $(-1, 2), (1, 2)$
Relative minimum: $(0, 0)$
Concave downward

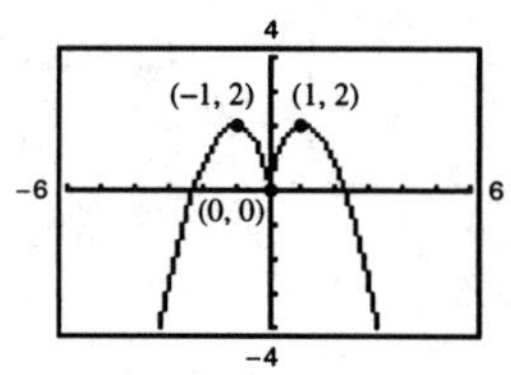

26. $y = (1-x)^{2/3}$

$y' = -\dfrac{2}{3}(1-x)^{-1/3} = -\dfrac{2}{3\sqrt[3]{1-x}}$

$y'' = -\dfrac{2}{9}(1-x)^{-4/3} = -\dfrac{2}{9(1-x)^{4/3}}$

Intercepts: $(0, 1), (1, 0)$
Relative minimum: $(1, 0)$
Concave downward

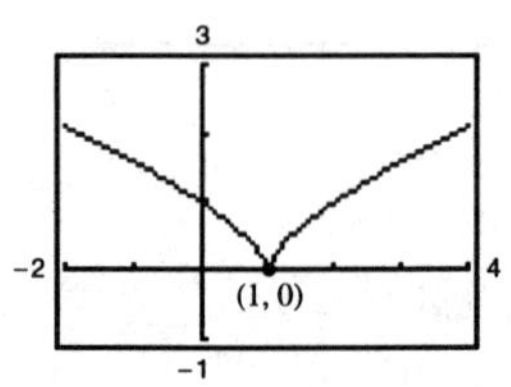

28. $y = x^{-1/3}$

$y' = -\frac{1}{3}x^{-4/3}$

$y'' = \frac{4}{9}x^{-7/3}$

No intercepts
No relative extrema
Decreasing
No points of inflection
Concave downward on $(-\infty, 0)$
Concave upward on $(0, \infty)$

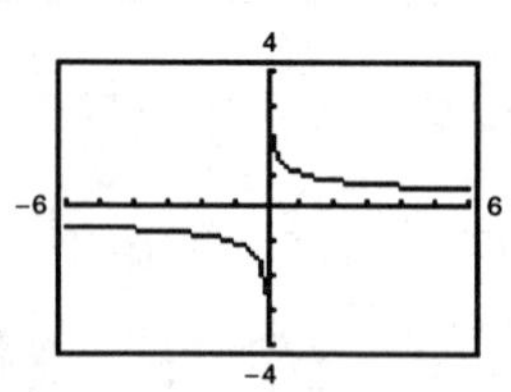

30. $y = x^{4/3}$

$y' = \frac{4}{3}x^{1/3}$

$y'' = \frac{4}{9}x^{-2/3}$

Intercept: $(0, 0)$
Relative minimum: $(0, 0)$
No points of inflection
Concave upward

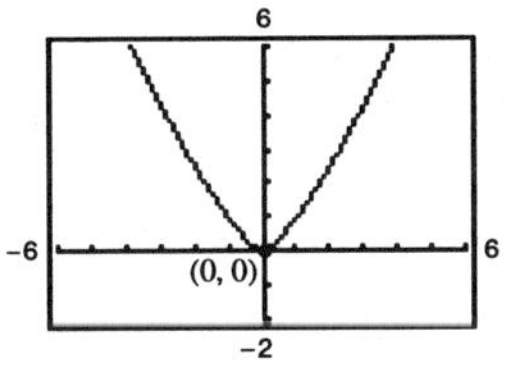

32. $y = \dfrac{x^2 + 1}{x^2 - 2}$

$y' = -\dfrac{6x}{(x^2 - 2)^2}$

$y'' = \dfrac{6(3x^2 + 2)}{(x^2 - 2)^3}$

Intercepts: $(0, -1/2)$
Relative maximum: $(0, -1/2)$
Vertical asymptotes: $x = \pm\sqrt{2}$
Horizontal asymptote: $y = 1$
Domain: $(-\infty, -\sqrt{2}), (-\sqrt{2}, \sqrt{2}), (\sqrt{2}, \infty)$

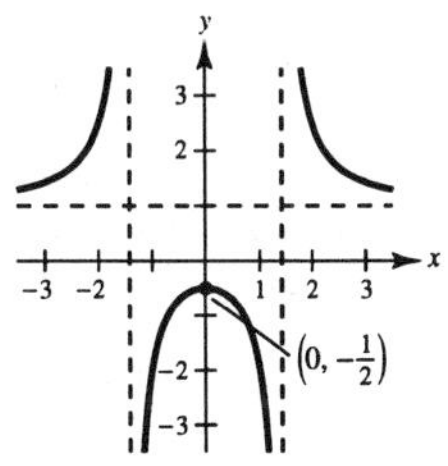

34. $y = \dfrac{x^2 - 6x + 12}{x - 4}$

$y' = \dfrac{x^2 - 8x + 12}{(x - 4)^2} = \dfrac{(x - 2)(x - 6)}{(x - 4)^2}$

$y'' = \dfrac{8}{(x - 4)^3}$

Intercept: $(0, -3)$
Relative maximum: $(6, 6)$
Relative minimum: $(2, -2)$
Vertical asymptote: $x = 4$
Domain: $(-\infty, 4), (4, \infty)$

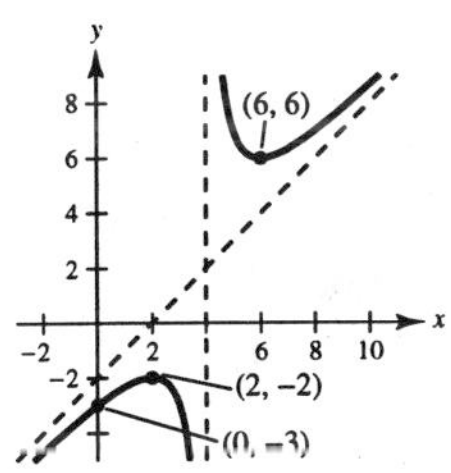

36. $y = x\sqrt{4 - x^2}$

$y' = \dfrac{2(2 - x^2)}{\sqrt{4 - x^2}}$

$y'' = \dfrac{2x(x^2 - 6)}{(4 - x^2)^{3/2}}$

Intercepts: $(0, 0), (2, 0), (-2, 0)$
Relative maximum: $(\sqrt{2}, 2)$
Relative minimum: $(-\sqrt{2}, -2)$
Point of inflection: $(0, 0)$
Domain: $[-2, 2]$

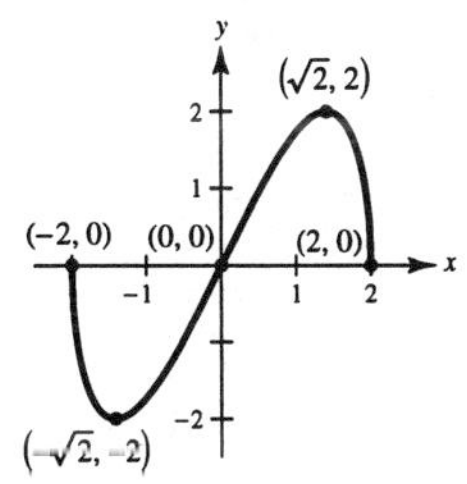

38. $y = x + \dfrac{32}{x^2}$

$y' = 1 - \dfrac{64}{x^3} = \dfrac{x^3 - 64}{x^3}$

$y'' = \dfrac{192}{x^4}$

Intercept: $(-2\sqrt[3]{4}, 0)$
Relative minimum: $(4, 6)$
Concave upward
Vertical asymptote: $x = 0$
Domain: $(-\infty, 0), (0, \infty)$

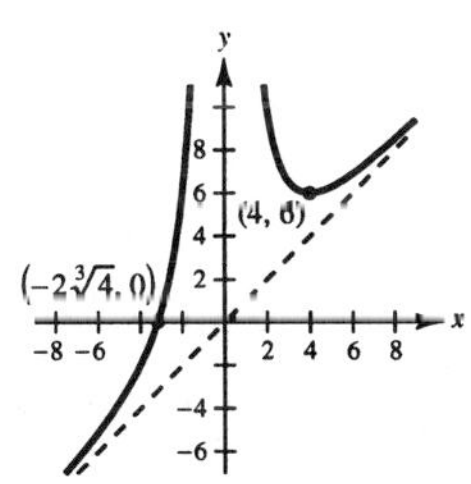

40. $y = x^4/(x^4 - 1)$

$y' = -4x^3/(x^4 - 1)^2$

$y'' = 4x^2(5x^4 + 3)/(x^4 - 1)^3$

Intercept: $(0, 0)$
Horizontal asymptote: $y = 1$
Vertical asymptotes: $x = \pm 1$
Domain: $(-\infty, -1), (-1, 1), (1, \infty)$
Relative maximum: $(0, 0)$

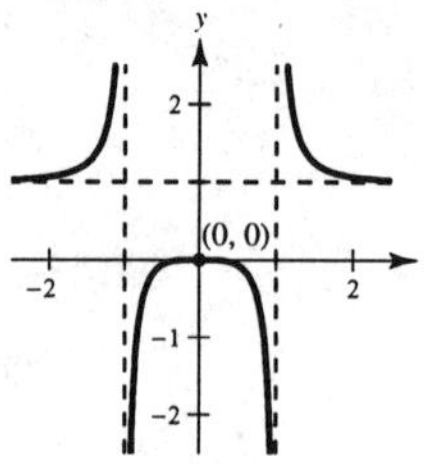

42. Since $\lim_{x \to \infty} f(x) = \infty$ and $\lim_{x \to -\infty} f(x) = -\infty$, we have $a > 0$.

$f(x) = x^3 + x^2 - x + 1$

(Solution not unique)

44. Since $\lim_{x \to \infty} f(x) = -\infty$ and $\lim_{x \to -\infty} f(x) = \infty$, we have $a < 0$.

$f(x) = -x^3 + 1$

(Solution not unique)

46. $f'(x) = -(x - 1)(x - 3) = -x^2 + 4x - 3$

$f(x) = -\frac{1}{3}x^3 + 2x^2 - 3x$

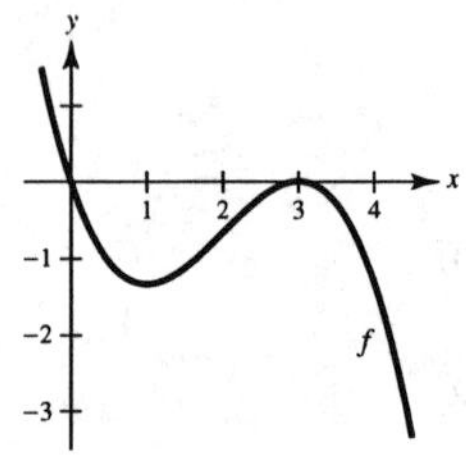

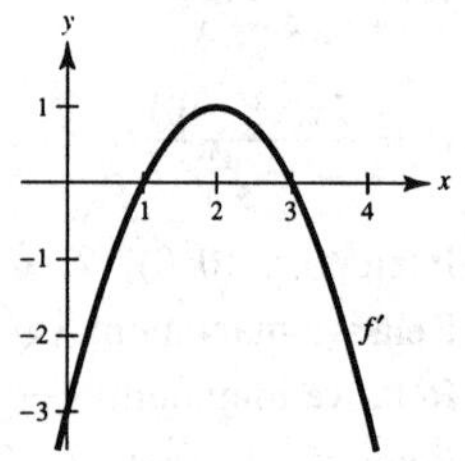

48. $f''(x) = 6x$

$f' = 3x^2$

$f(x) = x^3$

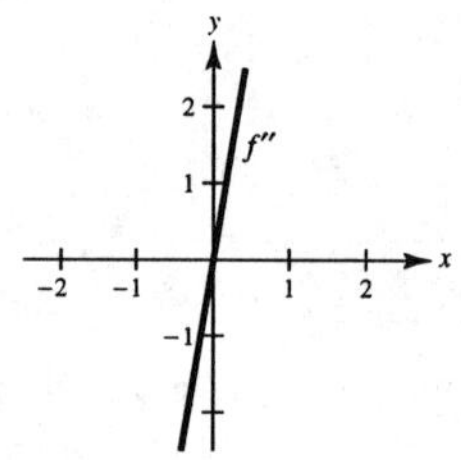

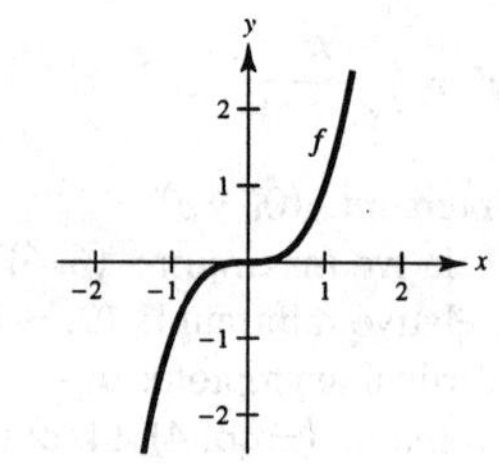

50. $f(4) = 2$

$f''(x) > 0$

$x \neq 4$

Concave upward

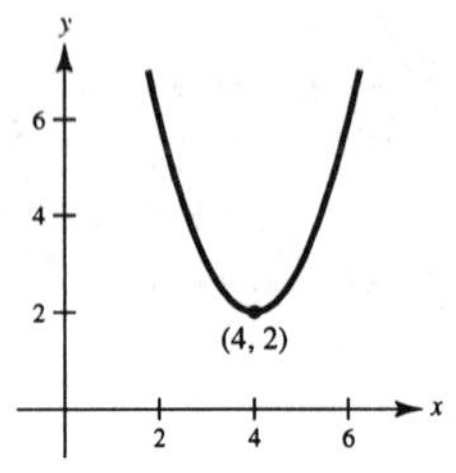

52. $P = 0.03t^2 - 0.01t + 3.39$ (Model I)

$P = 0.08t + 3.36$ (Model II)

$P = -0.07t^2 + 0.05t + 3.38$ (Model III)

(a)

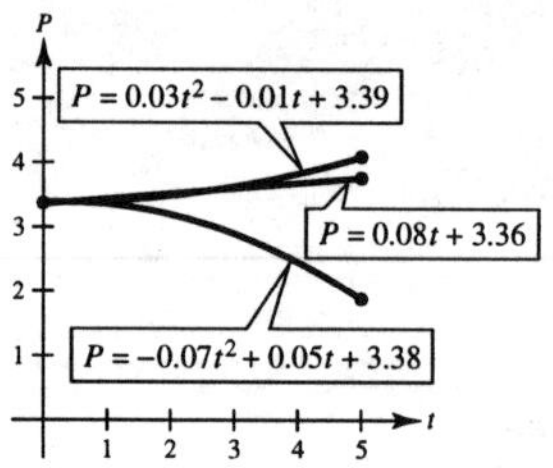

(b) I, II

(c) I most optimistic
III most pessimistic

Section 3.8 Differentials and Marginal Analysis

2. $dy = 3x^{1/2}dx = 3\sqrt{x}\,dx$

4. $dy = -2(1 - 2x^2)^{-3}(-4x)dx = \dfrac{8x}{(1 - 2x^2)^3}dx$

6. $dy = \dfrac{1}{3}(6x^2)^{-2/3}(12x)dx$

$= \dfrac{4x}{\sqrt[3]{36x^4}}dx = \dfrac{4}{\sqrt[3]{36x}}dx$

8. $f(x) = \sqrt{3x}, \quad x = 1, \quad \Delta x = 0.01$

$\Delta y = f(x + \Delta x) - f(x)$

$= \sqrt{3(1.01)} - \sqrt{3(1)} \approx 0.0086387$

10. $f(x) = \dfrac{x}{x^2 + 1}, \quad x = 1, \quad \Delta x = 0.01$

$\Delta y = f(x + \Delta x) - f(x)$

$= \dfrac{1.01}{(1.01)^2 + 1} - \dfrac{1}{1^2 + 1} \approx -0.0000247512$

12. $y = 1 - 2x^2, \quad x = 0, \quad \Delta x = dx = -0.1$

$dy = -4x\,dx = 0$

$\Delta y = 1 - 2(-0.1)^2 - 1 = -0.02$

14. $y = 2x + 1, \quad x = 2, \quad \Delta x = dx = 0.01$

$dy = 2dx = 0.02$

$\Delta y = \big(2(2.01) + 1\big) - \big(2(2) + 1\big) = 0.02$

16. $dy = -\dfrac{2}{x^3}dx, \quad \Delta y = \dfrac{1}{(x + \Delta x)^2} - \dfrac{1}{x^2}$

$dx = \Delta x$	dy	Δy	$\Delta y - dy$	$\dfrac{dy}{\Delta y}$
1.000	−0.2500	−0.1389	0.1111	1.7999
0.500	−0.1250	−0.0900	0.0350	1.3889
0.100	−0.0250	−0.0232	0.0018	1.0776
0.010	−0.0025	−0.0025	0.0000	1.0000
0.001	−0.0003	−0.0002	0.0001	1.5000

18. $dy = \dfrac{1}{2\sqrt{x}}dx, \quad \Delta y = \sqrt{x + \Delta x} - \sqrt{x}$

$dx = \Delta x$	dy	Δy	$\Delta y - dy$	$\dfrac{dy}{\Delta y}$
1.000	0.3536	0.3178	−0.0358	1.1126
0.500	0.1768	0.1669	−0.0099	1.0593
0.100	0.0354	0.0349	−0.0005	1.0143
0.010	0.0035	0.0035	0.0000	1.0000
0.001	0.0004	0.0004	0.0000	1.0000

20. $f(x) = \sqrt{25 - x^2}$

$f'(x) = \dfrac{1}{2}(25 - x^2)^{-1/2}(-2x) = \dfrac{-x}{\sqrt{25 - x^2}}$

$y - 4 = \dfrac{-3}{\sqrt{25 - 9}}(x - 3)$

$y - 4 = -\dfrac{3}{4}(x - 3)$

$y = \dfrac{3}{4}x + \dfrac{25}{4}$

$f(3 + 0.01) = 3.99248$

$y(3 + 0.01) = 3.9925$

$f(3 - 0.01) = 4.00748$

$y(3 - 0.01) = 4.0075$

22. $f(x) = 2x^{1/3} - 1$

$f'(x) = \dfrac{2}{3}x^{-2/3} = \dfrac{2}{3\sqrt[3]{x^2}}$

$y - 3 = \dfrac{2}{3}\left(\dfrac{1}{4}\right)(x - 8)$

$y - 3 = \dfrac{1}{6}(x - 8)$

$y = \dfrac{1}{6}x + \dfrac{5}{3}$

$f(8 + 0.01) = 3.001666$

$y(8 + 0.01) = 3.001667$

$f(8 - 0.01) = 2.9983326$

$y(8 - 0.01) = 2.9983333$

24. $N = \dfrac{10(5 + 3t)}{1 + 0.04t}$

$$dN = \frac{(1 + 0.04t)(30) - 10(5 + 3t)(0.04)}{(1 + 0.04t)^2}dt = \frac{28}{(1 + 0.04t)^2}dt$$

When $t = 5$ and $dt = 6 - 5 = 1$, we have the following.

$$dN = \frac{28}{[1 + 0.04(5)]^2}(1) = \frac{28}{1.44} \approx 19.44$$

The change in herd size will be approximately 19 deer.

26. $R = 30x - 0.15x^2, \quad x = 75, \quad dx = 1$

$$dR = (30 - 0.30x)dx$$
$$= [30 - 0.30(75)](1) = \$7.50$$

28. $P = -x^2 + 60x - 100, \quad x = 25, \quad dx = 1$

$$dP = (-2x + 60)dx = [-2(25) + 60](1) = \$10$$

30. $x = \dfrac{k}{p^2}$

$$2500 = \frac{k}{100}$$

$$250{,}000 = k$$

$$x = \frac{250{,}000}{p^2} \Rightarrow p^2 = \frac{250{,}000}{x} \Rightarrow p = \frac{500}{\sqrt{x}}$$

$$R = xp = x\left(\frac{500}{\sqrt{x}}\right) = 500\sqrt{x}$$

$$dR = \frac{250}{\sqrt{x}}dx = \frac{250}{\sqrt{3000}}(1) \approx \$4.56$$

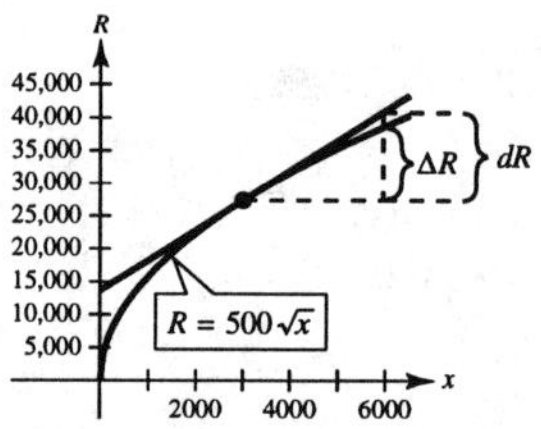

32. $C = 14.25x + 110{,}000$

$$dC = 14.25dx$$
$$= 14.25(1) = \$14.25$$

$\Delta C = dC$ because C is a linear function.

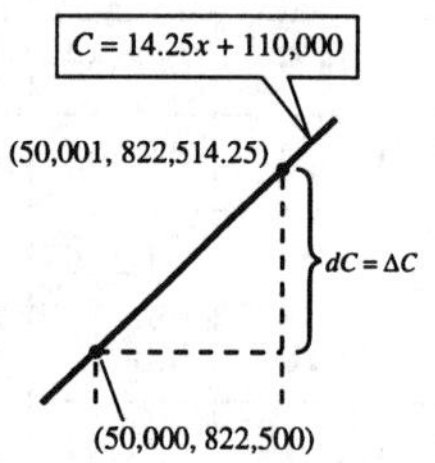

34. $A = x^2$

$$dA = 2x\,dx$$

When $x = 12$ and $dx = \pm 1/64$, we have

$$dA = 2(12)\left(\pm\frac{1}{64}\right) = \pm\frac{3}{8}\text{ in}^2.$$

When $A = 144$, the relative error is $\dfrac{dA}{A} = \dfrac{\pm 3/8}{144} \approx 0.0026$.

36. $V = x^3, \quad A = 6x^2$

(a) $dV = 3x^2dx$

$$\frac{dV}{V} = \frac{3x^2dx}{x^3} = \frac{3\,dx}{x}$$

When $x = 12$ and $dx = \pm 0.03$, we have

$$dV = 3(12)^2(\pm 0.03) = \pm 12.96\text{ in}^3.$$

$$\frac{dV}{V} = \frac{3(\pm 0.03)}{12} = \pm 0.0075$$

(b) $dA = 12x\,dx$

$$\frac{dA}{A} = \frac{12x\,dx}{6x^2} = \frac{2\,dx}{x}$$

$$dA = 12(12)(\pm 0.03) = \pm 4.32\text{ in}^2$$

$$\frac{dA}{A} = \frac{2(\pm 0.03)}{12} = \pm 0.005$$

38. $C = \dfrac{3t}{27 + t^3}$

$$dC = \frac{3(27 - 2t^3)}{(27 + t^3)^2}dt$$

When $t = 1$ and $dt = \frac{1}{2}$, we have $dC = \dfrac{3(25)}{(28)^2}\left(\dfrac{1}{2}\right) \approx 0.0478$.

Review Exercises for Chapter 3

2. $g(x) = (x - 1)^2(x - 3)$

$g'(x) = (x - 1)(3x - 7)$

Critical numbers: $x = 1, x = \frac{7}{3}$

4. $f(x) = (x + 1)^3$

$f'(x) = 3(x + 1)^2$

Critical number: $x = -1$

6. $g(x) = -x^2 + 7x - 12$

$g'(x) = -2x + 7$

Critical number: $x = \frac{7}{2}$
Increasing on $\left(-\infty, \frac{7}{2}\right)$
Decreasing on $\left(\frac{7}{2}, \infty\right)$

8. $f(x) = -x^3 + 6x^2 - 2$

$f'(x) = -3x^2 + 12x = -3x(x - 4)$

Critical numbers: $x = 0, 4$
Increasing on $(0, 4)$
Decreasing on $(-\infty, 0), (4, \infty)$

10. $N(t) = 0.0001t^4 - 0.01t^3 + 0.25t^2 - 1.22t + 26$
$0 \le t \le 25$

(a) Increasing on $(2.94, 19.91)$

(b) Decreasing on $(0, 2.94), (19.91, 25)$

(c) Subscriptions rose from 1973 to 1990, and decreased from 1970 to 1973, and from 1990 to 1995.

(d)

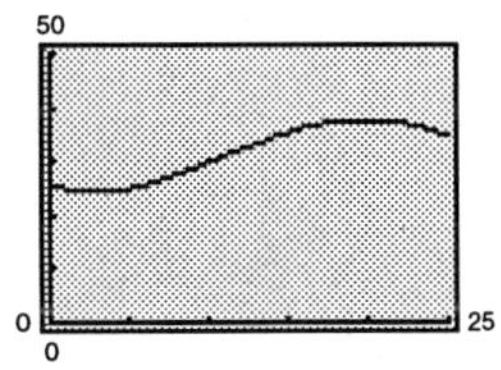

12. $f(x) = \frac{1}{4}x^4 - 8x$

$f'(x) = x^3 - 8$

Critical number: $x = 2$

Intervals	$(-\infty, 2)$	$(2, \infty)$
Sign of $f'(x)$	$-$	$+$
Conclusion	Decreasing	Increasing

Relative minimum: $(2, -12)$

14. $h(x) = 4 + 10x - x^2$

$h'(x) = 10 - 2x = 2(5 - x)$

Critical number: $x = 5$

Intervals	$(-\infty, 5)$	$(5, \infty)$
Sign of $f'(x)$	$+$	$-$
Conclusion	Increasing	Decreasing

Relative maximum: $(5, 29)$

16. $s(x) = x^4 - 8x^2 + 3$

$s'(x) = 4x^3 - 16x = 4x(x^2 - 4) = 4x(x - 2)(x + 2)$

Critical numbers: $x = 0, 2, -2$

Interval	$(-\infty, -2)$	$(-2, 0)$	$(0, 2)$	$(2, \infty)$
Sign of f'	$-$	$+$	$-$	$+$
Conclusion	Decreasing	Increasing	Decreasing	Increasing

Relative maxima: $(-2, -13), (2, -13)$

18. $f(x) = \dfrac{2}{x^2 - 1}$

$f'(x) = \dfrac{-4x}{(x^2 - 1)^2}$

Critical number: $x = 0$
Discontinuities: $x = \pm 1$

Interval	$(-\infty, -1)$	$(-1, 0)$	$(0, 1)$	$(1, \infty)$
Sign of f'	$+$	$+$	$-$	$-$
Conclusion	Increasing	Increasing	Decreasing	Decreasing

Relative maximum: $(0, -2)$

20. $g(x) = x - 6\sqrt{x}, \quad x > 0$

$$g'(x) = \frac{1-3}{\sqrt{x}} = \frac{\sqrt{x}-3}{\sqrt{x}}$$

Critical number: $x = 9$
Increasing on $(9, \infty)$
Decreasing on $(0, 9)$ Relative minimum: $(9, -9)$

Interval	$(0, 9)$	$(9, \infty)$
Sign of f'	$-$	$+$
Conclusion	Decreasing	Increasing

22. $f(x) = x^4 - 2x^3, \quad [0, 2]$

$f'(x) = 4x^3 - 6x^2 = 2x^2(2x - 3)$

Critical numbers: $x = 0, \frac{3}{2}$

x	$f(x)$	
0	0	maximum
$\frac{3}{2}$	-1.6875	minimum
2	0	maximum

24. $f(x) = \dfrac{8}{x} + x, \quad [1, 4]$

$$f'(x) = \frac{-8}{x^2} + 1 = \frac{x^2-8}{x^2}$$

Critical number: $x = 2\sqrt{2}$

x	$f(x)$	
1	9	maximum
4	6	
$2\sqrt{2}$	$4\sqrt{2}$	minimum

26. $f(x) = -x^4 + x^2 + 2, \quad [0, 2]$

$f'(x) = -4x^3 + 2x = 2x(1 - 2x^2)$

Critical numbers: $x = 0, \dfrac{\sqrt{2}}{2}$

x	$f(x)$	
0	2	minimum
2	-10	
$\frac{\sqrt{2}}{2}$	$\frac{9}{4}$	maximum

28. $f(x) = 4\sqrt{x} - x^2, \quad [0, 3]$

$$f'(x) = \frac{2}{\sqrt{x}} - 2x = \frac{2(1 - x^{3/2})}{\sqrt{x}}$$

Critical number: $x = 1$

x	$f(x)$	
0	0	
3	-2.072	minimum
1	3	maximum

30. $O = \dfrac{t^2 - t + 1}{t^2 + 1}, \quad 0 \le t$

The graph of O is:
(a) Lowest level is $\frac{1}{2}$ at $t = 1$.
(b) Highest level is 1 at $t = 0$.
(c) As $t \to \infty$, $O \to 1$.

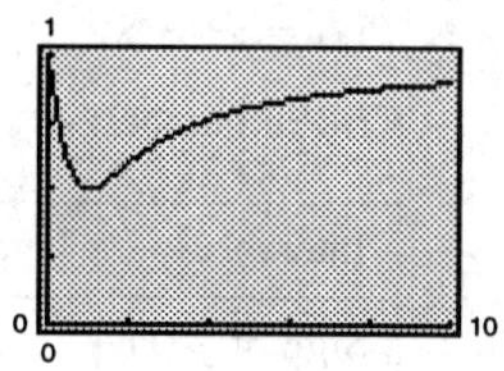

32. $h(x) = x^5 - 10x^2$

$h'(x) = 5x^4 - 20x$

$h''(x) = 20x^3 - 20 = 20(x - 1)(x^2 + x + 1)$

$f''(x) > 0$ for $x > 1$: concave upward on $(1, \infty)$

$f''(x) < 0$ for $x < 1$: concave downward on $(-\infty, 1)$

34. $h(x) = x^3 - 6x$

$h'(x) = 3x^2 - 6$

$h''(x) = 6x$

$h''(x) > 0$ for $x > 0$: concave upward on $(0, \infty)$

$h''(x) < 0$ for $x < 0$: concave downward on $(-\infty, 0)$

36. $f(x) = (x + 2)^2(x - 4) = x^3 - 12x - 16$

$f'(x) = 3x^2 - 12$

$f''(x) = 6x = 0$ when $x = 0$.

Since f changes concavity at $x = 0$, $(0, -16)$ is a point of inflection.

38. $f(x) = \dfrac{1}{4}x^4 - 2x^2 - x$

$f'(x) = x^3 - 4x - 1$

$f''(x) = 3x^2 - 4$

$f''(x) = 0$ when $x = \pm\dfrac{2}{\sqrt{3}}$.

Since f changes concavity at $x = \pm 2/\sqrt{3}$, the points of inflection are $\left(-2/\sqrt{3}, (-20/9) + \left(2/\sqrt{3}\right)\right)$ and $\left(2/\sqrt{3}, (-20/9) - \left(2/\sqrt{3}\right)\right)$.

40. $f(x) = x(x^2 - 3x - 9) = x^3 - 3x^2 - 9x$

$f'(x) = 3x^2 - 6x - 9 = 3(x^2 - 2x - 3) = 3(x - 3)(x + 1)$

$f''(x) = 6x - 6$

Critical numbers: $x = 3, -1$
$f''(3) > 0 \Rightarrow (3, -27)$ is a relative minimum.
$f''(-1) < 0 \Rightarrow (-1, 5)$ is a relative maximum.

42. $f(x) = \dfrac{(x - 2)^2}{4 - x}$

$f'(x) = \dfrac{(2 - x)(x - 6)}{(4 - x)^2}$

$f''(x) = 8/(4 - x)^3$

Critical numbers: $x = 2, 6$
$f''(2) > 0 \Rightarrow (2, 0)$ is a relative minimum.
$f''(6) < 0 \Rightarrow (6, -8)$ is a relative maximum.

44. $R = -\frac{2}{3}(x^3 - 12x^2 - 6), \quad 0 \le x \le 8$

$R' = -\frac{2}{3}(3x^2 - 24x)$

$R'' = -\frac{2}{3}(6x - 24)$

$R'' = 0$ when $x = 6$.

The point of diminshing return is $(6, 148)$.

46. Let x and y be the lengths shown in the figure. By similar triangles we have

$$\frac{5}{x} = \frac{y}{x + 4}$$

Thus, $y = 5 + (20/x)$. To minimize the length of the hypotenuse, we have the following.

$$z = \sqrt{(x + 4)^2 + y^2} = \sqrt{(x + 4)^2 + \left(5 + \frac{20}{x}\right)^2} = \sqrt{x^2 + 8x + 41 + \frac{200}{x} + \frac{400}{x^2}}$$

$$\frac{dz}{dx} = \frac{1}{2}\left[(x + 4)^2 + \left(5 + \frac{20}{x}\right)^2\right]^{-1/2}\left(2x + 8 - \frac{200}{x^2} - \frac{800}{x^3}\right)$$

$dz/dx = 0$ when

$$x + 4 - \frac{100}{x^2} - \frac{400}{x^3} = 0$$

$$x^4 + 4x^3 - 100x - 400 = 0$$

$$(x + 4)(x^3 - 100) = 0.$$

When $x = \sqrt[3]{100}$, $z \approx 12.7$ feet.

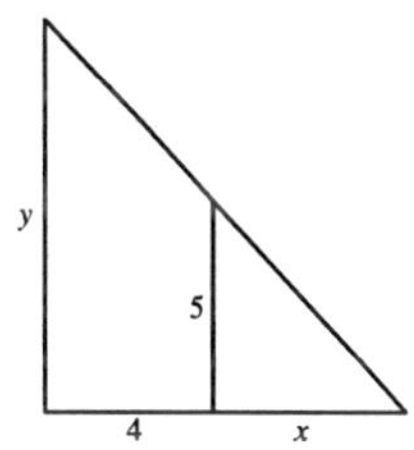

48. (a) $xy = 4800$

$$\text{Cost} = C = 3(x + 2y) + 4(x) = 7x + 6y = 7x + 6\left(\frac{4800}{x}\right)$$

$$C' = 7 - \frac{28{,}800}{x^2} = 0 \Rightarrow x \approx 64.14 \text{ feet}$$

$$y \approx 74.83 \text{ feet}$$

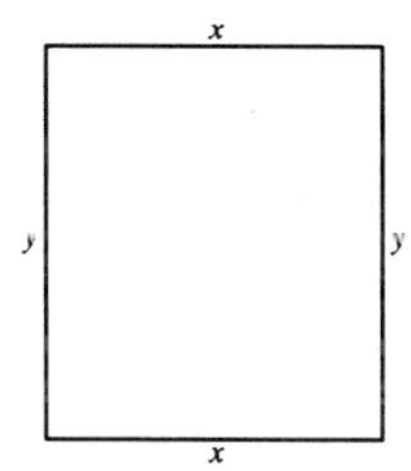

(b) If costs increase by \$1, then

$$C = 4(x + 2y) + 5(x) = 9x + 8y = 9x + 8\left(\frac{4800}{x}\right)$$

$$C' = 9 - \frac{38{,}400}{x^2} = 0 \Rightarrow x \approx 65.32 \text{ feet}$$

$$y \approx 73.48 \text{ feet.}$$

50. $N = -0.0135t^3 + 0.457t^2 - 4.98t + 24.5, \quad 7 \le t \le 14$

(a) N increasing on $(9.2, 14]$
N decreasing on $[7, 9.2)$

(b) Absolute minimum at $(9.2, 6.9)$
Absolute maximum at $(7, 7.4)$

(c) TV usage was a minimum in 1989 ($t = 9.2$), then rose afterwards.

52. (a) $P = R - C = xp - C = x(36 - 4x) - (2x^2 + 6)$

$= -6x^2 + 36x - 6$

$P'(x) = -12x + 36 = 0 \Rightarrow x = 3$ for maximum profit.

(b) $\overline{C} = \dfrac{C}{x} = \dfrac{2x^2 + 6}{x} = 2x + \dfrac{6}{x}$

$\overline{C}' = 2 - \dfrac{6}{x^2} = 0 \Rightarrow x^2 = 3 \Rightarrow x = \sqrt{3} \approx 1.73$ units

54. $F = ks^{3/2}$

When $s = 25$, $F = 50$ we have

$$50 = k(25)^{3/2} \Rightarrow k = \frac{2}{5}$$

$$F = \frac{2}{5}s^{3/2}.$$

The total cost per mile is

$$C = \frac{F}{s} + \frac{100}{s} = \frac{2}{5}s^{1/2} + \frac{100}{s}$$

$$\frac{dC}{ds} = \frac{1}{5}s^{-1/2} - \frac{100}{s^2}.$$

$\dfrac{dC}{ds} = 0$ when $s = 500^{2/3} \approx 63$ miles per hour.

56. $P = xp - C - xt$

$t = \$5$:

$P = x(600 - 3x) - (0.3x^2 + 6x + 600) - 5x$

$= -3.3x^2 + 589x - 600$

$\dfrac{dP}{dx} = -6.6x + 589$

$\dfrac{dP}{dx} = 0$ when $x = \dfrac{589}{6.6}$.

$P\left(\dfrac{589}{6.6}\right) \approx \$25{,}681.89$

$t = \$10$:

$P = x(600 - 3x) - (0.3x^2 + 6x + 600) - 10x$

$= -3.3x^2 + 584x - 600$

$\dfrac{dP}{dx} = -6.6x + 584$

$\dfrac{dP}{dx} = 0$ when $x = \dfrac{584}{6.6}$.

$P\left(\dfrac{584}{6.6}\right) \approx \$25{,}237.58$

$t = \$20$:

$P = x(600 - 3x) - (0.3x^2 + 6x + 600) - 20x$

$= -3.3x^2 + 574x - 600$

$\dfrac{dP}{dx} = -6.6x + 574$

$\dfrac{dP}{dx} = 0$ when $x = \dfrac{574}{6.6}$.

$P\left(\dfrac{574}{6.6}\right) \approx \$24{,}360.30$

58. $p = \sqrt{1800 - x}, \quad 0 \le x \le 1800$

$$\frac{dp}{dx} = \frac{-1}{2\sqrt{1800 - x}}$$

$$\eta = \frac{p/x}{dp/dx} = \frac{\sqrt{1800 - x}/x}{-1/2\sqrt{1800 - x}} = -2(1800 - x)/x$$

Elastic: $(0, 1200)$

Inelastic: $(1200, 1800)$

Demand is of unit elasticity when $x = 1200$.

60. $$g(x) = \frac{5x^2}{x^2 + 2}$$

$$\lim_{x\to\infty} \frac{5x^2}{x^2 + 2} = 5$$

Horizontal asymptote: $y = 5$
No vertical asymptotes

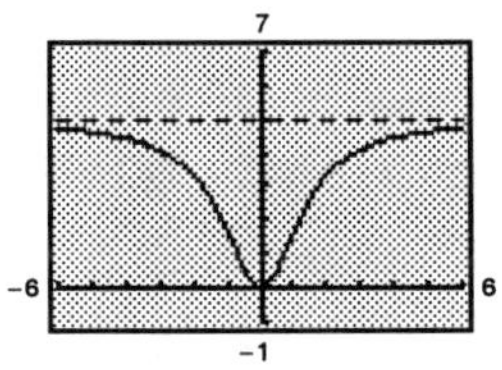

62. $$f(x) = \frac{3x}{\sqrt{x^2 + 2}}$$

$$\lim_{x\to\infty} \frac{3x}{\sqrt{x^2 + 2}} = 3$$

$$\lim_{x\to-\infty} \frac{3x}{\sqrt{x^2 + 2}} = -3$$

Horizontal asymptotes: $y = \pm 3$
No vertical asymptotes

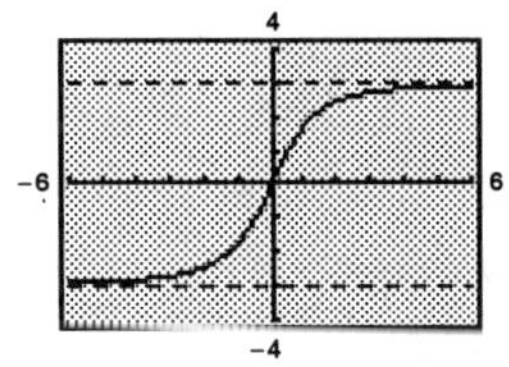

64. $h(x) = \dfrac{2x^2 + 3x - 5}{x - 1} = \dfrac{(x - 1)(2x + 5)}{x - 1} = 2x + 5, \quad x \ne 1$

There are no vertical asymptotes. There is a hole in the graph at $(1, 7)$.
There are no horizontal asymptotes since

$$\lim_{x\to\infty} h(x) = \infty \text{ and } \lim_{x\to-\infty} h(x) = -\infty.$$

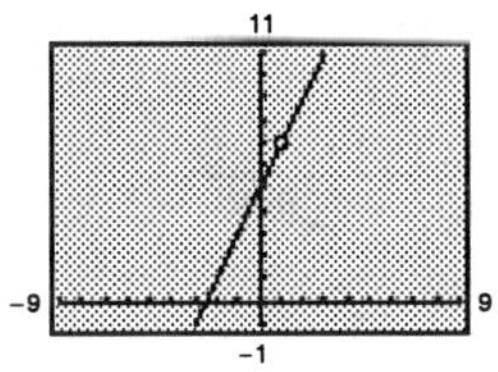

66. $$\lim_{x\to-1^+} \frac{x^2 - 2x + 1}{x + 1} = \infty$$

68. $$\lim_{x\to\infty} \frac{3x^2 - 2x + 3}{x + 1} = \lim_{x\to\infty} \frac{3x - 2 + (3/x)}{1 + (1/x)} = \infty$$

70. $$\lim_{x\to-\infty} \left(\frac{x}{x - 2} + \frac{2x}{x + 2}\right) = 1 + 2 = 3$$

72. $C = 10{,}000 + 48.9x, \quad R = 68.5x$

(a) $\overline{C} = \dfrac{C}{x} = \dfrac{10{,}000}{x} + 48.9$

(b) $\lim_{x\to\infty} \overline{C} = 48.9$

(c) $\overline{P} = \dfrac{P}{x} = \dfrac{R - C}{x} = 19.6 - \dfrac{10{,}000}{x}$

$\overline{P}(1{,}000{,}000) = 19.59$

$\overline{P}(2{,}000{,}000) = 19.595$

$\overline{P}(10{,}000{,}000) = 19.599$

(d) $\lim_{x\to\infty} \overline{P} = 19.6$

74. $f(x) = 4x^3 - x^4$

$f'(x) = 4x^2(3 - x)$

$f''(x) = 12x(2 - x)$

Domain: all real numbers
Range: all real numbers
Intercepts: $(0, 0)$, $(4, 0)$
Relative maximum: $(3, 27)$
Points of inflection: $(0, 0)$, $(2, 16)$

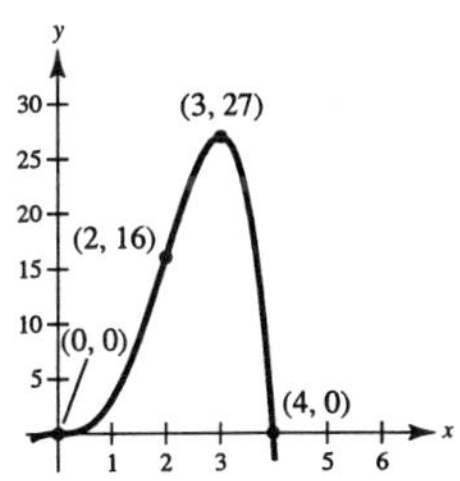

76. $f(x) = x + \dfrac{4}{x^2}$

$f'(x) = 1 - \dfrac{8}{x^3}$

$f''(x) = \dfrac{24}{x^4}$

Domain: all real numbers except 0
Range: all real numbers
Intercept: $\left(-\sqrt[3]{4}, 0\right)$
Vertical asymptote: $x = 0$
Relative minimum: $(2, 3)$
Concave upward on $(-\infty, 0), (0, \infty)$

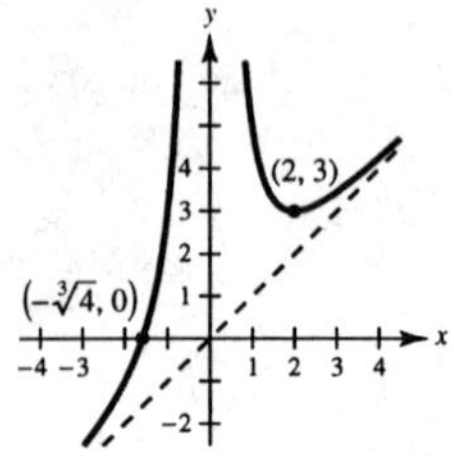

78. $f(x) = x^2 + \dfrac{2}{x}$

$f'(x) = 2x - \dfrac{2}{x^2} = \dfrac{2(x^3 - 1)}{x^2}$

$f''(x) = 2 + \dfrac{4}{x^3} = \dfrac{2(x^3 + 2)}{x^3}$

Domain: all real numbers except 0
Range: all real numbers
Intercept: $\left(-\sqrt[3]{2}, 0\right)$
Vertical asymptote: $x = 0$
Relative minimum: $(1, 3)$
Point of inflection: $\left(-\sqrt[3]{2}, 0\right)$

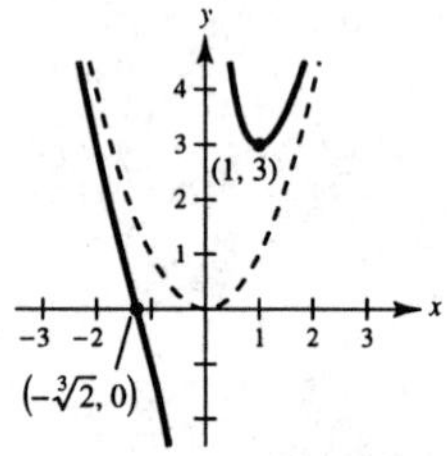

80. $f(x) = x^{4/5}$

$f'(x) = \dfrac{4}{5}x^{-1/5} = \dfrac{4}{5\sqrt[5]{x}}$

$f''(x) = -\dfrac{4}{25}x^{-6/5} = -\dfrac{4}{25x^{6/5}}$

Domain: $(-\infty, \infty)$
Range: $[0, \infty)$
Intercept: $(0, 0)$
Relative minimum: $(0, 0)$
Concave downward on $(-\infty, 0)$ and $(0, \infty)$

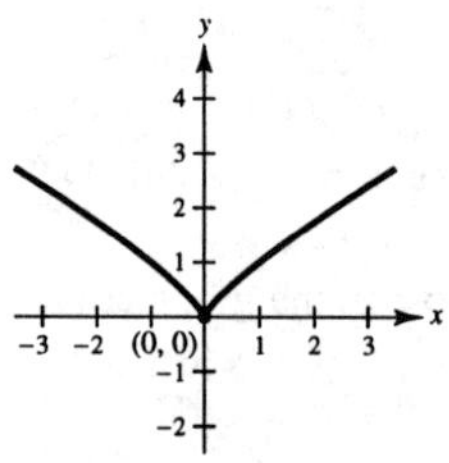

82. $y = (3x^2 - 2)^3$

$dy = 3(3x^2 - 2)^2(6x)dx$

$= 18x(3x^2 - 2)^2 dx$

84. $y = \dfrac{2 - x}{x + 5}$

$dy = \dfrac{(x + 5)(-1) - (2 - x)(1)}{(x + 5)^2}dx = \dfrac{-7}{(x + 5)^2}dx$

86. $C = 1.5\sqrt[3]{x} + 500, \quad x = 125$

$dC = 0.5x^{-2/3}dx = 0.02$

88. $P = 0.003x^2 + 0.019x - 1200, \quad x = 750$

$dP = (0.006x + 0.019)dx = 4.519$

90. $E = 22.5t + 7.5t^2 - 2.5t^3, \quad 0 \le t \le 4.5$

(a)

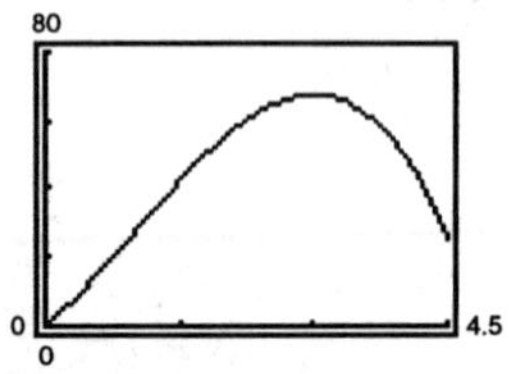

(b) $E'(t) = 0$ for $t = 3$

$E(3) = 67.5$

92. $p = 85 - 0.125x$

$\Delta p = [85 - 0.125(8)] - [85 - 0.125(7)] = -0.125$

$dp = -0.125dx = -0.125 \quad$ (since $dx = \Delta x = 8 - 7 = 1$)

CHAPTER 4
Exponential and Logarithmic Functions

CHAPTER 4
Exponential and Logarithmic Functions

Section 4.1 Exponential Functions

Solutions to Exercises

2. (a) $\left(\frac{1}{5}\right)^3 = \frac{1}{125}$

(b) $\left(\frac{1}{8}\right)^{1/3} = \sqrt[3]{\frac{1}{8}} = \frac{1}{2}$

(c) $64^{2/3} = \left(\sqrt[3]{64}\right)^2 = 4^2 = 16$

(d) $\left(\frac{5}{8}\right)^2 = \frac{25}{64}$

(e) $100^{3/2} = \left(\sqrt{100}\right)^3 = 10^3 = 1000$

(f) $4^{5/2} = \left(\sqrt{4}\right)^5 = 2^5 = 32$

4. (a) $\frac{5^3}{5^6} = \frac{1}{5^{6-3}} = \frac{1}{5^3} = \frac{1}{125}$

(b) $\left(\frac{1}{5}\right)^{-2} = \left(\frac{5}{1}\right)^2 = 25$

(c) $(8^{1/2})(2^{1/2}) = (8 \cdot 2)^{1/2} = 16^{1/2} = \sqrt{16} = 4$

(d) $(32^{3/2})\left(\frac{1}{2}\right)^{3/2} = \left(32 \cdot \frac{1}{2}\right)^{3/2}$

$= 16^{3/2} = \left(\sqrt{16}\right)^3 = 4^3 = 64$

6. (a) $(4^3)(4^2) = (64)(16) = 1024$

(b) $\left(\frac{1}{4}\right)^2(4^2) = \left(\frac{1}{4} \cdot 4\right)^2 = (1)^2 = 1$

(c) $(4^6)^{1/2} = 4^3 = 64$

(d) $[(8^{-1})(8^{2/3})]^3 = (8^{-1/3})^3 = 8^{-1} = \frac{1}{8}$

8. (a) $\left(\frac{1}{e}\right)^{-2} = \left(\frac{e}{1}\right)^2 = e^2$

(b) $\left(\frac{e^5}{e^2}\right)^{-1} = (e^3)^{-1} = e^{-3} = \frac{1}{e^3}$

(c) $\frac{e^5}{e^3} = e^2$

(d) $\frac{1}{e^{-3}} = e^3$

10. $5^{x+1} = 125$

$5^{x+1} = 5^3$

$x + 1 = 3$

$x = 2$

12. $\left(\frac{1}{5}\right)^{2x} = 625$

$5^{-2x} = 5^4$

$-2x = 4$

$x = -2$

14. $4^2 = (x + 2)^2$

$\pm 4 = x + 2$

$x = 2, -6$

16. $(x + 3)^{4/3} = 16$

$x + 3 = 16^{3/4}$

$x + 3 = 8$

$x = 5$

18. $e^x = 1$

$e^x = e^0$

$x = 0$

20. $e^{-1/x} = \sqrt{e} = e^{1/2}$

$\frac{-1}{x} = \frac{1}{2}$

$x = -2$

22. $\frac{x^2}{2} = e^2$

$x^2 = 2e^2$

$x = \pm\sqrt{2}e$

24. The graph of $f(x) = 3^{-x/2} = (1/3)^{x/2}$ is an exponential curve with the following characteristics.
Passes through $(0, 1)$, $\left(1, 1/\sqrt{3}\right)$, $(2, 1/3)$
Horizontal asymptote: $y = 0$
Therefore, it matches graph (c).

26. The graph of $f(x) = 3^{x-2}$ is an exponential curve with the following characteristics.
Passes through $(0, 1/9)$, $(2, 1)$, $(3, 3)$
Horizontal asymptote: $y = 0$
Therefore, it matches graph (f).

28. The graph of $f(x) = 3^x + 2$ is an exponential curve with the following characteristics.
Passes through $(0, 3)$, $(1, 5)$, $(-1, 7/3)$
Horizontal asymptote: $y = 2$
Therefore, it matches graph (b).

30. $f(x) = 4^x$

x	-2	-1	0	1	2
$f(x)$	$\frac{1}{16}$	$\frac{1}{4}$	1	4	16

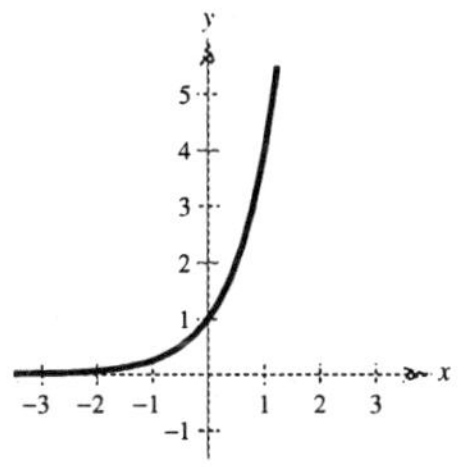

32. $f(x) = \left(\frac{1}{4}\right)^x$

x	-2	-1	0	1	2
$f(x)$	16	4	1	$\frac{1}{4}$	$\frac{1}{16}$

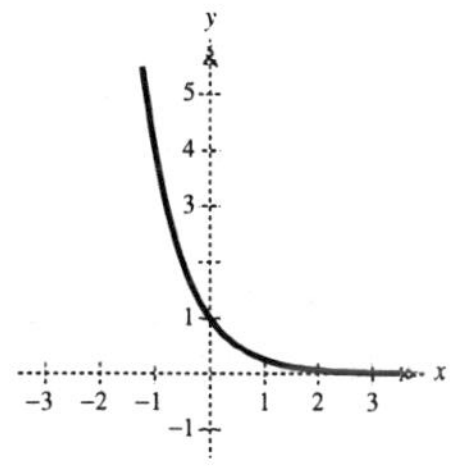

34. $y = 2^{-x^2}$

x	-2	-1	0	1	2
y	$\frac{1}{16}$	$\frac{1}{2}$	1	$\frac{1}{2}$	$\frac{1}{16}$

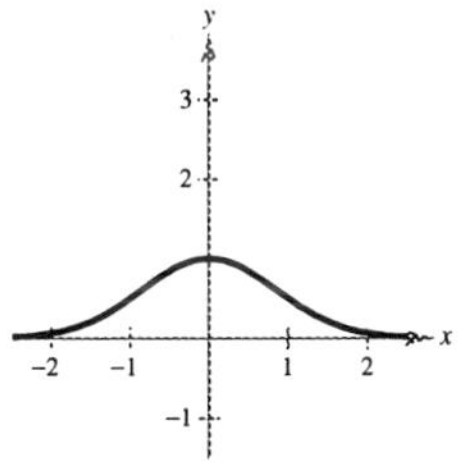

36. $y = 3^{|x|}$

x	-2	-1	0	1	2
y	9	3	1	3	9

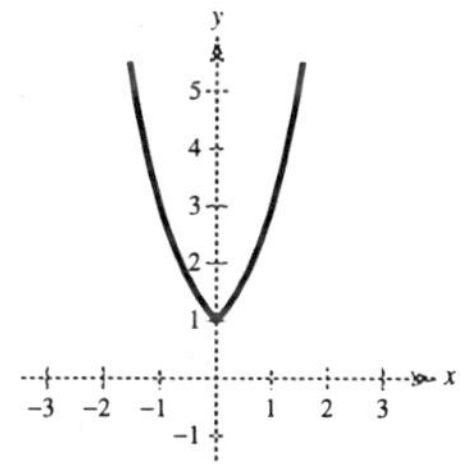

38. $s(t) = 2^{-t} + 3 = \left(\frac{1}{2}\right)^t + 3$

t	-2	-1	0	1	2
$s(t)$	7	5	4	$\frac{7}{2}$	$\frac{13}{4}$

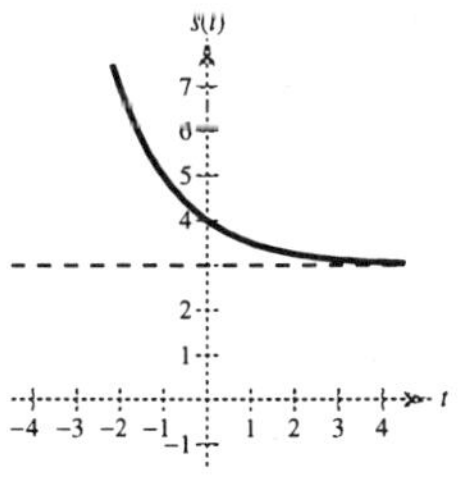

40. $f(x) = e^{2x}$

x	-1	0	$\frac{1}{2}$	1	2
$f(x)$	0.135	1	2.718	7.389	54.598

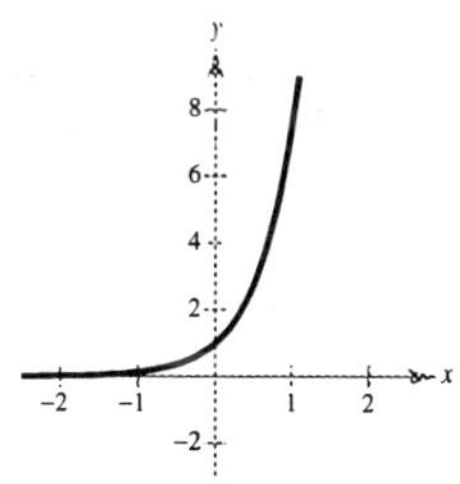

42. $A(t) = 500e^{0.15t}$

t	0	10	20	30
$A(t)$	500	2240.84	10,042.77	45,008.56

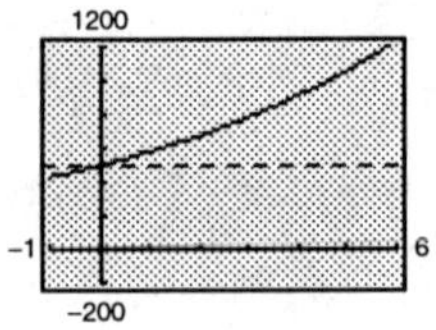

44. $g(x) = \dfrac{10}{1 + e^{-x}}$

x	−2	−1	0	1	2
$g(x)$	1.192	2.689	5	7.311	8.808

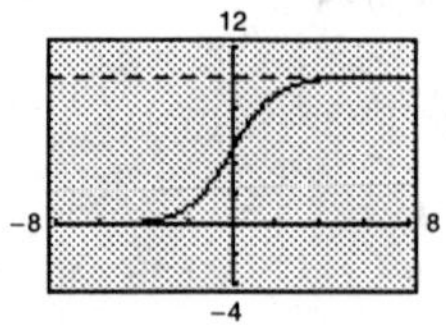

46. $A = P\left(1 + \dfrac{r}{n}\right)^{nt}, \quad r = 0.05, t = 20, P = 2500$

$= 2500\left(1 + \dfrac{0.05}{n}\right)^{20n}$

Continuous compounding: $A = Pe^{rt} = 2500e^{(0.05)20}$

n	1	2	4	12	365	continuous
A	6633.24	6712.66	6753.71	6781.60	6795.24	6795.70

48. $A = P\left(1 + \dfrac{r}{n}\right)^{nt}, \quad P = 2500, r = 0.05, t = 40$

$= 2500\left(1 + \dfrac{0.05}{n}\right)^{40n}$

Continuous compounding: $A = Pe^{rt} = 2500e^{(0.05)(40)}$

n	1	2	4	12	365	continuous
A	17,599.97	18,023.92	18,245.05	18,396.04	18,470.11	18,472.64

50. $A = Pe^{rt}, A = 100{,}000, r = 0.03 \Longrightarrow P = 100{,}000e^{-0.03t}$

t	1	10	20	30	40	50
P	97,044.55	74,081.82	54,881.16	40,656.97	30,119.42	22,313.02

52. $A = P\left(1 + \dfrac{r}{n}\right)^{nt}, A = 100{,}000, r = 0.06, n = 365 \Longrightarrow P = \dfrac{100{,}000}{\left(1 + \frac{0.06}{365}\right)^{365t}}$

t	1	10	20	30	40	50
P	94,176.92	54,883.87	30,112.39	16,532.33	9073.58	4979.93

54. (a) $p = 500 - 0.5e^{0.004(1000)} \approx \$472.70$

(b) $p = 500 - 0.5e^{0.004(1500)} \approx \$298.29$

56.

s	50	55	60	65	70
y	28.0	26.4	24.8	23.4	22.0

You can conclude that the miles per gallon decreases as speed increases.

58.

Interval	1	2	3	4	5	6	7	8	9	10
Number of cells	1	2	4	8	16	32	64	128	256	512

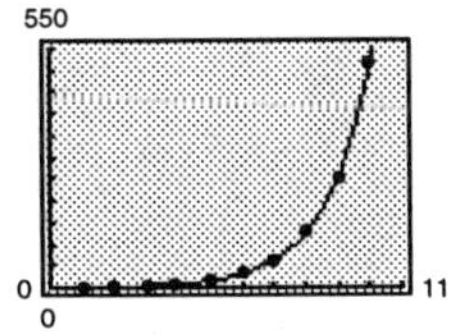

Model: $y = 2^{t-1}$ or $y = e^{(t-1)\ln 2}$

60. $N = \dfrac{95}{1 + 8.5e^{-0.12t}}$

(a) $t = 10 \Longrightarrow N = 26.68$ words/min

(b) $N = 70 \Longrightarrow t \approx 26.4$ weeks

(c) Yes, there is a limit as t increases without bound.

$$\lim_{t\to\infty} \frac{95}{1 + 8.5e^{-0.12t}} = \frac{95}{1+0} = 95 \text{ words/min}$$

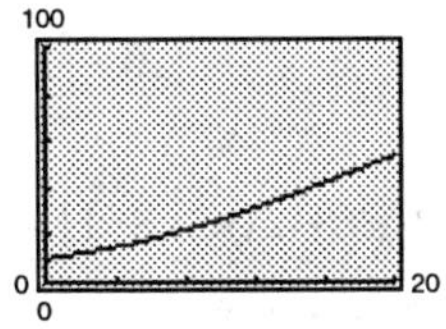

62.

x	-2	-1	0	1	2
y	-3.627	-1.175	0	1.175	3.627

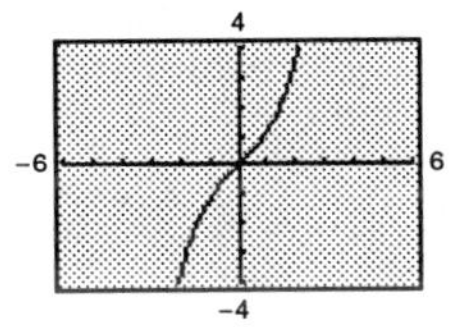

$$\lim_{x\to\infty} \frac{e^x - e^{-x}}{2} = \frac{\infty - 0}{2} = \infty$$

$$\lim_{x\to-\infty} \frac{e^x - e^{-x}}{2} = \frac{0 - \infty}{2} = -\infty$$

No horizontal asymptotes
Continuous on the entire real line

64.

x	-2	-1	0	1	2
y	0.502	0.581	0.667	0.758	0.854

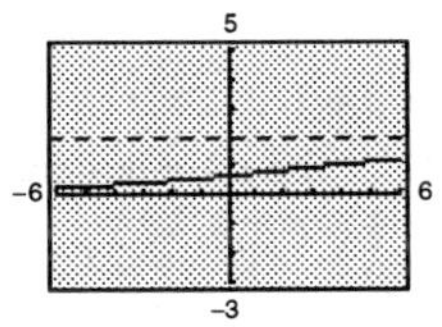

$$\lim_{x\to\infty} \frac{2}{1 + 2e^{-0.2x}} = \frac{2}{1+0} = 2$$

$$\lim_{x\to-\infty} \frac{2}{1 + 2e^{-0.2x}} = \frac{2}{\infty} = 0$$

Horizontal asymptotes: $y = 0$ and $y = 2$
Continuous on the entire real line

Section 4.2 Derivatives of Exponential Functions

2. $y' = 2e^{2x}$

$y'(0) = 2$

4. $y' = -2e^{-2x}$

$y'(0) = -2$

6. $y' = -e^{1-x}$

8. $f(x) = e^{1/x} = e^{x^{-1}}$

$$f'(x) = e^{1/x}\left(-\frac{1}{x^2}\right) = \frac{-e^{1/x}}{x^2}$$

10. $g(x) = e^{\sqrt{x}} = e^{x^{1/2}}$

$$g'(x) = e^{\sqrt{x}}\left(\frac{1}{2}x^{-1/2}\right) = \frac{e^{\sqrt{x}}}{2\sqrt{x}}$$

12. $y' = 2(10e^{-x})(-10e^{-x}) = -200e^{-2x}$

14. $f'(x) = \frac{1}{2}(e^x - e^{-x})$

16. $y' = x^2e^x + 2xe^x - 2xe^x - 2e^x + 2e^x = x^2e^x$

18. $g'(x) = e^{x^3}(3x^2)$

$g'(-1) = 3e^{-1}$

$y - \frac{1}{e} = \frac{3}{e}(x + 1)$

$y = \frac{3}{e}x + \frac{3}{e} + \frac{1}{e} = \frac{3}{e}x + \frac{4}{e}$

20. $y' = 3(e^{-x} + e^x)^2(-e^{-x} + e^x)$

$y'(0) = 3(4)(0) = 0$

$y - 8 = 0(x - 0)$

$y = 8$

22. $e^{xy} + x^2 - y^2 = 10$

$\left(y + x\frac{dy}{dx}\right)e^{xy} + 2x - 2y\frac{dy}{dx} = 0$

$\frac{dy}{dx}(xe^{xy} - 2y) = -ye^{xy} - 2x$

$$\frac{dy}{dx} = \frac{-ye^{xy} - 2x}{xe^{xy} - 2y} = \frac{-y(10 - x^2 + y^2) - 2x}{x(10 - x^2 + y^2) - 2y} = \frac{x^2y - y^3 - 2x - 10y}{xy^2 - x^3 + 10x - 2y}$$

24. $f'(x) = (1 + 2x)(4e^{4x}) + 2e^{4x} = 2e^{4x}[(1 + 2x)(2) + 1] = 2e^{4x}(4x + 3)$

$f''(x) = 2e^{4x}(4) + 8e^{4x}(4x + 3) = 8e^{4x}[1 + (4x + 3)] = 32e^{4x}(x + 1)$

26. $f'(x) = (3 + 2x)(-3e^{-3x}) + 2e^{-3x} = e^{-3x}(-9 - 6x + 2) = -e^{-3x}(6x + 7)$

$f''(x) = -e^{-3x}(6) + (6x + 7)(3e^{-3x}) = 3e^{-3x}(-2 + 6x + 7) = 3e^{-3x}(6x + 5)$

28. $f(x) = \frac{1}{2}(e^x - e^{-x})$

$f'(x) = \frac{1}{2}(e^x + e^{-x}) > 0$ for all x

$f'(x) \neq 0$ for any values of x. Thus, there are no relative extrema.

$f''(x) = \frac{1}{2}(e^x - e^{-x})$

$f''(x) = 0$ when $x = 0$.

We have a point of inflection at $(0, 0)$.

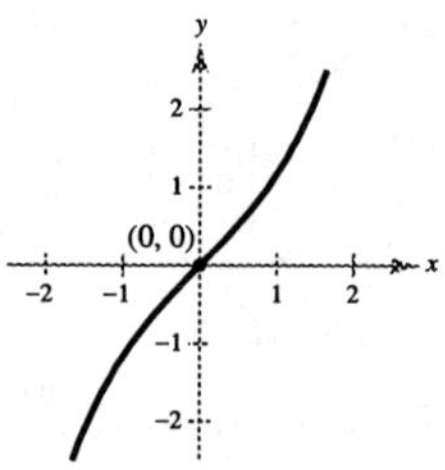

x	-2	-1	0	1	2
$f(x)$	-3.627	-1.175	0	1.175	3.627

30. $f(x) = xe^{-x}$

$f'(x) = -xe^{-x} + e^{-x} = e^{-x}(1 - x)$

$f'(x) = 0$ when $x = 1$.

$f''(x) = e^{-x}(-1) + (1 - x)(-e^{-x}) = -e^{-x}[1 + (1 - x)] = e^{-x}(x - 2)$

Since $f''(1) < 0$, we have a relative maximum at $(1, e^{-1})$.
$f''(x) = 0$ when $x = 2$ and we have a point of inflection at $(2, 2e^{-2})$.

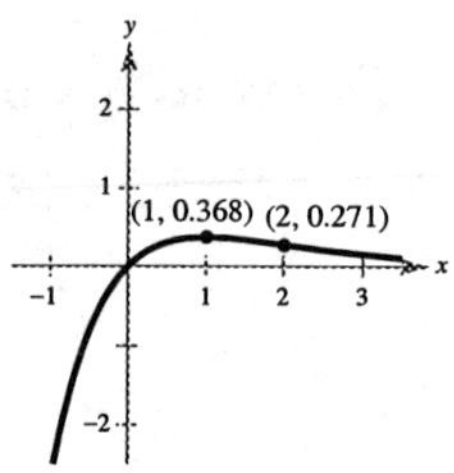

x	-1	0	1	2	3
$f(x)$	-2.718	0	0.368	0.271	0.149

Horizontal asymptote: $y = 0$

32. $g(x) = \dfrac{8}{1 + e^{-0.5/x}}$

Horizontal asymptote: $y = 4$
Vertical asymptote: $x = 0$ (from the left)

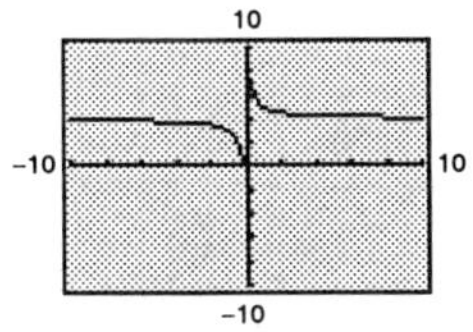

34. (a)

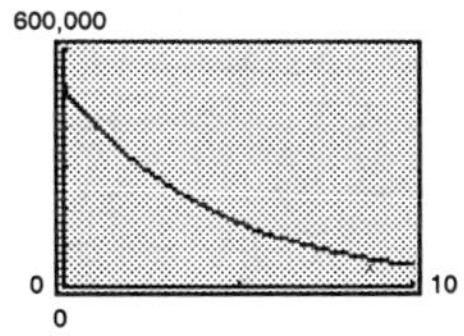

(b) $\dfrac{dV}{dt} = 500{,}000(-0.2231e^{-0.2231t}) = -111{,}550e^{-0.2231t}$

When $t = 1$, $dV/dt \approx -\$89{,}243.89$ per year.

(c) When $t = 5$, $dV/dt \approx -\$36{,}560.66$ per year.

(d) In this model, the initial rate of depreciation is greater than in a linear model.

36. $N = \dfrac{95}{1 + 8.5e^{-0.12t}}, \quad N' = \dfrac{96.9e^{-0.12t}}{(1 + 8.5e^{-0.12t})^2}$

(a) When $t = 5$, $N' = 1.66$ words/min/week.

(b) When $t = 10$, $N' = 2.30$ words/min/week.

(c) When $t = 30$, $N' = 1.74$ words/min/week.

38. $a = 20, \quad b = 0.5$

$p = 80e^{-0.5t} + 20$

$\dfrac{dp}{dt} = -40e^{-0.5t}$

When $t = 1$, $dp/dt \approx -24.3\%$.

When $t = 3$, $dp/dt \approx -8.9\%$.

40. $y = 8622.11 + 153.97t - 0.0001e^t$

(a)

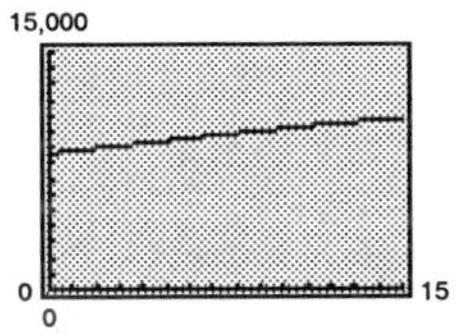

(b) For 1980, $t = 0$ and $y' \approx 154$ thousand/year
For 1990, $t = 10$ and $y' \approx 152$ thousand/year
For 1994, $t = 14$ and $y' \approx 34$ thousand/year

(c) $y' = 153.97 - 0.0001e^t$
$y'(0) \approx 153.97$ thousand/year
$y'(10) \approx 151.77$ thousand/year
$y'(14) \approx 33.71$ thousand/year

42. $f(x) = \dfrac{1}{\sigma\sqrt{2\pi}}e^{-(x-\mu)^2/2\sigma^2} = \dfrac{1}{\sqrt{2\pi}}e^{(x-\mu)^2/2}$

μ shifts the graph horizontally.

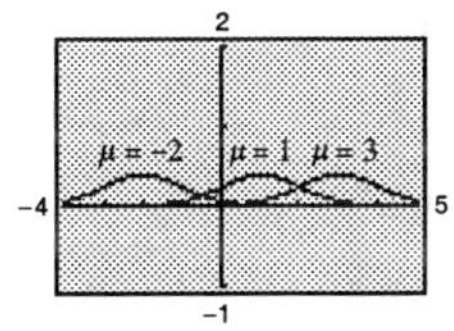

Section 4.3 Logarithmic Functions

2. $e^{2.128} \approx 8.4$

4. $e^{-2.2882\ldots} = 0.056$

6. $\ln 7.389\ldots = 2$

8. $\ln 1.284\ldots = 0.25$

10. The graph is a logarithmic curve that passes through the point (1, 0) with a vertical asymptote at $x = 0$. Therefore, it matches graph (d).

12. The graph is a logarithmic curve that passes through the point (2, 0) with a vertical asymptote at $x = 1$. Therefore, it matches graph (a).

14.

x	± 0.5	± 1	± 2
y	-0.69	0	0.69

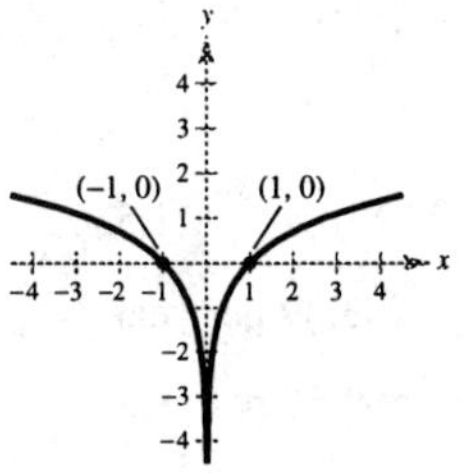

16.

x	0.5	1	2	3	4
y	4.31	5	5.69	6.10	6.39

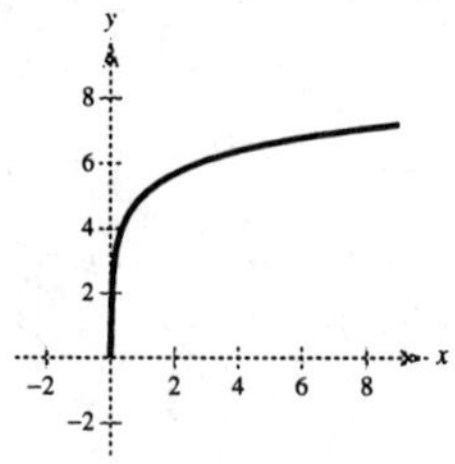

18.

x	0.5	1	2	3	4
y	-0.17	0	0.17	0.27	0.35

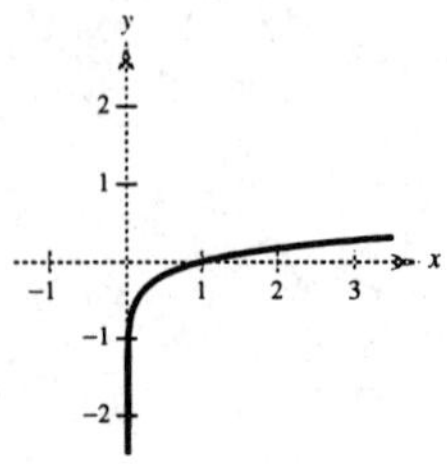

20. $f(g(x)) = f(\ln(x+1))$

$= e^{\ln(x+1)} - 1 = (x+1) - 1 = x$

$g(f(x)) = g(e^x - 1)$

$= \ln((e^x - 1) + 1) = \ln e^x = x$

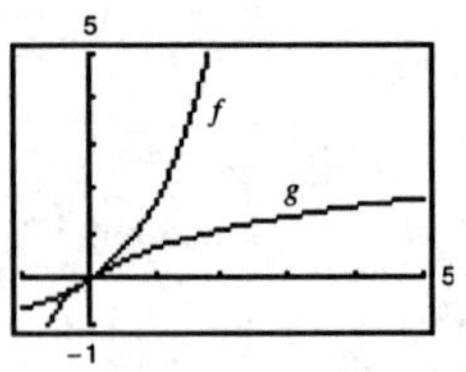

22. $f(g(x)) = f(\ln x^3)$

$= f(3 \ln x) = e^{3 \ln x/3} = e^{\ln x} = x$

$g(f(x)) = g(e^{x/3})$

$= \ln(e^{x/3})^3 = \ln e^x = x$

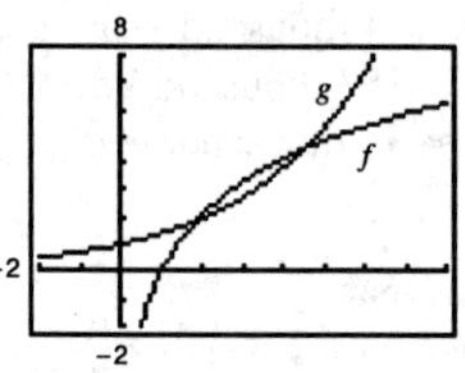

24. $\ln e^{2x-1} = 2x - 1$

26. $-1 + \ln e^{2x} = -1 + 2x$

$= 2x - 1$

28. $-8 + e^{\ln x^3} = -8 + x^3$

$= x^3 - 8$

30. (a) $\ln 0.25 = \ln \frac{1}{4} = \ln 1 - \ln 4 = \ln 1 - \ln 2^2 = \ln 1 - 2 \ln 2 = 0 - 2(0.6931) = -1.3862$

(b) $\ln 24 = \ln(3 \cdot 2^3) = \ln 3 + 3 \ln 2 = 1.0986 + 3(0.6931) = 3.1779$

(c) $\ln \sqrt[3]{12} = \frac{1}{3} \ln(3 \cdot 2^2) = \frac{1}{3}[\ln 3 + 2 \ln 2] = \frac{1}{3}[1.0986 + 2(0.6931)] \approx 0.8283$

(d) $\ln \frac{1}{72} = \ln 1 - \ln 72 = 0 - \ln(2^3 \cdot 3^2) = -[3 \ln 2 + 2 \ln 3] = -[3(0.6931) + 2(1.0986)] = -4.2765$

32. $\ln \frac{1}{5} = \ln 1 - \ln 5 = 0 - \ln 5 = -\ln 5$

34. $\ln \frac{xy}{z} = \ln xy - \ln z = \ln x + \ln y - \ln z$

36. $\ln\sqrt{2^3} = \ln 2^{3/2} = \dfrac{3}{2}\ln 2$

38. $\ln\dfrac{1}{e} = \ln 1 - \ln e = 0 - 1 = -1$

40. $\ln z(z-1)^2 = \ln z + \ln(z-1)^2 = \ln z + 2\ln(z-1)$

42. $\ln(2x+1) + \ln(2x-1) = \ln(2x+1)(2x-1)$

$= \ln(4x^2-1)$

44. $\dfrac{1}{3}[2\ln(x+3) + \ln x - \ln(x^2-1)] = \dfrac{1}{3}\ln\dfrac{(x+3)^2x}{(x^2-1)} = \ln\sqrt[3]{\dfrac{x(x+3)^2}{x^2-1}}$

46. $2\ln 3 - \dfrac{1}{2}\ln(x^2+1) = \ln 9 - \ln\sqrt{x^2+1}$

$= \ln\dfrac{9}{\sqrt{x^2+1}}$

48. $2\ln x + \frac{1}{2}\ln(x+1) = \ln x^2 + \ln\sqrt{x+1}$

$= \ln x^2\sqrt{x+1}$

50. $\frac{1}{2}\ln(x-1) + \frac{3}{2}\ln(x+2) = \ln(x-1)^{1/2} + \ln(x+2)^{3/2} = \ln[(x-1)^{1/2}(x+2)^{3/2}]$

52.
$$\begin{aligned} e^{\ln x^2} - 9 &= 0 \\ x^2 - 9 &= 0 \\ x^2 &= 9 \\ x &= \pm 3 \end{aligned}$$

54.
$$\begin{aligned} 2\ln x &= 4 \\ \ln x &= 2 \\ x &= e^2 \end{aligned}$$

56.
$$\begin{aligned} e^{-0.5x} &= 0.075 \\ -0.5x &= \ln 0.075 \\ x &= \frac{\ln 0.075}{-0.5} \\ &\approx 5.1805 \end{aligned}$$

58.
$$\begin{aligned} e^{-0.0174t} &= 0.5 \\ -0.0174t &= \ln 0.5 \\ t &= \frac{\ln 0.5}{-0.0174} \\ &\approx 39.8360 \end{aligned}$$

60.
$$\begin{aligned} 2^{1-x} &= 6 \\ \ln 2^{1-x} &= \ln 6 \\ (1-x)\ln 2 &= \ln 6 \\ 1 - x &= \frac{\ln 6}{\ln 2} \\ 1 - \frac{\ln 6}{\ln 2} &= x \\ x &\approx -1.5850 \end{aligned}$$

62.
$$\begin{aligned} 400(1.06)^t &= 1300 \\ (1.06)^t &= \frac{13}{4} \\ t\ln 1.06 &= \ln 13 - \ln 4 \\ t &= \frac{\ln 13 - \ln 4}{\ln 1.06} \\ &\approx 20.2279 \end{aligned}$$

64.
$$\begin{aligned} \frac{36}{1+e^{-t}} &= 20 \\ \frac{1+e^{-t}}{36} &= \frac{1}{20} \\ 1 + e^{-t} &= \frac{36}{20} = \frac{9}{5} \\ e^{-t} &= \frac{4}{5} \\ -t &= \ln\left(\frac{4}{5}\right) \\ t &= -\ln\left(\frac{4}{5}\right) = \ln\left(\frac{5}{4}\right) \approx 0.2231 \end{aligned}$$

66.
$$\begin{aligned} \ln\left(\frac{N}{N_0}\right) &= -kt \\ \ln\left(\frac{3.1}{13.6}\right) &= -\left(\frac{0.693}{5730}\right)t \\ -1.4787 &= -1.209 \times 10^{-4}t \\ t &\approx 12{,}226 \text{ years} \end{aligned}$$

68. $p = 250 - 0.8e^{0.005x}$

(a) $p = 200 = 250 - 0.8e^{0.005x}$

$$0.8e^{0.005x} = 50$$
$$e^{0.005x} = 62.5$$
$$0.005x = \ln 62.5$$
$$x = \frac{\ln 62.5}{0.005} \approx 827.03 \approx 827 \text{ units}$$

(b) $p = 125 = 250 - 0.8e^{0.005x}$

$$0.8e^{0.005x} = 125$$
$$e^{0.005x} = 156.25$$
$$0.005x = \ln 156.25$$
$$x = \frac{\ln 156.25}{0.005} \approx 1010.29 \approx 1010 \text{ units}$$

70. $P = 556{,}000e^{0.00733t} = 700{,}000$

$$e^{0.00733t} = \frac{700{,}000}{556{,}000}$$
$$0.00733t = \ln\left(\frac{700}{556}\right)$$
$$t \approx 31.4$$

$1970 + 31.4 = 2001.4$. In the year 2001.

72. $0.27 \times 10^{-12} = 10^{-12}\left(\frac{1}{2}\right)^{t/5700}$

$$0.27 = \left(\frac{1}{2}\right)^{t/5700}$$
$$\ln 0.27 = \frac{t}{5700} \ln \frac{1}{2}$$
$$t = \frac{5700 \ln 0.27}{\ln \frac{1}{2}} \approx 10{,}767 \text{ years}$$

74. $0.13 \times 10^{-12} = 10^{-12}\left(\frac{1}{2}\right)^{t/5700}$

$$0.13 = \left(\frac{1}{2}\right)^{t/5700}$$
$$\ln 0.13 = \frac{t}{5700} \ln \frac{1}{2}$$
$$t = \frac{5700 \ln 0.13}{\ln \frac{1}{2}} \approx 16{,}777 \text{ years}$$

76.

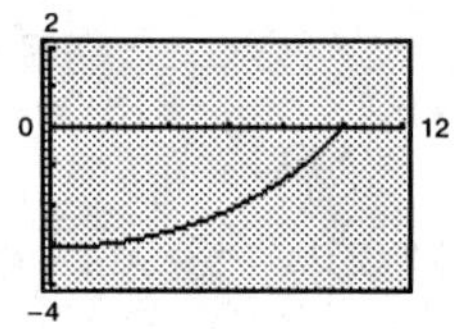

78. $f(x) = \dfrac{\ln x}{x}$

x	1	5	10	10^2	10^4	10^6
$f(x)$	0	0.3219	0.2303	0.0461	0.0009	0.00001

(a) $\displaystyle\lim_{x\to\infty} \frac{\ln x}{x} = 0$

(b)

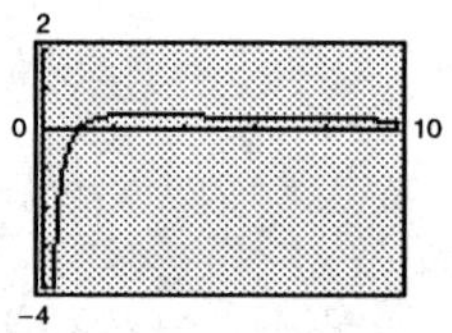

Relative maximum at (2.7183, 0.3679)
No relative minima

80.

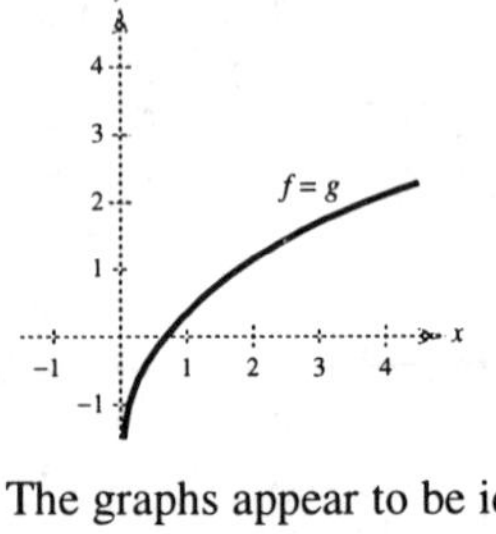

The graphs appear to be identical.

Section 4.4 Derivatives of Logarithmic Functions

2. $y = \ln x^{5/2}$

$= \dfrac{5}{2}\ln x$

$y' = \dfrac{5}{2x}$

$y'(1) = \dfrac{5}{2}$

4. $y = \ln x^{1/2}$

$= \dfrac{1}{2}\ln x$

$y' = \dfrac{1}{2x}$

$y'(1) = \dfrac{1}{2}$

6. $y = \ln(x^2 + 3)$

$y' = \dfrac{2x}{x^2 + 3}$

8. $y = \ln(1 - x^2)$

$y' = -\dfrac{2x}{1 - x^2}$

$= \dfrac{2x}{x^2 - 1}$

10. $y = \ln(1 - x)^{3/2} = \dfrac{3}{2}\ln(1 - x)$

$y' = \dfrac{3}{2}\left(\dfrac{-1}{1 - x}\right) = \dfrac{3}{2(x - 1)}$

12. $y = (\ln x^2)^2 = (2\ln x)^2 = 4(\ln x)^2$

$y' = 4\left[2(\ln x)\left(\dfrac{1}{x}\right)\right] = \dfrac{8\ln x}{x}$

14. $y = \ln\left(x\sqrt{x^2 - 1}\right) = \ln x + \dfrac{1}{2}\ln(x^2 - 1)$

$y' = \dfrac{1}{x} + \left(\dfrac{1}{2}\right)\dfrac{1}{x^2 - 1}(2x) = \dfrac{1}{x} + \dfrac{x}{x^2 - 1} = \dfrac{2x^2 - 1}{x(x^2 - 1)}$

16. $y = \ln\dfrac{x}{x^2 + 1} = \ln x - \ln(x^2 + 1)$

$y' = \dfrac{1}{x} - \dfrac{1}{x^2 + 1}(2x) = \dfrac{1}{x} - \dfrac{2x}{x^2 + 1} = \dfrac{1 - x^2}{x(x^2 + 1)}$

18. $y = \ln\sqrt{\dfrac{x + 1}{x - 1}} = \dfrac{1}{2}[\ln(x + 1) - \ln(x - 1)]$

$y' = \dfrac{1}{2}\left[\dfrac{1}{x + 1} - \dfrac{1}{x - 1}\right] = \dfrac{-1}{x^2 - 1} = \dfrac{1}{1 - x^2}$

20. $y = \dfrac{\ln x}{x^2}$

$y' = \dfrac{x^2(1/x) - 2x\ln x}{x^4} = \dfrac{x - 2x\ln x}{x^4} = \dfrac{1 - 2\ln x}{x^3}$

22. $y = \ln\left(x + \sqrt{4 + x^2}\right)$

$$y' = \frac{1 + \frac{1}{2}(4 + x^2)^{-1/2}(2x)}{x + \sqrt{4 + x^2}} = \frac{1 + \frac{x}{\sqrt{4 + x^2}}}{x + \sqrt{4 + x^2}} \cdot \frac{\sqrt{4 + x^2}}{\sqrt{4 + x^2}} = \frac{\sqrt{4 + x^2} + x}{\left(x + \sqrt{4 + x^2}\right)\sqrt{4 + x^2}} = \frac{1}{\sqrt{4 + x^2}}$$

24. $g(x) = \ln\dfrac{e^x + e^{-x}}{2} = \ln(e^x + e^{-x}) - \ln 2$

$g'(x) = \dfrac{e^x - e^{-x}}{e^x + e^{-x}}$

26. $f(x) = \ln\dfrac{1 + e^x}{1 - e^x} = \ln(1 + e^x) - \ln(1 - e^x)$

$f'(x) = \dfrac{e^x}{1 + e^x} - \dfrac{-e^x}{1 - e^x}$

$= \dfrac{e^x(1 - e^x) + e^x(1 + e^x)}{(1 + e^x)(1 - e^x)} = \dfrac{2e^x}{1 - e^{2x}}$

28. $3^x = e^{x(\ln 3)}$

30. $\log_3 x = \dfrac{1}{\ln 3}\ln x$

32. $\log_5 12 = \dfrac{1}{\ln 5}\ln 12 \approx 1.54396$ (calculator)

34. $\log_7 343 = 3$ because $7^3 = 343$.

36. $\log_9 3 = \frac{1}{2}$ because $9^{1/2} = 3$.

38. $y = \left(\frac{1}{4}\right)^x$

$y' = \left(\ln\frac{1}{4}\right)\left(\frac{1}{4}\right)^x = (-\ln 4)\left(\frac{1}{4}\right)^x$

40. $g(x) = \log_5 x$

$$g'(x) = \frac{1}{\ln 5} \cdot \frac{1}{x} = \frac{1}{x \ln 5}$$

42. $y = 6^{5x}$

$$y' = \ln 6 \cdot 6^{5x}(5) = 5 \ln 6 \cdot 6^{5x}$$

44. $f(x) = 10^{x^2}$

$$f'(x) = (\ln 10)10^{x^2}(2x) = 2x(\ln 10)10^{x^2}$$

46. $y = 3x3^x$

$$y' = 3[x(\ln 3)3^x + 3^x]$$
$$= 3[3^x(1 + x \ln 3)] = 3^{x+1}(1 + x \ln 3)$$

48. $y' = \dfrac{x(1/x) - \ln x}{x^2} = \dfrac{1 - \ln x}{x^2}$

$$y'(e) = 0$$
$$y - \frac{1}{e} = 0(x - e)$$
$$y = \frac{1}{e} \quad \text{Tangent line}$$

50. $g'(x) = (\ln 25)25^{2x^2}(4x)$

$$g'\left(-\tfrac{1}{2}\right) = (\ln 25)(5)(-2) = -10 \ln 25 = -20 \ln 5$$
$$y - 5 = -20 \ln 5 \left(x + \tfrac{1}{2}\right)$$
$$y = (-20 \ln 5)x + 5 - 10 \ln 5$$

52. $\ln xy + 5x = 30$

$$\ln x + \ln y + 5x = 30$$
$$\frac{1}{x} + \frac{1}{y}\frac{dy}{dx} + 5 = 0$$
$$\frac{dy}{dx} = \left(-\frac{1}{x} - 5\right)y$$
$$= -\frac{y}{x} - 5y = -\frac{y(1 + 5x)}{x}$$

54. $4xy + \ln(x^2y) = 7$

$$4xy + 2 \ln x + \ln y = 7$$
$$4xy' + 4y + \frac{2}{x} + \frac{y'}{y} = 0$$
$$\left(4x + \frac{1}{y}\right)y' = -4y - \frac{2}{x}$$
$$y' = \frac{-4y - (2/x)}{4x + (1/y)} = \frac{-4xy^2 - 2y}{4x^2y + x}$$

56. $f(x) = 2 \ln x + 3$

$$f'(x) = \frac{2}{x}$$
$$f''(x) = -\frac{2}{x^2}$$

58. $f(x) = \log_{10} x$

$$f'(x) = \frac{1}{\ln 10} \cdot \frac{1}{x} = \frac{1}{\ln 10}x^{-1}$$
$$f''(x) = \frac{-1}{\ln 10}x^{-2} = \frac{-1}{x^2 \ln 10}$$

60. $T = 87.97 + 34.96 \ln p + 7.91\sqrt{p}$

$$\frac{dT}{dp} = \frac{34.96}{p} + \frac{7.91}{2\sqrt{p}}$$

At $p = 60$, $\dfrac{dT}{dp} = \dfrac{34.96}{60} + \dfrac{7.91}{2\sqrt{60}} \approx 1.093$ degrees per pounds per square inch

62. $f(x) = 2 \ln x^3 = 6 \ln x$

$$f'(x) = \frac{6}{x}$$

At $(e, 6)$, the slope of the tangent line is $f'(e) = 6/e$.

Tangent line: $y - 6 = \dfrac{6}{e}(x - e)$

$$y = \frac{6}{e}x$$

64. $f(x) = \ln\left(x\sqrt{x + 3}\right) = \ln x + \dfrac{1}{2}\ln(x + 3)$

$$f'(x) = \frac{1}{x} + \frac{1}{2}\left(\frac{1}{x + 3}\right) = \frac{3(x + 2)}{2x(x + 3)}$$

At $(1.20, 0.90)$, the slope of the tangent line is $f'(1.20) = \frac{20}{21}$.

Tangent line: $y - 0.90 = \dfrac{20}{21}(x - 1.20)$

$$21y - 18.90 = 20x - 24$$
$$0 = 20x - 21y - 5.10$$
$$0 = 200x - 210y - 51$$

66. $f(x) = x^2 \log_3 x^2 = 2x^2 \log_3 x$

$$f'(x) = 2x^2 \frac{1}{\ln 3}\frac{1}{x} + 4x \log_3 x$$

$$= \frac{2x}{\ln 3} + 4x \log_3 x$$

At $(1, 0)$, the slope of the tangent line is $f'(1) = 2/(\ln 3)$.

Tangent line: $y - 0 = \dfrac{2}{\ln 3}(x - 1)$

$$y = \frac{2}{\ln 3}x - \frac{2}{\ln 3}$$

$$(\ln 3)y = 2x - 2$$

68. $y = \dfrac{x^2}{2} - \ln x$

$$y' = x - \frac{1}{x} = \frac{x^2 - 1}{x}$$

$y' = 0$ when $x = 1$.

$$y'' = 1 + \frac{1}{x^2} = \frac{x^2 + 1}{x^2}$$

[**Note:** $x = -1$ is not in the domain of the function.] Since $y''(1) = 2 > 0$, there is a relative minimum at $\left(1, \frac{1}{2}\right)$. Moreover, since $y'' > 0$ on $(0, \infty)$, it follows that the graph is concave upward in its domain and there are no inflection points.

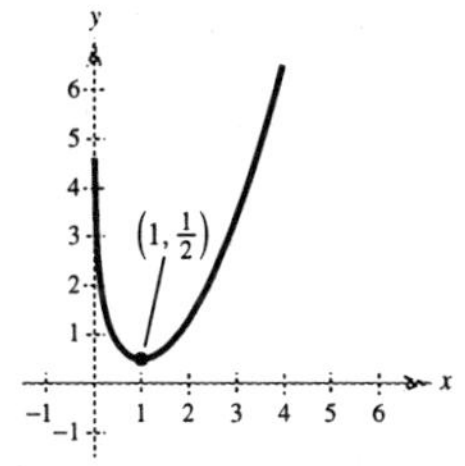

70. $y = x \ln x$

$$y' = x\left(\frac{1}{x}\right) + (\ln x)(1) = 1 + \ln x$$

$y' = 0$ when $x = e^{-1}$.

$$y'' = \frac{1}{x}$$

Since $y''(e^{-1}) > 0$, there is a relative minimum at $(e^{-1}, -e^{-1})$. Moreover, since $y'' > 0$ on $(0, \infty)$, it follows that the graph is concave upward in its domain and there are no inflection points.

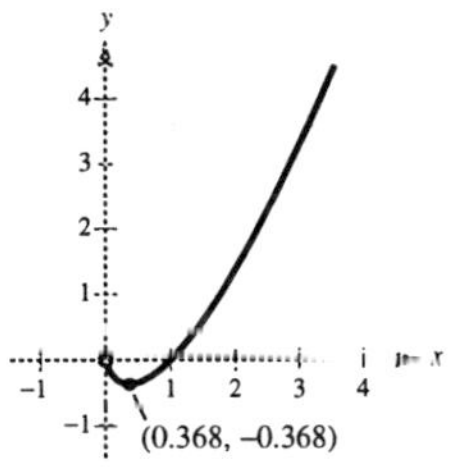

72. $y = (\ln x)^2$

$$y' = 2(\ln x)\left(\frac{1}{x}\right) = \frac{2 \ln x}{x}$$

$y' = 0$ when $x = 1$.

$$y'' = \frac{x(2/x) - (2 \ln x)(1)}{x^2} = \frac{2(1 - \ln x)}{x^2}$$

Since $y''(1) > 0$, there is a relative minimum at $(1, 0)$. Since $y'' = 0$ when $x = e$, it follows that there is an inflection point at $(e, 1)$.

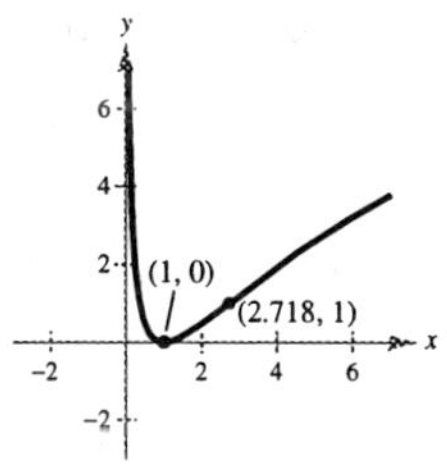

74. $x = 1000 - p \ln p$

$$\frac{dx}{dp} = 0 - \left[p\left(\frac{1}{p}\right) + (\ln p)(1)\right] = -(1 + \ln p)$$

When $p = 10$, $\dfrac{dx}{dp} = -(1 + \ln 10) \approx -3.3$.

76. $x = 300 - 50 \ln(\ln p)$

$$\frac{dx}{dp} = 0 - 50\frac{1/p}{\ln p} = -\frac{50}{p \ln p}$$

When $p = 10$, $\dfrac{dx}{dp} = \dfrac{-50}{10 \ln 10} \approx -2.171$.

78. $$x = \frac{500}{\ln(p^2 + 1)}$$

$$\ln(p^2 + 1) = \frac{500}{x}$$

$$p^2 + 1 = e^{500/x}$$

$$p = \sqrt{e^{500/x} - 1}$$

$$\frac{dp}{dx} = \frac{1}{2}(e^{500/x} - 1)^{-1/2}(e^{500/x})\left(\frac{-500}{x^2}\right)$$

$$= \frac{-250}{x^2} \cdot \frac{e^{500/x}}{\sqrt{e^{500/x} - 1}}$$

When $p = 10$,

$$x = \frac{500}{\ln(101)} \approx 108.34 \text{ and } \frac{dp}{dx} = -0.215.$$

Note that $dp/dx = 1/[dx/dp]$ from Exercise 75.

80. $A = \left(\frac{1}{2}\right)^{t/276}$

$$\frac{dA}{dt} = \left(\ln\frac{1}{2}\right)\left(\frac{1}{2}\right)^{t/276}\left(\frac{1}{276}\right)$$

When $t = 500$,

$$\frac{dA}{dt} = \left(\ln\frac{1}{2}\right)\left(\frac{1}{2}\right)^{500/276}\left(\frac{1}{276}\right) \approx -7.15 \times 10^{-4}.$$

82. Answers will vary.

Section 4.5 Exponential Growth and Decay

2. Since $y = \frac{1}{2}$ when $t = 0$, it follows that $C = \frac{1}{2}$. Moreover, since $y = 5$ when $t = 5$, we have $5 = \frac{1}{2}e^{5k}$ and

$k = \frac{1}{5}\ln 10 \approx 0.4605.$

Thus, $y = \frac{1}{2}e^{0.4605t}$.

4. Since $y = 2$ when $t = 0$, it follows that $C = 2$. Moreover, since $y = 1$ when $t = 5$, we have $1 = 2e^{5k}$ and

$k = \frac{1}{5}\ln\frac{1}{2} \approx -0.1386.$

Thus, $y = 2e^{-0.1386t}$.

6. Using the fact that $y = \frac{1}{2}$ when $t = 3$ and $y = 5$ when $t = 4$, we have $\frac{1}{2} = Ce^{3k}$ and $5 = Ce^{4k}$. From these two equations we have

$$\frac{1/2}{e^{3k}} = \frac{5}{e^{4k}}, \; \frac{1}{2}e^{4k} = 5e^{3k}, \text{ and } e^k = 10.$$

Thus, $k = \ln 10$ and we have $y = Ce^{t\ln 10}$. Since $5 = Ce^{4\ln 10}$, it follows that

$$C = \frac{5}{e^{4\ln 10}} = \frac{1}{2000}.$$

Therefore,

$$y = \frac{1}{2000}e^{t\ln 10} \approx \frac{1}{2000}e^{2.3026t}.$$

8. $\frac{dy}{dt} = -\frac{2}{3}y$

$y = 20$ when $t = 0$: $y = 20e^{-(2/3)t}$

$$\frac{dy}{dt} = 20\left(-\frac{2}{3}\right)e^{-(2/3)t} = -\frac{2}{3}[20e^{-(2/3)t}] = -\frac{2}{3}y$$

Exponential decay

10. $\frac{dy}{dt} = 5.2y$

$y = 18$ when $t = 0$: $y = 18e^{5.2t}$

$$\frac{dy}{dt} = 18(5.2)e^{5.2t} = 5.2(18e^{5.2t}) = 5.2y$$

Exponential growth

12. Since $y = Ce^{[\ln(1/2)/1620]t}$, we have $1.5 = Ce^{[\ln(1/2)/1620](1000)} \Longrightarrow C \approx 2.301$ which implies that the initial quantity is 2.301 grams. When $t = 10{,}000$, we have $y = 2.301e^{[\ln(1/2)/1620](10{,}000)} \approx 0.032$ gram.

14. Since the initial quantity is 3 grams, we have $y = 3e^{[\ln(1/2)/5730]t}$. When $t = 1000$, $y = 3e^{[\ln(1/2)/5730](1000)} \approx 2.658$ grams. When $t = 10{,}000$, $y = 3e^{[\ln(1/2)/5730](10{,}000)} \approx 0.895$ gram.

16. Since $y = Ce^{[\ln(1/2)/24{,}360]t}$, we have $0.4 = Ce^{[\ln(1/2)/24{,}360](10{,}000)} \Longrightarrow C \approx 0.532$ which implies that the initial quantity is 0.532 gram. When $t = 1000$, we have $y = 0.532e^{[\ln(1/2)/24{,}360](1000)} \approx 0.517$ gram.

18. When $t = 1$, we have $0.9957\,C = Ce^k \Longrightarrow k = \ln 0.9957$. If we let h = half-life, we have

$$k = \frac{\ln(1/2)}{h}.$$

Thus,

$$\ln 0.9957 = \frac{\ln(1/2)}{h} \Longrightarrow h = \frac{\ln(1/2)}{\ln 0.9957} \approx 160.850 \text{ years.}$$

20. $y = Ce^{[\ln(1/2)/5730]t}$

Since 30% of the radioactive carbon remains, we have the following.

$$0.30C = Ce^{[\ln(1/2)/5730]t}$$

$$\ln 0.30 = \left[\frac{\ln(1/2)}{5730}\right]t$$

Thus, $t = \dfrac{5730 \ln 0.30}{\ln(1/2)} \approx 9952.813$ years.

22. (a) Let $y = Ce^{kt}$, with $t = 0$ corresponding to 1980.

Hence, $C = 311.9$ (billion)

$$459.3 = 311.9e^{k(12)}$$

$$e^{12k} = \frac{459.3}{311.9}$$

$$k = \frac{1}{12}\ln\left(\frac{459.3}{311.9}\right) \approx 0.03225$$

$$y = 311.9e^{0.03225t}$$

1990: $y = 311.9e^{0.03225(10)} \approx 430.6$ billion

(b) $623.8 = 311.9e^{kt}$

$$2 = e^{kt}$$

$$t = \frac{1}{k}\ln 2 \approx 21.49 \text{ years to double}$$

(c) From year t to $t + 1$ the increase is

$$311.9e^{k(t+1)} - 311.9e^{kt} = 311.9e^{kt}(e^k - 1)$$

The relative increase is $\dfrac{311.9e^{kt}(e^k - 1)}{311.9e^{kt}} = e^k - 1 \approx 3\%$.

24. Since $A = 20{,}000e^{0.105t}$, the time to double is given by the following.

$$40{,}000 = 20{,}000e^{0.105t}$$

$$\ln 2 = 0.105t$$

$$t = \frac{\ln 2}{0.105} \approx 6.601 \text{ years}$$

Amount after 10 years:

$$A = 20{,}000e^{0.105(10)} \approx \$57{,}153.02$$

Amount after 25 years:

$$A = 20{,}000e^{0.105(25)} \approx \$276{,}091.48$$

26. Since $A = 10{,}000e^{rt}$ and $A = 20{,}000$ when $t = 5$, we have the following.

$$20{,}000 = 10{,}000e^{5r}$$

$$r = \frac{\ln 2}{5} \approx 0.1386 = 13.86\%$$

Amount after 10 years:

$$A = 10{,}000e^{[(\ln 2)/5](10)} = \$40{,}000$$

Amount after 25 years:

$$A = 10{,}000e^{[(\ln 2)/5](25)} = \$320{,}000$$

28. Since $A = 2000e^{rt}$ and $A = 6008.33$ when $t = 25$, we have the following.

$$6008.33 = 2000e^{25r}$$
$$\ln 3.004165 = 25r$$
$$r \approx 0.044 = 4.4\%$$

The time to double is given by

$$4000 = 2000e^{0.044t}$$
$$\ln 2 = 0.044t$$
$$t \approx 15.753 \text{ years.}$$

Amount after 10 years:

$$A = 2000e^{0.044(10)} \approx \$3105.41$$

30. (a) By equating $A = P(1 + i)^t$ and $A = Pe^{rt}$ we have the following.

$$P(1 + i)^t = Pe^{rt}$$
$$(1 + i)^t = (e^r)^t$$
$$1 + i = e^r$$

Therefore, $i = e^r - 1$.

(b) If $r = 0.06$, then $i = e^{0.06} - 1 \approx 0.0618$ or 6.18%.

32.

Number of Compoundings per Year	4	12	365	Continuous
Effective Yield	7.714%	7.763%	7.788%	7.788%

$n = 4$: $\quad i = \left(1 + \frac{0.075}{4}\right)^4 - 1 \approx 0.07714 \approx 7.714\%$

$n = 12$: $\quad i = \left(1 + \frac{0.075}{12}\right)^{12} - 1 \approx 0.07763 \approx 7.763\%$

$n = 365$: $\quad i = \left(1 + \frac{0.075}{365}\right)^{365} - 1 \approx 0.07788 \approx 7.788\%$

Continuous: $\quad i = e^{0.075} - 1 \approx 0.07788 \approx 7.788\%$

34. (a) For $r = 10\%$, the approximate time necessary for the investment to double is $\frac{70}{10} = 7$ years.

(b) For $r = 7\%$, the approximate time necessary for the investment to double is $\frac{70}{7} = 10$ years.

36. Let $t = 0$ correspond to 1987.

(a) $y = Ce^{kt} = 538.7e^{kt}$

$$1361.7 = 538.7e^{k9}$$
$$k = \frac{1}{9}\ln\left(\frac{1361.7}{538.7}\right) \approx 0.10304$$

In 2000, $t = 13$ and $y \approx \$2056$ million.

(b) (0, 538.7), (9, 1361.7)

$$y - 538.7 = \frac{1361.7 - 538.7}{9 - 0}(x - 0)$$
$$y = 91.44x + 538.7 \quad \text{linear model}$$
$$y(13) \approx \$1727 \text{ million}$$

(c)

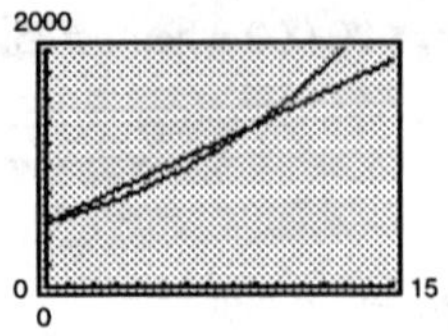

38. $S = 30(1 - 3^{kt})$

(a) Since $S = 5$ when $t = 1$, we have

$$5 = 30(1 - e^k)$$
$$e^k = 1 - \tfrac{1}{6}$$
$$k = \ln \tfrac{5}{6}.$$

[**Note:** S is in thousands of units.]
Therefore, $S = 30(1 - e^{[\ln(5/6)]t})$.

(b) $\lim_{t\to\infty} 30(1 - e^{[\ln(5/6)]t}) = 30$

The saturation point for the market is 30,000 units.

(c) When $t = 5$, we have

$$S = 30(1 - e^{[\ln(5/6)](5)}) \approx 17.944.$$

Thus, $S \approx 17{,}944$ units.

(d)

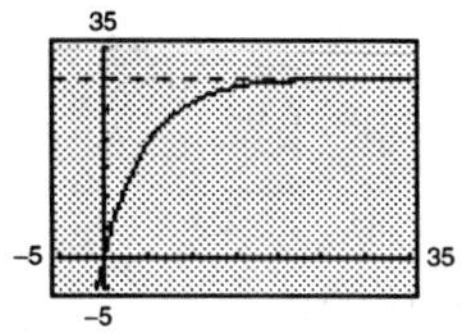

40. (a) Since $N = 20$ when $t = 30$, we have

$$20 = 30(1 - e^{30k})$$
$$k = \frac{\ln(1/3)}{30} \approx -0.0366$$
$$N = 30(1 - e^{-0.0366t}).$$

(b) $25 = 30(1 - e^{-0.0366t})$

$$t = \frac{\ln(1/6)}{-0.0366} \approx 49 \text{ days}$$

42. (a) Since $p = Ce^{kx}$ and $p = 5$ when $x = 300$, and $p = 4$ when $x = 400$, we have the following.

$$5 = Ce^{300k} \text{ and } 4 = Ce^{400k}$$
$$C = \frac{5}{e^{300k}} \text{ and } C = \frac{4}{e^{400k}}$$
$$\frac{5}{e^{300k}} = \frac{4}{e^{400k}}$$
$$5e^{400k} = 4e^{300k}$$
$$e^{100k} = \frac{4}{5}$$
$$k = \frac{1}{100}\ln\frac{4}{5}$$

Therefore, $C = \dfrac{5}{e^{300[(1/100)\ln(4/5)]}} = \dfrac{5}{e^{3\ln(4/5)}} = \dfrac{625}{64}$ and $p = \dfrac{625}{64}e^{[\ln(4/5)/100]x} \approx 9.7656e^{-0.0022x}$.

(b) $$R = xp = \frac{625}{64}xe^{[\ln(4/5)/100]x}$$

$$R' = \frac{625}{64}\left\{x\left[\frac{\ln(4/5)}{100}e^{[\ln(4/5)/100]x}\right] + e^{[\ln(4/5)/100]x}\right\} = \frac{625}{64}e^{[\ln(4/5)/100]x}\left[\frac{\ln(4/5)}{100}x + 1\right]$$

$$R' = 0 \text{ when } x = \frac{100}{\ln(4/5)} \approx 448 \text{ units.}$$

$p(448) \approx \$3.59$

44. $A = Ve^{-0.04t} = 100{,}000e^{0.6\sqrt{t}}e^{-0.04t} = 100{,}000e^{(0.6\sqrt{t}-0.04t)}$

$$A'(t) = 100{,}000\left(\frac{0.6}{2\sqrt{t}} - 0.04\right)e^{(0.6\sqrt{t}-0.04t)} = 0$$

$$\frac{0.6}{2\sqrt{t}} = 0.04$$

$$\sqrt{t} = \frac{0.6}{(0.04)2} = 7.5$$

$$t = 56.25 \approx 56$$

The timber should be harvested in 2046.

Review Exercises for Chapter 4

2. $16^{3/2} = \left(\sqrt{16}\right)^3 = 4^3 = 64$

4. $\left(\frac{27}{8}\right)^{-1/3} = \left(\frac{8}{27}\right)^{1/3} = \frac{2}{3}$

6. $(9^{1/3})(3^{1/3}) = (9 \cdot 3)^{1/3} = 27^{1/3} = 3$

8. $\frac{1}{4}\left(\frac{1}{2}\right)^{-3} = \frac{1}{4}(2)^3 = \frac{8}{4} = 2$

10. $x^{3/4} = 8$

$x^{1/4} = 2$

$x = 2^4 = 16$

12. $\frac{x^3}{3} = e^3$

$x^3 = 3e^3$

$x = e\sqrt[3]{3}$

14. $p = 5.331(1.0396)^t, \quad 0 \le t \le 34$

1960: $t = 0 \Longrightarrow p \approx 5.33\%$

1970: $t = 10 \Longrightarrow p \approx 7.86\%$

1988: $t = 28 \Longrightarrow p \approx 15.81\%$

16. $f(t) = \left(\frac{1}{6}\right)^t$

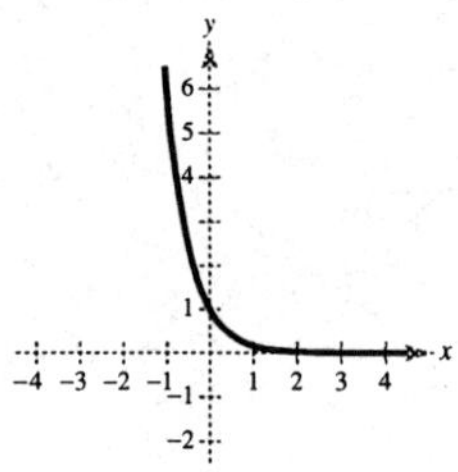

18. $g(x) = \frac{1}{4}(2)^{-3x}$

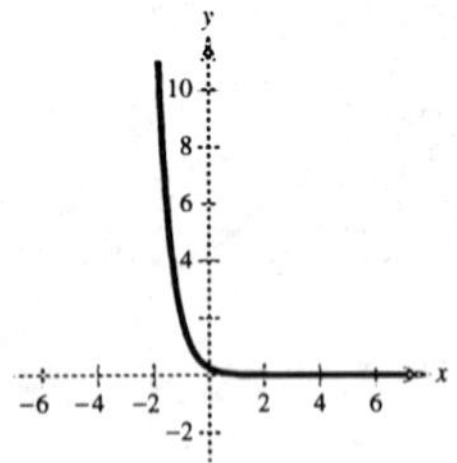

20. $y = 1096e^{-0.39t}$

If $t = 20$, $y = 0.449 < 1$, which indicates that the species is endangered.

22. $f(t) = e^{4t} - 1$

$f(0) = e^0 - 1 = 0$

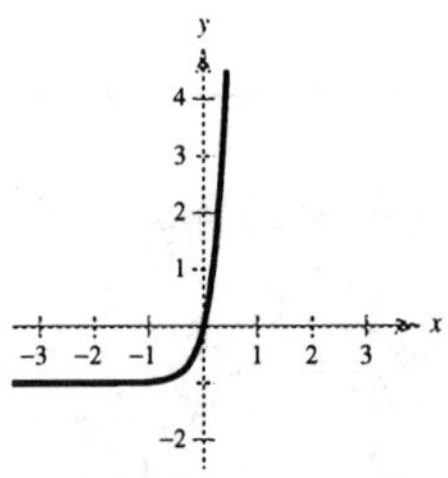

24. $g(x) = \dfrac{24}{1 + e^{-0.3x}}$

$g(300) \approx 24$

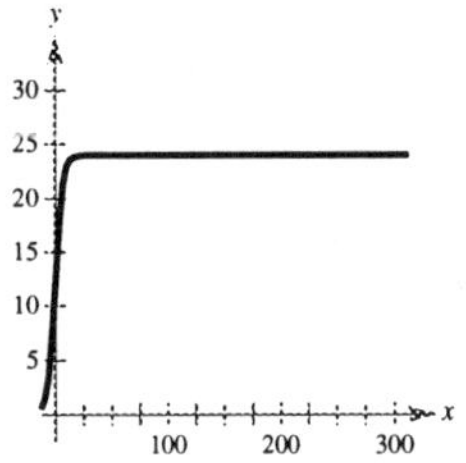

26. (a) $P = \dfrac{5000}{1 + 4999e^{-0.8t}}, \quad 0 \le t$

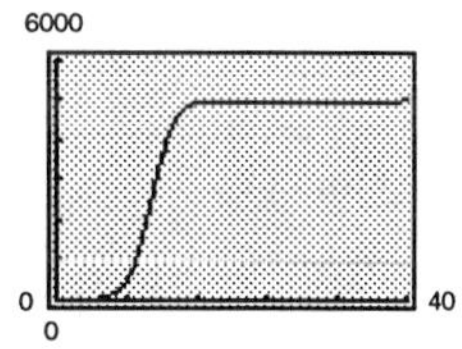

(b) When $t = 5$, $P = 54$ students.

(c) Yes, as $t \to \infty$, $P \to 5000$.

28.

n	1	2	4	12	365	Continuous
A	16,035.68	16,310.19	16,453.31	16,551.02	16,598.95	16,600.58

$A = P\left(1 + \dfrac{r}{n}\right)^{nt} = 5000\left(1 + \dfrac{0.06}{n}\right)^{20n}$

$A = Pe^{rt} = 5000e^{0.06(20)} \approx 16{,}600.58$

(Continuous)

30. (a) $A = P\left(1 + \dfrac{r}{n}\right)^{nt} = 2000\left(1 + \dfrac{0.065}{4}\right)^{4(10)} \approx \$3811.12$

(b) $A = Pe^{rt} = 2000e^{0.0625(10)} \approx \$3736.49$

Account (a) will be greater.

32. $R = 979.39 + 2.82e^{0.3645t}$

1987: $t = 7$, $P \approx \$1015.56$ million

1990: $t = 10$, $P \approx \$1087.35$ million

1996: $t = 16$, $P \approx \$1941.12$ million

34. $y = x^2e^x$

$y' = x^2e^x + 2xe^x = (x^2 + 2x)e^x$

36. $y = \dfrac{x - 1}{e^x + 2x^2}$

$y' = \dfrac{(e^x + 2x^2)(1) - (x - 1)(e^x + 4x)}{(e^x + 2x^2)^2}$

$= \dfrac{e^x + 2x^2 - xe^x + e^x - 4x^2 + 4x}{(e^x + 2x^2)^2} = \dfrac{2e^x - xe^x - 2x^2 + 4x}{(e^x + 2x^2)^2} = \dfrac{(2 - x)(2x + e^x)}{(e^x + 2x^2)^2}$

38. $y = \sqrt[3]{xe^{3x}} = (xe^{3x})^{1/3} = x^{1/3}e^x$

$y' = \dfrac{1}{3}x^{-2/3}e^x + x^{1/3}e^x = \dfrac{e^x(1 + 3x)}{3x^{2/3}}$

40. $y = \dfrac{e^x}{1 - xe^x}$

$y' = \dfrac{(1 - xe^x)e^x + e^x(xe^x + e^x)}{(1 - xe^x)^2} = \dfrac{e^x(1 + e^x)}{(1 - xe^x)^2}$

42. $f(x) = x^3e^x$

$f'(x) = x^2e^x(3 + x)$

$f''(x) = xe^x(x^2 + 6x + 6)$

$(-3, -1.344)$ is a relative minimum.

There are inflection points at $(0, 0)$, $\left(-3 + \sqrt{3}, -0.574\right)$, and $\left(-3 - \sqrt{3}, -0.933\right)$.
$y = 0$ is a horizontal asymptote.

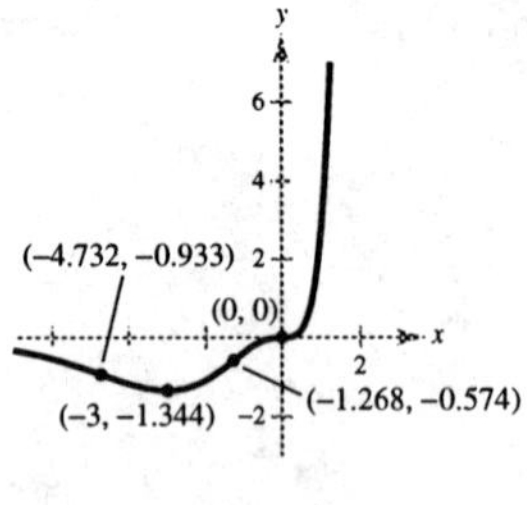

44. $f(x) = \dfrac{1}{xe^x}$

$$f'(x) = \frac{-x - 1}{x^2e^x}$$

$$f''(x) = \frac{x^2 + 2x + 2}{x^3e^x}$$

$(-1, -2.718)$ is a relative maximum.
$y = 0$ is a horizontal asymptote.
$x = 0$ is a vertical asymptote.

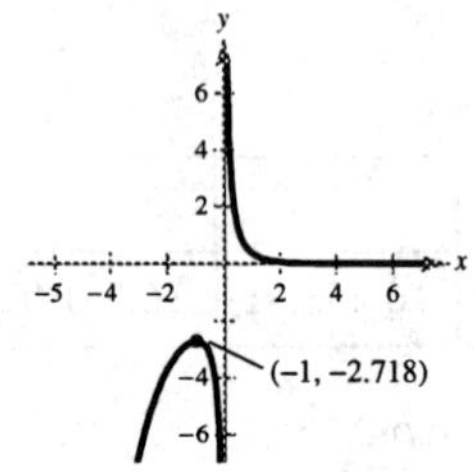

46. $\ln 0.6 \approx -0.5108$

$e^{-0.5108} \approx 0.6$

48. $e^{-4} \approx 0.0183$

$\ln 0.0183 \approx -4$

50. $y = 5 + \ln x$

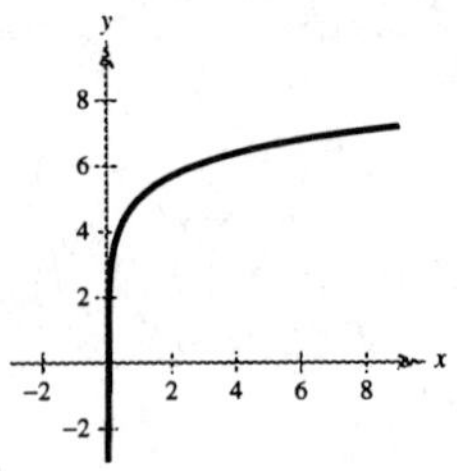

52. $y = -\frac{5}{6}\ln x$

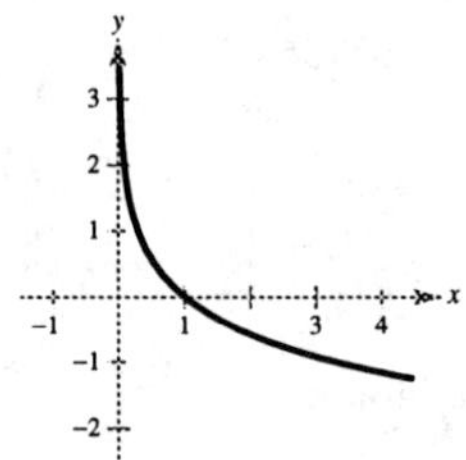

54.
$$\ln\frac{x^2}{(x + 1)^3} = \ln x^2 - \ln(x + 1)^3$$
$$= 2\ln x - 3\ln(x + 1)$$

56.
$$\ln\frac{e^4}{5x} = \ln e^4 - \ln 5x$$
$$= 4\ln e - \ln 5 - \ln x$$
$$= 4 - \ln 5 - \ln x$$

58. $\ln x = 3e^{-1} = \dfrac{3}{e}$

$e^{3/e} = x \approx 3.015$

60. $e^{2x-1} - 6 = 0$

$$e^{2x-1} = 6$$
$$2x - 1 = \ln 6$$
$$x = \frac{1 + \ln 6}{2} \approx 1.396$$

62. $e^{-1.386x} = 0.25$

$$-1.386x = \ln 0.25$$
$$x = \frac{\ln 0.25}{-1.386} \approx 1.0002$$

64. $100(1.21)^x = 110$

$$1.21^x = \frac{110}{100} = 1.1$$
$$x\ln 1.21 = \ln 1.1$$
$$x = \frac{\ln 1.1}{\ln 1.21} \approx 0.5$$

66. $w = 2.17 + \dfrac{1}{0.099 + 0.75e^{-0.17t}}$

(a)

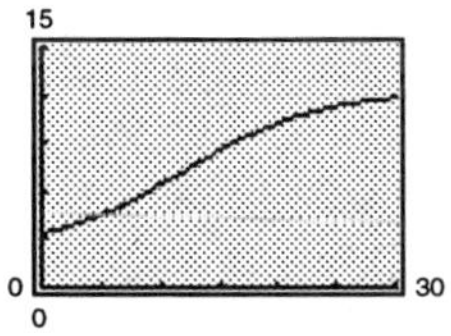

(b) The curve intersects $w = 10$ at $t = 19.2$, or 1989.

(c) After a while the equation will be nearly constant, which means it might not be a good model.

68. $y = \ln \dfrac{x(x-1)}{x-2} = \ln x + \ln(x-1) - \ln(x-2)$

$y' = \dfrac{1}{x} + \dfrac{1}{x-1} - \dfrac{1}{x-2}$

70. $f(x) = \ln\sqrt{4x} = \dfrac{1}{2}\ln 4x = \dfrac{1}{2}(\ln 4 + \ln x)$

$f'(x) = \dfrac{1}{2x}$

72. $y = \ln(x^2-2)^{2/3} = \dfrac{2}{3}\ln(x^2-2)$

$y' = \dfrac{2}{3} \cdot \dfrac{2x}{x^2-2} = \dfrac{4x}{3(x^2-2)}$

74. $y = \ln \dfrac{e^x}{1+e^x} = \ln e^x - \ln(1+e^x) = x - \ln(1+e^x)$

$y' = 1 - \dfrac{e^x}{1+e^x} = \dfrac{1}{1+e^x}$

76. $y = \dfrac{8\ln x}{x^2}$

$y' = 8\left(\dfrac{1-2\ln x}{x^3}\right)$

$y'' = 8\left(\dfrac{6\ln x - 5}{x^4}\right)$

$(e^{1/2}, 1.472)$ is a relative maximum.
$(e^{5/6}, 1.259)$ is a point of inflection.

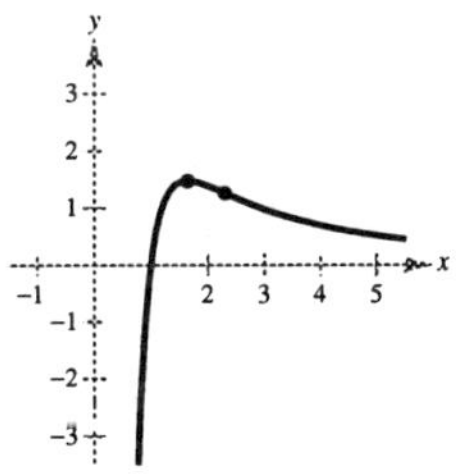

78. $y = \ln \dfrac{x^2}{9-x^2} = 2\ln x - \ln(9-x^2)$

No relative extrema
Inflection points: $(\sqrt{3}, -0.693), (-\sqrt{3}, -0.693)$

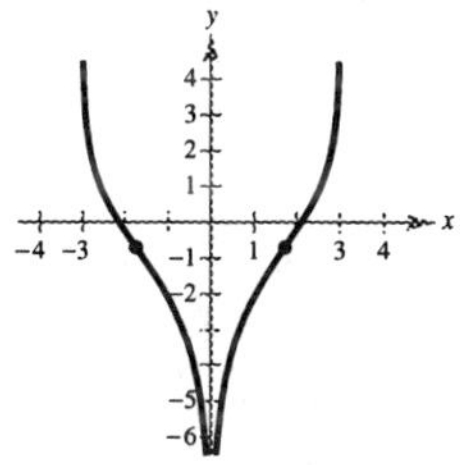

80. $\log_2 32 = \log_2 2^5 = 5\log_2 2 = 5$

82. $\log_4 \frac{1}{64} = \log_4 4^{-3} = -3\log_4 4 = -3$

84. $\log_4 12 = \dfrac{\ln 12}{\ln 4} \approx 1.7925$

86. $\log_{10} 125 = \dfrac{\ln 125}{\ln 10} \approx 2.0969$

88. $y = \log_{10}\dfrac{3}{x} = \log_{10} 3 - \log_{10} x$

$y' = \dfrac{-1}{\ln 10} \cdot \dfrac{1}{x} = \dfrac{-1}{x\ln 10}$

90. $y = \log_{16}(x^2 - 3x)$

$y' = \dfrac{1}{\ln 16} \cdot \dfrac{2x-3}{x^2-3x} = \dfrac{2x-3}{(x^2-3x)\ln 16}$

92. $C = P(1.05)^t$

(a) $C = 24.95(1.05)^{10} \approx \$40.64$

(b) $C'(t) = P \ln(1.05)(1.05)^t$

$= P \ln(1.05)(1.05) \approx 0.0512P$

94. $P = P_0 e^{0.025t}$, P_0 = initial population

(a) $2\,P_0 = P_0 e^{0.025t}$

$2 = e^{0.025t}$

$\ln 2 = 0.025t$

$t = \dfrac{\ln 2}{0.025} \approx 27.7$ years

(b) $3\,P_0 = P_0 e^{0.025t}$

$3 = e^{0.025t}$

$\ln 3 = 0.025t$

$t = \dfrac{\ln 3}{0.025} \approx 43.9$ years

96. $\dfrac{1}{2} = e^{k(5.2)}$

$k = \dfrac{\ln(1/2)}{5.2} \approx -0.1333$

$0.1 = 0.5e^{-0.1333t}$

$0.2 = e^{-0.1333t}$

$t = \dfrac{\ln(0.2)}{-0.1333} \approx 12.1$ years

98. $y = 1135.3\, e^{kt}$

$4190.6 = 1135.3\, e^{k(9)}$

$k = \dfrac{1}{9} \ln\left(\dfrac{4190.6}{1135.3}\right) \approx 0.1451$

For $t = 13$, $y = 1135.3e^{k(13)} \approx \$7487.7$ million in the year 2000.

CHAPTER 5
Integration and Its Applications

CHAPTER 5
Integration and Its Applications

Section 5.1 Antiderivatives and Indefinite Integrals

Solutions to Exercises

2. $\frac{d}{dx}(8\sqrt{x} + C) = 8 \cdot \frac{1}{2}x^{-1/2} = \frac{4}{\sqrt{x}}$

4. $\frac{d}{dx}(x - 3\sqrt[3]{x} + C) = \frac{d}{dx}(x - 3x^{1/3} + C)$

$= 1 - x^{-2/3} = 1 - \frac{1}{\sqrt[3]{x^2}}$

6. $\frac{d}{dx}\left(\frac{x^3}{3} - 4x + C\right) = x^2 - 4$

$= (x - 2)(x + 2)$

8. $\frac{d}{dx}\left(\frac{3(x+1)}{\sqrt[3]{x}} + C\right) = \frac{d}{dx}(3x^{2/3} + 3x^{-1/3} + C)$

$= 2x^{-1/3} - x^{-4/3} = \frac{2x - 1}{x^{4/3}}$

10. $\int -4\,dx = -4x + C$

$\frac{d}{dx}[-4x + C] = -4$

12. $\int t^4\,dt = \frac{t^5}{5} + C$

$\frac{d}{dt}\left[\frac{t^5}{5} + C\right] = t^4$

14. $\int 4y^{-3}dy = 4\left(\frac{y^{-2}}{-2}\right) + C = -\frac{2}{y^2} + C$

$\frac{d}{dy}[-2y^{-2} + C] = 4y^{-3}$

16. $\int e\,dt = et + C$

$\frac{d}{dt}[et + C] = e$

18. $\int v^{-1/2}dv = \frac{v^{1/2}}{1/2} + C = 2\sqrt{v} + C$

$\frac{d}{dv}[2v^{1/2} + C] = v^{-1/2}$

	Given	Rewrite	Integrate	Simplify
20.	$\int \frac{1}{x^2}dx$	$\int x^{-2}dx$	$\frac{x^{-1}}{-1} + C$	$-\frac{1}{x} + C$
22.	$\int x(x^2 + 3)dx$	$\int (x^3 + 3x)dx$	$\frac{x^4}{4} + \frac{3x^2}{2} + C$	$\frac{x^2}{4}(x^2 + 6) + C$
24.	$\int \frac{1}{(2x)^3}dx$	$\int \frac{1}{8}x^{-3}dx$	$\frac{1}{8}\left(\frac{x^{-2}}{-2}\right) + C$	$-\frac{1}{16x^2} + C$

26. If $f'(x) = x$, then $f(x) = \dfrac{x^2}{2} + C$.

For example, $f(x) = \dfrac{x^2}{2}$ or $f(x) = \dfrac{x^2}{2} + 3$.

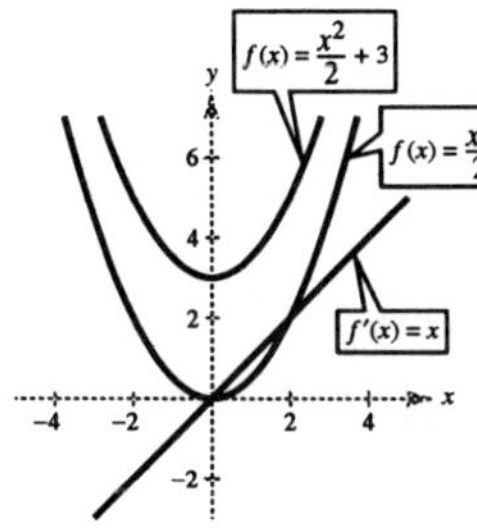

28. $$\int (x^2 - 2x + 3)dx = \frac{x^3}{3} - x^2 + 3x + C$$

$$\frac{d}{dx}\left[\frac{x^3}{3} - x^2 + 3x + C\right] = x^2 - 2x + 3$$

30. $$\int \left(\sqrt{x} + \frac{1}{2\sqrt{x}}\right)dx = \int \left(x^{1/2} + \frac{1}{2}x^{-1/2}\right)dx = \frac{2}{3}x^{3/2} + x^{1/2} + C = \frac{1}{3}\sqrt{x}(2x + 3) + C$$

$$\frac{d}{dx}\left[\frac{2}{3}x^{3/2} + x^{1/2} + C\right] = x^{1/2} + \frac{1}{2}x^{-1/2} = \sqrt{x} + \frac{1}{2\sqrt{x}}$$

32. $$\int \left(\sqrt[4]{x^3} + 1\right)dx = \int (x^{3/4} + 1)dx = \frac{4}{7}x^{7/4} + x + C$$

$$\frac{d}{dx}\left[\frac{4}{7}x^{7/4} + x + C\right] = x^{3/4} + 1 = \sqrt[4]{x^3} + 1$$

34. $$\int \frac{1}{4x^2}dx = \frac{1}{4}\int x^{-2}dx = \frac{1}{4}\left(\frac{x^{-1}}{-1}\right) + C = \frac{-1}{4x} + C$$

$$\frac{d}{dx}\left[\frac{-1}{4x} + C\right] = \frac{1}{4x^2}$$

36. $$\int \frac{t^2 + 2}{t^2}dt = \int (1 + 2t^{-2})dt = t + \frac{2t^{-1}}{-1} + C = t - \frac{2}{t} + C$$

$$\frac{d}{dt}\left[t - \frac{2}{t} + C\right] = 1 + \frac{2}{t^2} = \frac{t^2 + 2}{t^2}$$

38. $$\int \sqrt{x}(x + 1)dx = \int (x^{3/2} + x^{1/2})dx = \frac{2}{5}x^{5/2} + \frac{2}{3}x^{3/2} + C = \frac{2}{15}x^{3/2}(3x + 5) + C$$

$$\frac{d}{dx}\left[\frac{2}{5}x^{5/2} + \frac{2}{3}x^{3/2} + C\right] = x^{3/2} + x^{1/2} = \sqrt{x}(x + 1)$$

40. $$\int (2t^2 - 1)^2dt = \int (4t^4 - 4t^2 + 1)dt = \frac{4}{5}t^5 - \frac{4}{3}t^3 + t + C$$

$$\frac{d}{dt}\left[\frac{4}{5}t^5 - \frac{4}{3}t^3 + t + C\right] = 4t^4 - 4t^2 + 1 = (2t^2 - 1)^2$$

42. $$\int (1 + 3t)t^2dt = \int (t^2 + 3t^3)dt = \frac{1}{3}t^3 + \frac{3}{4}t^4 + C = \frac{t^3}{12}(4 + 9t) + C$$

$$\frac{d}{dt}\left[\frac{1}{3}t^3 + \frac{3}{4}t^4 + C\right] = t^2 + 3t^3 = (1 + 3t)t^2$$

44. $$f(x) = \int \left(\frac{1}{5}x - 2\right)dx = \frac{1}{10}x^2 - 2x + C$$

$$f(10) = -10 = \frac{1}{10}(100) - 2(10) + C = -10 + C \Rightarrow C = 0$$

$$f(x) = \frac{1}{10}x^2 - 2x$$

46. $f(x) = \int (2x - 3)(2x + 3)dx = \int (4x^2 - 9)dx$

$= \frac{4}{3}x^3 - 9x + C$

$f(3) = 0 = \frac{4}{3}(27) - 9(3) + C = 9 + C \Rightarrow C = -9$

$f(x) = \frac{4}{3}x^3 - 9x - 9$

48. $f'(x) = \frac{-5}{x^2} = -5x^{-2},\ x > 0,\ f(1) = 2$

$f(x) = -5\frac{x^{-1}}{-1} + C = \frac{5}{x} + C$

$f(1) = 5 + C = 2 \Rightarrow C = -3$

$f(x) = \frac{5}{x} - 3$

50. $y = \int 2(x - 1)dx = \int (2x - 2)dx = x^2 - 2x + C$

At $(3, 2)$, $2 = 9 - 6 + C$ which implies that $C = -1$.
Thus, $y = x^2 - 2x - 1$.

52. $f(x) = \int \frac{6}{x^2}dx = 6\int x^{-2}dx = 6\left(\frac{x^{-1}}{-1}\right) + C = -\frac{6}{x} + C$

At $(2, 5)$, $5 = -\frac{6}{2} + C$ which implies that $C = 8$.

Thus, $f(x) = -\frac{6}{x} + 8$.

54. $f'(x) = \int x^2 dx = \frac{1}{3}x^3 + C_1$

Since $f'(0) = 0 + C_1 = 6$, we know that $C_1 = 6$. Thus,

$f'(x) = \frac{1}{3}x^3 + 6$

$f(x) = \int \left(\frac{1}{3}x^3 + 6\right)dx = \frac{1}{12}x^4 + 6x + C_2.$

Since $f(0) = C_2 = 3$, we have $f(x) = \frac{1}{12}x^4 + 6x + 3$.

56. $f'(x) = \int x^{-3/2}dx = -2x^{-1/2} + C_1$

Since $f'(1) = -2 + C_1 = 2$, we know that $C_1 = 4$. Thus,

$f'(x) = -2x^{-1/2} + 4$

$f(x) = \int (-2x^{-1/2} + 4)dx = -4x^{1/2} + 4x + C_2.$

Since $f(9) = -12 + 36 + C_2 = -4$, we know that $C_2 = -28$. Therefore,

$f(x) = -4\sqrt{x} + 4x - 28 = -4(\sqrt{x} - x + 7).$

58. $C = \int \left(\frac{1}{50}x + 10\right)dx = \frac{1}{100}x^2 + 10x + C_1$

Since $C = 1000$ when $x = 0$, we have $C_1 = 1000$.
Thus, $C = \frac{1}{100}x^2 + 10x + 1000$.

60. $C = \int \left(\frac{1}{10}x^{1/4} + 10\right)dx = \frac{2}{25}x^{5/4} + 10x + C_1$

Since $C = 2300$ when $x = 0$, we have $C_1 = 2300$.
Thus, $C = \frac{2}{25}x^{5/4} + 10x + 2300$.

62. $R = \int (100 - 6x - 2x^2)dx = 100x - 3x^2 - \frac{2}{3}x^3 + C$

Since $R = 0$ when $x = 0$, it follows that $C = 0$.
Thus, $R = 100x - 3x^2 - \frac{2}{3}x^3$ and the demand function is

$p = \frac{R}{x} = 100 - 3x - \frac{2}{3}x^2.$

64. $P = \int (-40x + 250)dx = -20x^2 + 250x + C$

Since $P = 650$ when $x = 5$, we have $C = -100$.
Thus, $P = -20x^2 + 250x - 100$.

66. $a(t) = -32$

$v(t) = -32t + C_1 = -32t$ (since rock is dropped)

$s(t) = -16t^2 + C_2 = -16t^2 + 6000$ (since 6000 feet deep)

$s(t) = 0 = -16t^2 + 6000$

$16t^2 = 6000$

$t^2 = \frac{6000}{16} = 375$

$t = \sqrt{375} \approx 19.36$ seconds

68. (a) $v(t) = \int -32\,dt = -32t + C_1$

Since $v(0) = 16$, we have $v(t) = -32t + 16$.

$$s(t) = \int (-32t + 16)dt = -16t^2 + 16t + C_2$$

Since $s(0) = 64$, we have $s(t) = -16t^2 + 16t + 64$. The sandbag will hit the ground when $s = 0$.

$$-16t^2 + 16t + 64 = 0$$

$$-16(t^2 - t - 4) = 0$$

By the Quadratic Formula or a graphing utility, $t = \dfrac{1 + \sqrt{17}}{2} \approx 2.56$ sec.

(b) $v\left(\dfrac{1 + \sqrt{17}}{2}\right) = -32\left(\dfrac{1 + \sqrt{17}}{2}\right) + 16 \approx -65.97$ ft/sec.

70. $P = \int 500t^{1.06}dt = \dfrac{500}{2.06}t^{2.06} + C$

Since $P = 50{,}000$ when $t = 0$, we have $P = \frac{500}{2.06}t^{2.06} + 50{,}000$. When $t = 10$, $P(10) \approx 77{,}868$.

72.

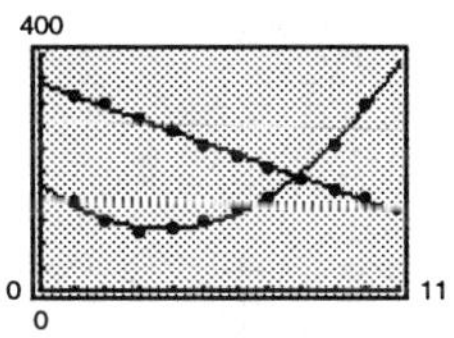

For marginal benefit, $y' = -19.88x + 351.33$

For marginal cost, $y' = 5.38x^2 - 40.62x + 182.33$

Integrating benefit: $y = -9.94x^2 + 351.33x + C_1$

$$y(0) = 0 \Rightarrow C_1 = 0$$

$$y = -9.94x^2 + 351.33x$$

Integrating cost: $y = 1.793x^3 - 20.31x^2 + 182.33x + C_2$

$$y(0) = 425 \Rightarrow C_2 = 425$$

$$y = 1.793x^3 - 20.31x^2 + 182.33x + 425$$

Graphing the benefit and cost equations together, you see that benefit exceeds cost on (2.3, 11].

Section 5.2 The General Power Rule

	$\int u^n \frac{du}{dx}dx$	u	$\frac{du}{dx}$
2.	$\int (3 - 4x^2)^3(-8x)dx$	$3 - 4x^2$	$-8x$
4.	$\int 3x^2\sqrt{x^3 + 1}\,dx$	$x^3 + 1$	$3x^2$
6.	$\int \frac{1}{(1 + 2x)^2}(2)dx$	$1 + 2x$	2

8. $\displaystyle\int (x^2-1)^3(2x)dx = \frac{(x^2-1)^4}{4} + C$

10. $\displaystyle\int \sqrt{3-x^3}(3x^2)dx = -\int (3-x^3)^{1/2}(-3x^2)dx = -\frac{2}{3}(3-x^3)^{3/2} + C$

12. $\displaystyle\int (x-3)^{5/2}dx = \frac{2}{7}(x-3)^{7/2} + C$

14. $\displaystyle\int x(1-2x^2)^3dx = -\frac{1}{4}\int (1-2x^2)^3(-4x)dx = -\frac{1}{4}\cdot\frac{(1-2x^2)^4}{4} + C = -\frac{1}{16}(1-2x^2)^4 + C$

16. $\displaystyle\int \frac{x^2}{(x^3-1)^2}dx = \frac{1}{3}\int (x^3-1)^{-2}(3x^2)dx = \frac{1}{3}\left[\frac{(x^3-1)^{-1}}{-1}\right] + C = -\frac{1}{3(x^3-1)} + C$

18. $\displaystyle\int \frac{6x}{(1+x^2)^3}dx = 3\int (1+x^2)^{-3}(2x)dx = 3\left[\frac{(1+x^2)^{-2}}{-2}\right] + C = -\frac{3}{2(1+x^2)^2} + C$

20. $\displaystyle\int \frac{4x+6}{(x^2+3x+7)^3}dx = 2\int (x^2+3x+7)^{-3}(2x+3)dx = 2\left[\frac{(x^2+3x+7)^{-2}}{-2}\right] + C = -\frac{1}{(x^2+3x+7)^2} + C$

22. $\displaystyle\int u^3\sqrt{u^4+2}\,du = \frac{1}{4}\int (u^4+2)^{1/2}(4u^3)du = \frac{1}{4}\cdot\frac{2}{3}(u^4+2)^{3/2} + C = \frac{1}{6}(u^4+2)^{3/2} + C$

24. $\displaystyle\int \frac{x^2}{\sqrt{1-x^3}}dx = -\frac{1}{3}\int (1-x^3)^{-1/2}(-3x^2)dx = -\frac{1}{3}(2)(1-x^3)^{1/2} + C = -\frac{2}{3}\sqrt{1-x^3} + C$

26. $\displaystyle\int \frac{t+2t^2}{\sqrt{t}}dt = \int (t^{1/2}+2t^{3/2})dt = \frac{2}{3}t^{3/2} + 2\left(\frac{2}{5}t^{5/2}\right) + C = \frac{2}{3}t^{3/2} + \frac{4}{5}t^{5/2} + C = \frac{2}{15}t^{3/2}(5+6t) + C$

28. $\displaystyle\int \left(1+\frac{1}{t}\right)^3\left(\frac{1}{t^2}\right)dt = -\int (1+t^{-1})^3(-t^{-2})dt = -\frac{1}{4}\left(1+\frac{1}{t}\right)^4 + C$

30. $\displaystyle\int \frac{1}{3x^2}dx = \int \frac{1}{3}x^{-2}dx = \frac{1}{3}\left(\frac{x^{-1}}{-1}\right) + C = -\frac{1}{3x} + C$

32. $\displaystyle\int (3-2x-4x^2)(1+4x)dx = -\frac{1}{2}\int (3-2x-4x^2)(-2-8x)dx$

$$= \left(-\frac{1}{2}\right)\frac{(3-2x-4x^2)^2}{2} + C = -\frac{1}{4}(3-2x-4x^2)^2 + C$$

34. Let $u = 1-x^3$, then $du = -3x^2dx$ and $x^2dx = -\frac{1}{3}du$.

$$\int x^2(1-x^3)^2dx = \int (1-x^3)^2x^2dx = \int u^2\left(-\frac{1}{3}du\right) = -\frac{1}{3}\left(\frac{u^3}{3}\right) + C = -\frac{1}{9}u^3 + C = -\frac{1}{9}(1-x^3)^3 + C$$

36. Let $u = t^2+1$, then $du = 2t\,dt$ and $t\,dt = \frac{1}{2}\,du$.

$$\int t\sqrt{t^2+1}\,dt = \int (t^2+1)^{1/2}t\,dt = \int u^{1/2}\left(-\frac{1}{2}du\right) = \frac{1}{2}\left(\frac{2}{3}u^{3/2}\right) + C = \frac{1}{3}(t^2+1)^{3/2} + C$$

38. Let $u = 2x + 1$, then $du = 2\,dx$ and $3\,dx = \frac{3}{2}\,du$.

$$\int \frac{3}{\sqrt{2x+1}}\,dx = \int (2x+1)^{-1/2}(3)\,dx = \int u^{-1/2}\left(\frac{3}{2}\,du\right) = \frac{3}{2}(2\,u^{1/2}) + C = 3\sqrt{2x+1} + C$$

40. Let $u = 4 - x^{3/2}$, then $du = -\frac{3}{2}x^{1/2}dx$ and $x^{1/2}dx = -\frac{2}{3}du$.

$$\int \sqrt{x}(4 - x^{3/2})^2\,dx = \int (4 - x^{3/2})^2 x^{1/2}dx = \int u^2\left(-\frac{2}{3}\,du\right) = -\frac{2}{3}\left(\frac{u^3}{3}\right) + C = -\frac{2}{9}(4 - x^{3/2})^3 + C$$

42. $$\int (3 - 2x)^2\,dx = -\frac{1}{2}\int (3 - 2x)^2(-2\,dx)$$

$$= -\frac{1}{2}(3 - 2x)^3/3 + C_1 = -\frac{1}{6}(3 - 2x)^3 + C_1 = -\frac{1}{6}(27 - 54x + 36x^2 - 8x^3) + C_1$$

$$\int (3 - 2x)^2\,dx = \int (9 - 12x + 4x^2)\,dx = 9x - 6x^2 + \frac{4}{3}x^3 + C_2$$

The two answers differ by a constant.

44. $$\int x(2x^2 + 1)^2\,dx = \frac{1}{4}\int (2x^2 + 1)^2(4x\,dx) = \frac{1}{4}\frac{(2x^2+1)^3}{3} + C_1 = \frac{1}{12}(8x^6 + 12x^4 + 6x^2 + 1) + C_1$$

$$\int x(2x^2 + 1)^2\,dx = \int (4x^5 + 4x^3 + x)\,dx = \frac{4x^6}{6} + x^4 + \frac{x^2}{2} + C_2$$

The two answers differ by a constant.

46. $f(x) = \int x\sqrt{1 - x^2}\,dx = -\frac{1}{3}(1 - x^2)^{3/2} + C$ (See Exercise 45.)

Since $f(0) = \frac{7}{3}$, it follows that $C = \frac{8}{3}$, and we have

$$f(x) = -\frac{1}{3}(1 - x^2)^{3/2} + \frac{8}{3} = \frac{1}{3}[8 - (1 - x^2)^{3/2}].$$

48. (a) $$C = \int \frac{12}{\sqrt[3]{12x + 1}}\,dx$$

$$= \int (12x + 1)^{-1/3}(12)\,dx$$

$$= \frac{3}{2}(12x + 1)^{2/3} + K$$

Since $C(13) = 100$, it follows that $K \approx 56.35$.

$$C = \frac{3}{2}(12x + 1)^{2/3} + 56.35$$

(b) 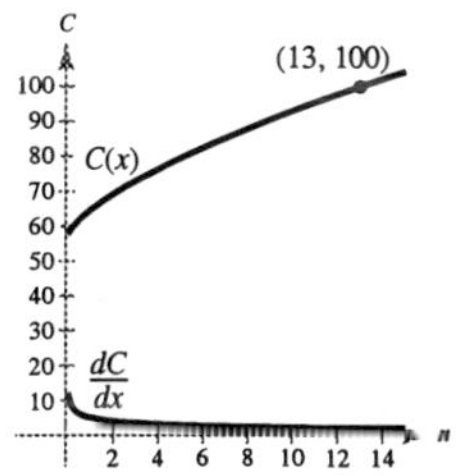

50. $$x = \int \frac{10}{\sqrt{p - 3}}\,dp = 10\int (p - 3)^{-1/2}dp = 10(2)(p - 3)^{1/2} + C = 20\sqrt{p - 3} + C$$

Since $x = 100$ when $p = 3$, it follows that $C = 100$. Therefore, we have $x = 20\sqrt{p - 3} + 100$.

52. $$x = \int \frac{-400}{(0.02p - 1)^3}\,dp = -400\left(\frac{1}{0.02}\right)\int (0.02p - 1)^{-3}(0.02)\,dp = -20{,}000\left[\frac{(0.02p - 1)^{-2}}{-2}\right] + C = \frac{10{,}000}{(0.02p - 1)^2} + C$$

Since $x = 10{,}000$ when $p = 100$, it follows that $C = 0$. Therefore, we have $x = \dfrac{10{,}000}{(0.02p - 1)^2}$.

54. $\dfrac{dQ}{dt} = k(100 - t)^2, \quad 0 \le t \le 100$

$$Q = \int k(100 - t)^2\,dt = \frac{-k}{3}(100 - t)^3 + C$$

When $t = 100$, $Q = 0 \Rightarrow C = 0$.

When $t = 0$, $Q = 2{,}000{,}000 \Rightarrow 2{,}000{,}000 = \dfrac{-k}{3}(100)^3$

$$-6{,}000{,}000 = 1{,}000{,}000k$$

$$k = -6$$

Thus, $Q = 2(100 - t)^3$. When $t = 50$, we have $Q = 2(100 - 50)^3 = \$250{,}000$ remaining to be disbursed.

56. (a) $Q = \displaystyle\int \frac{0.93}{(x - 19{,}999)^{0.07}}\,dx = (x - 19{,}999)^{0.93} + C$

Since $Q = 20{,}000$ when $x = 20{,}000$,

$$20{,}000 = (20{,}000 - 19{,}999)^{0.93} + C = 1 + C$$

$$C = 19{,}999.$$

Thus $Q = (x - 19{,}999)^{0.93} + 19{,}999$.

(c)

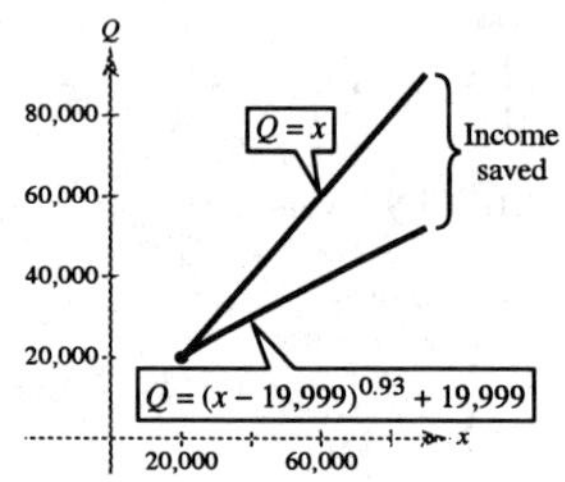

(b)

x	20,000	50,000	100,000	150,000
Q	20,000	34,578.29	56,296.67	77,011.54
$x - Q$	0	15,421.71	43,703.33	72,998.46

58. $\displaystyle\int \frac{x}{\sqrt{3x + 2}}\,dx = \frac{2}{27}(3x + 2)^{3/2} - \frac{4}{9}\sqrt{3x + 2} + C$

Section 5.3 Exponential and Logarithmic Integrals

2. $\displaystyle\int e^{-3x}(-3)\,dx = e^{-3x} + C$

4. $\displaystyle\int e^{-0.25x}\,dx = -4\int e^{-0.25x}(-0.25)\,dx = -4e^{-0.25x} + C$

6. $\displaystyle\int 3xe^{0.5x^2}\,dx = 3\int e^{0.5x^2}\,x\,dx = 3e^{0.5x^2} + C$

8. $\displaystyle\int (2x + 1)e^{x^2 + x}\,dx = \int e^{x^2 + x}(2x + 1)\,dx = e^{x^2 + x} + C$

10. $\displaystyle\frac{3}{2}\int e^{x^2 - 8x}(2x - 8)\,dx = \frac{3}{2}e^{x^2 - 8x} + C$

12. $\displaystyle -6\int e^{-(x+1)/2}\left(-\frac{1}{2}\right)dx = -6e^{-(x+1)/2} + C$

14. $\displaystyle\int \frac{1}{x - 5}\,dx = \ln|x - 5| + C$

16. $\displaystyle\int \frac{1}{6x - 5}\,dx = \frac{1}{6}\int \frac{6}{6x - 5}\,dx = \frac{1}{6}\ln|6x - 5| + C$

18. $\displaystyle\int \frac{x^2}{3 - x^3}\,dx = -\frac{1}{3}\int \frac{-3x^2}{3 - x^3}\,dx = -\frac{1}{3}\ln|3 - x^3| + C$

20. $\displaystyle\int \frac{x}{x^2 + 4}\,dx = \frac{1}{2}\int \frac{2x}{x^2 + 4}\,dx = \frac{1}{2}\ln(x^2 + 4) + C$

22. $\displaystyle\int \frac{x^2 + 2x + 3}{x^3 + 3x^2 + 9x + 1}\,dx = \frac{1}{3}\int \frac{3(x^2 + 2x + 3)}{x^3 + 3x^2 + 9x + 1}\,dx = \frac{1}{3}\int \frac{3x^2 + 6x + 9}{x^3 + 3x^2 + 9x + 1}\,dx = \frac{1}{3}\ln|x^3 + 3x^2 + 9x + 1| + C$

24. $\int \frac{1}{1 + e^x} e^x \, dx = \ln(1 + e^x) + C$

26. $-2 \int e^{1/(4x^2)} \left(-\frac{1}{2}x^{-3}\right) dx = -2e^{1/(4x^2)} + C$

28. $\int (e^x - e^{-x})^2 \, dx = \int (e^{2x} - 2 + e^{-2x}) dx$

$= \frac{1}{2}e^{2x} - 2x - \frac{1}{2}e^{-2x} + C$

30. $3 \int \frac{1}{2 + e^x} e^x \, dx = 3 \ln(2 + e^x) + C$

32. $\frac{-1}{3} \int \frac{1}{2 - e^{3x}} (-3e^{3x}) \, dx = -\frac{1}{3} \ln |2 - e^{3x}| + C$

34. $\int (3x^2 + e^x)^{1/2} (6x + e^x) \, dx = \frac{2}{3}(3x^2 + e^x)^{3/2} + C$

36. $\int \frac{2(e^x - e^{-x})}{(e^x + e^{-x})^2} dx = 2 \int (e^x + e^{-x})^{-2}(e^x - e^{-x}) dx = (2)\frac{(e^x + e^{-x})^{-1}}{-1} + C = -\frac{2}{e^x + e^{-x}} + C$

38. $\int \frac{1}{\sqrt{x + 1}} dx = \int (x + 1)^{-1/2} \, dx = 2\sqrt{x + 1} + C$

40. $\int \frac{x + 5}{x} dx = \int \left[1 + 5\left(\frac{1}{x}\right)\right] dx = x + 5 \ln |x| + C$

42. $\int \frac{x - 1}{4x} dx = \frac{1}{4} \int \left(1 - \frac{1}{x}\right) dx = \frac{1}{4}(x - \ln |x|) + C$

44. $\int \frac{3}{1 + e^{-3x}} dx = \int \frac{1}{e^{3x} + 1} 3e^{3x} \, dx = \ln(e^{3x} + 1) + C$

46. $\int \frac{x - 3}{x + 3} dx = \int \left(1 + \frac{-6}{x + 3}\right) dx = x - 6 \ln |x + 3| + C$

48. $\int \frac{5}{e^{-5x} + 7} dx = \frac{1}{7} \int \frac{1}{7e^{5x} + 1} 35e^{5x} \, dx = \frac{1}{7} \ln(7e^{5x} + 1) + C$

50. $f'(x) = \frac{x^3 - 4x^2 + 3}{x - 3}$; $(4, -1)$

$f(x) = \int \frac{x^3 - 4x^2 + 3}{x - 3} dx = \int \left(x^2 - x - 3 - \frac{6}{x - 3}\right) dx = \frac{x^3}{3} - \frac{x^2}{2} - 3x - 6 \ln |x - 3| + C$

$-1 = \frac{64}{3} - \frac{16}{2} - 12 - 6 \ln (1) + C$

$C = -\frac{7}{3}$

$f(x) = \frac{x^3}{3} - \frac{x^2}{2} - 3x - 6 \ln |x - 3| - \frac{7}{3}$

52. (a) $P = \int -125e^{-t/20} \, dt$

$= (-125)e^{-t/20}(-20) + C$

$= 2500e^{-t/20} + C$

Since $P = 0$ when $t = 0$, $C = 0$.
Hence, $P = 2500e^{-t/20}$.

(b) $P(15) = 2500e^{-15/20} \approx 1181$ fish

(c) We solve for t in the equation $P = 1$:

$1 = 2500e^{-t/20}$

$\frac{1}{2500} = e^{-t/20}$

$\ln\left(\frac{1}{2500}\right) = \frac{-t}{20}$

$t = 20 \ln 2500 \approx 156.5$ days

54. (a) $R = \int \left(50 - 0.02x + \frac{100}{x+1}\right) dx$

$= 50x - 0.01x^2 + 100 \ln(x+1) + C$

Since $R = 0$ when $x = 0$, we have $C = 0$ and $R = 50x - 0.01x^2 + 100 \ln(x+1)$.

(b)

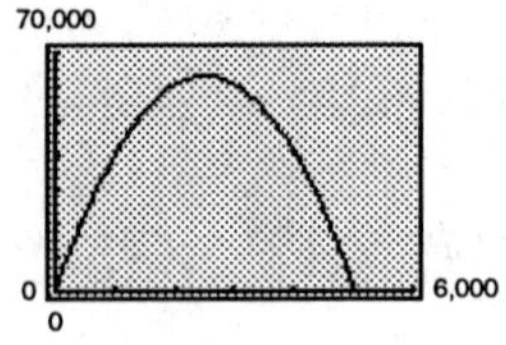

(c) When $x = 1500$, $R \approx \$53{,}231.39$.

(d) When $R = 60{,}230$, $x \approx 1949$ or $x \approx 3054$ units.

56. (a) $\frac{ds}{dt} = 58.17t^2 + \frac{26{,}272.58}{t}$

$s = 19.39t^3 + 26{,}272.58 \ln t + C$

1992: $55{,}484 = 19.39(12)^3 + 26{,}272.58 \ln(12) + C$

$C = -43{,}306.83$

$s(t) = 19.39t^3 + 26{,}272.58 \ln t - 43{,}306.83$

(b) 1994: $s(14) \approx \$79{,}234$ million

Section 5.4 Area and the Fundamental Theorem of Calculus

2. $\int_0^3 2x\,dx$

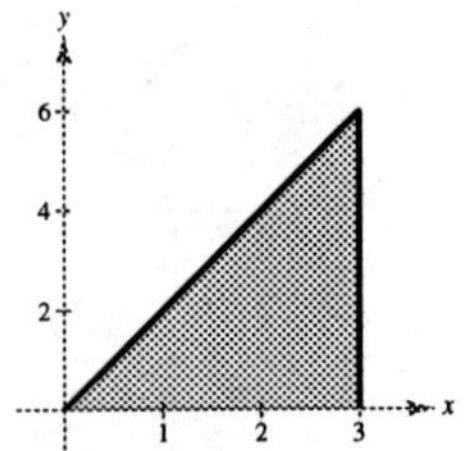

$\text{Area} = \frac{1}{2}(\text{Base})(\text{Height})$

$= \frac{1}{2}(3)(6) = 9$

4. $\int_{-3}^3 \sqrt{9 - x^2}\,dx$

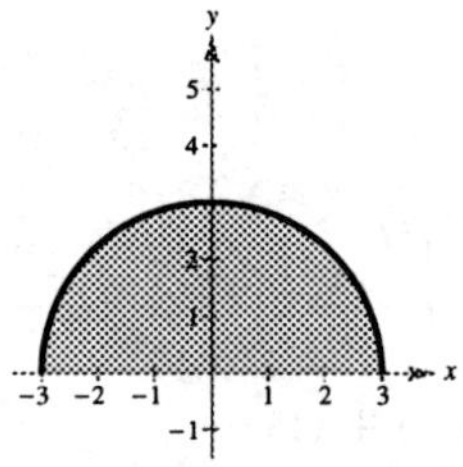

$\text{Area} = \frac{1}{2}\pi r^2$

$= \frac{1}{2}\pi(3)^2 = \frac{9}{2}\pi$

6. $A = \int_1^2 \frac{1}{x^2} dx = \int_1^2 x^{-2}\,dx = \frac{x^{-1}}{-1}\Big]_1^2 = -\frac{1}{x}\Big]_1^2 = -\frac{1}{2} + 1 = \frac{1}{2}$

8. $A = \int_0^4 3e^{-x/2}\,dx = 3(-2)\int_0^4 e^{-x/2}\left(-\frac{1}{2}\right)dx = -6e^{-x/2}\Big]_0^4 = -6(e^{-2} - 1) = 6(1 - e^{-2}) \approx 5.188$

10. $A = \int_1^4 \frac{x^2 + 4}{x}\,dx = \int_1^4 \left[x + 4\left(\frac{1}{x}\right)\right]dx = \left[\frac{x^2}{2} + 4 \ln |x|\right]_1^4 = (8 + 4 \ln 4) - \left(\frac{1}{2} + 4 \ln 1\right) = \frac{15}{2} + 8 \ln 2 \approx 13.045$

12. $\int_2^7 3\,dv = 3v\Big]_2^7 = 21 - 6 = 15$

14. $\int_2^5 (-3x + 4)dx = \left[-3\left(\frac{x^2}{2}\right) + 4x\right]_2^5 = \left(-\frac{75}{2} + 20\right) - (-6 + 8) = -\frac{39}{2}$

16. $\int_0^3 (3x^2 + x - 2)dx = \left[x^3 + \frac{x^2}{2} - 2x\right]_0^3 = \left(27 + \frac{9}{2} - 6\right) = \frac{51}{2}$

18. $\int_2^2 (x - 3)^4\, dx = \frac{(x - 3)^5}{5}\Big]_2^2 = -\frac{1}{5} + \frac{1}{5} = 0$

20. $\int_1^4 \sqrt{\frac{2}{x}}\, dx = \int_1^4 \sqrt{2}x^{-1/2}\, dx = \sqrt{2}(2x^{1/2})\Big]_1^4 = 2\sqrt{2x}\Big]_1^4 = 2\sqrt{8} - 2\sqrt{2} = 4\sqrt{2} - 2\sqrt{2} = 2\sqrt{2}$

22. $\int_0^1 \frac{x - \sqrt{x}}{3}\, dx = \frac{1}{3}\left(\frac{x^2}{2} - \frac{2}{3}x^{3/2}\right)\Big]_0^1 = \frac{1}{3}\left(\frac{1}{2} - \frac{2}{3}\right) - 0 = -\frac{1}{18}$

24. $\int_0^4 (x^{1/2} + x^{1/4})\, dx = \left[\frac{2}{3}x^{3/2} + \frac{4}{5}x^{5/4}\right]_0^4 = \frac{2}{3}(8) + \frac{4}{5}(4\sqrt[4]{4}) = \frac{16}{3} + \frac{16\sqrt{2}}{5} = \frac{16}{15}(5 + 3\sqrt{2})$

26. $\int_0^2 \frac{x}{\sqrt{1 + 2x^2}}\, dx = \frac{1}{4}\int_0^2 (1 + 2x^2)^{-1/2}(4x)\, dx = \frac{1}{4}(2)(1 + 2x^2)^{1/2}\Big]_0^2 = \frac{1}{2}\sqrt{1 + 2x^2}\Big]_0^2 = \frac{3}{2} - \frac{1}{2} = 1$

28. $\int_1^2 e^{1-x}\, dx = -\int_1^2 e^{1-x}(-1)\, dx = -e^{1-x}\Big]_1^2 = 1 - e^{-1} = \frac{e - 1}{e} \approx 0.632$

30. $\int_{-1}^1 (e^x - e^{-x})\, dx = \left[e^x + e^{-x}\right]_{-1}^1 = (e + e^{-1}) - (e^{-1} + e) = 0$

32. $\int_0^3 |2x - 3|\, dx = \int_0^{3/2} -(2x - 3)\, dx + \int_{3/2}^3 (2x - 3)\, dx$

$= \left[-x^2 + 3x\right]_0^{3/2} + \left[x^2 - 3x\right]_{3/2}^3 = \left(-\frac{9}{4} + \frac{9}{2}\right) + (9 - 9) - \left(\frac{9}{4} - \frac{9}{2}\right) = \frac{9}{2}$

34. $\int_{-4}^4 (4 - |x|)\, dx = \int_{-4}^0 (4 + x)\, dx + \int_0^4 (4 - x)\, dx$

$= \left[4x + \frac{x^2}{2}\right]_{-4}^0 + \left[4x - \frac{x^2}{2}\right]_0^4 = 0 - (-16 + 8) + (16 - 8) - 0 = 16$

36. $\int_2^3 \frac{x + 1}{x^2 + 2x - 3}\, dx = \left[\frac{1}{2}\ln|x^2 + 2x - 3|\right]_2^3 = \frac{1}{2}\ln 12 - \frac{1}{2}\ln 5 \approx 0.438$

38. $\int_1^2 \frac{(2 + \ln x)^3}{x}\, dx = \left[\frac{(2 + \ln x)^4}{4}\right]_1^2 = \frac{1}{4}(2 + \ln 2)^4 - 4 \approx 9.152$

40. $\int_0^2 (x + 4)dx = \left[\frac{x^2}{2} + 4x\right]_0^2 = 10$

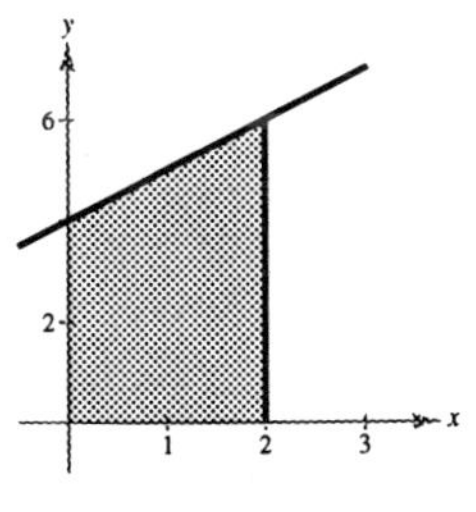

42. $\displaystyle\int_0^1 \sqrt{x}(1-x)dx = \int_0^1 (x^{1/2} - x^{3/2})dx$

$$= \left[\frac{2}{3}x^{3/2} - \frac{2}{5}x^{5/2}\right]_0^1$$

$$= \left(\frac{2}{3} - \frac{2}{5}\right) = \frac{4}{15}$$

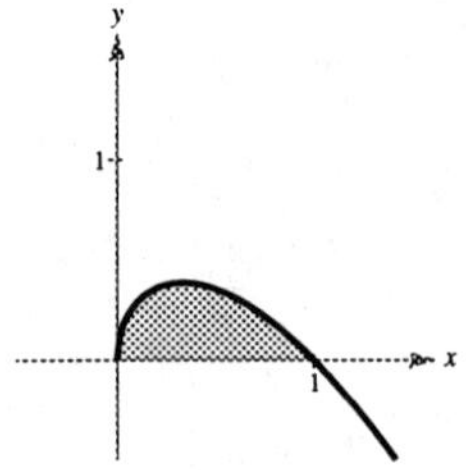

44. $\displaystyle\int_0^{\ln 6} \frac{e^x}{2}dx = \left[\frac{e^x}{2}\right]_0^{\ln 6} = 3 - \frac{1}{2} = \frac{5}{2}$

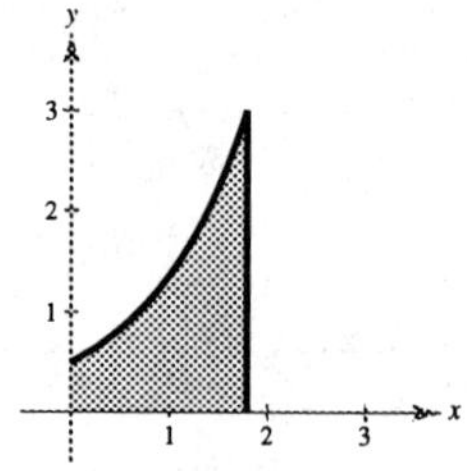

46. $\displaystyle A = \int_0^4 \left(1 + \sqrt{x}\right)dx = \left[x + \frac{2}{3}x^{3/2}\right]_0^4 = 4 + \frac{16}{3} = \frac{28}{3}$

48. $\displaystyle A = \int_{-2}^1 3e^x\,dx = 3e^x\Big]_{-2}^1 = 3(e - e^{-2}) \approx 7.75$

50. (a) $\displaystyle\int_0^5 2g(x)\,dx = 2\int_0^5 g(x)\,dx = 2(3) = 6$

(b) $\displaystyle\int_5^0 f(x)\,dx = -\int_0^5 f(x)dx = -8$

(c) $\displaystyle\int_5^5 f(x)\,dx = 0$

(d) $\displaystyle\int_0^5 [f(x) - f(x)]\,dx = \int_0^5 0\,dx = 0$

52. Average value $\displaystyle = \frac{1}{10-0}\int_0^{10} 5e^{0.2(x-10)}\,dx$

$$= \frac{1}{10}(25 - 25e^{-2})$$

$$= \frac{5}{2} - \frac{5}{2}e^{-2}$$

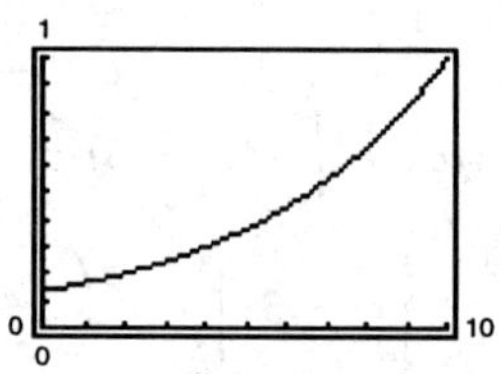

To find the x-values for which $f(x) = \frac{5}{2} - \frac{5}{2}e^{-2}$, we let $5e^{0.2(x-10)} = \frac{5}{2} - \frac{5}{2}e^{-2}$ and solve for x. Using a graphing utility, we obtain $x \approx 5.807$.

54. Average value $\displaystyle = \frac{1}{7-0}\int_0^7 \frac{5x}{x^2+1}dx$

$$= \frac{1}{7}\cdot\frac{5}{2}\ln(x^2+1)\Big]_0^7$$

$$= \frac{5}{14}\ln 50$$

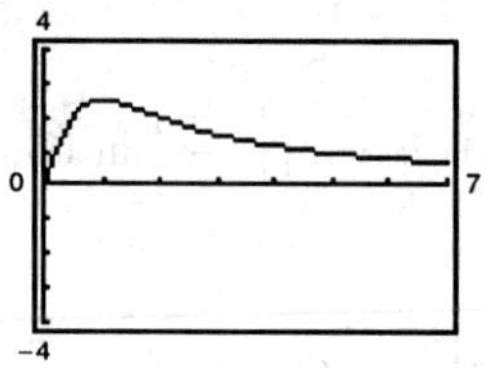

To find the x-values for which $f(x) = \frac{5}{14}\ln 50$, we solve for x as follows:

$$\frac{5x}{x^2+1} = \frac{5}{14}\ln 50$$

Using a graphing utility, we obtain $x \approx 0.3055$.

56. Average value $= \dfrac{1}{2-0}\displaystyle\int_0^2 \frac{1}{(x-3)^2}\,dx = \frac{1}{2}\int_0^2 (x-3)^{-2}\,dx = -\frac{1}{2(x-3)}\bigg]_0^2 = \frac{1}{2} - \frac{1}{6} = \frac{1}{3}$

To find the x-value(s) for which $f(x) = \frac{1}{3}$, we let

$$\frac{1}{(x-3)^2} = \frac{1}{3}$$

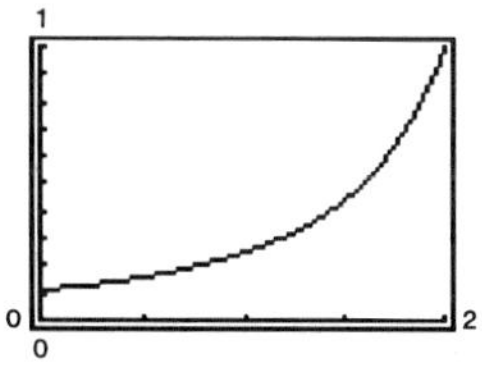

and solve for x to obtain the following.

$$3 = (x-3)^2$$

$$\pm\sqrt{3} = x - 3$$

$$3 \pm \sqrt{3} = x$$

$$x = 3 - \sqrt{3} \approx 1.268 \text{ in the interval } (0, 2).$$

58. $g(x) = x^3 - 2x$ is odd.

$[g(x) = -g(-x)]$

60. $f(t) = 5t^4 + 1$ is even.

$[f(t) = f(-t)]$

62. (a) $\displaystyle\int_{-2}^0 x^3\,dx = -\int_0^2 x^3\,dx = -4$

(b) $\displaystyle\int_{-2}^2 x^3\,dx = 0$ since $f(x) = x^3$ is an odd function.

(c) $\displaystyle\int_0^2 3x^3\,dx = 3\int_0^2 x^3\,dx = 3(4) = 12$

64. $\displaystyle\Delta C = \int_{10}^{13} \frac{20{,}000}{x^2}\,dx$

$\displaystyle = -\frac{20{,}000}{x}\bigg]_{10}^{13}$

$\displaystyle = -20{,}000\left(\frac{1}{13} - \frac{1}{10}\right) \approx \$461.54$

66. $\displaystyle\Delta R = \int_{500}^{503} 75\left(20 + \frac{900}{x}\right)dx$

$\displaystyle = \Big[1500x + 67{,}500\ln x\Big]_{500}^{503}$

$\approx \$4903.79$

68. $\displaystyle\Delta P = \int_{125}^{128} \frac{25}{2}\left(40 - 3\sqrt{x}\right)dx$

$\displaystyle = \frac{25}{2}\left[40x - 2x^{3/2}\right]_{125}^{128} = \frac{25}{2}[40(128) - 2(128)^{3/2}] - \frac{25}{2}[40(125) - 2(125)^{3/2}] \approx \$234.69$

70. Capital accumulation $\displaystyle = \int_0^5 500\sqrt{t+1}\,dt = \frac{1000}{3}(t+1)^{3/2}\bigg]_0^5 = \frac{1000}{3}(6)^{3/2} - \frac{1000}{3} \approx \$4565.65$

72. Capital accumulation $\displaystyle = \int_0^5 \frac{12{,}000t}{(t^2+2)^2}\,dt = -\frac{6000}{t^2+2}\bigg]_0^5 = 6000\left(-\frac{1}{27} + \frac{1}{2}\right) \approx \$2777.78$

74. $\displaystyle V = \int_0^3 10{,}000(t-6)\,dt = 10{,}000\left(\frac{t^2}{2} - 6t\right)\bigg]_0^3 = 10{,}000\left(\frac{9}{2} - 18\right) = -\$135{,}000.00$

76. $\displaystyle \frac{1}{R-0}\int_0^R k(R^2 - r^2)\,dr = \frac{k}{R}\left(R^2 r - \frac{r^3}{3}\right)\bigg]_0^R = \frac{2kR^2}{3}$

78. (a) $$\frac{dN}{dt} = k[650 - N(t)]$$

$$\frac{dN}{650 - N} = k\,dt$$

$$\int \frac{dN}{650 - N} = \int k\,dt$$

$$-\ln|650 - N| = kt + C_1$$

$$\ln|650 - N| = -kt - C_1$$

$$650 - N = e^{-kt - C_1} = Ce^{-kt}$$

$$N = 650 - Ce^{-kt}$$

$$N(0) = 300 \Rightarrow 300 = 650 - C \Rightarrow C = 350$$

$$N(2) = 500 \Rightarrow 500 = 650 - 350e^{-2k}$$

$$\Rightarrow \frac{150}{350} = e^{-2k}$$

$$\Rightarrow k = -\frac{1}{2}\ln\frac{3}{7} = \frac{1}{2}\ln\frac{7}{3} \approx 0.42365$$

$$N(t) = 650 - 350e^{-0.42365t}$$

(b) $$\frac{1}{5 - 0}\int_0^5 N(t)\,dt = \frac{1}{5}\int_0^5 (650 - 350e^{-0.42365t})dt \approx 504 \text{ coyotes}$$

80. $$\int_{1/2}^{1} (x + 1)\sqrt{1 - x}\,dx = \frac{17}{60}\sqrt{2}$$

82. $$\int_0^1 x^3(x^3 + 1)^3\,dx = \frac{1921}{1820}$$

Section 5.5 The Area of a Region Bounded by Two Graphs

2. $$A = \int_{-2}^{2} [(2x + 5) - (x^2 + 2x + 1)]\,dx = \int_{-2}^{2} (-x^2 + 4)\,dx = \left[-\frac{x^3}{3} + 4x\right]_{-2}^{2} = \frac{32}{3}$$

4. $$A = \int_0^1 (x^2 - x^3)\,dx = \left[\frac{x^3}{3} - \frac{x^4}{4}\right]_0^1 = \frac{1}{12}$$

6. $$A = 2\int_0^1 [(x - 1)^3 - (x - 1)]dx$$

$$= 2\left[\frac{(x - 1)^4}{4} - \frac{(x - 1)^2}{2}\right]_0^1 = \frac{1}{2}$$

8. $$A = \int_1^2 \left[(-x + 3) - \frac{2}{x}\right] dx$$

$$= \left[-\frac{x^2}{2} + 3x - 2\ln x\right]_1^2$$

$$= (4 - 2\ln 2) - \left(\frac{1}{2}\right)$$

$$= \frac{3}{2} - 2\ln 2$$

10. The region is bounded by the graphs of $y = 1 - x^2$ and $y = x^2 - 1$, from $x = -1$ to $x = 1$, as shown in the figure.

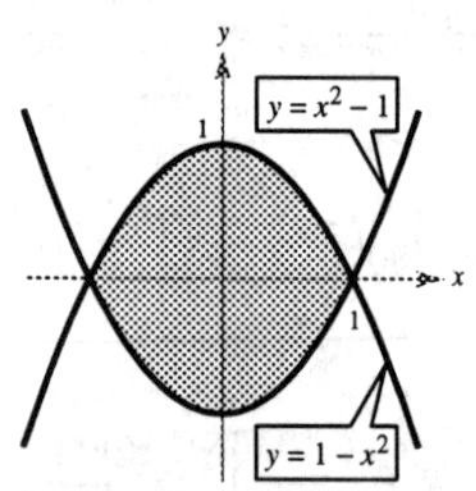

12. The region is bounded by the graphs of $y = x - 6$ and $y = x^2 + 5x - 6$ from $x = -4$ to $x = 0$, as shown in the figure.

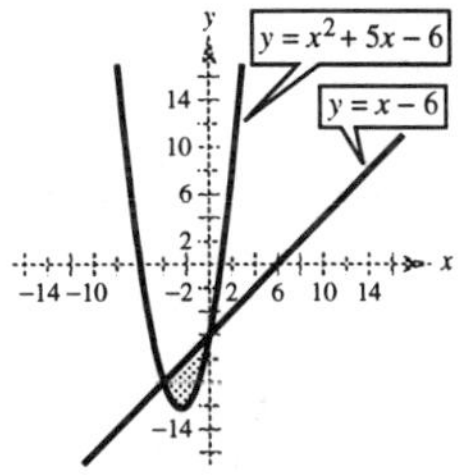

14. The region is bounded by the graphs of $x = y + 6$ and $x = y^2$ from $y = -2$ to $y = 3$, as shown in the figure.

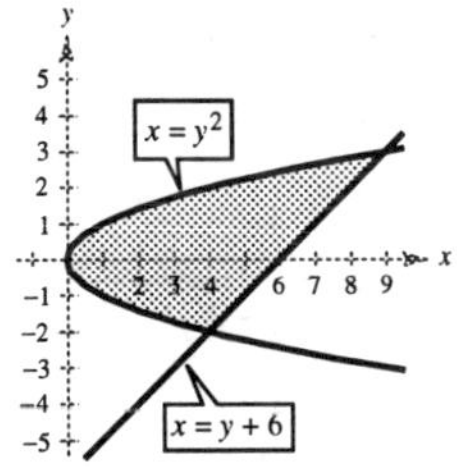

16. The points of intersection are found by evaluating $x = 1$ and $x = 5$ for the function $y = \dfrac{1}{x^2}$.

$$A = \int_1^5 \frac{1}{x^2}\,dx = -\frac{1}{x}\Big]_1^5 = \frac{4}{5}$$

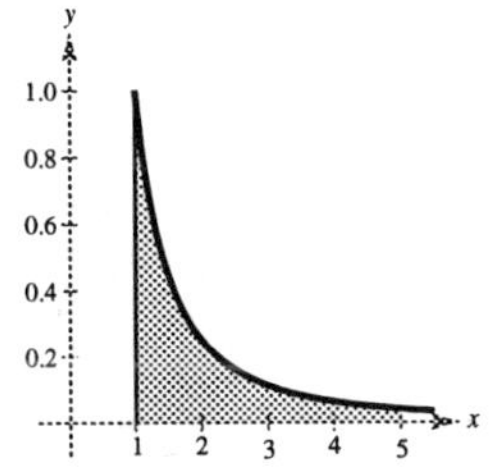

18. The point of intersection of the two graphs is found by equating y-values and solving for x.

$$x^3 - 2x + 1 = -2x$$
$$x^3 + 1 = 0$$
$$x = -1$$

$$A = \int_{-1}^{1} [(x^3 - 2x + 1) - (-2x)]\,dx$$
$$= \int_{-1}^{1} (x^3 + 1)\,dx = \left[\frac{x^4}{4} + x\right]_{-1}^{1} = 2$$

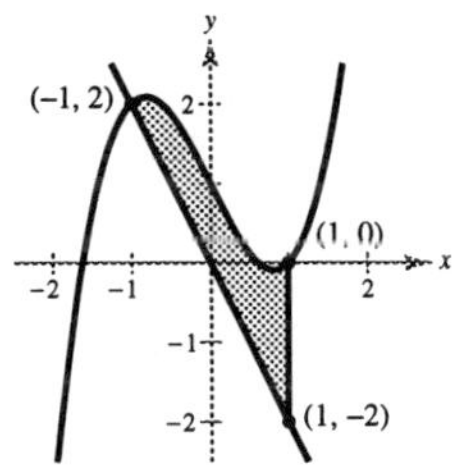

20. The points of intersection of f and g are found by setting $f(x) = g(x)$ and solving for x.

$$\sqrt{3x} + 1 = x + 1$$
$$\sqrt{3x} = x$$
$$3x = x^2$$
$$x^2 - 3x = 0$$
$$x(x - 3) = 0$$
$$x = 0,\ 3$$

$$A = \int_0^3 \left[\left(\sqrt{3x} + 1\right) - (x + 1)\right]dx = \int_0^3 \left(\sqrt{3x} - x\right)dx$$
$$= \left[\frac{2}{9}(3x)^{3/2} - \frac{x^2}{2}\right]_0^3 = \frac{3}{2}$$

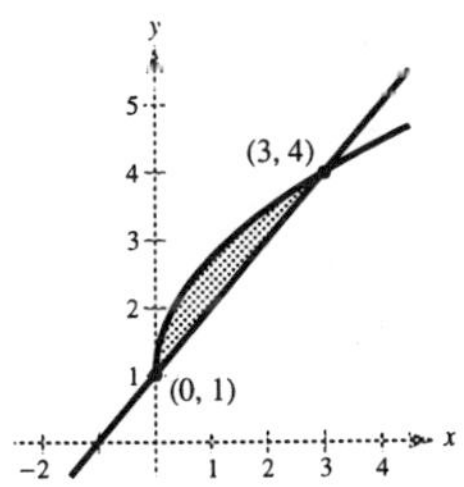

22. The points of intersection are found by setting $f(y)$ equal to $g(y)$ and solving for y.

$$y^2 = y + 2$$
$$y^2 - y - 2 = 0$$
$$(y + 1)(y - 2) = 0$$
$$y = -1, 2$$

$$A = \int_{-1}^{2} [(y + 2) - y^2]\,dy = \left[\frac{y^2}{2} + 2y - \frac{y^3}{3}\right]_{-1}^{2}$$
$$= \left(2 + 4 - \frac{8}{3}\right) - \left(\frac{1}{2} - 2 + \frac{1}{3}\right) = \frac{9}{2}$$

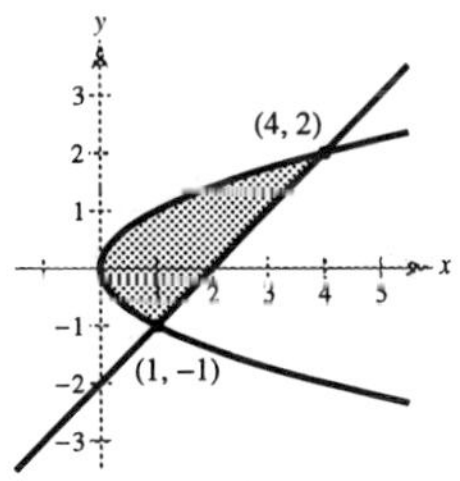

24. $A = \int_0^9 \sqrt{y}\,dy = \frac{2}{3}y^{3/2}\Big]_0^9 = 18$

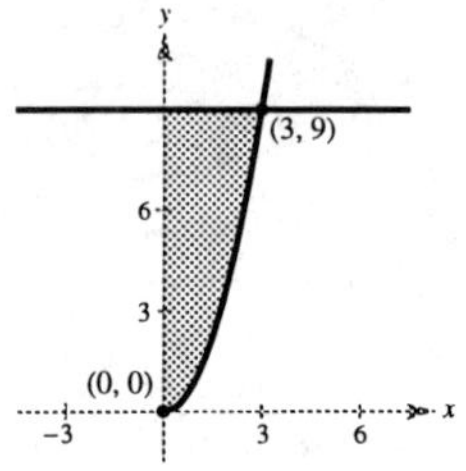

26. $A = \int_{1/2}^{1}\left[\frac{1}{x} - (-e^x)\right]dx = \ln x + e^x\Big]_{1/2}^{1}$

$= e - \left(\ln\frac{1}{2} + e^{1/2}\right) = e + \ln 2 - e^{1/2} \approx 1.763$

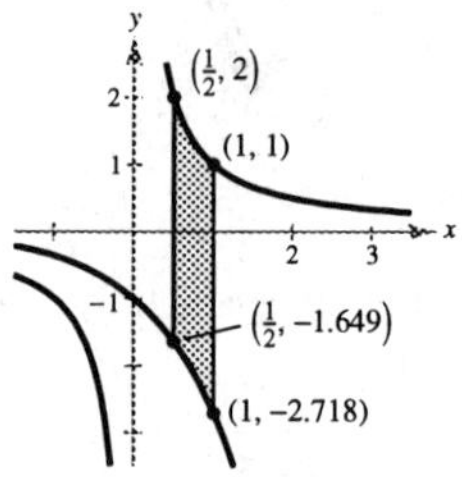

28. $A = \int_1^2 x^2 dx + \int_2^4 \frac{8}{x}dx = \frac{x^3}{3}\Big]_1^2 + 8\ln x\Big]_2^4$

$= \left(\frac{8}{3} - \frac{1}{3}\right) + 8(\ln 4 - \ln 2) = \frac{7}{3} + 8\ln 2$

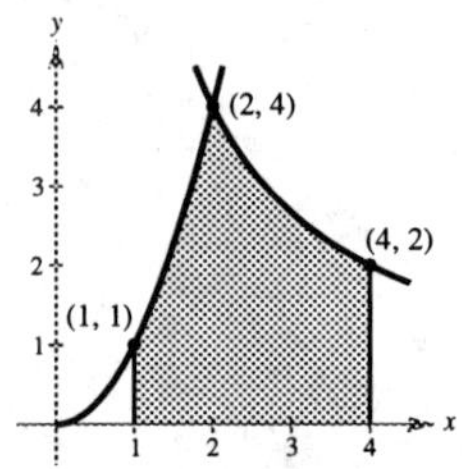

30. $A = \int_1^3 \frac{e^{1/x}}{x^2}dx = -e^{1/x}\Big]_1^3 = -e^{1/3} + e^1 \approx 1.323$

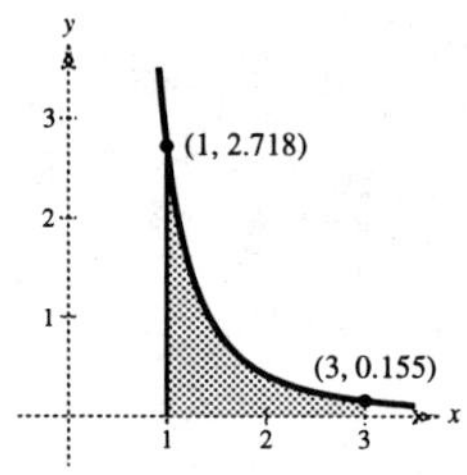

32. The points of intersection of f and g are found by setting $f(x) = g(x)$ and solving for x.

$$3 - 2x - x^2 = 0$$
$$(3 + x)(1 - x) = 0$$
$$x = -3,\ 1$$

$$A = \int_{-3}^{1}(3 - 2x - x^2)\,dx = \left(3x - x^2 - \frac{x^3}{3}\right)\Big]_{-3}^{1}$$
$$= \left(3 - 1 - \frac{1}{3}\right) - (-9 - 9 + 9) = \frac{32}{3}$$

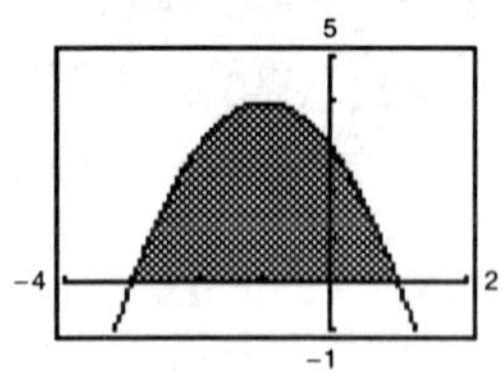

34. The points of intersection of f and g are found by setting $f(x) = g(x)$ and solving for x.

$$-x^2 + 4x + 2 = x + 2$$
$$0 = x^2 - 3x = x(x - 3)$$
$$x = 0,\ 3$$

$$A = \int_0^3 [(-x^2 + 4x + 2) - (x + 2)]dx$$
$$= \int_0^3 (-x^2 + 3x)\,dx = \left[-\frac{x^3}{3} + \frac{3x^2}{2}\right]_0^3 = \frac{9}{2}$$

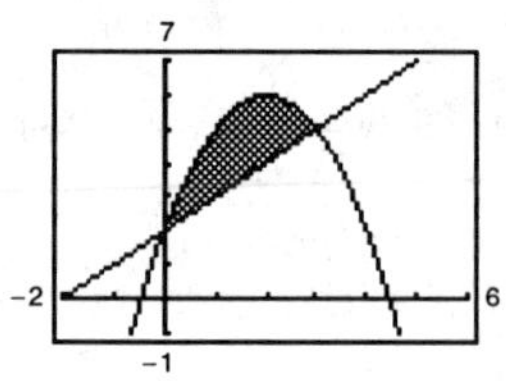

36. The line through (6, 4) and (0, 0) is $y = \frac{2}{3}x$ and the line through (6, 4) and (4, 0) is $y = 2x - 8$. Therefore, the area is given by the following.

$$A = \int_0^4 \left[\frac{y+8}{2} - \frac{3}{2}y\right] dy = \int_0^4 (4 - y)dy = \left[4y - \frac{y^2}{2}\right]_0^4 = 8$$

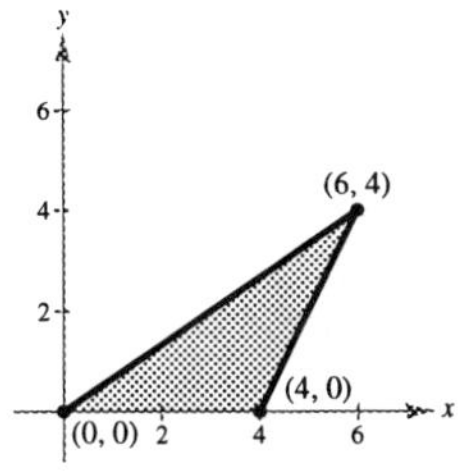

38. The point of equilibrium is found by equating $300 - x = 100 + x$ to obtain $x = 100$ and $p = 200$.

$$CS = \int_0^{100} [(300 - x) - 200]dx = \left[100x - \frac{x^2}{2}\right]_0^{100} = 5000$$

$$PS = \int_0^{100} [200 - (100 + x)]\,dx = \left[100x - \frac{x^2}{2}\right]_0^{100} = 5000$$

40. The point of equilibrium is found by equating $1000 - 0.4x^2 = 42x$ to obtain $x = 20$ and $p = 840$.

$$CS = \int_0^{20} [(1000 - 0.4x^2) - 840]dx = \left[160x - \frac{0.4x^3}{3}\right]_0^{20} \approx 2133.33$$

$$PS = \int_0^{20} [840 - 42x]dx = \left[840x - 21x^2\right]_0^{20} = 8400$$

42. The point of equilibrium is found by equating the supply and demand functions.

$$\begin{aligned}
\sqrt{25 - 0.1x} &= \sqrt{9 + 0.1x} - 2 \\
25 - 0.1x &= 9 + 0.1x - 4\sqrt{9 + 0.1x} + 4 \\
12 - 0.2x &= -4\sqrt{9 + 0.1x} \\
0.05x - 3 &= \sqrt{9 + 0.1x} \\
0.0025x^2 - 0.3x + 9 &= 9 + 0.1x \\
0.0025x^2 - 0.4x &= 0 \\
x &= 160 \text{ and } p = 3
\end{aligned}$$

$$CS = \int_0^{160} \left[\sqrt{25 - 0.1x} - 3\right] dx = \left[-\frac{20}{3}\left(\sqrt{25 - 0.1x}\right)^3 - 3x\right]_0^{160} \approx 173.33$$

$$PS = \int_0^{160} \left[3 - \left(\sqrt{9 + 0.1x} - 2\right)\right] dx = \left[5x - \frac{20}{3}\left(\sqrt{9 + 0.1x}\right)^3\right]_0^{160} \approx 146.67$$

44. If the demand and supply functions do not intersect, then there would be no intersection point for which supply = demand.

46. The model R_1 projects greater revenue than R_2. The difference in total revenue is

$$\int_6^{10} (R_1 - R_2)\,dt = \int_6^{10} [(7.21 + 0.26t + 0.02t^2) - (7.21 + 0.1t + 0.01t^2)]dt = \int_6^{10} (0.16t + 0.01t^2)\,dt \approx \$7.73 \text{ billion.}$$

48. $\displaystyle\int_{25}^{50} [N_1(t) - N_2(t)]\,dt \approx 4531$ individuals

50. Demand: (6000, 331), (8000, 303)

$$m = \frac{303 - 331}{8000 - 6000} = \frac{-7}{500}$$

$$p - 331 = -\frac{7}{500}(x - 6000)$$

$$p = -\frac{7x}{500} + 415$$

Point of equilibrium: $-\frac{7x}{500} + 415 = \frac{11x}{400}$

$$x = 10{,}000$$

$$p = 275$$

$$CS = \int_0^{10{,}000} \left[\left(-\frac{7x}{500} + 415\right) - 275\right]dx = \int_0^{10{,}000} \left(-\frac{7x}{500} + 140\right)dx = \left[-\frac{7x^2}{1000} + 140x\right]_0^{10{,}000} = \$700{,}000$$

$$PS = \int_0^{10{,}000} \left(275 - \frac{11x}{400}\right)dx = \left[275x - \frac{11x^2}{800}\right]_0^{10{,}000} = \$1{,}375{,}000$$

52. $\int_0^{10} [R(t) - C(t)]\,dt = \int_0^{10} [100 - (60 + 0.2t^2)]\,dt = \int_0^{10} (40 - 0.2t^2)\,dt = \left[40t - \frac{2}{30}t^3\right]_0^{10} = 333.33$ million dollars

54. Income inequality $= \int_0^{100} [x - f(x)]\,dx = \int_0^{10} (x - 0.0614x^{3/2} - 0.3867x + 0.0387)\,dx \approx 614.37$

Section 5.6 The Definite Integral as the Limit of a Sum

2. The midpoints of the four intervals are $\frac{1}{4}, \frac{3}{4}, \frac{5}{4}$, and $\frac{7}{4}$. The approximate area is

$$A \approx \frac{2-0}{4}\left[\left(\sqrt{\frac{1}{4}} + 1\right) + \left(\sqrt{\frac{3}{4}} + 1\right) + \left(\sqrt{\frac{5}{4}} + 1\right) + \left(\sqrt{\frac{7}{4}} + 1\right)\right]$$

$$= \frac{1}{2}\left[4 + \frac{1}{2} + \frac{\sqrt{3}}{2} + \frac{\sqrt{5}}{2} + \frac{\sqrt{7}}{2}\right] \approx 3.903.$$

The exact area is $A = \int_0^2 (\sqrt{x} + 1)dx = \left[\frac{2}{3}x^{3/2} + x\right]_0^2 \approx 3.886.$

4. The midpoints of the four intervals are $-\frac{3}{4}, -\frac{1}{4}, \frac{1}{4}$, and $\frac{3}{4}$. The approximate area is

$$A \approx \frac{1-(-1)}{4}\left[f\left(-\frac{3}{4}\right) + f\left(-\frac{1}{4}\right) + f\left(\frac{1}{4}\right) + f\left(\frac{3}{4}\right)\right] = \frac{1}{2}\left[\frac{7}{16} + \frac{15}{16} + \frac{15}{16} + \frac{7}{16}\right] = \frac{11}{8} = 1.375.$$

The exact area is $A = \int_{-1}^1 (1 - x^2)\,dx = \left[x - \frac{x^3}{3}\right]_{-1}^1 = \frac{4}{3} \approx 1.333.$

6. The midpoints of the four intervals are $\frac{19}{8}, \frac{25}{8}, \frac{31}{8}$, and $\frac{37}{8}$. The approximate area is

$$A \approx \frac{5-2}{4}\left[\frac{25}{8} + \frac{43}{8} + \frac{61}{8} + \frac{79}{8}\right] = 19.5.$$

The exact area is $A = \int_2^5 (3x - 4)\,dx = \left[\frac{3}{2}x^2 - 4x\right]_2^5 = 19.5.$

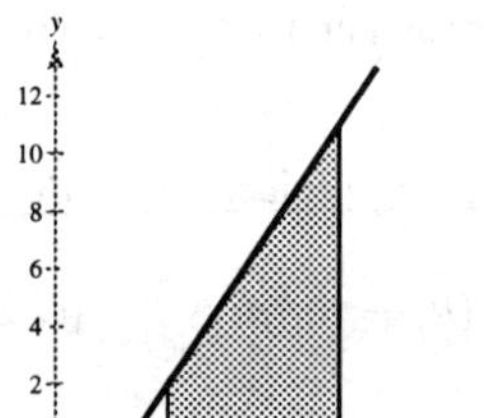

8. The midpoints of the four intervals are $\frac{1}{8}, \frac{3}{8}, \frac{5}{8}$, and $\frac{7}{8}$. The approximate area is

$$A \approx \frac{1-0}{4}\left[\frac{127}{512} + \frac{357}{512} + \frac{515}{512} + \frac{553}{512}\right] \approx 0.758.$$

The exact area is $A = \displaystyle\int_0^1 (2x - x^3)\,dx = \left[x^2 - \frac{x^4}{4}\right]_0^1 = \frac{3}{4} = 0.750.$

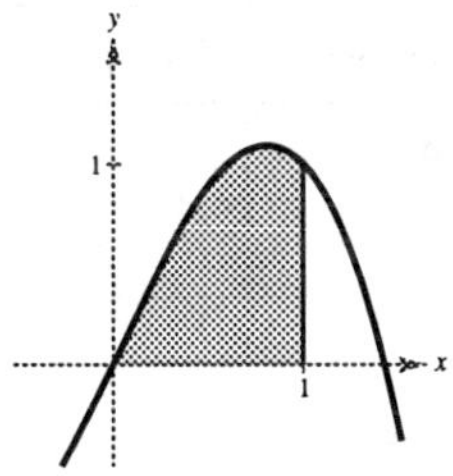

10. The midpoints of the four intervals are $-\frac{7}{8}, -\frac{5}{8}, -\frac{3}{8}$, and $-\frac{1}{8}$. The approximate area is

$$A \approx \frac{0-(-1)}{4}\left[\frac{735}{512} + \frac{325}{512} + \frac{99}{512} + \frac{9}{512}\right] = \frac{73}{128} \approx 0.5703.$$

The exact area is $A = \displaystyle\int_{-1}^0 (x^2 - x^3)\,dx = \left[\frac{x^3}{3} - \frac{x^4}{4}\right]_{-1}^0 = \frac{7}{12} \approx 0.5833.$

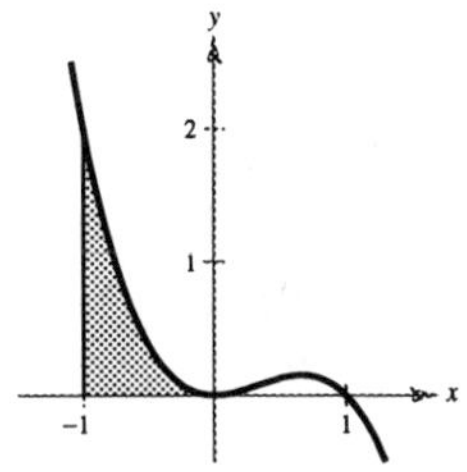

12. The midpoints of the four intervals are $\frac{3}{8}, \frac{9}{8}, \frac{15}{8}$, and $\frac{21}{8}$. The approximate area is

$$A \approx \frac{3-0}{4}\left[\frac{189}{512} + \frac{1215}{512} + \frac{2025}{512} + \frac{1323}{512}\right] \approx 6.961.$$

The exact area is $A = \displaystyle\int_0^3 (3x^2 - x^3)\,dx = \left[x^3 - \frac{x^4}{4}\right]_0^3 = \frac{27}{4} = 6.750.$

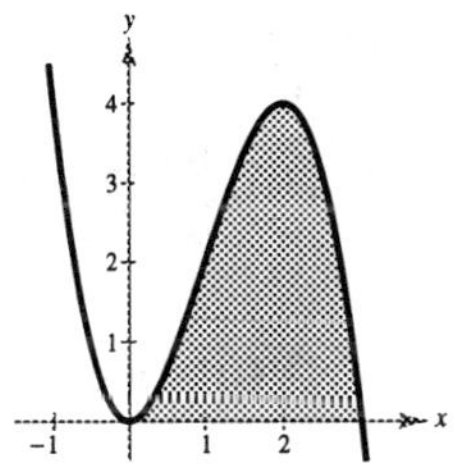

14. $\displaystyle\int_0^4 (2x^3 + 3)\,dx = 140$

Using the Midpoint Rule with $n = 80$, we get 139.99.

16. $\displaystyle\int_1^2 (x^3 - 1)\,dx = 2.75$

Using the Midpoint Rule with $n = 7$, we get 2.74.

18. The midpoints of the four intervals are $\frac{1}{2}, \frac{3}{2}, \frac{5}{2}$, and $\frac{7}{2}$. The approximate area is

$$A \approx \frac{4-0}{4}\left[\frac{7}{4} + \frac{15}{4} + \frac{15}{4} + \frac{7}{4}\right] = 11.$$

The exact area is $A = \displaystyle\int_0^4 (4y - y^2)\,dy = \left[2y^2 - \frac{y^3}{3}\right]_0^4 = \frac{32}{3} \approx 10.667.$

20. For $n = 8$, we have $x_0 = 1, x_1 = \frac{5}{4}, x_2 = \frac{3}{2}, x_3 = \frac{7}{4}, x_4 = 2, x_5 = \frac{9}{4}, x_6 = \frac{5}{2}, x_7 = \frac{11}{4}, x_8 = 3$, and the approximation is

$$A \approx \frac{3-1}{2(8)}\left[1 + 2\left(\frac{16}{25}\right) + 2\left(\frac{4}{9}\right) + 2\left(\frac{16}{49}\right) + 2\left(\frac{1}{4}\right) + 2\left(\frac{16}{81}\right) + 2\left(\frac{4}{25}\right) + 2\left(\frac{16}{121}\right) + \frac{1}{9}\right] \approx 0.677.$$

The exact area is $A = \displaystyle\int_1^3 \frac{1}{x^2}\,dx = -\frac{1}{x}\Big]_1^3 = \frac{2}{3} \approx 0.667.$

The midpoint approximation is 0.662, which is better.

22. For $n = 4$, we have $x_0 = 0, x_1 = 1, x_2 = 2, x_3 = 3, x_4 = 4$, and the approximation is

$$\int_0^4 \sqrt{1 + x^2}\,dx \approx \frac{4-0}{2(4)}\left[1 + 2\sqrt{2} + 2\sqrt{5} + 2\sqrt{10} + \sqrt{17}\right] \approx 9.374.$$

24. For $n = 4$, we have $x_0 = 1, x_1 = 2, x_2 = 3, x_3 = 4, x_4 = 5$, and the approximation is

$$\int_1^5 \frac{\sqrt{x-1}}{x}\,dx \approx \frac{5-1}{2(4)}\left[0 + 2\left(\frac{1}{2}\right) + 2\left(\frac{\sqrt{2}}{3}\right) + 2\left(\frac{\sqrt{3}}{4}\right) + \frac{2}{5}\right] \approx 1.604.$$

26.

n	Midpoint Rule	Trapezoidal Rule
4	5.459	5.433
8	5.452	5.446
12	5.451	5.448
16	5.450	5.449
20	5.450	5.449

28. $A = \int_0^4 x\sqrt{\frac{4-x}{4+x}}\,dx$

$\approx \frac{4-0}{2(10)}\left[f(0) + 2f\left(\frac{2}{5}\right) + 2f\left(\frac{4}{5}\right) + 2f\left(\frac{6}{5}\right) + 2f\left(\frac{8}{5}\right) + 2f(2) + 2f\left(\frac{12}{5}\right) + 2f\left(\frac{14}{5}\right) + 2f\left(\frac{16}{5}\right) + 2f\left(\frac{18}{5}\right) + f(4)\right] \approx 3.347$

30. For the Midpoint Rule, use the average length.

(a) Midpoint Rule: $\frac{160}{8}(25 + 52 + 68 + 82 + 77.5 + 74 + 77.5 + 40) \approx 9920$ sq. ft.

(b) Trapezoidal Rule: $\frac{160}{2(8)}[0 + 2(50) + 2(54) + 2(82) + 2(82) + 2(73) + 2(75) + 2(80) + 0] \approx 9920$ sq. ft.

Section 5.7 Volumes of Solids of Revolution

2. $V = \pi\int_0^1 (x^2)^2\,dx = \pi\left(\frac{x^5}{5}\right)\Big]_0^1 = \frac{\pi}{5}$

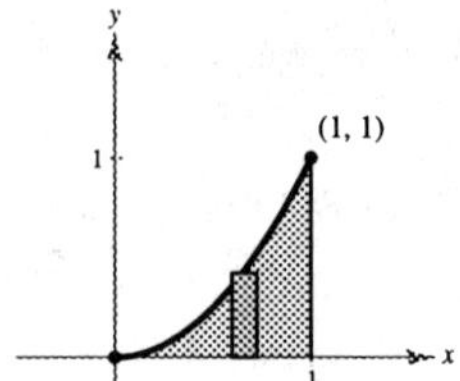

4. $V = \pi\int_{-2}^2 \left(\sqrt{4-x^2}\right)^2 dx = \pi\int_{-2}^2 (4-x^2)\,dx$

$= \pi\left[4x - \frac{x^3}{3}\right]_{-2}^2 = \frac{32\pi}{3}$

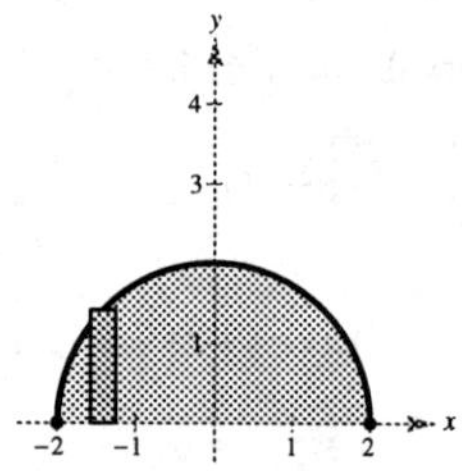

6. $V = \pi\int_0^4 x^2\,dx = \pi\left(\frac{x^3}{3}\right)\Big]_0^4 = \frac{64\pi}{3}$

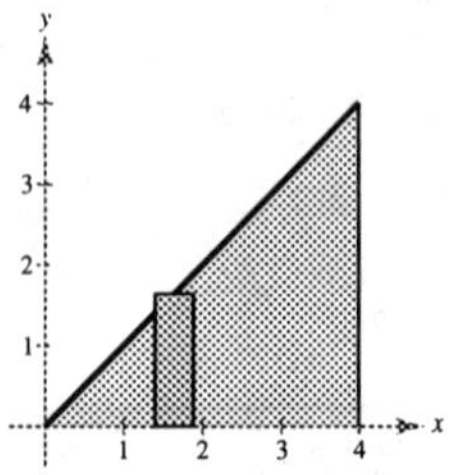

8. The points of intersection of the two graphs occur when $x = \pm 2$.

$$V = \pi \int_{-2}^{2} [(5)^2 - (x^2 + 1)^2]\,dx$$

$$= 2\pi \int_{0}^{2} (24 - 2x^2 - x^4)\,dx$$

$$= 2\pi\left[24x - \frac{2}{3}x^3 - \frac{1}{5}x^5\right]_0^2$$

$$= \frac{1088\pi}{15}$$

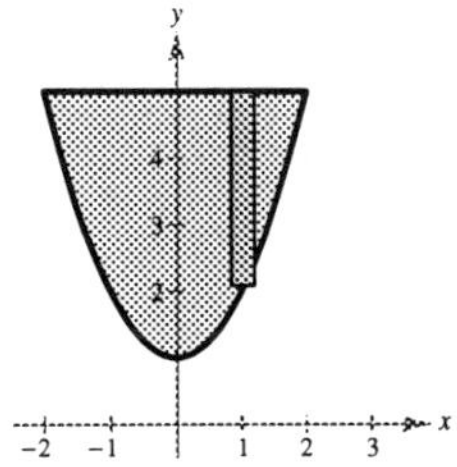

10. $V = \pi \int_{0}^{1} [(e^{x-1})^2 - x^2]\,dx$

$$= \pi \int_{0}^{1} (e^{2x-2} - x^2)\,dx$$

$$= \pi\left[\frac{1}{2}e^{2x-2} - \frac{x^3}{3}\right]_0^1$$

$$= \pi\left[\left(\frac{1}{2} - \frac{1}{3}\right) - \left(\frac{1}{2}e^{-2}\right)\right]$$

$$= \pi\left(\frac{1}{6} - \frac{1}{2e^2}\right) \approx 0.311$$

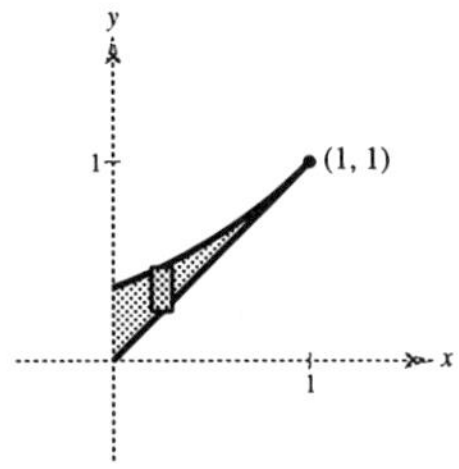

12. $V = \pi \int_{0}^{4} (\sqrt{x})^2\,dx$

$$= \pi \int_{0}^{4} x\,dx$$

$$= \pi\left(\frac{x^2}{2}\right)\Bigg]_0^4 = 8\pi$$

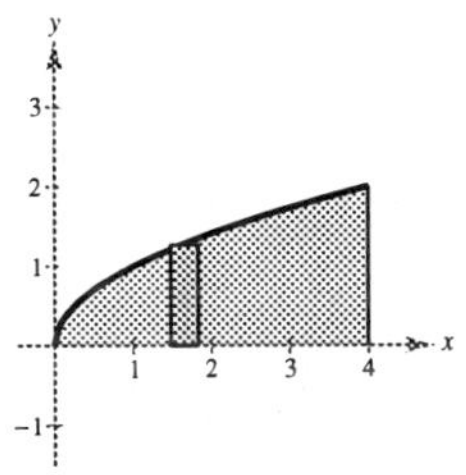

14. $V = \pi \int_{1}^{3} \left(\frac{1}{x}\right)^2\,dx$

$$= \pi\left(-\frac{1}{x}\right)\Bigg]_1^3$$

$$= \pi\left(-\frac{1}{3} + 1\right) = \frac{2}{3}\pi$$

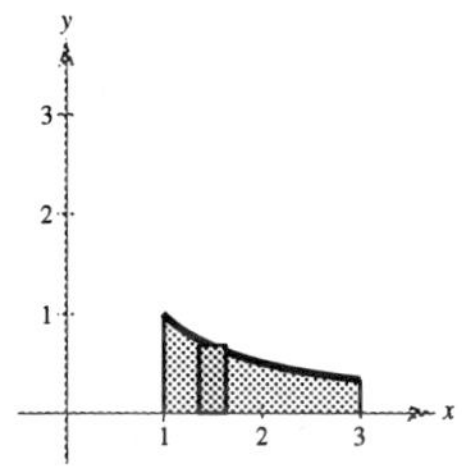

16. The points of intersection of the two graphs occur when $x = 0$ and $x = 2$.

$$V = \pi \int_{0}^{2} [(4x - x^2)^2 - (x^2)^2]\,dx$$

$$= \pi \int_{0}^{2} (16x^2 - 8x^3)\,dx$$

$$= \pi\left[\frac{16}{3}x^3 - 2x^4\right]_0^2 = \frac{32\pi}{3}$$

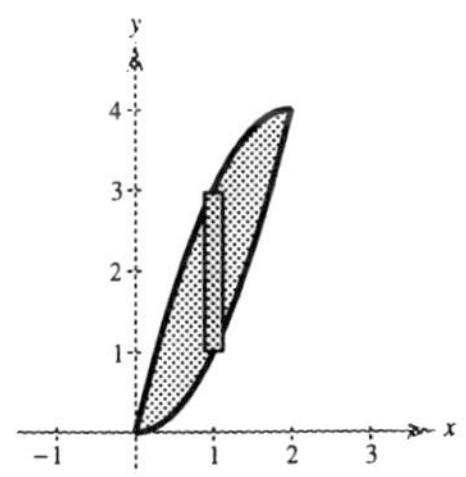

18. $V = \pi \int_{0}^{4} (\sqrt{16 - y^2})^2\,dy$

$$= \pi \int_{0}^{4} (16 - y^2)\,dy$$

$$= \pi\left[16y - \frac{1}{3}y^3\right]_0^4 = \frac{128\pi}{3}$$

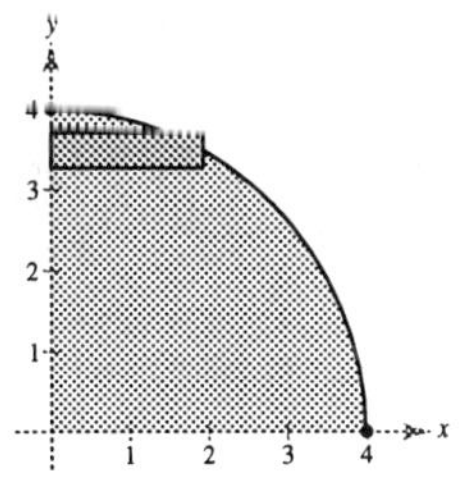

20. The points of intersection occur when $y = 0$ and $y = 1$.

$$V = \pi \int_0^1 [y(y-1)]^2\,dy$$

$$= \pi \int_0^1 (y^4 - 2y^3 + y^2)\,dy$$

$$= \pi\left[\frac{1}{5}y^5 - \frac{1}{2}y^4 + \frac{1}{3}y^3\right]_0^1$$

$$= \frac{\pi}{30}$$

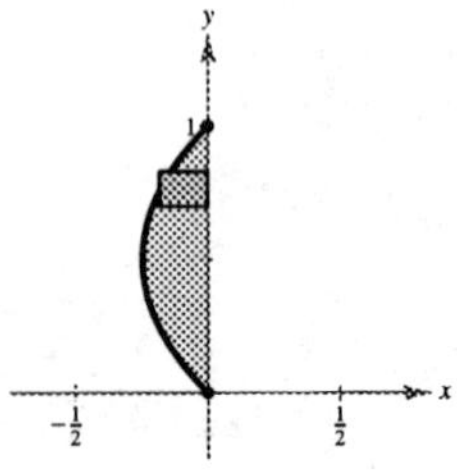

22. $V = \pi \displaystyle\int_1^4 (-y^2 + 4y)^2\,dy$

$$= \pi \int_1^4 (y^4 - 8y^3 + 16y^2)\,dy$$

$$= \pi\left[\frac{1}{5}y^5 - 2y^4 + \frac{16}{3}y^3\right]_1^4$$

$$= \frac{153\pi}{5}$$

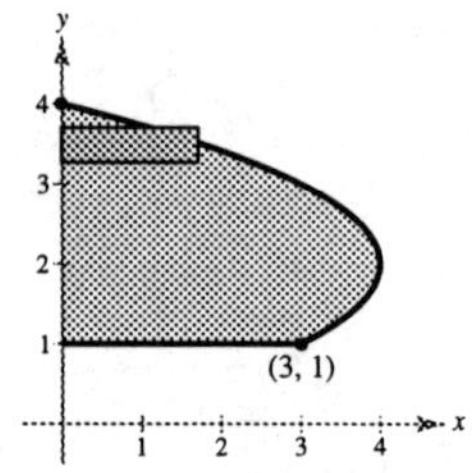

24. The region is a rectangle with vertices $(0, 0)$, $(2, 0)$, $(0, 4)$, and $(2, 4)$.

$$V = \pi \int_0^4 (2)^2\,dy$$

$$= \pi(4y)\Big]_0^4$$

$$= 16\pi$$

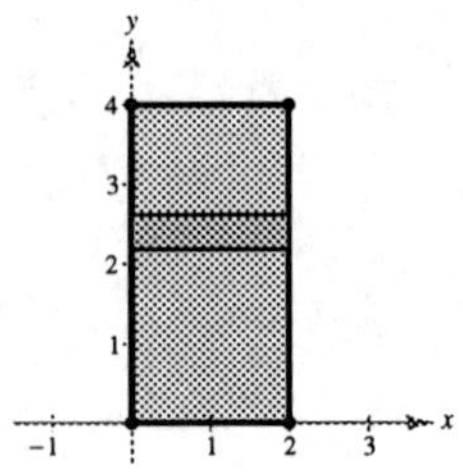

26. A right circular cone can be formed be revolving the region bounded by $y = (r/h)x$, $x = h$, and $y = 0$ about the x-axis.

$$V = \pi \int_0^h \left(\frac{r}{h}x\right)^2 dx$$

$$= \pi\left(\frac{r^2}{h^2}\cdot\frac{x^3}{3}\right)\Big]_0^h$$

$$= \pi\frac{r^2}{h^2}\cdot\frac{h^3}{3}$$

$$= \frac{1}{3}\pi r^2 h$$

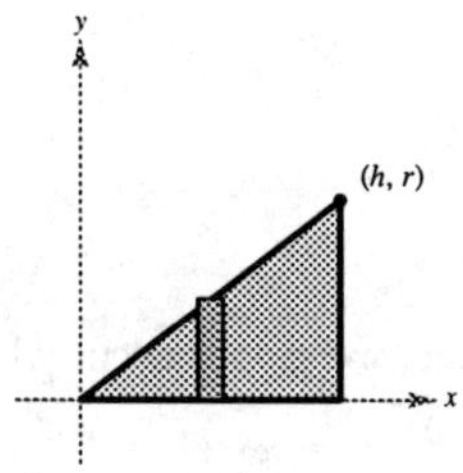

28. The right half of the ellipse is given by $x = \sqrt{25 - (25/9)y^2}$.

$$V = \pi \int_{-3}^3 \left(25 - \frac{25}{9}y^2\right) dy$$

$$= 50\pi \int_0^3 \left(1 - \frac{1}{9}y^2\right) dy$$

$$= 50\pi\left(y - \frac{1}{27}y^3\right)\Big]_0^3$$

$$= 100\pi$$

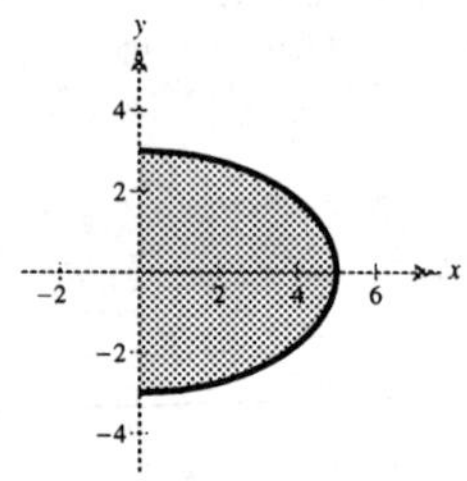

30. The graph has x-intercepts at $(0, 0)$ and $(2, 0)$.

$$V = \pi \int_0^2 \left[\frac{1}{8}x^2\sqrt{2-x}\right]_0^2 dx$$

$$= \frac{\pi}{64}\int_0^2 (2x^4 - x^5)\,dx$$

$$= \frac{\pi}{64}\left[\frac{2x^5}{5} - \frac{x^6}{6}\right]_0^2$$

$$= \frac{\pi}{30}$$

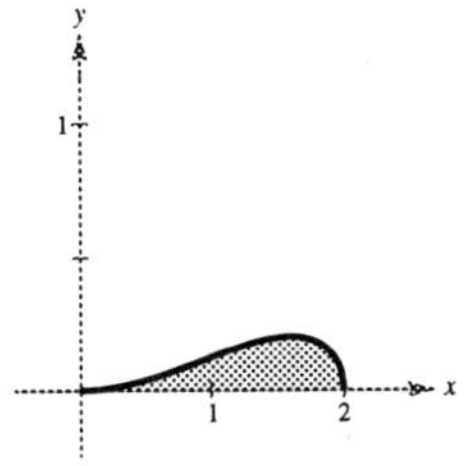

32. (a)

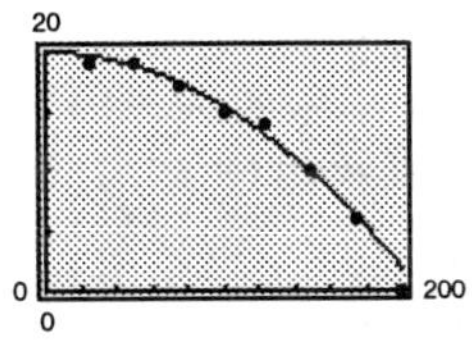

(b) $y = 20 - 0.00045x^2$

$$x = \sqrt{\frac{20-y}{0.00045}}$$

$$V \approx \int_0^{20} \left(\frac{20-y}{0.00045}\right)\pi\,dy \approx 1{,}396{,}263 \text{ cubic feet}$$

(c) $(1,396{,}263)(7.48) \approx 10{,}444{,}050$ gallons

34. For $n = 4$, the midpoints are $\frac{3}{8}, \frac{9}{8}, \frac{15}{8}$, and $\frac{21}{8}$.

$$V = \pi \int_0^3 \left[\frac{10}{x^2+1}\right]^2 dx$$

$$= 100\pi \int_0^3 \frac{1}{x^4 + 2x^2 + 1}\,dx \approx 100\pi\left(\frac{3-0}{4}\right)(0.76862 + 0.19482 + 0.04904 + 0.01606) \sim 242.344$$

Review Exercises for Chapter 5

2. $\displaystyle\int \frac{3}{5}x\,dx = \frac{3}{5}\cdot\frac{x^2}{2} + C = \frac{3x^2}{10} + C$

4. $\displaystyle\int (5 - 6x^2)dx = 5x - 3x^3 + C$

6. $\displaystyle\int 6x^2\sqrt{x}\,dx = \int 6x^{5/2}\,dx = \frac{12}{7}x^{7/2} + C$

8. $\displaystyle\int \left(\frac{4}{\sqrt{x}} + \sqrt{x}\right)dx = \int (4x^{-1/2} + x^{1/2})\,dx$

$$= 8x^{1/2} + \frac{2}{3}x^{3/2} + C$$

10. $\displaystyle f(x) = \int (x^{-1/3} - 1)dx = \frac{3}{2}x^{2/3} - x + C$

$f(8) = 4 = \frac{3}{2}(4) - 8 + C = -2 \Rightarrow C = 6$

Hence, $f(x) = \frac{3}{2}x^{2/3} - x + 6$.

12. $\displaystyle f'(x) = \int (6x^{-1/2} + 3)\,dx = 12x^{1/2} + 3x + C_1$

$f'(1) = 12 = 12 + 3 + C_1 \Rightarrow C_1 = -3$

$f'(x) = 12x^{1/2} + 3x - 3$

$\displaystyle f(x) = \int (12x^{1/2} + 3x - 3)\,dx = 8x^{3/2} + \frac{3}{2}x^2 - 3x + C_2$

$f(4) = 56 = 64 + 24 - 12 + C_2 \Rightarrow C_2 = -20$

$f(x) = 8x^{3/2} + \frac{3}{2}x^2 - 3x - 20$

14. (a) $R = \int 0.675t^{3/2}\,dt = 0.27t^{5/2} + C$

$= 0.27t^{5/2}$ since $R = 0$ when $t = 0$.

(b) $R = 27{,}000 = 0.27t^{5/2}$

$100{,}000 = t^{5/2}$

$t = 400$ weeks

16. $\int (x-6)^{4/3}\,dx = \frac{3}{7}(x-6)^{7/3} + C$

18. $\int \frac{4x}{\sqrt{1-3x^2}}\,dx = -\frac{2}{3}\int (1-3x^2)^{-1/2}(-6x)\,dx$

$= -\frac{2}{3}(2)(1-3x^2)^{1/2} + C$

$= -\frac{4}{3}\sqrt{1-3x^2} + C$

20. $\int \frac{x^2}{(x^3-4)^2}\,dx = \frac{1}{3}\int (x^3-4)^{-2}(3x^2)\,dx$

$= \frac{1}{3}(x^3-4)^{-1} + C$

$= \frac{1}{3(x^3-4)} + C$

22. $\int \frac{\sqrt{x}}{(1-x^{3/2})^3}\,dx = -\frac{2}{3}\int (1-x^{3/2})^{-3}\left(-\frac{3}{2}x^{1/2}\right)dx = -\frac{2}{3}(1-x^{3/2})^{-2}\left(\frac{1}{-2}\right) + C = \frac{1}{3}(1-x^{3/2})^{-2} + C$

24. $C = \int \frac{5x}{\sqrt{x^2+1000}}\,dx = \frac{5}{2}\int (x^2+1000)^{-1/2}(2x)\,dx = 5(x^2+1000)^{1/2} + C_1$

$C(225) = 1136.06 = 5(225^2 + 1000)^{1/2} + C_1 \Longrightarrow C_1 = 0$

$C = 5\sqrt{x^2+1000}$

(a) $C(500) \approx \$2505.00$

(b) $C(1000) \approx \$5002.50$

26. $\int (2t-1)e^{t^2-t}\,dt = e^{t^2-t} + C$

28. $\int \frac{4}{6x-1}\,dx = \frac{2}{3}\int \frac{1}{6x-1}(6)\,dx = \frac{2}{3}\ln|6x-1| + C$

30. $\int \frac{x-4}{x^2-8x}\,dx = \frac{1}{2}\int \frac{1}{x^2-8x}(2x-8)\,dx$

$= \frac{1}{2}\ln|x^2-8x| + C$

32. $\int \frac{e^{5x}}{5+e^{5x}}\,dx = \frac{1}{5}\int \frac{1}{5+e^{5x}}(5e^{5x})\,dx$

$= \frac{1}{5}\ln(5+e^{5x}) + C$

34. $A = \int_{-2}^{2} (4-x^2)\,dx = 2\int_0^2 (4-x^2)\,dx$

$= 2\left[4x - \frac{1}{3}x^3\right]_0^2 = \frac{32}{3}$

36. $A = 2\int_0^3 \sqrt{9-x^2}\,dx$

$= 2\left[\frac{9}{2}\arcsin\left(\frac{x}{3}\right) + \frac{1}{2}x\sqrt{9-x^2}\right]_0^3 = \frac{9\pi}{2}$

38. $A = 8 - \int_0^2 2xe^{x^2-4}\,dx = 8 - \left[e^{x^2-4}\right]_0^2$

$= 8 - (1 - e^{-4}) = 7 + e^{-4}$

40. $\int_{-1}^{1} (t^2+2)\,dt = \left[\frac{t^3}{3} + 2t\right]_{-1}^{1} = \frac{14}{3}$

42. $\int_1^4 2x\sqrt{x}\,dx = \int_1^4 2x^{3/2}\,dx = \frac{4}{5}x^{5/2}\Big]_1^4 = \frac{4}{5}(4^{5/2} - 1) = \frac{4}{5}(32-1) = \frac{124}{5}$

44. $\int_3^6 \frac{x}{3\sqrt{x^2-8}}\,dx = \frac{1}{3}\left(\frac{1}{2}\right)\int_3^6 (x^2-8)^{-1/2}(2x)\,dx = \frac{1}{6}(2)(x^2-8)^{1/2}\Big]_3^6 = \frac{1}{3}\sqrt{x^2-8}\Big]_3^6 = \frac{1}{3}(2\sqrt{7} - 1)$

46. $\displaystyle\int_0^1 x^2(x^3+1)^3\,dx = \frac{1}{3}\int_0^1 (x^3+1)^3(3x^2)\,dx = \left(\frac{1}{3}\right)\frac{(x^3+1)^4}{4}\Bigg]_0^1 = \frac{1}{12}(16-1) = \frac{5}{4}$

48. $\displaystyle\int_0^{\ln 5} e^{x/5}\,dx = 5e^{x/5}\Big]_0^{\ln 5} = 5e^{1/5\ln 5} - 5 = 5^{6/5} - 5 \approx 1.899$

50. $\displaystyle\int_1^3 \frac{1}{x(\ln x+2)^2}\,dx = \int_1^3 (\ln x+2)^{-2}\frac{1}{x}\,dx = -(\ln x+2)^{-1}\Big]_1^3 = \frac{-1}{\ln x+2}\Bigg]_1^3 = \frac{-1}{\ln 3+2} + \frac{1}{2} \approx 0.177$

52. $\displaystyle P = \int 0.4\left(1 - \frac{5000}{x}\right)dx = 0.4x - 2000\ln|x| + C$

$P(100{,}000) - P(75{,}000) \approx \$9424.64$

54. $\displaystyle\text{Average value} = \frac{1}{10-2}\int_2^{10} \frac{20\ln x}{x}\,dx$

$\displaystyle= \frac{5}{4}(\ln x)^2\Big]_2^{10}$

$\displaystyle= \frac{5}{4}(\ln 10)^2 - \frac{5}{4}(\ln 2)^2 \approx 6.0268$

To find the value of x for which $f(x) = 6.0268$, we solve

$$\frac{20\ln x}{x} = 6.0268.$$

Using a graphing utility, $x \approx 5.877$.

56. $\displaystyle\text{Average value} = \frac{1}{2-0}\int_0^2 x^3\,dx = \left(\frac{1}{2}\right)\left(\frac{x^4}{4}\right)\Bigg]_0^2 = 2$

To find the x-value for which $f(x) = 2$, we let $x^3 = 2$ and solve for x.

$x^3 = 2 \Longrightarrow x = \sqrt[3]{2}$

58. $\displaystyle C = \frac{15{,}000}{33}\int_0^5 (1.00 + 0.1t + 0.02t^2)\,dt = \frac{15{,}000}{33}\left[1.00t + 0.05t^2 + \frac{0.02}{3}t^3\right]_0^5 = \$3219.70$

60. (a)

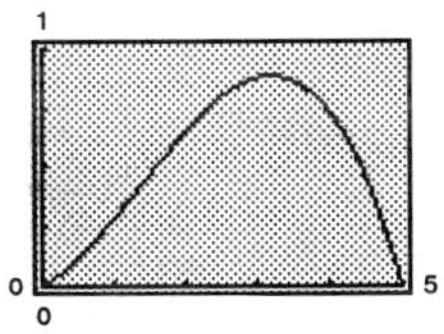

(b) $V'(t) = 0.1729 + 0.3044t - 0.1122t^2$

$V'(t) > 0$ on $[0, 3.2)$ Increasing

$V'(t) < 0$ on $(3.2, 5]$ Decreasing

(c) Maximum volume is 0.88 at $t = 3.2$.

(d) $\displaystyle\text{Average} = \frac{1}{5-0}\int_0^5 V(t)dt \approx 0.53$ liters

(e) Answers vary.

62. $\displaystyle\int_{-3}^3 (x^4 + x^2)\,dx = 2\int_0^3 (x^4 + x^2)\,dx = 2(57.6) = 115.2$ (even function)

64. $\displaystyle\int_{-1}^0 (x^3 - x)\,dx = -\int_0^1 (x^3 - x)\,dx = \frac{1}{4}$ (odd function)

66.

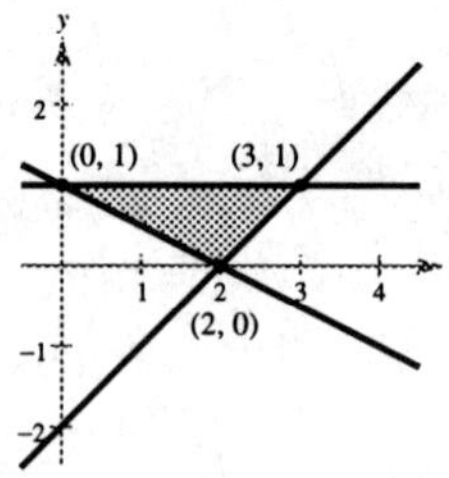

$$A = \int_0^1 [(y + 2) - (2 - 2y)]\,dy$$

$$= \int_0^1 3y\,dy = \frac{3}{2}y^2\Big]_0^1 = \frac{3}{2}$$

68.

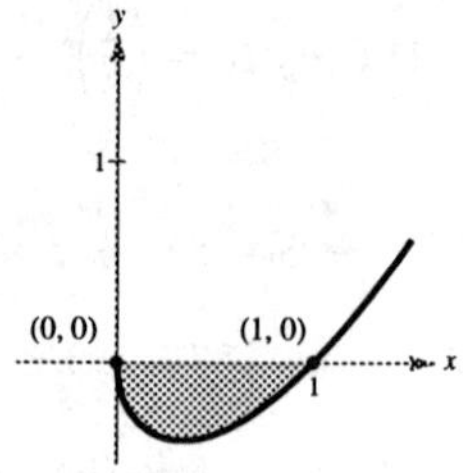

$$A = \int_0^1 [0 - (x^{3/2} - x^{1/2})]\,dx = \int_0^1 (-x^{3/2} + x^{1/2})\,dx$$

$$= \left[-\frac{2}{5}x^{5/2} + \frac{2}{3}x^{3/2}\right]_0^1 = \frac{4}{15}$$

70.

$$4 - x = x^2 - 5x + 8$$

$$x^2 - 4x + 4 = 0$$

$$(x - 2)^2 = 0$$

$$x = 2$$

$$A = \int_0^2 [(x^2 - 5x + 8) - (4 - x)]\,dx$$

$$= \int_0^2 (x^2 - 4x + 4)\,dx = \frac{x^3}{3} - 2x^2 + 4x\Big]_0^2 = \frac{8}{3}$$

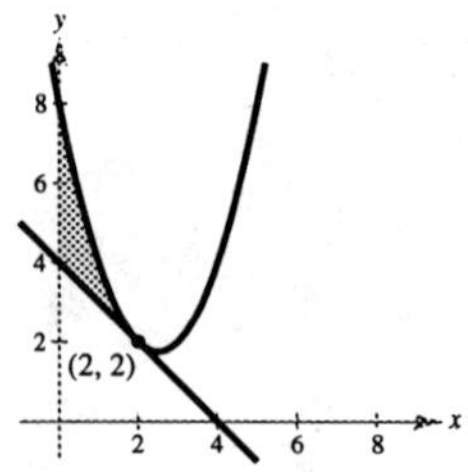

72. $A = 2\displaystyle\int_0^1 (x - x^5)\,dx$

$$= 2\left[\frac{1}{2}x^2 - \frac{1}{6}x^6\right]_0^1$$

$$= \frac{2}{3}$$

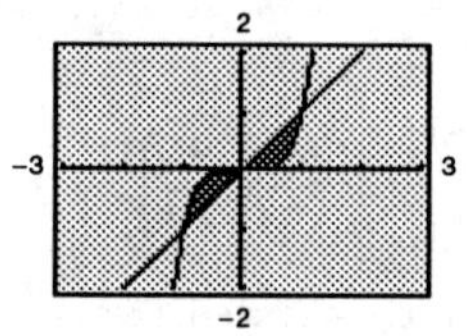

74. Demand function = Supply function

$$\sqrt{100{,}000 - 0.15x^2} = \sqrt{0.01x^2 + 36{,}000}$$

$$64{,}000 = 0.16x^2$$

$$x^2 = 400{,}000$$

$$x \approx 632.46 \implies \text{price} = 200$$

Using a graphing utility, we find that

Consumer surplus $= \displaystyle\int_0^{632.46} (\text{demand} - \text{price})\,dx \approx 51{,}146.5$, and

Producer surplus $= \displaystyle\int_0^{632.46} (\text{price} - \text{supply})\,dx \approx 4304.5$.

76. (a) $\displaystyle\int_3^6 [(25.18t + 81.79) - (14.23t + 72.14)]\,dt \approx 176.8$ million

(b) $\displaystyle\int_3^6 [(25.18t + 81.79) - (1.49t + 19.77)]\,dt \approx 505.9$ million

78. Total sleep, and especially REM sleep, diminish with age. To estimate the amount of non-REM sleep, you would integrate the function representing the curve from 0 to 50.

80. $\int_{-1}^{1} \sqrt{1-x^2}\,dx$

$n = 4$: 1.6297

$n = 20$: $1.5762 \left(\text{exact } \frac{\pi}{2} \approx 1.5708\right)$

82. $\int_{-1}^{1} e^{3-x^2}\,dx$

$n = 4$: 30.3130

$n = 20$: 30.0130

84. $V = \pi \int_0^{16} \left(\sqrt{16-x}\right)^2 dx = \pi\left[16x - \frac{x^2}{2}\right]_0^{16} = 128\pi \approx 402.124$

86. $y = \frac{1}{x}, \quad 1 \le x \le 2$

$$\text{Volume} = \pi \int_1^2 \left(\frac{1}{x}\right)^2 dx = \pi\left[\frac{-1}{x}\right]_1^2 = \pi\left[\frac{-1}{2} + 1\right] = \frac{\pi}{2}$$

88. $V = \pi \int_0^4 \left[2^2 - \left(\sqrt{x}\right)^2\right] dx = \pi \int_0^4 (4-x)\,dx = \pi\left[4x - \frac{x^2}{2}\right]_0^4 = 8\pi$

90.
$$V = \pi \int_{-2}^{2} \left[\left(4 - \frac{1}{2}x^2\right)^2 - 2^2\right] dx$$
$$= 2\pi \int_0^2 \left[12 - 4x^2 + \frac{1}{4}x^4\right] dx$$
$$= 2\pi\left[12x - \frac{4x^3}{3} + \frac{1}{20}x^5\right]_0^2$$
$$= \frac{448}{15}\pi$$

92.
$$V = \pi \int_0^3 (3-y)^2\,dy$$
$$= \pi \int_0^3 (9 - 6y + y^2)\,dy$$
$$= \pi\left[9y - 3y^2 + \frac{y^3}{3}\right]_0^3$$
$$= 9\pi$$

$\left(\text{Volume of cone} = \frac{1}{3}\pi r^2 h = \frac{1}{3}\pi(3)^2 3 = 9\pi\right)$

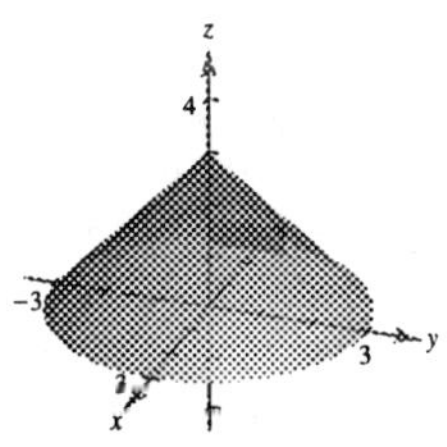

CHAPTER 6
Techniques of Integration

CHAPTER 6
Techniques of Integration

Section 6.1 Integration by Substitution

Solutions to Exercises

2. $\int (x + 5)^{3/2}\,dx = \frac{2}{5}(x + 5)^{5/2} + C$

4. $\int \frac{2t - 1}{t^2 - t + 2}\,dt = \ln(t^2 - t + 2) + C$

6. $\int (3 + x)^5\,dx = \frac{(3 + x)^6}{6} + C$

8. $\int \frac{6x^2 + 2}{x^3 + x}\,dx = 2\int \frac{3x^2 + 1}{x^3 + x}dx = 2\ln|x^3 + x| + C$

$= \ln(x^3 + x)^2 + C$

10. $\int \frac{1}{(3x + 1)^2}\,dx = \frac{1}{3}\int (3x + 1)^{-2}(3)\,dx$

$= \frac{-1}{3}(3x + 1)^{-1} + C$

$= \frac{-1}{3(3x + 1)} + C$

12. $\int \frac{1}{\sqrt{5x + 1}}\,dx = \frac{1}{5}\int (5x + 1)^{-1/2}(5)\,dx$

$= \frac{1}{5}(2)(5x + 1)^{1/2} + C$

$= \frac{2}{5}\sqrt{5x + 1} + C$

14. $\int 2x(e^{x^2})^3\,dx = \frac{1}{3}\int e^{3x^2}\,6x\,dx = \frac{1}{3}e^{3x^2} + C$

16. $\int \frac{2x}{x - 4}\,dx = 2\int \left[1 + \frac{4}{x - 4}\right]dx$

$= 2[x + 4\ln|x - 4|] + C$

18. $\int \frac{t}{\sqrt{1 - t^2}}\,dt = -\frac{1}{2}\int (1 - t^2)^{-1/2}(-2t)dt$

$= -\frac{1}{2}(2)(1 - t^2)^{1/2} + C$

$= -\sqrt{1 - t^2} + C$

20. $\int \frac{e^x}{1 + e^x}\,dx = \ln(1 + e^x) + C$

22. Let $u = x + 1$, then $x = u - 1$ and $dx = du$.

$$\int \frac{x^2}{(x + 1)^3}\,dx = \int \frac{(u - 1)^2}{u^3}\,du = \int \frac{u^2 - 2u + 1}{u^3}\,du$$

$$= \int (u^{-1} - 2u^{-2} + u^{-3})du = \ln|u| + \frac{2}{u} - \frac{1}{2u^2} + C = \ln|x + 1| + \frac{2}{x + 1} - \frac{1}{2(x + 1)^2} + C$$

24. Let $u = x - 4$, then $x = u + 4$ and $dx = du$.

$$\int \frac{5x}{(x - 4)^3}\,dx = \int \frac{5(u + 4)}{u^3}\,du = 5\int (u^{-2} + 4u^{-3})\,du$$

$$= 5\left[-\frac{1}{u} - \frac{2}{u^2}\right] + C = -5\left[\frac{u + 2}{u^2}\right] + C = -\frac{5(x - 2)}{(x - 4)^2} + C$$

26. Let $u = 1 + \sqrt{x}$, then $x = (u - 1)^2$ and $dx = 2(u - 1)du$.

$$\int \frac{1}{1 + \sqrt{x}}\,dx = \int \left(\frac{1}{u}\right)2(u - 1)du = 2\int \left[1 - \frac{1}{u}\right]du$$

$$= 2[u - \ln|u|] + C_1 = 2\left[1 + \sqrt{x} - \ln\left|1 + \sqrt{x}\right|\right] + C_1 = 2\sqrt{x} - 2\ln\left|1 + \sqrt{x}\right| + C$$

28. $$\int \frac{6x + \sqrt{2x}}{x}\,dx = \int \left(6 + \sqrt{2}x^{-1/2}\right)dx = 6x + 2\sqrt{2}x^{1/2} + C$$

30. $$\int_2^4 \sqrt{4x + 1}\,dx = \frac{1}{4}\int_2^4 (4x + 1)^{1/2}(4)\,dx = \frac{1}{6}(4x + 1)^{3/2}\Big]_2^4 = \frac{1}{6}(17^{3/2} - 27)$$

32. $$\int_0^2 e^{-2x}\,dx = \frac{-1}{2}\int_0^2 e^{-2x}(-2)\,dx = \frac{-1}{2}e^{-2x}\Big]_0^2 = \frac{-1}{2}(e^{-4} - 1) = \frac{1}{2}(1 - e^{-4})$$

34. Let $u = x + 5$, then $x = u - 5$ and $dx = du$. $u = 5$ when $x = 0$ and $u = 6$ when $x = 1$.

$$\int_0^1 x(x + 5)^4\,dx = \int_5^6 (u - 5)u^4\,du = \int_5^6 (u^5 - 5u^4)\,du$$

$$= \left[\frac{u^6}{6} - u^5\right]_5^6 = \frac{u^5}{6}\Big[u - 6\Big]_5^6 = 0 - \frac{3125}{6}(-1) = \frac{3125}{6}$$

36. Let $u = 1 - x$, then $x = 1 - u$ and $dx = -du$. $u = 1$ when $x = 0$ and $u = 0.5$ when $x = 0.5$.

$$\int_0^{0.5} x^2(1 - x)^3\,dx = \int_1^{0.5} -(1 - u)^2u^3\,du = \int_1^{0.5} (-u^5 + 2u^4 - u^3)\,du$$

$$= \left[-\frac{u^6}{6} + \frac{2u^5}{5} - \frac{u^4}{4}\right]_1^{0.5} = -\frac{u^4}{60}\Big[10u^2 - 24u + 15\Big]_1^{0.5} = \frac{7}{640}$$

38. Let $u = 2x + 1$, then $x = \dfrac{u - 1}{2}$ and $dx = \dfrac{1}{2}du$. When $x = 0$, $u = 1$, and when $x = 4$, $u = 9$.

$$\int_0^4 x\sqrt{2x + 1}\,dx = \int_1^9 \left(\frac{u - 1}{2}\right)\sqrt{u}\left(\frac{1}{2}\right)du = \int_1^9 \frac{1}{4}(u^{3/2} - u^{1/2})\,du$$

$$= \frac{1}{4}\left[\frac{2}{5}u^{5/2} - \frac{2}{3}u^{3/2}\right]_1^9 = \frac{1}{4}\left[\frac{2}{5}(3^5) - \frac{2}{3}(3^3)\right] - \frac{1}{4}\left[\frac{2}{5} - \frac{2}{3}\right] = \frac{298}{15} \approx 19.867$$

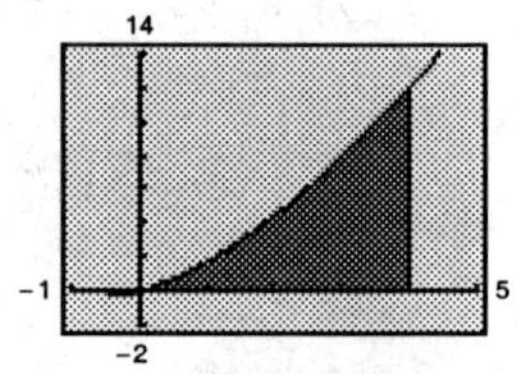

40. Let $u = x + 2$, then $x = u - 2$, and $du = dx$. When $x = 0$, $u = 2$, and when $x = 7$, $u = 9$.

$$\int_0^7 x^2\sqrt{x + 2}\,dx = \int_2^9 (u - 2)^2\sqrt{u}\,du = \int_2^9 (u^{5/2} - 4u^{3/2} + 4u^{1/2})\,du$$

$$= \left[\frac{2}{7}u^{7/2} - \frac{8}{5}u^{5/2} + \frac{8}{3}u^{3/2}\right]_2^9 \approx 308.057 - 1.724 = 306.333$$

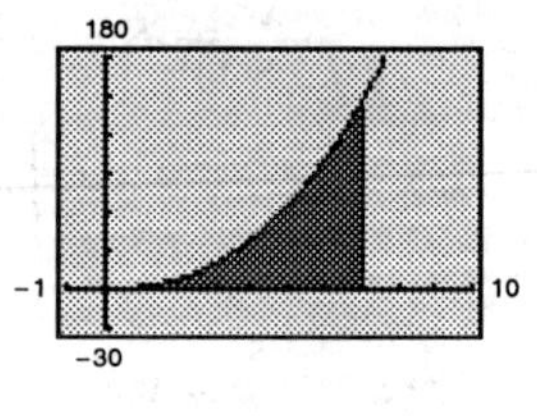

42. Let $u = x + 3$, then $x = u - 3$ and $dx = du$. When $x = \frac{1}{2}$, $u = \frac{7}{2}$ and when $x = 6$, $u = 9$.

$$\int_{1/2}^6 \frac{2x - 1}{\sqrt{x + 3}}\,dx = \int_{7/2}^9 \frac{2(u - 3) - 1}{\sqrt{u}}\,du = \int_{7/2}^9 (2u^{1/2} - 7u^{-1/2})\,du$$

$$= \left[\frac{4}{3}u^{3/2} - 14u^{1/2}\right]_{7/2}^9 \approx (-6) - (-17.461) = 11.461$$

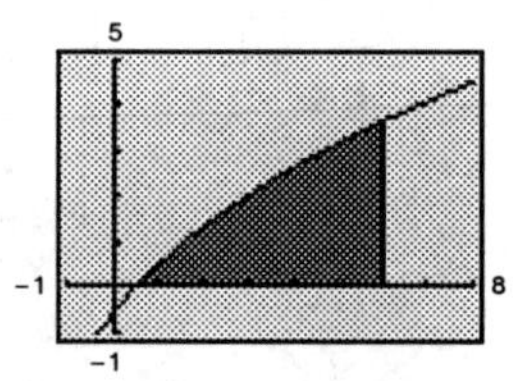

44. Let $u = x - 2$, then $x = u + 2$ and $du = dx$. When $x = 2$, $u = 0$, and when $x = 10$, $u = 8$.

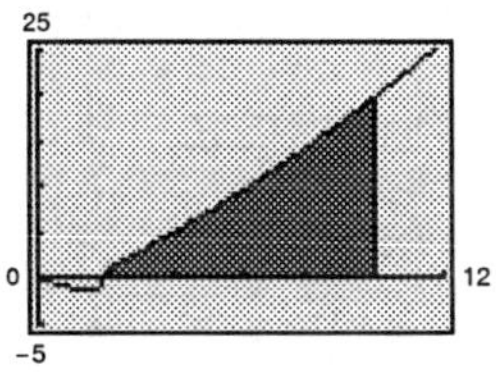

$$\int_2^{10} x\sqrt[3]{x-2}\,dx = \int_0^8 (u+2)u^{1/3}\,du = \int_0^8 (u^{4/3} + 2u^{1/3})\,du$$

$$= \left[\frac{3}{7}u^{7/3} + \frac{3}{2}u^{4/3}\right]_0^8 \approx 78.857$$

46. Let $u = 1 - x$, then $x = 1 - u$ and $dx = -du$. $u = 1$ when $x = 0$ and $u = 0$ when $x = 1$.

$$A = \int_0^1 x\sqrt[3]{1-x}\,dx = \int_1^0 -(1-u)\sqrt[3]{u}\,du = \int_1^0 (u^{4/3} - u^{1/3})\,du$$

$$= \left[\frac{3}{7}u^{7/3} - \frac{3}{4}u^{4/3}\right]_1^0 = \frac{9}{28} \text{ square unit}$$

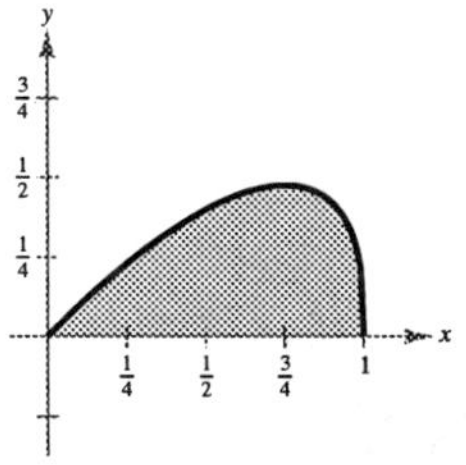

48. Let $u = 1 + \sqrt{x}$, then $x = (u-1)^2$ and $dx = 2(u-1)\,du$. $u = 1$ when $x = 0$ and $u = 3$ when $x = 4$.

$$A = \int_0^4 \frac{1}{1+\sqrt{x}}\,dx = \int_1^3 \frac{2(u-1)}{u}\,du = 2\int_1^3 \left(1 - \frac{1}{u}\right)du$$

$$= 2\Big[u - \ln|u|\Big]_1^3 = 4 - 2\ln 3 \approx 1.803 \text{ square units}$$

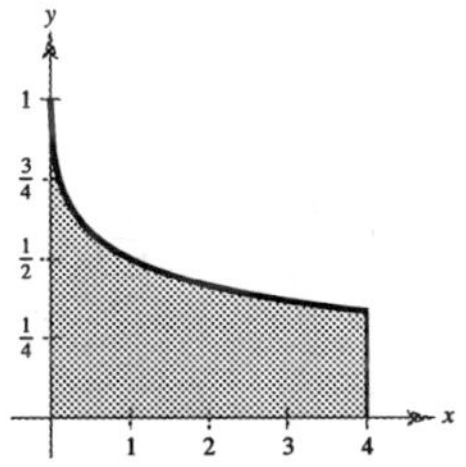

50. $V = \pi\displaystyle\int_0^1 \left[\sqrt{x}(1-x)^2\right]^2 dx = \pi\int_0^1 x(1-x)^4\,dx$

Let $u = 1 - x$, then $x = 1 - u$ and $dx = -du$. $u = 1$ when $x = 0$ and $u = 0$ when $x = 1$.

$$V = \pi\int_1^0 -(1-u)u^4\,du = \pi\int_1^0 (u^5 - u^4)\,du = \pi\left[\frac{u^6}{6} - \frac{u^5}{5}\right]_1^0 = \frac{\pi}{30} \text{ cubic units}$$

52. $\displaystyle\frac{1}{2-0}\int_0^2 [f(x) - g(x)]\,dx = \frac{1}{2}\int_0^2 \left[x\sqrt{4x+1} - 2\sqrt{x^3}\right]dx = \frac{1}{2}\int_0^2 x\sqrt{4x+1}\,dx - \int_0^2 x^{3/2}\,dx$

For the first integral, let $u = 4x + 1$, then $x = \dfrac{u-1}{4}$ and $dx = \dfrac{1}{4}du$. $u = 1$ when $x = 0$ and $u = 9$ when $x = 2$.

$$\frac{1}{2}\int_1^9 \left(\frac{u-1}{4}\right)\sqrt{u}\left(\frac{1}{4}du\right) - \int_0^2 x^{3/2}\,dx = \frac{1}{32}\int_1^9 (u^{3/2} - u^{1/2})\,du - \left[\frac{2}{5}x^{5/2}\right]_0^2$$

$$= \frac{1}{32}\left[\frac{2}{5}u^{5/2} - \frac{2}{3}u^{3/2}\right]_1^9 - \frac{8\sqrt{2}}{5} = \frac{149}{60} - \frac{8\sqrt{2}}{5} \approx 0.221$$

54. Let $u = 1 - x$, then $x = 1 - u$ and $dx = -du$.

$$\int x^3(1-x)^{3/2}dx = \int (1-u)^3 u^{3/2}\,du = \int (u^3 - 3u^2 + 3u - 1)u^{3/2}\,du = \int (u^{9/2} - 3u^{7/2} + 3u^{5/2} - u^{3/2})du$$

$$= \frac{2}{11}u^{11/2} - 3\left(\frac{2}{9}u^{9/2}\right) + 3\left(\frac{2}{7}u^{7/2}\right) - \frac{2}{5}u^{5/2} + C$$

$$= \frac{2}{1155}u^{5/2}(105u^3 - 385u^2 + 495u - 231) + C$$

$$= \frac{2}{1155}(1-x)^{5/2}[105(1 - 3x + 3x^2 - x^3) - 385(1 - 2x + x^2) + 495(1 - x) - 231] + C$$

$$= \frac{2}{1155}(1-x)^{5/2}(-105x^3 - 70x^2 - 40x - 16) + C$$

$$= -\frac{2}{1155}(1-x)^{5/2}(105x^3 + 70x^2 + 40x + 16) + C$$

Therefore, we have the following.

$$\int_a^b \frac{1155}{32}x^3(1-x)^{3/2}dx = \left[-\frac{1}{16}(1-x)^{5/2}(105x^3 + 70x^2 + 40x + 16)\right]_a^b$$

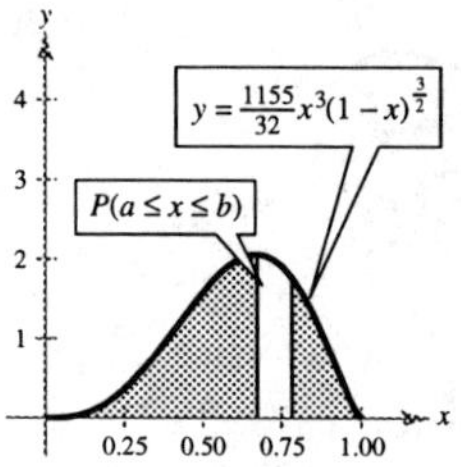

(a) $a = 0$ and $b = 0.25$
$P(0 \le x \le 0.25) \approx 0.0252 = 2.52\%$

(b) $a = 0.50$ and $b = 1$
$P(0.50 \le x \le 1) \approx 0.7361 = 73.61\%$

56. The average daily revenue is given by the following.

$$\frac{1}{365 - 0}\int_0^{365} [0.06t^2(365 - t)^{1/2} + 1250]\,dt = \frac{1}{365}\int_0^{365} [0.06t^2(365 - t)^{1/2}]dt + \frac{1}{365}\int_0^{365} 1250\,dt$$

$$= \frac{1}{365}\int_{\sqrt{365}}^{0} 0.06(365 - u^2)^2 u(-2u)\,du + 1250 \approx \$24{,}520.95$$

58. $A = \int_0^4 \sqrt[3]{x}\sqrt{4 - x}\,dx \approx \frac{4 - 0}{10}\left[f\left(\frac{1}{5}\right) + f\left(\frac{3}{5}\right) + f(1) + f\left(\frac{7}{5}\right) + f\left(\frac{9}{5}\right) + f\left(\frac{11}{5}\right) + f\left(\frac{13}{5}\right) + f(3) + f\left(\frac{17}{5}\right) + f\left(\frac{19}{5}\right)\right] \approx 5.885$

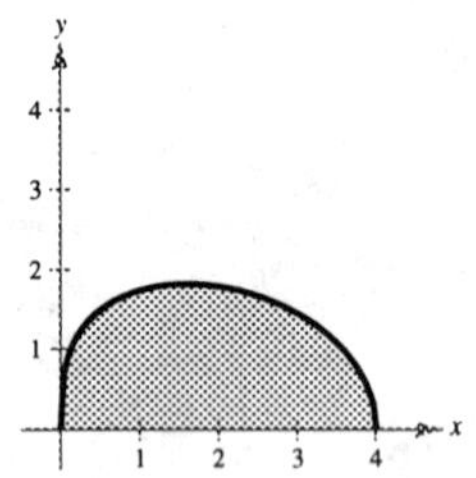

Section 6.2 Integration by Parts and Present Value

2. Let $u = x$ and $dv = e^{-x}\,dx$, then $du = dx$ and $v = -e^{-x}$.

$$\int xe^{-x}\,dx = -xe^{-x} - \int -e^{-x}\,dx = -xe^{-x} - e^{-x} + C$$

4. Let $u = x^2$ and $dv = e^{2x}dx$, then $du = 2x\,dx$ and $v = \frac{1}{2}e^{2x}$.

$$\int x^2e^{2x}\,dx = \frac{1}{2}x^2e^{2x} - \int xe^{2x}\,dx$$

Let $u = x$ and $dv = e^{2x}\,dx$, then $du = dx$ and $v = \frac{1}{2}e^{2x}$.

$$\int x^2e^{2x}\,dx = \frac{1}{2}x^2e^{2x} - \left[\frac{1}{2}xe^{2x} - \int \frac{1}{2}e^{2x}\,dx\right]$$

$$= \frac{1}{2}x^2e^{2x} - \frac{1}{2}xe^{2x} + \frac{1}{4}e^{2x} + C = \frac{1}{4}e^{2x}(2x^2 - 2x + 1) + C$$

6. $\displaystyle\int \ln x^2\,dx = 2\int \ln x\,dx$

Let $u = \ln x$ and $dv = dx$, then $du = \frac{1}{x}\,dx$ and $v = x$.

$$\int \ln x^2\,dx = 2\left[x\ln x - \int dx\right] = 2x\ln x - 2x + C = 2x(\ln x - 1) + C$$

8. $\displaystyle\int e^{-2x}\,dx = -\frac{1}{2}\int e^{-2x}(-2)\,dx = -\frac{1}{2}e^{-2x} + C$

10. Let $u = x$ and $dv = e^{-2x}\,dx$, then $du = dx$ and $v = -\frac{1}{2}e^{-2x}$.

$$\int xe^{-2x}\,dx = -\frac{1}{2}xe^{-2x} - \int -\frac{1}{2}e^{-2x}\,dx = -\frac{1}{2}xe^{-2x} - \frac{1}{4}e^{-2x} + C = -\frac{1}{4}e^{-2x}(2x + 1) + C$$

12. $\displaystyle\int x^2e^{x^3}\,dx = \frac{1}{3}\int e^{x^3}(3x^2)\,dx = \frac{1}{3}e^{x^3} + C$

14. $\displaystyle\int \frac{x}{e^x}\,dx = \int xe^{-x}\,dx$

Let $u = x$ and $dv = e^{-x}\,dx$, then $du = dx$ and $v = -e^{-x}$.

$$\int \frac{x}{e^x}\,dx = -xe^{-x} - \int -e^{-x}\,dx = -xe^{-x} - e^{-x} + C = -e^{-x}(x + 1) + C = -\frac{x+1}{e^x} + C$$

16. Let $u = \ln x$ and $dv = x^3\,dx$. Then $du = \frac{1}{x}\,dx$ and $v = \frac{x^4}{4}$.

$$\int x^3\ln x\,dx = \frac{x^4}{4}\ln x - \int \frac{x^4}{4}\left(\frac{1}{x}\right)dx = \frac{x^4}{4}\ln x - \int \frac{x^3}{4}\,dx = \frac{x^4}{4}\ln x - \frac{x^4}{16} + C$$

18. $\displaystyle\int \frac{1}{x(\ln x)^3}\,dx = \int (\ln x)^{-3}\frac{1}{x}\,dx = \frac{(\ln x)^{-2}}{-2} + C = -\frac{1}{2(\ln x)^2} + C$

20. Let $u = \ln 3x$ and $dv = dx$, then $du = \frac{1}{x}\,dx$ and $v = x$.

$$\int \ln(3x)\,dx = x\ln(3x) - \int dx = x\ln(3x) - x + C = x[\ln(3x) - 1] + C$$

22. Let $u = x - 1$, then $x = u + 1$ and $du = dx$.

$$\int x\sqrt{x-1}\,dx = \int (u+1)u^{1/2}\,du = \int (u^{3/2} + u^{1/2})\,du = \frac{2}{5}u^{5/2} + \frac{2}{3}u^{3/2} + C = \frac{2}{5}(x-1)^{5/2} + \frac{2}{3}(x-1)^{3/2} + C$$

24. Let $u = x$ and $dv = (2+3x)^{-1/2}\,dx$, then $du = dx$ and $v = \frac{2}{3}(2+3x)^{1/2}$.

$$\int \frac{x}{\sqrt{2+3x}}\,dx = \frac{2}{3}x(2+3x)^{1/2} - \frac{2}{3}\int (2+3x)^{1/2}\,dx = \frac{2}{3}x(2+3x)^{1/2} - \frac{4}{27}(2+3x)^{3/2} + C$$

$$= \frac{2}{27}(2+3x)^{1/2}[9x - 2(2+3x)] + C = \frac{2}{27}\sqrt{2+3x}\,(3x-4) + C$$

26. Let $u = x^2e^{x^2}$ and $dv = x(x^2+1)^{-2}\,dx$, then $du = 2xe^{x^2}(x^2+1)\,dx$ and $v = -\dfrac{1}{2(x^2+1)}$.

$$\int \frac{x^3e^{x^2}}{(x^2+1)^2}\,dx = \int (x^2e^{x^2})\left[\frac{x}{(x^2+1)^2}\right]dx = -\frac{x^2e^{x^2}}{2(x^2+1)} + \int xe^{x^2}\,dx$$

$$= -\frac{x^2e^{x^2}}{2(x^2+1)} + \frac{1}{2}e^{x^2} + C = \frac{-x^2e^{x^2} + e^{x^2}(x^2+1)}{2(x^2+1)} + C = \frac{e^{x^2}}{2(x^2+1)} + C$$

28. Area $= -\displaystyle\int_{-1}^{1} (x^2-1)e^x\,dx = \int_{-1}^{1} (1-x^2)e^x\,dx$

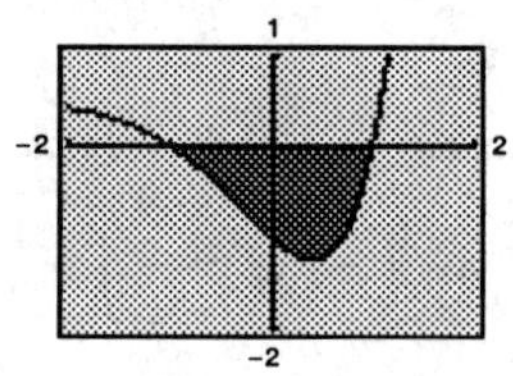

Let $u = 1 - x^2$, $dv = e^x\,dx$, $du = -2x\,dx$, and $v = e^x$.

$$\int (1-x^2)e^x\,dx = (1-x^2)e^x - \int -2xe^x\,dx = (1-x^2)e^x + 2\int xe^x\,dx$$

$$= (1-x^2)e^x + 2(xe^x - e^x) \qquad \text{(Example 1)}$$

Hence,

$$\text{Area} = \int_{-1}^{1} (1-x^2)e^x\,dx = \Big[(1-x^2)e^x + 2xe^x - 2e^x\Big]_{-1}^{1}$$

$$= (0 + 2e - 2e) - (-2e^{-1} - 2e^{-1}) = \frac{4}{e} \approx 1.472$$

30. Area $= \displaystyle\int_{1}^{e} \frac{\ln x}{x^2}\,dx$

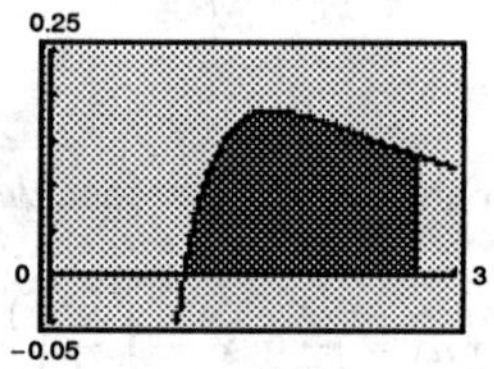

Let $u = \ln x$, $dv = \dfrac{1}{x^2}\,dx$, $du = \dfrac{1}{x}\,dx$ and $v = -\dfrac{1}{x}$.

$$\int \frac{\ln x}{x^2}\,dx = -\frac{1}{x}\ln x - \int \frac{-1}{x^2}\,dx = -\frac{1}{x}\ln x - \frac{1}{x}$$

Hence,

$$\text{Area} = \int_{1}^{e} \frac{\ln x}{x^2}\,dx = \left[-\frac{1}{x}\ln x - \frac{1}{x}\right]_{1}^{e} = \left(-\frac{1}{e} - \frac{1}{e}\right) - (-1) = 1 - \frac{2}{e} \approx 0.264$$

32. Let $u = x^2$ and $dv = e^{-x}\,dx$, then $du = 2x\,dx$ and $v = -e^{-x}$.

$$\int_0^2 \frac{x^2}{e^x}\,dx = -x^2e^{-x}\Big]_0^2 + 2\int_0^2 xe^{-x}\,dx$$

Let $u = x$ and $dv = e^{-x}\,dx$, then $du = dx$ and $v = -e^{-x}$.

$$\int_0^2 \frac{x^2}{e^x}\,dx = -4e^{-2} + 2\left\{\Big[-xe^{-x}\Big]_0^2 - \int_0^2 -e^{-x}\,dx\right\} = -4e^{-2} - 4e^{-2} - 2\Big[e^{-x}\Big]_0^2 = -10e^{-2} + 2 \approx 0.647$$

34. Let $u = \ln(1 + 2x)$ and $dv = dx$, then $du = \dfrac{2}{1 + 2x}\,dx$ and $v = x$.

$$\int_0^1 \ln(1 + 2x)\,dx = x\ln(1 + 2x)\Big]_0^1 - \int_0^1 \frac{2x}{1 + 2x}\,dx = \ln 3 - \int_0^1 \left[1 - \frac{1}{2x + 1}\right]dx$$

$$= \ln 3 - \left[x - \frac{1}{2}\ln(2x + 1)\right]_0^1 = \ln 3 - 1 + \frac{1}{2}\ln 3 = \frac{3}{2}\ln 3 - 1 \approx 0.648$$

36. (a) Let $u = x$ and $dv = \sqrt{4 + x}\,dx$, then $du = dx$ and $v = \frac{2}{3}(4 + x)^{3/2}$.

$$\int x\sqrt{4 + x}\,dx = \frac{2}{3}x(4 + x)^{3/2} - \frac{2}{3}\int (4 + x)^{3/2}\,dx = \frac{2}{3}x(4 + x)^{3/2} - \frac{4}{15}(4 + x)^{5/2} + C$$

$$= \frac{2}{15}(4 + x)^{3/2}[5x - 2(4 + x)] + C = \frac{2}{15}(4 + x)^{3/2}(3x - 8) + C$$

(b) Let $u = \sqrt{4 + x}$, then $x = u^2 - 4$ and $dx = 2u\,du$.

$$\int x\sqrt{4 + x}\,dx = \int (u^2 - 4)u(2u)\,du = 2\int (u^4 - 4u^2)\,du = 2\left[\frac{u^5}{5} - \frac{4u^3}{3}\right] + C$$

$$= \frac{2}{15}u^3(3u^2 - 20) + C = \frac{2}{15}\left(\sqrt{4 + x}\right)^3[3(4 + x) - 20] + C = \frac{2}{15}(4 + x)^{3/2}(3x - 8) + C$$

38. (a) Let $u = x$ and $dv = \sqrt{4 - x}\,dx$, then $du = dx$ and $v = -\frac{2}{3}(4 - x)^{3/2}$.

$$\int x\sqrt{4 + x}\,dx = -\frac{2}{3}x(4 - x)^{3/2} + \frac{2}{3}\int (4 - x)^{3/2}\,dx = -\frac{2}{3}x(4 - x)^{3/2} - \frac{4}{15}(4 - x)^{5/2} + C$$

$$= -\frac{2}{15}(4 - x)^{3/2}[5x + 2(4 - x)] + C = -\frac{2}{15}(4 - x)^{3/2}(3x + 8) + C$$

(b) Let $u = \sqrt{4 - x}$, then $x = 4 - u^2$ and $dx = -2u\,du$.

$$\int x\sqrt{4 - x}\,dx = \int (4 - u^2)u(-2u)\,du = -2\int (4u^2 - u^4)\,du = -2\left(\frac{4u^3}{3} - \frac{u^5}{5}\right) + C$$

$$= -\frac{2}{15}u^3(20 - 3u^2) + C = -\frac{2}{15}\left(\sqrt{4 - x}\right)^3[20 - 3(4 - x)] + C = -\frac{2}{15}(4 - x)^{3/2}(3x + 8) + C$$

40. Let $u = x^n$ and $dv = e^{ax}\,dx$, then $du = nx^{n-1}\,dx$ and $v = \dfrac{1}{a}e^{ax}$.

$$\int x^n e^{ax}\,dx = \frac{x^n e^{ax}}{a} - \frac{n}{a}\int x^{n-1}e^{ax}\,dx$$

42. Using $n = 1$ and $a = -3$, we have the following.

$$\int xe^{-3x}\,dx = \frac{xe^{-3x}}{-3} + \frac{1}{3}\int e^{-3x}\,dx = -\frac{1}{3}xe^{-3x} - \frac{1}{9}e^{-3x} + C = -\frac{1}{9}e^{-3x}(3x + 1) + C$$

44. Using $n = \frac{1}{2}$, Exercise 39 yields

$$\int x^{1/2}\ln x\,dx = \frac{x^{1/2+1}}{(\frac{1}{2} + 1)^2}\left[-1 + \left(\frac{1}{2} + 1\right)\ln x\right] + C = \frac{4}{9}x^{3/2}\left[-1 + \frac{3}{2}\ln x\right] + C$$

46. Let $u = x$ and $dv = e^{-x/3}\,dx$, then $du = dx$ and $v = -3e^{-x/3}$.

$$A = \int_0^3 \frac{1}{9}xe^{-x/3}\,dx = \frac{1}{9}\int_0^3 xe^{-x/3}\,dx$$

$$= \frac{1}{9}\left\{\left[-3xe^{-x/3}\right]_0^3 + 3\int_0^3 e^{-x/3}\,dx\right\} = \frac{1}{9}\left\{-9e^{-1} - 9\left[e^{-x/3}\right]_0^3\right\} = -2e^{-1} + 1 = 1 - \frac{2}{e} \approx 0.264$$

48. Using Exercise 40 twice with $n = 2$ and $a = 2$ and with $n = 1$ and $a = 2$, we have the following.

$$V = \pi \int_0^1 x^2 e^{2x}\,dx$$

$$= \pi\left\{\left[\frac{x^2e^{2x}}{2}\right]_0^1 - \int_0^1 xe^{2x}\,dx\right\} = \pi\left\{\frac{e^2}{2} - \left[\frac{xe^{2x}}{2}\right]_0^1 + \frac{1}{2}\int_0^1 e^{2x}\,dx\right\} = \frac{\pi}{2}\left[\frac{e^{2x}}{2}\right]_0^1 = \frac{\pi}{4}(e^2 - 1) \approx 5.018$$

50. $\displaystyle\int_1^4 \ln x(x^2+4)\,dx = \frac{112}{3}\ln 4 - 19 \approx 32.755$

52. $\displaystyle\int_1^e x^9 \ln x\,dx = \frac{9}{100}e^{10} + \frac{1}{100} \approx 1982.392$

54. (a)

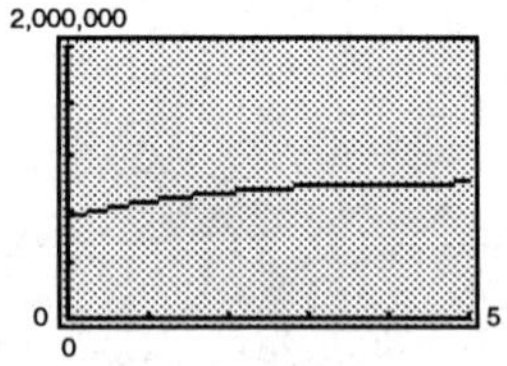

Since the graph of I is increasing on $[0, 5]$, or equivalently, $\frac{dI}{dt} > 0$ on $[0, 5]$, the gift income is expected to increase.

(b) $$\text{Incoming during 5 years} = \int_0^5 2000(375 + 68t\,e^{-0.2t})dt = 2000\Big[375t\Big]_0^5 + 136{,}000\int_0^5 te^{-0.2t}\,dt$$

$$= 3{,}750{,}000 + 136{,}000\left\{\left[\frac{te^{-0.2t}}{-0.2}\right]_0^5 + \frac{1}{0.2}\int_0^5 e^{-0.2t}\,dt\right\}$$

$$= 3{,}750{,}000 + 136{,}000\left\{-9.196986 - \Big[25e^{-0.2t}\Big]_0^5\right\}$$

$$= 3{,}750{,}000 + 136{,}000[-9.196986 - 25(e^{-1} - 1)] \approx \$4{,}648{,}419.80$$

(c) $\text{Average} = \dfrac{4{,}648{,}419.80}{5} = \$929{,}683.96$

$I(3) = \$973{,}915.15$, which is close to the average.

56. Using Exercise 40 twice with $n = 2$ and $a = -\frac{1}{30}$ and with $n = 1$ and $a = -\frac{1}{30}$, we have the following.

$$\int (410.5t^2e^{-t/30} + 25{,}000)\,dt = 410.5\left[-30t^2e^{-t/30} + 60\int te^{-t/30}\,dt\right] + 25{,}000t$$

$$= 25{,}000t + 410.5\left[-30t^2e^{-t/30} - 1800te^{-t/30} + 1800\int e^{-t/30}\,dt\right]$$

$$= 25{,}000t + 410.5[-30t^2e^{-t/30} - 1800t\,e^{-t/30} - 54{,}000e^{-t/30}] + C$$

$$= 25{,}000t - 12{,}315e^{-t/30}(t^2 + 60t + 1800) + C$$

Therefore, we have the following.

(a) $\displaystyle\frac{1}{91-0}\int_0^{91}(410.5t^2e^{-t/30} + 25{,}000)\,dt = \frac{1}{91}\Big[25{,}000t - 12{,}315e^{-t/30}(t^2 + 60t + 1800)\Big]_0^{91} \approx \$167{,}316.12$

(b) $\displaystyle\frac{1}{365-274}\int_{274}^{365}(410.5t^2e^{-t/30} + 25{,}000)\,dt = \frac{1}{91}\Big[25{,}000t - 12{,}315e^{-t/30}(t^2 + 60t + 1800)\Big]_{274}^{365} \approx \$26{,}253.48$

(c) $\displaystyle\int_0^{365}(410.5t^2e^{-t/30} + 25{,}000)dt \approx 31{,}281{,}948.97$

58. $\displaystyle V = \int_0^{t_1} c(t)e^{-rt}\,dt = \int_0^{10} 450e^{-0.04t}\,dt = \frac{450}{-0.04}e^{-0.04t}\Big]_0^{10} \approx \$3708.90$

60. $V = \int_0^{t_1} c(t)e^{-rt}\,dt = \int_0^{6} (30{,}000 + 500t)e^{-0.07t}\,dt \approx \$153{,}816.01$

62. $V = \int_0^{t_1} c(t)e^{-rt}\,dt = \int_0^{10} (5000 + 25t\,e^{t/10})e^{-0.06t}\,dt \approx \$39{,}238.17$

64. (a) $\int_0^{3} (300{,}000 + 125{,}000t)dt = 300{,}000t + 62{,}500t^2\Big]_0^3 = \$1{,}462{,}500$

(b) $\int_0^{3} (300{,}000 + 125{,}000t)e^{-0.05t}\,dt \approx 1{,}345{,}043.50$

66. $V = \pi\int_1^4 \left(\frac{10}{\sqrt{x}\,e^x}\right)^2 dx = 100\pi\int_1^4 \frac{1}{xe^{2x}}dx$

$$\approx 100\pi\left(\frac{4-1}{12}\right)\left[f\left(\frac{9}{8}\right) + f\left(\frac{11}{8}\right) + f\left(\frac{13}{8}\right) + f\left(\frac{15}{8}\right) + f\left(\frac{17}{8}\right) + f\left(\frac{19}{8}\right) + f\left(\frac{21}{8}\right) + f\left(\frac{23}{8}\right) + f\left(\frac{25}{8}\right) + f\left(\frac{27}{8}\right) + f\left(\frac{29}{8}\right) + f\left(\frac{31}{8}\right)\right] \approx 15.026$$

68. Future value $= e^{rt_1}\int_0^{t_1} f(t)e^{-rt}\,dt = e^{(0.1)5}\int_0^{5} (3000e^{0.05t})e^{-0.1t}\,dt \approx \$21{,}881.75$

70. Future value $= e^{(0.1)18}\int_0^{18} (300t)e^{-0.1t}\,dt \approx \$97{,}489.42$

Four years at Notre Dame costs 4(23,660) = \$94,640, so there would be enough money.

Section 6.3 Partial Fractions and Logistics Growth

2. $\frac{10x + 3}{x(x + 1)} = \frac{A}{x} + \frac{B}{x + 1}$

Basic equation: $10x + 3 = A(x + 1) + Bx$
When $x = 0$: $3 = A$.
When $x = -1$: $-7 = -B$, $B = 7$.

$$\frac{10x + 3}{x^2 + x} = \frac{3}{x} + \frac{7}{x + 1}$$

4. $\frac{3x + 11}{(x - 3)(x + 1)} = \frac{A}{x - 3} + \frac{B}{x + 1}$

Basic equation: $3x + 11 = A(x + 1) + B(x - 3)$
When $x = 3$: $20 = 4A$, $A = 5$.
When $x = -1$: $8 = -4B$, $B = -2$.

$$\frac{3x + 11}{x^2 - 2x - 3} = \frac{5}{x - 3} - \frac{2}{x + 1}$$

6. $\frac{7x + 5}{6(2x + 1)(x + 1)} = \frac{1}{6}\left(\frac{A}{2x + 1} + \frac{B}{x + 1}\right)$

Basic equation: $7x + 5 = A(x + 1) + B(2x + 1)$
When $x = -\frac{1}{2}$: $\frac{3}{2} = \frac{1}{2}A$, $A = 3$.
When $x = -1$: $-2 = -B$, $B = 2$.

$$\frac{7x + 5}{6(2x^2 + 3x + 1)} = \frac{1}{6}\left(\frac{3}{2x + 1} + \frac{2}{x + 1}\right) = \frac{1}{2(2x + 1)} + \frac{1}{3(x + 1)}$$

8. $\frac{3x^2 - x + 1}{x(x + 1)^2} = \frac{A}{x} + \frac{B}{x + 1} + \frac{C}{(x + 1)^2}$

Basic equation: $3x^2 - x + 1 = A(x + 1)^2 + Bx(x + 1) + Cx$
When $x = 0$: $1 = A$.
When $x = -1$: $5 = -C$, $C = -5$.
When $x = 1$: $3 = 4A + 2B + C$, $B = 2$.

$$\frac{3x^2 - x + 1}{x(x + 1)^2} = \frac{1}{x} + \frac{2}{x + 1} - \frac{5}{(x + 1)^2}$$

10. $\dfrac{3x-4}{(x-5)^2} = \dfrac{A}{x-5} + \dfrac{B}{(x-5)^2}$

Basic equation: $3x - 4 = A(x-5) + B$
When $x = 5$: $11 = B$.
When $x = 0$: $-4 = -5A + 11 \Rightarrow A = 3$.

$$\frac{3x-4}{(x-5)^2} = \frac{3}{x-5} + \frac{11}{(x-5)^2}$$

12. $\dfrac{6x^2-5x}{(x+2)^3} = \dfrac{A}{x+2} + \dfrac{B}{(x+2)^2} + \dfrac{C}{(x+2)^3}$

Basic equation: $6x^2 - 5x = A(x+2)^2 + B(x+2) + C$
When $x = -2$: $34 = C$.
When $x = 0$: $0 = 4A + 2B + 34$.
When $x = 1$: $1 = 9A + 3B + 34$.
Solving these equations yields $A = 6$ and $B = -29$.

$$\frac{6x^2-5x}{(x+2)^3} = \frac{6}{x+2} - \frac{29}{(x+2)^2} + \frac{34}{(x+2)^3}$$

14. $\dfrac{9}{(x-3)(x+3)} = \dfrac{A}{x-3} + \dfrac{B}{x+3}$

Basic equation: $9 = A(x+3) + B(x-3)$
When $x = 3$: $9 = 6A$, $A = \frac{3}{2}$.
When $x = -3$: $9 = -6B$, $B = -\frac{3}{2}$.

$$\int \frac{9}{x^2-9}\,dx = \frac{3}{2}\int \frac{1}{x-3}\,dx - \frac{3}{2}\int \frac{1}{x+3}\,dx = \frac{3}{2}\ln|x-3| - \frac{3}{2}\ln|x+3| + C = \frac{3}{2}\ln\left|\frac{x-3}{x+3}\right| + C$$

16. $\dfrac{-4}{(x+2)(x-2)} = \dfrac{A}{x+2} + \dfrac{B}{x-2}$

Basic equation: $-4 = A(x-2) + B(x+2)$
When $x = -2$: $-4 = -4A$, $A = 1$.
When $x = 2$: $-4 = 4B$, $B = -1$.

$$\int \frac{-4}{x^2-4}\,dx = \int \frac{1}{x+2}\,dx - \int \frac{1}{x-2}\,dx = \ln|x+2| - \ln|x-2| + C = \ln\left|\frac{x+2}{x-2}\right| + C$$

18. $\dfrac{3}{x(x-3)} = \dfrac{A}{x} + \dfrac{B}{x-3}$

Basic equation: $3 = A(x-3) + Bx$
When $x = 0$: $3 = -3A$, $A = -1$.
When $x = 3$: $3 = 3B$, $B = 1$.

$$\int \frac{3}{x^2-3x}\,dx = -\int \frac{1}{x}\,dx + \int \frac{1}{x+3}\,dx = -\ln|x| + \ln|x+3| + C = \ln\left|\frac{x+3}{x}\right| + C$$

20. $\dfrac{5}{(x-2)(x+3)} = \dfrac{A}{x-2} + \dfrac{B}{x+3}$

Basic equation: $5 = A(x+3) + B(x-2)$
When $x = 2$: $5 = 5A$, $A = 1$.
When $x = -3$: $5 = -5B$, $B = -1$.

$$\int \frac{5}{x^2+x-6}\,dx = \int \frac{1}{x-2}\,dx - \int \frac{1}{x+3}\,dx = \ln|x-2| - \ln|x+3| + C = \ln\left|\frac{x-2}{x+3}\right| + C$$

22. $\dfrac{1}{(2x-3)(2x+3)} = \dfrac{A}{2x-3} + \dfrac{B}{2x+3}$

Basic equation: $1 = A(2x+3) + B(2x-3)$
When $x = \frac{3}{2}$: $1 = 6A$, $A = \frac{1}{6}$.
When $x = -\frac{3}{2}$: $1 = -6B$, $B = -\frac{1}{6}$.

$$\int \frac{1}{4x^2-9}\,dx = \frac{1}{6}\int \frac{1}{2x-3}\,dx - \frac{1}{6}\int \frac{1}{2x+3}\,dx = \frac{1}{12}\ln|2x-3| - \frac{1}{12}\ln|2x+3| + C = \frac{1}{12}\ln\left|\frac{2x-3}{2x+3}\right| + C$$

24. $\displaystyle\int \frac{x+1}{(x+1)(x+3)}\,dx = \int \frac{1}{x+3}\,dx = \ln|x+3| + C$

26. $\dfrac{3x^2-7x-2}{x(x+1)(x-1)} = \dfrac{A}{x} + \dfrac{B}{x+1} + \dfrac{C}{x-1}$

Basic equation: $3x^2 - 7x - 2 = A(x+1)(x-1) + Bx(x-1) + Cx(x+1)$
When $x = 0$: $-2 = -A$, $A = 2$.
When $x = -1$: $8 = 2B$, $B = 4$.
When $x = 1$: $-6 = 2C$, $C = -3$.

$$\int \frac{3x^2-7x-2}{x^3-x}\,dx = 2\int \frac{1}{x}\,dx + 4\int \frac{1}{x+1}\,dx - 3\int \frac{1}{x-1}\,dx = 2\ln|x| + 4\ln|x+1| - 3\ln|x-1| + C$$

28. $\dfrac{4x^2+2x-1}{x^2(x+1)} = \dfrac{A}{x} + \dfrac{B}{x^2} + \dfrac{C}{x+1}$

Basic equation: $4x^2 + 2x - 1 = Ax(x+1) + B(x+1) + Cx^2$

When $x = 0$: $-1 = B$, $B = -1$.

When $x = -1$: $1 = C$, $C = 1$.

When $x = 1$: $5 = 2A + 2B + C$, $A = 3$.

$$\int \frac{4x^2+2x-1}{x^3+x^2}\,dx = 3\int \frac{1}{x}\,dx - \int \frac{1}{x^2}\,dx + \int \frac{1}{x+1}\,dx$$

$$= 3\ln|x| + \frac{1}{x} + \ln|x+1| + C = \frac{1}{x} + \ln|x^3| + \ln|x+1| + C = \frac{1}{x} + \ln|x^4+x^3| + C$$

30. $\dfrac{x^4}{x^3-3x^2+3x-1} = x + 3 + \dfrac{6x^2-8x+4}{(x-1)^3}$

$$\frac{6x^2-8x+3}{(x-1)^3} = \frac{A}{x-1} + \frac{B}{(x-1)^2} + \frac{C}{(x-1)^3}$$

Basic equation: $6x^2 - 8x + 3 = A(x-1)^2 + B(x-1) + C = Ax^2 + (B-2A)x + (A-B+C)$
Therefore, $A = 6$, $B - 2A = -8$, and $A - B + C = 3$.
Solving these equations yields $A = 6$, $B = 4$, and $C = 1$.

$$\int \frac{x^4}{(x-1)^3}\,dx = \int (x+3)\,dx + 6\int \frac{1}{x-1}\,dx + 4\int \frac{1}{(x-1)^2}\,dx + \int \frac{1}{(x-1)^3}\,dx$$

$$= \frac{x^2}{2} + 3x + 6\ln|x-1| - \frac{4}{x-1} - \frac{1}{2(x-1)^2} + C$$

32. $\dfrac{3x}{(x-3)^2} = \dfrac{A}{x-3} + \dfrac{B}{(x-3)^2}$

Basic equation: $3x = A(x-3) + B = Ax + (B-3A)$
Therefore, $A = 3$ and $B - 3A = 0$. Solving these equations yields $A = 3$ and $B = 9$.

$$\int \frac{3x}{x^2-6x+9}\,dx = 3\int \frac{1}{x-3}\,dx + 9\int \frac{1}{(x-3)^2}\,dx = 3\ln|x-3| - \frac{9}{x-3} + C$$

34. $\dfrac{3}{(2x+1)(x+2)} = \dfrac{A}{2x+1} + \dfrac{B}{x+2}$

Basic equation: $3 = A(x+2) + B(2x+1)$

When $x = -\frac{1}{2}$: $3 = \frac{3}{2}A$, $A = 2$.

When $x = -2$: $3 = -3B$, $B = -1$.

$$\int_0^1 \frac{3}{2x^2+5x+2}\,dx = 2\int_0^1 \frac{1}{2x+1}\,dx - \int_0^1 \frac{1}{x+2}\,dx = \ln|2x+1|\Big]_0^1 - \ln|x+2|\Big]_0^1$$

$$= \ln 3 - (\ln 3 - \ln 2) = \ln 2 \approx 0.693$$

36. $\displaystyle\int_0^1 \frac{x^2-x}{x^2+x+1}\,dx = \int_0^1 \left[1 - \frac{2x+1}{x^2+x+1}\right]dx = \left[x - \ln(x^2+x+1)\right]_0^1 = 1 - \ln 3 \approx -0.099$

38. $\dfrac{x^3-1}{x^2-4} = x + \dfrac{4x-1}{x^2-4}$

$\dfrac{4x-1}{x^2-4} = \dfrac{A}{x-2} + \dfrac{B}{x+2}$

Basic equation: $4x - 1 = A(x+2) + B(x-2)$

When $x = 2$: $7 = 4A \Rightarrow A = \dfrac{7}{4}$

When $x = -2$: $-9 = -4B \Rightarrow B = \dfrac{9}{4}$

$$\int_0^1 \frac{x^3-1}{x^2-4} = \int_0^1 \left(x + \frac{7}{4}\cdot\frac{1}{x-2} + \frac{9}{4}\cdot\frac{1}{x+2}\right)dx = \frac{x^2}{2} + \frac{7}{4}\ln|x-2| + \frac{9}{4}\ln|x+2|\Big]_0^1$$

$$= \frac{1}{2} + \frac{9}{4}\ln 3 = \frac{7}{4}\ln 2 + \frac{9}{4}\ln 2 = \frac{1}{2} - \frac{7}{4}\ln 2 + \frac{9}{4}\ln\left(\frac{3}{2}\right) \approx 0.199$$

40. $\dfrac{x^4-4}{x^2-1} = x^2 + 1 + \dfrac{-3}{x^2-1}$

$\dfrac{-3}{x^2-1} = \dfrac{A}{x-1} + \dfrac{B}{x+1}$

Basic equation: $-3 = A(x+1) + B(x-1)$

When $x = -1$: $-3 = -2B \Rightarrow B = \dfrac{3}{2}$

When $x = 1$: $-3 = 2A \Rightarrow A = \dfrac{-3}{2}$

$$\int_2^4 \frac{x^4-4}{x^2-1}\,dx = \int_2^4 \left(x^2 + 1 - \frac{3}{2}\cdot\frac{1}{x-1} + \frac{3}{2}\cdot\frac{1}{x+1}\right)dx$$

$$= \left[\frac{x^3}{3} + x - \frac{3}{2}\ln|x-1| + \frac{3}{2}\ln|x+1|\right]_2^4$$

$$= \frac{62}{3} - 3\ln 3 + \frac{3}{2}\ln 5 \approx 19.785$$

42. $\displaystyle A = \int_{-1}^2 \frac{-4}{(x+2)(x-3)}\,dx = \int_{-1}^2 \left(\frac{4}{5}\cdot\frac{1}{x+2} - \frac{4}{5}\cdot\frac{1}{x-3}\right)dx$

$$= \frac{4}{5}\Big[\ln|x+2| - \ln|x-3|\Big]_{-1}^2 = \frac{4}{5}(\ln 4 + \ln 4)$$

$$= \frac{8}{5}\ln 4 \approx 2.2181$$

44. $A = \int_{-1}^{0} \frac{x^2+2x-1}{x^2-4}\,dx = \int_{-1}^{0} \left(1 + \frac{7}{4}\cdot\frac{1}{x-2} + \frac{1}{4}\cdot\frac{1}{x+2}\right) dx$

$= \left[x + \frac{7}{4}\ln|x-2| + \frac{1}{4}\ln|x+2|\right]_{-1}^{0}$

$= \left(\frac{7}{4}\ln 2 + \frac{1}{4}\ln 2\right) - \left(-1 + \frac{7}{4}\ln 3\right)$

$= 1 + 2\ln 2 = \frac{7}{4}\ln 3 \approx 0.4637$

46. $\frac{1}{x(x+a)} = \frac{A}{x} + \frac{B}{x+a}$

Basic equation: $1 = A(x+a) + Bx$

When $x = 0$: $1 = aA, \quad A = \frac{1}{a}$.

When $x = -a$: $1 = -aB, \quad B = -\frac{1}{a}$.

$\frac{1}{x(x+a)} = \frac{1}{a}\left(\frac{1}{x} - \frac{1}{x+a}\right)$

48. $\frac{1}{(x+1)(a-x)} = \frac{A}{x+1} + \frac{B}{a-x}$

Basic equation: $1 = A(a-x) + B(x+1)$

When $x = a$: $1 = B(a+1) \Longrightarrow B = \frac{1}{a+1}$

When $x = -1$: $1 = A(a+1) \Longrightarrow A = \frac{1}{a+1}$

$\frac{1}{(x+1)(a-x)} = \frac{1/(a+1)}{x+1} + \frac{1/(a+1)}{a-x}$

50. $V = \pi\int_{0}^{3} \frac{16}{(x+1)^2(x-4)^2}\,dx = \frac{16\pi}{125}\int_{0}^{3}\left[\frac{2}{x+1} - \frac{2}{x-4} + \frac{5}{(x+1)^2} + \frac{5}{(x-4)^2}\right] dx$

$= \frac{16\pi}{125}\left[2\ln\left|\frac{x+1}{x-4}\right| - \frac{5(2x-3)}{(x+1)(x-4)}\right]_{0}^{3} = \frac{16\pi}{125}\left[\left(2\ln 4 + \frac{15}{4}\right) - \left(2\ln\frac{1}{4} - \frac{15}{4}\right)\right]$

$= \frac{16\pi}{125}\left(2\ln 16 + \frac{15}{2}\right) \approx 5.246$

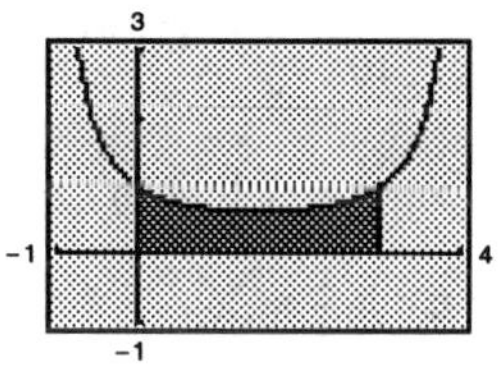

52. $V = \pi\int_{-2}^{0}\left(\frac{25x}{x^2+x-6}\right)^2 dx$

$\frac{(25x)^2}{(x+3)^2(x-2)^2} = \frac{A}{x+3} + \frac{B}{(x+3)^3} + \frac{C}{x-2} + \frac{D}{(x-2)^2}$

Basic equation: $625x^2 = A(x+3)(x-2)^2 + B(x-2)^2 + C(x-2)(x+3)^2 + D(x+3)^2$

When $x = 2$: $2500 = 25D \Longrightarrow D = 100$

When $x = -3$: $5625 = 25B \Longrightarrow B = 225$

When $x = 0$: $0 = 12A + 900 - 18C + 900 - 900 = 6A - 9B$

When $x = 1$: $625 = 4A + 225 - 16C + 1600 - 300 = A - 4C$

Hence, $A = -60$ and $C = 60$.

$V = \pi\int_{-2}^{0}\left[\frac{-60}{x+3} + \frac{225}{(x+3)^2} + \frac{60}{x-2} + \frac{100}{(x-2)^2}\right] dx$

$= \pi\left[-60\ln|x+3| - \frac{225}{x+3} + 60\ln|x-2| - \frac{100}{x-2}\right]_{-2}^{0}$

$= \pi\left[\left(-60\ln 3 - \frac{225}{3} + 60\ln 2 + 50\right) - (-225 + 60\ln 4 + 25)\right]$

≈ 212.04

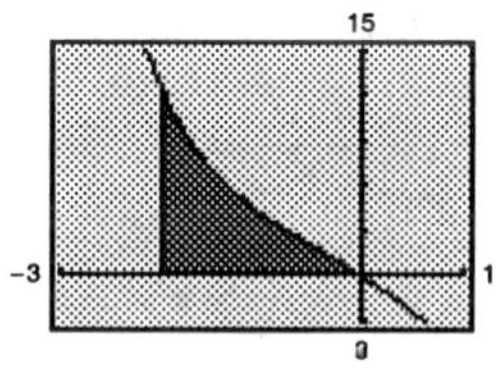

54. $t = 5010\int \frac{1}{(x+1)(500-x)}\,dx$

$= \int\left(\frac{10}{x+1} - \frac{10}{x-500}\right)dx$

$t = 10\ln|x+1| - 10\ln|500-x| + C$

Since $x = 1$ when $t = 0$:

$0 = 10\ln 2 - 10\ln 499 + C \Rightarrow C \approx 55.1946.$

Thus:

$t = 10\ln(x+1) - 10\ln(500-x) + 55.1946.$

(a) If $x = (0.75)(500) = 375$, then $t \approx 66.2$ hours.

(b) When $t = 100$, you can use a graphing utility to solve for x:

$100 = 10 - \ln(x+1) - 10\ln(500-x) + 55.1946 \Rightarrow x \approx 494$ individuals.

56. (a) $kt = \int \frac{1}{y(10-y)}\,dy = \frac{1}{10}\int\left(\frac{1}{y} + \frac{1}{10-y}\right)dy = \frac{1}{10}(\ln y - \ln|10-y|) + C_1 = \frac{1}{10}\ln\left(\frac{y}{10-y}\right) + C_1$

$10kt - C_2 = \ln\left(\frac{y}{10-y}\right)$

$\frac{y}{10-y} = Ce^{10kt}$

Since $y = 1$ when $t = 0$, $\frac{1}{9} = C$ and $\frac{y}{10-y} = \frac{1}{9}e^{10kt}$.

Solving for y,

$\frac{9y}{10-y} = e^{10kt}$

$9y = 10e^{10kt} - ye^{10kt}$

$y(9 + e^{10kt}) = 10e^{10kt}$

$y = \frac{10e^{10kt}}{9 + e^{10kt}} = \frac{10}{1 + 9e^{-10kt}}$

(b) Since $y = 2$ when $t = 2$,

$\frac{9(2)}{10-2} = d^{10k(2)}$

$\frac{9}{4} = e^{20k} \Rightarrow k = 0.04055$

58. Both species have positive growth rates. The rate of growth is increasing on $[0, 3]$ and decreasing on $[3, \infty)$ for P. aurelia. The rate of growth is increasing on $[0, 2]$ and decreasing on $[2, \infty]$ for P. caudatum. Hence, the former species has a greater limiting population.

Section 6.4 Integration Tables and Completing the Square

2. Formula 11: $u = x,\quad du = dx,\quad a = 2,\quad b = 3$

$$\int \frac{1}{x(2+3x)^2}\,dx = \frac{1}{2}\left[\frac{1}{2+3x} + \frac{1}{2}\ln\left|\frac{x}{2+3x}\right|\right] + C$$

4. Formula 29: $u = x, \quad du = dx, \quad a = 3$

$$4\int \frac{1}{x^2 - 9}\,dx = \frac{4}{2(3)}\ln\left|\frac{x-3}{x+3}\right| + C = \frac{2}{3}\ln\left|\frac{x-3}{x+3}\right| + C$$

6. Formula 22: $u = x, \quad du = dx, \quad a = 3$

$$\int x^2\sqrt{x^2+9}\,dx = \frac{1}{8}\left[x(2x^2+9)\sqrt{x^2+9} - 81\ln\left|x + \sqrt{x^2+9}\right|\right] + C$$

8. Formula 37: $u = x^2, \quad du = 2x\,dx$

$$\int \frac{x}{1+e^{x^2}}\,dx = \frac{1}{2}\int \frac{2x}{1+e^{x^2}}\,dx = \frac{1}{2}[x^2 - \ln(1+e^{x^2})] + C$$

10. Formula 11: $u = x, \quad du = dx, \quad a = b = 1$

$$\int \frac{1}{x(1+x)^2}\,dx = \frac{1}{1+x} + \ln\left|\frac{x}{1+x}\right| + C$$

12. Formula 25: $u = x, \quad du = dx, \quad a = 1$

$$\int \frac{1}{\sqrt{x^2-1}}\,dx = \ln\left|x + \sqrt{x^2-1}\right| + C$$

14. Formula 24: $u = x, \quad du = dx, \quad a = 3$

$$\int \frac{\sqrt{x^2-9}}{x^2}\,dx = -\frac{\sqrt{x^2-9}}{x} + \ln\left|x + \sqrt{x^2-9}\right| + C$$

16. Formula 42: $u = x^3, \quad du = 3x^2dx$

$$\int x^2(\ln x^3)^2\,dx = \frac{1}{3}\int (\ln x^3)^2(3x^2)\,dx = \frac{x^3}{3}[2 - 2\ln x^3 + (\ln x^3)^2] + C$$

18. Formula 37: $u = x, \quad du = dx$

$$\int \frac{1}{1+e^x}\,dx = x - \ln(1+e^x) + C$$

20. Formula 29: $u = x^2, \quad du = 2x\,dx, \quad a = 3$

$$\int \frac{x}{x^4-9}\,dx = \frac{1}{2}\int \frac{2x}{(x^2)^2-9}\,dx = \frac{1}{12}\ln\left|\frac{x^2-3}{x^2+3}\right| + C$$

22. Formula 17: $u = t, \quad du = dt, \quad a = 3, \quad b = 4$

$$\int \frac{\sqrt{3+4t}}{t}\,dt = 2\sqrt{3+4t} + 3\int \frac{1}{t\sqrt{3+4t}}\,dt$$

Formula 15: $u = t, \quad du = dt, \quad a = 3, \quad b = 4$

$$\int \frac{\sqrt{3+4t}}{t}\,dt = 2\sqrt{3+4t} + \frac{3}{\sqrt{3}}\ln\left|\frac{\sqrt{3+4t}-\sqrt{3}}{\sqrt{3+4t}+\sqrt{3}}\right| + C = 2\sqrt{3+4t} + \sqrt{3}\ln\left|\frac{\sqrt{3+4t}-\sqrt{3}}{\sqrt{3+4t}+\sqrt{3}}\right| + C$$

24. Formula 21: $u = x, \quad du = dx, \quad a = \sqrt{3}$

$$\int \sqrt{3+x^2}\,dx = \frac{1}{2}\left[x\sqrt{3+x^2} + 3\ln\left|x + \sqrt{3+x^2}\right|\right] + C$$

26. Formula 28: $u = x, \quad du = dx, \quad a = 2$

$$\int \frac{1}{x^2\sqrt{x^2-4}}\,dx = \frac{\sqrt{x^2-4}}{4x} + C$$

28. Formula 4: $u = x, \quad du = dx, \quad a = 1, \quad b = -3$

$$2\int \frac{x}{(1-3x)^2}\,dx = \frac{2}{9}\left[\frac{1}{1-3x} + \ln|1-3x|\right] + C$$

30. Formula 34: $u = x^2, \quad du = 2x\,dx$

$$\int xe^{x^2}\,dx = \frac{1}{2}\int e^{x^2}(2x)\,dx = \frac{1}{2}e^{x^2} + C$$

32. Formula 13: $u = x, \quad du = dx, \quad a = -1, \quad b = 2$

$$\frac{1}{2}\int \frac{1}{x^2(2x-1)^2}\,dx = -\frac{1}{2}\left[\frac{4x-1}{x(2x-1)} - 4\ln\left|\frac{x}{2x-1}\right|\right] + C$$

34. Formula 43: $u = x, \quad du = dx, \quad n = 3$
Formula 42: $u = x, \quad du = dx$

$$\int (\ln x)^3\,dx = x(\ln x)^3 - 3\int (\ln x)^2\,dx = x(\ln x)^3 - 3x[2 - 2\ln x + (\ln x)^2] + C$$

$$= x[(\ln x)^3 - 3(\ln x)^2 + 6\ln x - 6] + C$$

36. $\displaystyle\int_0^1 \frac{2}{1+e^{4x}}\,dx = 2\left[x - \frac{1}{4}\ln(1+e^{4x})\right]_0^1 = 2\left[1 - \frac{1}{4}\ln(1+e^4) + \frac{1}{4}\ln 2\right] \approx 0.3375$

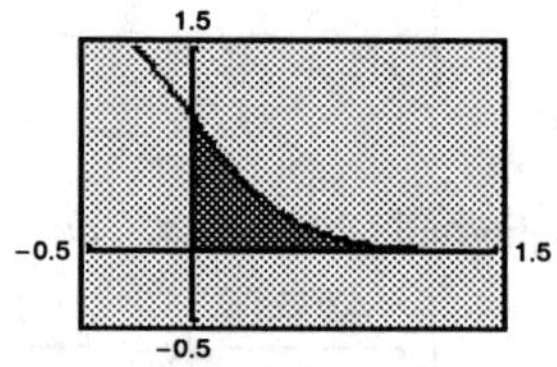

38. Formula 21: $u = e^x, \quad du = e^x\,dx, \quad a = 1$

$$\text{Area} = \int_1^2 \frac{e^x}{1-e^{2x}}\,dx = -\int_1^2 \frac{1}{(e^x)^2 - 1}(e^x\,dx) = \left[\frac{-1}{2}\ln\left|\frac{e^x-1}{e^x+1}\right|\right]_1^2$$

$$= \frac{-1}{2}\ln\left(\frac{e^2-1}{e^2+1}\right) + \frac{1}{2}\ln\left(\frac{e-1}{e+1}\right) \approx 0.25$$

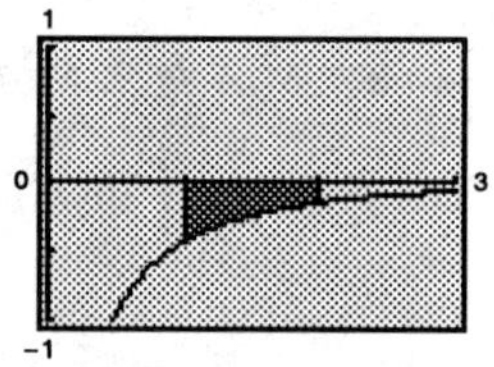

40. Formula 2: $u = 1 + 2\sqrt{x}, \quad du = \dfrac{1}{\sqrt{x}}\,dx$

$$\text{Area} = \int_1^4 \frac{1}{\sqrt{x}(1+2\sqrt{x})}\,dx = \int_1^4 \frac{1}{1+2\sqrt{x}}\left(\frac{1}{\sqrt{x}}\right)dx$$

$$= \left[\ln\left(1+2\sqrt{x}\right)\right]_1^4 = \ln 5 - \ln 3 \approx 0.5108$$

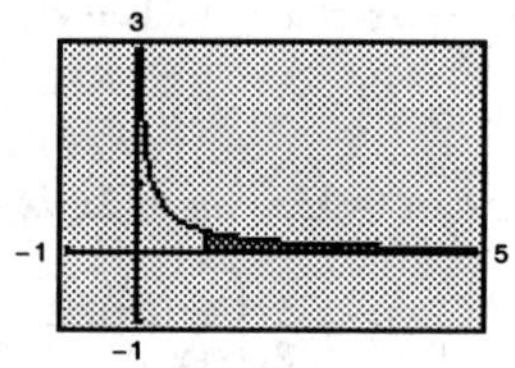

42. Formula 4: $u = x, \quad du = dx, \quad a = 4, \quad b = 1$

$$\int_0^5 \frac{x}{(4+x)^2}\,dx = \left[\frac{4}{4+x} + \ln|4+x|\right]_0^5 = \left(\frac{4}{9} + \ln 9\right) - (1 + \ln 4) = -\frac{5}{9} + \ln\frac{9}{4} \approx 0.255$$

44. Formula 40: $u = x, \quad du = dx$

$$\int_1^4 x\ln x\,dx = \left[\frac{x^2}{4}(-1 + 2\ln x)\right]_1^4$$

$$= (4)(-1 + 2\ln 4) - \left(\frac{1}{4}\right)(-1 + 2\ln 1)$$

$$= -\frac{15}{4} + 8\ln 4 \approx 7.340$$

46. (a) Formula 41: $u = x, \quad du = dx, \quad n = 4$

$$\int x^4\ln x\,dx = \frac{x^5}{25}(-1 + 5\ln x) + C$$

(b) Let $u = \ln x$ and $dv = x^4\,dx$, then $du = \dfrac{1}{x}\,dx$ and $v = \dfrac{x^5}{5}$.

$$\int x^4\ln x\,dx = \frac{x^5}{5}\ln x - \frac{1}{5}\int x^4\,dx$$

$$= \frac{x^5}{5}\ln x - \frac{x^5}{25} + C$$

$$= \frac{x^5}{25}(-1 + 5\ln x) + C$$

48. (a) Formula 29: $u = x, \quad du = dx, \quad a = \sqrt{75} = 5\sqrt{3}$

$$\int \frac{1}{x^2 - 75}\,dx = \frac{1}{10\sqrt{3}} \ln\left|\frac{x - 5\sqrt{3}}{x + 5\sqrt{3}}\right| + C$$

(b) $\dfrac{1}{(x - 5\sqrt{3})(x + 5\sqrt{3})} = \dfrac{A}{x - 5\sqrt{3}} + \dfrac{B}{x + 5\sqrt{3}}$

Basic equation: $1 = A(x + 5\sqrt{3}) + B(x - 5\sqrt{3})$

When $x = 5\sqrt{3}$: $1 = 10\sqrt{3}A, \ A = \dfrac{1}{10\sqrt{3}}$.

When $x = -5\sqrt{3}$: $1 = -10\sqrt{3}B, \ B = -\dfrac{1}{10\sqrt{3}}$.

$$\int \frac{1}{x^2 - 75}\,dx = \frac{1}{10\sqrt{3}}\left(\int \frac{1}{x - 5\sqrt{3}}\,dx - \int \frac{1}{x + 5\sqrt{3}}\,dx\right)$$

$$= \frac{1}{10\sqrt{3}}\left(\ln\left|x - 5\sqrt{3}\right| - \ln\left|x + 5\sqrt{3}\right|\right) + C = \frac{1}{10\sqrt{3}} \ln\left|\frac{x - 5\sqrt{3}}{x + 5\sqrt{3}}\right| + C$$

50. (a) $2x^2 + 12x + 14 = 2(x^2 + 6x + 7)$

$= 2(x^2 + 6x + 9 - 9 + 7)$

$= 2[(x + 3)^2 - 2$

(b) $3x^2 - 12x - 9 = 3(x^2 - 4x - 3)$

$= 3(x^2 - 4x + 4 - 4 - 3)$

$= 3[(x - 2)^2 - 7]$

(c) $x^2 - 2x = x^2 - 2x + 1 - 1$

$= (x - 1)^2 - 1$

(d) $9 + 8x - x^2 = -(x^2 - 8x - 9)$

$= -(x^2 - 8x + 16 - 16 - 9)$

$= -[(x - 4)^2 - 25]$

$= 25 - (x - 4)^2$

52. Formula 29: $u = x + 2, \quad du = dx, \quad a = 3$

$$\int \frac{1}{x^2 + 4x - 5}\,dx = \int \frac{1}{(x + 2)^2 - 9}\,dx = \frac{1}{6}\ln\left|\frac{(x + 2) - 3}{(x + 2) + 3}\right| + C = \frac{1}{6}\ln\left|\frac{x - 1}{x + 5}\right| + C$$

54. Formula 21: $u = x - 3, \quad du = dx, \quad a = 3$

$$\int \sqrt{x^2 - 6x}\,dx = \int \sqrt{(x - 3)^2 - 9}\,dx = \frac{1}{2}\left[(x - 3)\sqrt{x^2 - 6x} - 9\ln\left|(x - 3) + \sqrt{x^2 - 6x}\right|\right] + C$$

56. Formula 31: $u = x + 3, \quad du = dx, \quad a = 4$

$$\int \frac{\sqrt{7 - 6x - x^2}}{x + 3}\,dx = \int \frac{\sqrt{16 - (x + 3)^2}}{x + 3}\,dx = \sqrt{7 - 6x - x^2} - 4\ln\left|\frac{4 + \sqrt{7 - 6x - x^2}}{x + 3}\right| + C$$

58. Formula 23: $u = x^2 + 2, \quad du = 2x\,dx, \quad a = 1$

$$\int \frac{x\sqrt{x^4 + 4x^2 + 5}}{x^2 + 2}\,dx = \frac{1}{2}\int \frac{\sqrt{(x^2 + 2)^2 + 1}}{x^2 + 2}(2x)\,dx = \frac{1}{2}\left(\sqrt{x^4 + 4x^2 + 5} - \ln\left|\frac{1 + \sqrt{x^4 + 4x^2 + 5}}{x^2 + 2}\right|\right) + C$$

60. Formula 37: $u = 4.20 - 0.25t, \quad du = -0.25\,dt$

$$\text{Average} = \frac{1}{28 - 21}\int_{21}^{28} \frac{375}{1 + e^{4.20 - 0.25t}}\,dt = \frac{375}{7(-0.25)}\int_{21}^{28} \frac{-0.25}{1 + e^{4.20 - 0.25t}}\,dt$$

$$= -\frac{1500}{7}\Big[4.20 - 0.25t - \ln(1 + e^{4.20 - 0.25t})\Big]_{21}^{28} \approx 323.352$$

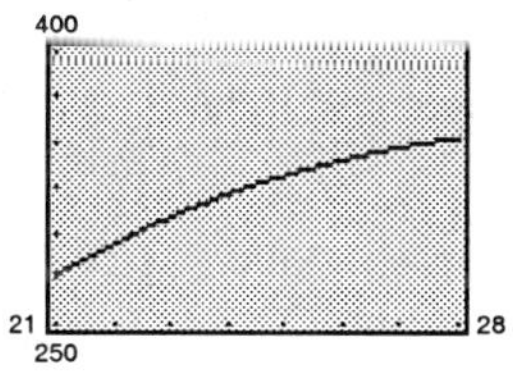

62. Equating the supply and demand functions produces $x = 12$, $p = 4$ as the point of equilibrium.

Consumer surplus: Formula 25: $u = x$, $du = dx$, $a = 9$

$$\int_0^{12} \left(\frac{60}{\sqrt{x^2+81}} - 4\right) dx = \left[60 \ln\left|x + \sqrt{x^2+81}\right| - 4x\right]_0^{12} \approx 17.92$$

Producer surplus:

$$\int_0^{12} \left(4 - \frac{x}{3}\right) dx = \left(4x - \frac{x^2}{6}\right)\Big]_0^{12} = 24$$

Section 6.5 Numerical Integration

2. Exact: $\displaystyle\int_0^1 \left(\frac{x^2}{2} + 1\right) dx = \left[\frac{x^3}{6} + x\right]_0^1 = \frac{7}{6} \approx 1.1667$

Trapezoidal Rule: $\displaystyle\int_0^1 \left(\frac{x^2}{2} + 1\right) dx \approx \frac{1}{8}\left[1 + 2\left(\frac{1}{32} + 1\right) + 2\left(\frac{1}{8} + 1\right) + 2\left(\frac{9}{32} + 1\right) + \frac{3}{2}\right] \approx 1.1719$

Simpson's Rule: $\displaystyle\int_0^1 \left(\frac{x^2}{2} + 1\right) dx \approx \frac{1}{12}\left[1 + 4\left(\frac{1}{32} + 1\right) + 2\left(\frac{1}{8} + 1\right) + 4\left(\frac{9}{32} + 1\right) + \frac{3}{2}\right] \approx 1.1667$

4. Exact: $\displaystyle\int_1^2 \frac{1}{x}\, dx = \ln|x|\Big]_1^2 = \ln 2 \approx 0.6931$

Trapezoidal Rule: $\displaystyle\int_1^2 \frac{1}{x}\, dx \approx \frac{1}{8}\left[1 + 2\left(\frac{4}{5}\right) + 2\left(\frac{2}{3}\right) + 2\left(\frac{4}{7}\right) + \frac{1}{2}\right] \approx 0.6970$

Simpson's Rule: $\displaystyle\int_1^2 \frac{1}{x}\, dx \approx \frac{1}{12}\left[1 + 4\left(\frac{4}{5}\right) + 2\left(\frac{2}{3}\right) + 4\left(\frac{4}{7}\right) + \frac{1}{2}\right] \approx 0.6933$

6. Exact: $\displaystyle\int_1^2 \frac{1}{x}\, dx = \ln|x|\Big]_1^2 = \ln 2 \approx 0.6931$

Trapezoidal Rule: $\displaystyle\int_1^2 \frac{1}{x}\, dx \approx \frac{1}{16}\left[1 + 2\left(\frac{8}{9}\right) + 2\left(\frac{4}{5}\right) + 2\left(\frac{8}{11}\right) + 2\left(\frac{2}{3}\right) + 2\left(\frac{8}{13}\right) + 2\left(\frac{4}{7}\right) + 2\left(\frac{8}{15}\right) + \frac{1}{2}\right] \approx 0.6941$

Simpson's Rule: $\displaystyle\int_1^2 \frac{1}{x}\, dx \approx \frac{1}{24}\left[1 + 4\left(\frac{8}{9}\right) + 2\left(\frac{4}{5}\right) + 4\left(\frac{8}{11}\right) + 2\left(\frac{2}{3}\right) + 4\left(\frac{8}{13}\right) + 2\left(\frac{4}{7}\right) + 4\left(\frac{8}{15}\right) + \frac{1}{2}\right] \approx 0.6932$

8. Exact: $\displaystyle\int_0^4 \sqrt{x}\, dx = \frac{2}{3}x^{3/2}\Big]_0^4 = \frac{16}{3} \approx 5.3333$

Trapezoidal Rule: $\displaystyle\int_0^4 \sqrt{x}\, dx \approx \frac{4}{16}\left(0 + 2\sqrt{\tfrac{1}{2}} + 2 + 2\sqrt{\tfrac{3}{2}} + 2\sqrt{2} + 2\sqrt{\tfrac{5}{2}} + 2\sqrt{3} + 2\sqrt{\tfrac{7}{2}} + 2\right) \approx 5.2650$

Simpson's Rule: $\displaystyle\int_0^4 \sqrt{x}\, dx \approx \frac{4}{24}\left(0 + 4\sqrt{\tfrac{1}{2}} + 2 + 4\sqrt{\tfrac{3}{2}} + 2\sqrt{2} + 4\sqrt{\tfrac{5}{2}} + 2\sqrt{3} + 4\sqrt{\tfrac{7}{2}} + 2\right) \approx 5.3046$

10. Exact: $\displaystyle\int_0^2 x\sqrt{x^2+1}\, dx = \frac{1}{2}\int_0^2 (x^2+1)^{1/2}(2x)\, dx = \frac{1}{3}(x^2+1)^{3/2}\Big]_0^2 = \frac{1}{3}(5\sqrt{5} - 1) \approx 3.3934$

Trapezoidal Rule: $\displaystyle\int_0^2 x\sqrt{x^2+1}\, dx \approx \frac{2}{8}\left[0 + 2\left(\frac{1}{2}\right)\sqrt{\tfrac{5}{4}} + 2(1)\sqrt{2} + 2\left(\frac{3}{2}\right)\sqrt{\tfrac{13}{4}} + 2\sqrt{5}\right] \approx 3.4567$

Simpson's Rule: $\displaystyle\int_0^2 x\sqrt{x^2+1}\, dx \approx \frac{2}{12}\left[0 + 4\left(\frac{1}{2}\right)\sqrt{\tfrac{5}{4}} + 2\sqrt{2} + 4\left(\frac{3}{2}\right)\sqrt{\tfrac{13}{4}} + 2\sqrt{5}\right] \approx 3.3922$

12. (a) Trapezoidal Rule: $\int_0^2 \frac{1}{\sqrt{1+x^3}}\,dx \approx \frac{2}{8}\left[1 + 2\left(\frac{1}{\sqrt{9/8}}\right) + 2\left(\frac{1}{\sqrt{2}}\right) + 2\left(\frac{1}{\sqrt{35/8}}\right) + \frac{1}{3}\right] \approx 1.3973 \approx 1.40$

(b) Simpson's Rule: $\int_0^2 \frac{1}{\sqrt{1+x^3}}\,dx \approx \frac{2}{12}\left[1 + 4\left(\frac{1}{\sqrt{9/8}}\right) + 2\left(\frac{1}{\sqrt{2}}\right) + 4\left(\frac{1}{\sqrt{35/8}}\right) + \frac{1}{3}\right] \approx 1.4052 \approx 1.41$

14. $\int_0^1 \sqrt{1-x^2}\,dx, \quad n = 8$

Trapezoidal Rule: 0.772

Simpson's Rule: 0.780

(*Note*: This is an approximation of $\pi/4 \approx 0.785$.)

16. (a) Trapezoidal Rule: $\int_0^2 e^{-x^2}\,dx \approx \frac{2}{8}(1 + 2e^{-1/4} + 2e^{-1} + 2e^{-9/4} + e^{-4}) \approx 0.881$

(b) Simpson's Rule: $\int_0^2 e^{-x^2}\,dx \approx \frac{2}{12}(1 + 4e^{-1/4} + 2e^{-1} + 4e^{-9/4} + e^{-4}) \approx 0.882$

18. (a) Trapezoidal Rule: $\int_0^3 \frac{x}{2+x+x^2}dx \approx \frac{3}{12}\left[0 + 2\left(\frac{2}{11}\right) + 2\left(\frac{1}{4}\right) + 2\left(\frac{6}{23}\right) + 2\left(\frac{1}{4}\right) + 2\left(\frac{10}{43}\right) + \frac{3}{14}\right] \approx 0.641$

(b) Simpson's Rule: $\int_0^3 \frac{x}{2+x+x^2}dx \approx \frac{3}{18}\left[0 + 4\left(\frac{2}{11}\right) + 2\left(\frac{1}{4}\right) + 4\left(\frac{6}{23}\right) + 2\left(\frac{1}{4}\right) + 4\left(\frac{10}{43}\right) + \frac{3}{14}\right] \approx 0.653$

20. $V = \int_0^{t_1} c(t)e^{-rt}\,dt = \int_0^8 \left(200{,}000 + 15{,}000\sqrt[3]{t}\right)e^{-0.10t}\,dt$

$$\approx \frac{8}{24}[200{,}000 + 4(215{,}000)e^{-0.10} + 2(200{,}000 + 15{,}000\sqrt[3]{2})e^{-0.20} + 4(200{,}000 + 15{,}000\sqrt[3]{3})e^{-0.30}$$

$$+ 2(200{,}000 + 15{,}000\sqrt[3]{4})e^{-0.40} + 4(200{,}000 + 15{,}000\sqrt[3]{5})e^{-0.50}$$

$$+ 2(200{,}000 + 15{,}000\sqrt[3]{6})e^{-0.60} + 4(200{,}000 + 15{,}000\sqrt[3]{7})e^{-0.70}$$

$$+ 230{,}000e^{-0.80}] \approx \$1{,}215{,}971.94$$

22. $\Delta R \approx \int_{14}^{16} 50\sqrt{x}\sqrt{20-x}\,dx \approx \frac{2}{12}[50\sqrt{14}\sqrt{6} + 200\sqrt{14.5}\sqrt{5.5} + 100\sqrt{15}\sqrt{5} + 200\sqrt{15.5}\sqrt{4.5} + 50(4)(2)] \approx \$863.44$

24. $P(0 \le x \le 2) = \int_0^2 \frac{1}{\sqrt{2\pi}}e^{-x^2/2}\,dx \approx \frac{1}{\sqrt{2\pi}}\left(\frac{2}{18}\right)(1 + 4e^{-1/18} + 2e^{-2/9} + 4e^{-1/2} + 2e^{-8/9} + 4e^{-25/18} + e^{-2})$

$\approx 0.4772 = 47.72\%$

26. $A \approx \frac{120}{3(12)}[75 + 4(81) + 2(84) + 4(76) + 2(67) + 4(68) + 2(69) + 4(72) + 2(68) + 4(56) + 2(42) + 4(23) + 0]$

≈ 7463.3333 square feet

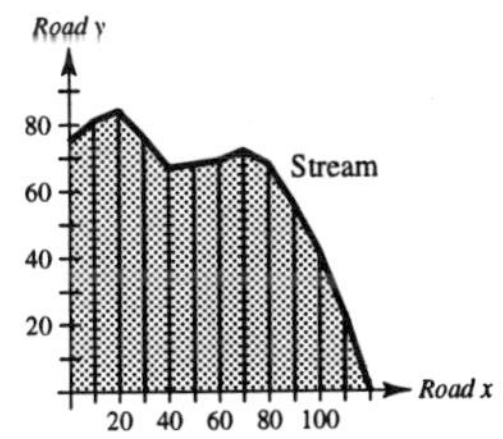

28. $f(x) = \dfrac{1}{x+1}$

$f'(x) = -\dfrac{1}{(x+1)^2}$

$f''(x) = \dfrac{2}{(x+1)^3}$

$f'''(x) = -\dfrac{6}{(x+1)^4}$

$f^{(4)}(x) = \dfrac{24}{(x+1)^5}$

(a) Trapezoidal Rule: Since $f''(x)$ is maximum in $[0, 1]$ when $x = 0$, we have

$$|\text{Error}| \le \frac{(1-0)^3}{12(4)^2}(2) = \frac{1}{96} \approx 0.01.$$

(b) Simpson's Rule: Since $f^{(4)}(x)$ is maximum in $[0, 1]$ when $x = 0$, we have

$$|\text{Error}| \le \frac{(1-0)^5}{180(4)^4}(24) \approx 0.00052.$$

30. $f(x) = e^{-x^2}$

$f'(x) = -2xe^{-x^2}$

$f''(x) = -2(-2x^2 + 1)e^{-x^2}$

$f'''(x) = 4(-2x^3 + 3x)e^{-x^2}$

$f^{(4)}(x) = -2(-8x^4 + 24x^2 - 6)e^{-x^2}$

(a) Trapezoidal Rule: Since $|f''(x)|$ is maximum in $[0, 1]$ when $x = 0$, we have

$$|\text{Error}| \le \frac{(1-0)^3}{12(4)^2}(2) = \frac{1}{96} \approx 0.01.$$

(b) Simpson's Rule: Since $|f^{(4)}(x)|$ is maximum in $[0, 1]$ when $x = 0$, we have

$$|\text{Error}| \le \frac{(1-0)^5}{180(4)^4}(12) = \frac{1}{3840} \approx 0.00026.$$

32. $f(x) = \dfrac{1}{x}$

$f'(x) = -\dfrac{1}{x^2}$

$f''(x) = \dfrac{2}{x^3}$

$f'''(x) = -\dfrac{6}{x^4}$

$f^{(4)}(x) = \dfrac{24}{x^5}$

(a) Trapezoidal Rule: Since $f''(x)$ is maximum in $[1, 3]$ when $x = 1$ and $f''(1) = 2$, we have the following.

$$|\text{Error}| \le \frac{(3-1)^3}{12n^2}(2) < 0.0001$$

$$n^2 > 13{,}333.33$$

$$n > 115.47$$

Let $n = 116$.

(b) Simpson's Rule: Since $f^{(4)}(x)$ is maximum in $[1, 3]$ when $x = 1$ and $f^{(4)}(1) = 24$, we have the following.

$$|\text{Error}| \le \frac{(3-1)^5}{180n^4}(24) < 0.0001$$

$$n^4 > 42{,}666.67$$

$$n > 14.4$$

Let $n = 16$. (n must be even.)

34. $f(x) = \ln x$

$f'(x) = \dfrac{1}{x}$

$f''(x) = \dfrac{-1}{x^2}$

$f'''(x) = \dfrac{2}{x^3}$

$f^{(4)}(x) = \dfrac{-6}{x^4}$

(a) Trapezoidal Rule: Since $|f''(x)|$ is maximum in $[3, 5]$ when $x = 3$, and $|f''(3)| = \frac{1}{9}$, we have

$$|\text{Error}| \le \frac{(5-3)^3}{12n^2}\left(\frac{1}{9}\right) < 0.0001$$

$$n^2 > 740.74$$

$$n > 27.2$$

Take $n = 28$.

(b) Simpson's Rule: Since $|f^{(4)}(x)|$ is maximum in $[3, 5]$ when $x = 3$, and $|f^{(4)}(3)| = \left|-\frac{6}{81}\right| = \frac{2}{27}$, we have

$$|\text{Error}| \le \frac{(5-3)^5}{180n^4}\left(\frac{2}{27}\right) < 0.0001$$

$$n^4 > 131.69$$

$$n > 3.39$$

Take $n = 4$.

36. $\int_1^4 x^2\sqrt{x+4}\,dx \approx 55.6246 \quad (n = 100)$

38. $\int_2^5 10x^2e^{-x}\,dx \approx 11.0405 \quad (n = 100)$

40. $$V = \pi\int_0^4 x^2(x+4)^{2/3}dx$$
$$\approx \pi\left(\frac{4-0}{3(4)}\right)[0 + 4(5)^{2/3} + 2(4)(6)^{2/3} + 4(9)(7)^{2/3} + 16(4)] \approx 244.883 \text{ cubic units}$$

42. $$y = \frac{1}{3}(x^{3/2} - 3x^{1/2} + 2)$$
$$y' = \frac{1}{3}\left(\frac{3}{2}x^{1/2} - \frac{3}{2}x^{-1/2}\right) = \left(\frac{1}{2}\right)\left(\frac{x-1}{\sqrt{x}}\right)$$
$$1 + (y')^2 = 1 + \frac{(x-1)^2}{4x} = \frac{(x+1)^2}{4x}$$
$$S = \int_0^1 \sqrt{\frac{(x+1)^2}{4x}}\,dx = \int_0^1 \frac{x+1}{2\sqrt{x}}\,dx = \frac{1}{2}\int_0^1 (x^{1/2} + x^{-1/2})\,dx = \frac{1}{2}\left[\frac{2}{3}x^{3/2} + 2x^{1/2}\right]_0^1 = \frac{4}{3} \text{ units}$$

Simpson's Rule is inappropriate for this problem since the integral is undefined when $x = 0$.

44. $\frac{1}{6}\int_0^6 [12 - 4\ln(t^2 - 4t + 6)]\,dt \approx 5.781$ grams

(Simpson's Rule with $n = 100$)

Section 6.6 Improper Integrals

2. This integral converges since $\int_{-\infty}^0 e^{2x}\,dx = \lim_{b\to-\infty}\left[\frac{1}{2}e^{2x}\right]_b^0 = \frac{1}{2} - 0 = \frac{1}{2}$.

4. This integral diverges since $\int_1^\infty \frac{1}{\sqrt{x}}\,dx = \lim_{b\to\infty}\left[2\sqrt{x}\right]_1^b = \infty - 2 = \infty$.

6. This integral converges since $\int_0^\infty \frac{5}{e^{2x}}\,dx = -\frac{5}{2}\int_0^\infty e^{-2x}(-2)\,dx = -\frac{5}{2}\lim_{b\to\infty}\left[e^{-2x}\right]_0^b = -\frac{5}{2}(0-1) = \frac{5}{2}$.

8. This integral diverges since

$$\int_{1/2}^\infty \frac{1}{\sqrt{2x-1}}\,dx = \int_{1/2}^1 \frac{1}{\sqrt{2x-1}}\,dx + \int_1^\infty \frac{1}{\sqrt{2x-1}}\,dx$$
$$= \lim_{b\to(1/2)^+}\left[\sqrt{2x-1}\right]_b^1 + \lim_{c\to\infty}\left[\sqrt{2x-1}\right]_1^c = (1-0) + (\infty - 1) = \infty.$$

The second integral diverges.

10. This integral converges since $\int_{-\infty}^1 \frac{1}{x^2}\,dx = \lim_{b\to-\infty}\left[-\frac{1}{x}\right]_b^1 = 1 + 0 = 1$.

12. This integral diverges since

$$\int_{-\infty}^\infty \frac{|x|}{x^2+1}\,dx = \int_{-\infty}^0 \frac{-x}{x^2+1}\,dx + \int_0^\infty \frac{x}{x^2+1}\,dx$$
$$= \lim_{b\to-\infty}\left[-\frac{1}{2}\ln(x^2+1)\right]_b^0 + \lim_{c\to\infty}\left[\frac{1}{2}\ln(x^2+1)\right]_0^c = \infty + \infty = \infty.$$

14. This integral diverges since

$$\int_{-\infty}^{\infty} x^2 e^{-x^3}\,dx = \int_{-\infty}^{0} x^2 e^{-x^3}\,dx + \int_{0}^{\infty} e^{-x^3}\,dx$$

$$= \lim_{b\to-\infty}\left[-\frac{1}{3}e^{-x^3}\right]_{-\infty}^{0} + \lim_{c\to\infty}\left[-\frac{1}{3}e^{-x^3}\right]_{0}^{c} = \left(-\frac{1}{3}+\infty\right)+\left(0+\frac{1}{3}\right) = \infty.$$

16. This integral converges since $\displaystyle\int_3^4 \frac{1}{\sqrt{x-3}}\,dx = \lim_{b\to3^+}\left[2\sqrt{x-3}\right]_b^4 = 2.$

18. This integral diverges since

$$\int_0^2 \frac{1}{(x-1)^2}\,dx = \int_0^1 \frac{1}{(x-1)^2}\,dx + \int_1^2 \frac{1}{(x-1)^2}\,dx$$

$$= \lim_{b\to1^-}\left[-\frac{1}{x-1}\right]_0^b + \lim_{c\to1^+}\left[-\frac{1}{x-1}\right]_c^2 = (\infty-1)+(-1+\infty) = \infty.$$

20. This integral converges since $\displaystyle\int_0^{27}\frac{5}{\sqrt[3]{x}}\,dx = 5\int_0^{27} x^{-1/3}\,dx = 5\lim_{b\to0^+}\left[\frac{3}{2}x^{2/3}\right]_b^{27} = 5\left(\frac{27}{2}-0\right) = \frac{135}{2}.$

22. This integral converges since

$$\int_0^2 \frac{x}{\sqrt{4-x^2}}\,dx = \lim_{b\to2^-}\left[-\sqrt{4-x^2}\right]_0^b = 0+2 = 2.$$

24. This integral diverges since

$$\int_0^1 \frac{1}{x}\,dx = \lim_{b\to0^+}\left[\ln|x|\right]_b^1 = 0-(-\infty) = \infty.$$

26. This integral diverges since

$$\int_0^2 \frac{1}{(x-1)^{4/3}}\,dx = \int_0^1 \frac{1}{(x-1)^{4/3}}\,dx + \int_1^2 \frac{1}{(x-1)^{4/3}}\,dx$$

$$= \lim_{b\to1^-}\left[-\frac{3}{\sqrt[3]{x-1}}\right]_0^b + \lim_{c\to1^+}\left[-\frac{3}{\sqrt[3]{x-1}}\right]_c^2 = (\infty-3)+(-3+\infty) = \infty.$$

28. This integral converges since $\displaystyle\int_3^5 \frac{1}{x^2\sqrt{x^2-9}}\,dx = \lim_{b\to3^+}\left[\frac{\sqrt{x^2-9}}{9x}\right]_b^5 = \frac{4}{45}.$

30. $a = 2,\ n = 4$: $\displaystyle\lim_{x\to\infty} x^4e^{-2x}$

x	1	10	25	50
x^4e^{-2x}	0.1353	2.061×10^{-5}	7.534×10^{-17}	2.325×10^{-37}

32. $a = \frac{1}{2},\ n = 5$: $\displaystyle\lim_{x\to\infty} x^5e^{-(1/2)x}$

x	1	10	25	50
$x^5e^{-(1/2)x}$	0.6065	673.7947	36.3931	0.0043

34. $\displaystyle\int_0^\infty (x-1)e^{-x}\,dx = \lim_{b\to\infty}\left[-xe^{-x}\right]_0^b = 0$

36. $\displaystyle\int_0^\infty xe^{-x}\,dx = \lim_{b\to\infty}\left[-(x+1)e^{-x}\right]_0^b = 0+1 = 1$

38. (a) Present value $\displaystyle = \int_0^{20} 75{,}000e^{-0.08t}\,dt = \left[-\frac{75{,}000}{0.08}e^{-0.08t}\right]_0^{20} \approx \$748{,}222.01$

(b) Present value $\displaystyle = \int_0^{\infty} 75{,}000e^{-0.08t}\,dt = \lim_{b\to\infty}\left[-\frac{75{,}000}{0.08}e^{-0.08t}\right]_0^{b} = \$937{,}500.00$

40. $C = 650{,}000 + \int_0^n 25{,}000(1 + 0.08t)e^{-0.12t}\,dt = 650{,}000 + 25{,}000\left[-\frac{1}{9}(125 + 6t)e^{-0.12t}\right]_0^n$

(a) For $n = 5$, we have $C = 650{,}000 + 25{,}000\left(-\frac{155}{9}e^{-0.6} + \frac{125}{9}\right) \approx \$760{,}928.32$.

(b) For $n = 10$, we have $C = 650{,}000 + 25{,}000\left(-\frac{185}{9}e^{-1.2} + \frac{125}{9}\right) \approx \$842{,}441.86$.

(c) For $n = \infty$, we have $C = 650{,}000 + 25{,}000 \lim_{n\to\infty}\left[-\frac{1}{9}(125 + 6t)e^{-0.12t}\right]_0^n = 650{,}000 + 25{,}000\left(0 + \frac{125}{9}\right) \approx \$997{,}222.22$.

42. (a) $A = \int_0^\infty e^{-x}\,dx = \lim_{b\to\infty}\left[-e^{-x}\right]_0^b = 1$

(b) $V = \pi\int_0^\infty (e^{-x})^2\,dx = \pi \lim_{b\to\infty}\left[-\frac{1}{2}e^{-2x}\right]_0^b = \frac{\pi}{2}$

44. $P(a \le x \le b) = \int_a^b \frac{1}{\sigma\sqrt{2\pi}}e^{-(x-\mu)^2/2\sigma^2}\,dx = \int_a^b h(x)\,dx$

$\mu = 36, \quad \sigma = 0.2$ Use numerical integration.

(a) $P(35.5 \le x < \infty) = \int_{35.5}^\infty h(x)\,dx \approx 0.9938$

(b) $P(35.9 \le x < \infty) = \int_{35.9}^\infty h(x)dx \approx 0.6915$

Review Exercises for Chapter 6

2. $\int (x^2 + 2x - 1)\,dx = \frac{x^3}{3} + x^2 - x + C$

4. $\int \frac{2}{(x-1)^2}\,dx = \int 2(x-1)^{-2}\,dx = \frac{2(x-1)^{-1}}{-1} + C = \frac{-2}{x-1} + C$

6. $\int 3xe^{-x^2}\,dx = -\frac{3}{2}e^{-x^2} + C$

8. $\int \frac{2x^3 - x}{x^4 - x^2 + 1}\,dx = \frac{1}{2}\int \frac{4x^3 - 2x}{x^4 - x^2 + 1}\,dx = \frac{1}{2}\ln|x^4 - x^2 + 1| + C$

10. $\int \frac{1}{\sqrt{2x-9}}\,dx = \frac{1}{2}\int (2x-9)^{-1/2}(2)\,dx = \sqrt{2x-9} + C$

12. $\int (x^2 - 1)e^{x^3-3x}\,dx = \frac{1}{3}e^{x^3-3x} + C$

14. Use substitution and let $u = 1 - x$, $x = 1 - u$, $du = -dx$.

$$\int x(1-x)^2\,dx = \int (1-u)u^2(-du) - \int (u^3 - u^2)du = \frac{u^4}{4} - \frac{u^3}{3} + C = \frac{(1-x)^4}{4} - \frac{(1-x)^3}{3} + C$$

16. Use substitution and let $u = 1 + x$, then $x = u - 1$ and $dx = du$.

$$\int x^2\sqrt{1+x}\,dx = \int (u-1)^2\sqrt{u}\,du = \int (u^{5/2} - 2u^{3/2} + u^{1/2})\,du$$

$$= \frac{2}{7}u^{7/2} - \frac{4}{5}u^{5/2} + \frac{2}{3}u^{3/2} + C = \frac{2}{105}u^{3/2}(15u^2 - 42u + 35) + C$$

$$= \frac{2}{105}(1+x)^{3/2}[15(1 + 2x + x^2) - 42(1+x) + 35] + C$$

$$= \frac{2}{105}(1+x)^{3/2}(15x^2 - 12x + 8) + C$$

18. Use substitution and let $u = \sqrt{x}$, then $x = u^2$ and $dx = 2u\,du$.

$$\int \frac{\sqrt{x}}{1+\sqrt{x}}\,dx = \int \frac{u}{1+u}(2u)du = 2\int \frac{u^2}{1+u}\,du$$

$$= 2\int \left(u - 1 + \frac{1}{1+u}\right)du = 2\left(\frac{u^2}{2} - u + \ln|1+u|\right) + C = x - 2\sqrt{x} + 2\ln\left|1+\sqrt{x}\right| + C$$

20. Use substitution and let $u = x - 1,\ x = u + 1,\ du = dx$.

$$\int \frac{x}{x-1}dx = \int \frac{u+1}{u}\,du = \int \left(1 + \frac{1}{u}\right)du = u + \ln|u| + C = x - 1 + \ln|x-1| + C$$

22. Use substitution and let $u = x - 2,\ x = u + 2,\ du = dx$.

$$\int_2^3 x^2\sqrt{x-2}\,dx = \int_0^1 (u+2)^2u^{1/2}\,du = \int_0^1 (u^{5/2} + 4u^{3/2} + 4u^{1/2})\,du$$

$$= \left[\frac{2}{7}u^{7/2} + \frac{8}{5}u^{5/2} + \frac{8}{3}u^{3/2}\right]_0^1 = \frac{478}{105} \approx 4.552$$

24. Use substitution and let $u = x + 3,\ x = u - 3,\ du = dx$.

$$\int_{-3}^0 x(x+3)^4\,dx = \int_0^3 (u-3)u^4\,du = \int_0^3 (u^5 - 3u^4)\,du = \left[\frac{u^6}{6} - \frac{3u^5}{5}\right]_0^3 = \frac{243}{2} - \frac{3}{5}(243) = -\frac{243}{10}$$

26. (a) $P(0.40 \le x \le 0.60) = \int_{0.4}^{0.6} \frac{140}{27}x^2(1-x)^{1/3}\,dx$

$u = 1 - x,\ x = 1 - u,\ du = -dx$

$$\int_{0.6}^{0.4} \frac{140}{27}(1-u)^2u^{1/3}(-du) = \frac{-140}{27}\int_{0.6}^{0.4} (u^{1/3} - 2u^{4/3} + u^{7/3})\,du$$

$$= \frac{-140}{27}\left[\frac{3}{4}u^{4/3} - \frac{6}{7}u^{7/3} + \frac{3}{10}u^{10/3}\right]_{0.6}^{0.4} \approx 0.206$$

(b) $P(0 \le x \le 0.50) = \int_0^{0.5} \frac{140}{27}x^2(1-x)^{1/3}\,dx$

As in part (a), this probability is

$$\frac{140}{27}\left[\frac{3}{4}u^{4/3} - \frac{6}{7}u^{7/3} + \frac{3}{10}u^{10/3}\right]_1^{0.5} \approx 0.184$$

28. (a)

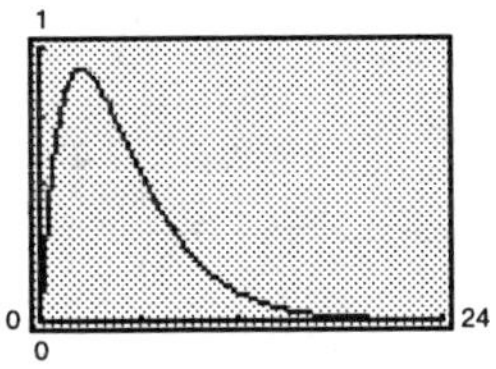

(b) $E'(t) = e^{-0.4t} + t(-0.4)e^{-0.4t} = e^{-0.4t}(1 - 0.4t)$

$E'(t) = 0$ when $t = 2.5$.
Increasing on $[0, 2.5]$
Decreasing on $[2.5, 24]$

(c) At $t = 2.5$ hours

(d) Average $= \dfrac{1}{24}\displaystyle\int_0^{24} E(t)\,dt \approx 0.26$ or 26%

(e) $E(t) = 0.26$ for $t = 0.27$ and $t = 9.08$.

30. Use integration by parts and let $u = \ln x$ and $dv = \sqrt{x}\,dx$, then $du = (1/x)dx$ and $v = (2/3)x^{3/2}$.

$$\int \sqrt{x}\ln x\,dx = \frac{2}{3}x^{3/2}\ln x - \int \frac{2}{3}x^{3/2}\left(\frac{1}{x}\right)dx$$

$$= \frac{2}{3}x^{3/2}\ln x - \frac{2}{3}\int x^{1/2}\,dx = \frac{2}{3}x^{3/2}\ln x - \frac{4}{9}x^{3/2} + C = \frac{2}{9}x^{3/2}[(3\ln x) - 2] + C$$

32. Use integration by parts and let $u = \ln\left(\dfrac{x}{x+1}\right) = \ln x - \ln(x+1)$ and $dv = dx$. Then $du = \left(\dfrac{1}{x} - \dfrac{1}{x+1}\right)dx$ and $v = x$.

$$\int \ln\left(\frac{x}{x+1}\right)dx = x\ln\left(\frac{x}{x+1}\right) - \int\left(1 - \frac{x}{x+1}\right)dx = x\ln\left(\frac{x}{x+1}\right) - \int \frac{1}{x+1}\,dx$$

$$= x\ln\left(\frac{x}{x+1}\right) - \ln|x+1| + C = x\ln|x| - x\ln|x+1| - \ln|x+1| + C$$

$$= x\ln|x| - (x+1)\ln|x+1| + C$$

34. $u = (\ln x)^3$, $dv = dx$, $du = 3(\ln x)^2\dfrac{1}{x}\,dx$, $v = x$

$$\int (\ln x)^3\,dx = x(\ln x)^3 - \int x(3)(\ln x)^2\frac{1}{x}\,dx = x(\ln x)^3 - \int 3(\ln x)^2\,dx$$

Using parts again: $u = (\ln x)^2$, $dv = 3\,dx$, $du = 2(\ln x)\dfrac{1}{x}$, $v = 3x$

$$\int (\ln x)^3\,dx = x(\ln x)^3 - \left[3x(\ln x)^2 - \int 6\ln x\,dx\right]$$

Using parts a third time: $u = \ln x$, $dv = 6\,dx$, $du = \dfrac{1}{x}\,dx$, $v = 6x$

$$\int (\ln x)^3\,dx = x(\ln x)^3 - 3x(\ln x)^2 + \left[6x\ln x - \int 6x\left(\frac{1}{x}\,dx\right)\right]$$

$$= x(\ln x)^3 - 3x(\ln x)^2 + 6x\ln x - 6x + C$$

36. Present value $= \displaystyle\int_0^{t_1} c(t)e^{-rt}\,dt$

$$= \int_0^{10} (20{,}000 + 1500t)e^{-0.06t}\,dt$$

$$= (-750{,}000 - 25{,}000t)e^{-0.06t}\Big]_0^{10}$$

$$= \$201{,}188.36$$

38. Present value $= \displaystyle\int_0^{t_1} c(t)e^{-rt}\,dt$

$$= \int_0^{5} (10{,}000 + 100e^{t/2})e^{-0.05t}\,dt$$

$$= \frac{-200{,}000}{e^{0.05t}} + 222.22e^{0.45t}\Big]_0^{5}$$

$$= \$46{,}126.01$$

40. $\displaystyle\int_0^3 1000e^{-0.06t}\,dt \approx \$2745.50$

42. $\displaystyle 50{,}000 = \int_0^{t_0} 6000e^{-0.06t}\,dt = -100{,}000e^{-0.06t}\Big]_0^{t_0} = -100{,}000(e^{-0.06t_0} - 1)$

$$\Rightarrow \frac{-1}{2} = e^{-0.06t_0} - 1$$

$$\frac{1}{2} = e^{-0.06t_0}$$

$$\ln\frac{1}{2} = -0.06t_0 \Rightarrow t_0 \approx 11.55 \text{ years}$$

44. Use partial fractions.

$$\frac{2x-1}{(x-1)^2} = \frac{A}{x-1} + \frac{B}{(x-1)^2}$$

Basic equation: $2x - 1 = A(x-1) + B$
When $x = 1$: $1 = B$.
When $x = 2$: $3 = A + B, A = 2$.

$$\int \frac{4x-2}{3(x-1)^2}\,dx = \frac{2}{3}\int \frac{2x-1}{(x-1)^2}\,dx = \frac{2}{3}\int\left[\frac{2}{x-1} + \frac{1}{(x-1)^2}\right]dx = \frac{2}{3}\left[2\ln|x-1| - \frac{1}{x-1}\right] + C$$

46. Use partial fractions.

$$\frac{4x^2-x-5}{x^2(x+5)} = \frac{A}{x} + \frac{B}{x^2} + \frac{C}{x+5}$$

Basic equation: $4x^2 - x - 5 = Ax(x+5) + B(x+5) + Cx^2$
When $x = 0$: $-5 = 5B$, $B = -1$.
When $x = -5$: $100 = 25C$, $C = 4$.
When $x = 1$: $-2 = 6A + 6B + C$, $A = 0$.

$$\int \frac{4x^2-x-5}{x^2(x+5)}\,dx = \int\left(-\frac{1}{x^2} + \frac{4}{x+5}\right)dx = \frac{1}{x} + 4\ln|x+5| + C$$

48. $\displaystyle\frac{x^2+2x-12}{x^2+3x} = 1 + \frac{-x-12}{x^2+3x}$

Partial fractions: $\displaystyle\frac{-x-12}{x^2+3x} = \frac{A}{x} + \frac{B}{x+3}$

Basic equation: $-x - 12 = A(x+3) + Bx$

When $x = 0$: $-12 = 3A, A = -4$.

When $x = -3$: $-9 = -3B, B = 3$.

$$\int \frac{x^2+2x-12}{x(x+3)}\,dx = \int\left(1 + \frac{-4}{x} + \frac{3}{x+3}\right)dx$$

$$= x - 4\ln|x| + 3\ln|x+3| + C$$

50. (a) $\displaystyle y = \frac{L}{1+be^{-kt}} = \frac{2700}{1+be^{-kt}}$

Since $y = 300$ when $t = 0$, $\displaystyle\frac{2700}{1+b} = 300 \Rightarrow b = 8$.

Since $y = 966$ when $t = 5$, we have

$$966 = \frac{2700}{1+8e^{-5k}}.$$

Solving for k,

$$1 + 8e^{-5k} = 2.795031$$

$$8e^{-5k} = 1.795031$$

$$k \approx 0.29888.$$

Thus, $\displaystyle y = \frac{2700}{1+8e^{-0.29888}}.$

(b) When $t = 4$, $y \approx 789$ pheasants.

(c) $y = 1750$ when $t \approx 9$ years.

52. Use Formula 10 from the tables and let $u = x, a = 4, b = 3$.

$$\int \frac{1}{x(4+3x)}\,dx = \frac{1}{4}\ln\left|\frac{x}{4+3x}\right| + C$$

54. Use Formula 42 from the tables and let $u = x^2$ and $du = 2x\,dx$.

$$\int x(\ln x^2)^2\,dx = \frac{1}{2}\int (\ln x^2)^2(2x)\,dx$$

$$= \frac{x^2}{2}[2 - 2\ln x^2 + (\ln x^2)^2] + C$$

56. Use Formula 33 from the tables and let $u = x, a = 4$, and $du = dx$.

$$\int_1^3 \frac{1}{x^2\sqrt{16-x^2}}\,dx = -\frac{\sqrt{16-x^2}}{16x}\Bigg]_1^3 = -\frac{\sqrt{7}}{48} + \frac{\sqrt{15}}{16} = \frac{1}{48}\left(3\sqrt{15} - \sqrt{7}\right) \approx 0.187$$

58. Use Formula 30 from the tables and let $u = x, a = 3, n = 2$:

$$\int \frac{1}{(x^2-9)^2}\,dx = \frac{-1}{18}\left[\frac{x}{x^2-9} + \int \frac{1}{x^2-9}\,dx\right]$$

$$= \frac{-1}{18}\left[\frac{x}{x^2-9} + \frac{1}{6}\ln\left|\frac{x-3}{x+3}\right|\right] + C \qquad \text{(Formula 29)}$$

60. Use Formula 43 from the tables and let $u = x, n = 4$:

$$\int (\ln x)^4\,dx = x(\ln x)^4 - 4\int (\ln x)^3\,dx$$

$$= x(\ln x)^4 - 4\left[x(\ln x)^3 - 3\int (\ln x)^2\,dx\right]$$

$$= x(\ln x)^4 - 4x(\ln x)^3 + 12x[2 - 2\ln x + (\ln x)^2] + C \qquad \text{(Formula 42)}$$

62.

$$\frac{1}{x^2-8x-52} = \frac{1}{x^2-8x+16-68} = \frac{1}{(x-4)^2-68}$$

$$\int \frac{1}{(x-4)^2-\left(2\sqrt{17}\right)^2}\,dx = \frac{1}{4\sqrt{17}}\ln\left|\frac{x-4-2\sqrt{17}}{x-4+2\sqrt{17}}\right| + C \qquad \text{(Formula 29)}$$

64.

$$x^4 + 6x^2 + 10 = (x^4 + 6x^2 + 9) + 1 = (x^2+3)^2 + 1$$

$$\frac{1}{2}\int \frac{2x}{\sqrt{(x^2+3)^2+1}}\,dx = \frac{1}{2}\ln\left|x^2 + 3 + \sqrt{(x^2+3)^2+1}\right| + C \qquad \text{(Formula 25)}$$

66. $$\int_0^2 (x^2+1)\,dx \approx \frac{2-0}{2(4)}\left[1 + 2\left(\frac{5}{4}\right) + 2(2) + 2\left(\frac{13}{4}\right) + 5\right] = 4.75$$

68. $$\int_0^2 \frac{1}{\sqrt{1+x^3}}\,dx \approx \frac{1}{8}\left[1 + \frac{2}{\sqrt{1+(1/4)^3}} + \frac{2}{\sqrt{1+(1/2)^3}} + \frac{2}{\sqrt{1+(3/4)^3}} + \frac{2}{\sqrt{2}} + \frac{2}{\sqrt{1+(5/4)^3}}\right.$$

$$\left. + \frac{2}{\sqrt{1+(3/2)^3}} + \frac{2}{\sqrt{1+(7/4)^3}} + \frac{1}{3}\right] \approx 1.4010$$

70. $$\int_1^2 x^3\,dx \approx \frac{2-1}{3(4)}\left[1 + 4\left(\frac{5}{4}\right)^3 + 2\left(\frac{3}{2}\right)^3 + 4\left(\frac{7}{4}\right)^3 + 8\right] = 3.75 \qquad \text{(exact!)}$$

72. $\displaystyle\int_0^1 e^{x^2}\,dx \approx \frac{1}{18}(1 + 4e^{1/36} + 2e^{1/9} + 4e^{1/4} + 2e^{4/9} + 4e^{25/36} + e) \approx 1.4629$

74. Following Exercise 67, use $n = 8$:

$$|E| \le \frac{2 - 0^3}{12(8)^2}(219) \le 2.28125$$

76. Following Exercise 69, with $n = 8$:

$$|E| \le \frac{(4 - 2)^5}{180(8)^4}(24) \approx 0.001$$

78. $\displaystyle\int_{-\infty}^{0} \frac{3}{(1 - 3x)^{2/3}}\,dx = \lim_{b\to-\infty} -\int_b^0 (1 - 3x)^{-2/3}(-3\,dx) = \lim_{b\to-\infty} -\Big[3(1 - 3x)^{1/3}\Big]_b^0$ diverges

80. $\displaystyle\int_0^{\infty} 2x^2e^{-x^3}\,dx = \lim_{b\to\infty} -\frac{2}{3}\int_0^b e^{-x^3}(-3x^2)\,dx = \lim_{b\to\infty} -\frac{2}{3}\Big[e^{-x^3}\Big]_0^b = \frac{2}{3}$

82. $\displaystyle\int_1^2 \frac{x}{16(x - 1)^2}\,dx = \lim_{a\to1^+} \frac{1}{16}\int_a^2 \frac{x}{(x - 1)^2}\,dx = \lim_{a\to1^+} \frac{1}{16}\left[\frac{-1}{x - 1} + \ln(x - 1)\right]_a^2$ diverges

84. $\displaystyle\int_0^2 \frac{x + 2}{(x - 1)^2}\,dx = \lim_{b\to1^-} \int_0^b \frac{x + 2}{(x - 1)^2}\,dx + \lim_{c\to1^+} \int_c^2 \frac{x + 2}{(x - 1)^2}\,dx$

$\displaystyle= \lim_{b\to1^-} \left[\ln|x - 1| - \frac{3}{x - 1}\right]_0^b + \lim_{c\to1^+}\left[\ln|x - 1| - \frac{3}{x - 1}\right]_c^2$ diverges

86. (a) $\displaystyle C = C_0 + \int_0^n c(t)e^{-rt}\,dt = 1.5 \times 10^6 + \int_0^{20} 50{,}000e^{-0.06t}\,dt \approx 1.5 \times 10^6 + 582{,}388.16 = \$2{,}082{,}388.16$

(b) $\displaystyle C = 1.5 \times 10^6 + \int_0^{\infty} 50{,}000e^{-0.06t}\,dt \approx 1.5 \times 10^6 + 833{,}333.33 = 2{,}333{,}333.33$ (See Exercise 79.)

CHAPTER 7
Functions of Several Variables

CHAPTER 7
Functions of Several Variables

Section 7.1 The Three-Dimensional Coordinate System

2. (a) See graph.

(b) See graph.

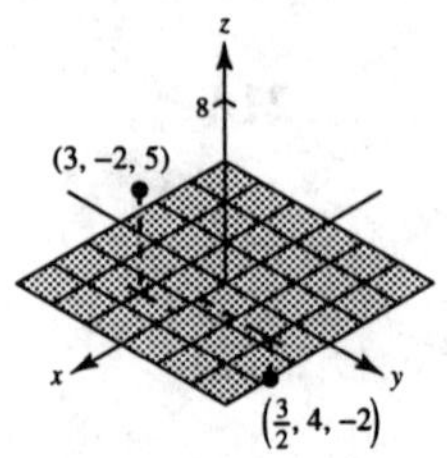

4. (a) See graph.

(b) See graph.

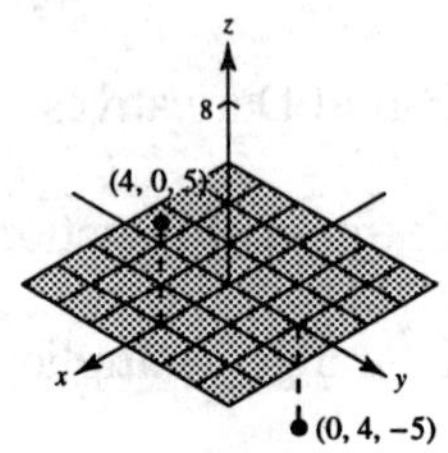

6. $d = \sqrt{(2+4)^2 + (-1+1)^2 + (5-1)^2}$

$= \sqrt{52} = 2\sqrt{13}$

8. $d = \sqrt{(8-8)^2 + (-2+2)^2 + (4-2)^2} = 2$

10. Midpoint $= \left(\dfrac{4+8}{2}, \dfrac{0+8}{2}, \dfrac{-6+20}{2}\right) = (6, 4, 7)$

12. Midpoint $= \left(\dfrac{0+4}{2}, \dfrac{-2+2}{2}, \dfrac{5+7}{2}\right) = (2, 0, 6)$

14. $(1, 0, 0) = \left(\dfrac{x+0}{2}, \dfrac{y+(-2)}{2}, \dfrac{z+1}{2}\right)$

$1 = \dfrac{x}{2}$ $\qquad 0 = \dfrac{y-2}{2}$ $\qquad 0 = \dfrac{z+1}{2}$ $\qquad (x, y, z) = (2, 2, -1)$

$x = 2$ $\qquad y = 2$ $\qquad z = -1$

16. $(0, 1, 1) = \left(\dfrac{x+3}{2}, \dfrac{y+3}{2}, \dfrac{z+0}{2}\right)$

$0 = \dfrac{x+3}{2}$ $\qquad 1 = \dfrac{y+3}{2}$ $\qquad 1 = \dfrac{z}{2}$ $\qquad (x, y, z) = (-3, -1, 2)$

$x = -3$ $\qquad y = -1$ $\qquad z = 2$

18. Let $A = (5, 3, 4)$, $B = (7, 1, 3)$, and $C = (3, 5, 3)$. Then we have the following.

$d(AB) = \sqrt{(7-5)^2 + (1-3)^2 + (3-4)^2} = 3$

$d(AC) = \sqrt{(3-5)^2 + (5-3)^2 + (3-4)^2} = 3$

$d(BC) = \sqrt{(3-7)^2 + (5-1)^2 + (3-3)^2} = 4\sqrt{2}$

Since $d(AB) = d(AC)$, the triangle is isosceles.

20. Let $A = (5, 0, 0)$, $B = (0, 2, 0)$, and $C = (0, 0, -3)$. Then we have the following.

$d(AB) = \sqrt{(0-5)^2 + (2-0)^2 + (0-0)^2} = \sqrt{29}$

$d(AC) = \sqrt{(0-5)^2 + (0-0)^2 + (-3-0)^2} = \sqrt{34}$

$d(BC) = \sqrt{(0-0)^2 + (0-2)^2 + (-3-0)^2} = \sqrt{13}$

The triangle is neither right nor isosceles.

22. $(x-2)^2 + (y-3)^2 + (z-1)^2 = 9$

24. The midpoint of the diameter is the center.

$$\text{Center} = \left(\frac{-1+0}{2}, \frac{-2+3}{2}, \frac{1+3}{2}\right) = \left(\frac{-1}{2}, \frac{1}{2}, 2\right)$$

The radius is the distance from the center to either endpoint.

$$\text{Radius} = \sqrt{\left(\frac{-1}{2}\right)^2 + \left(\frac{1}{2} - 3\right)^2 + (2-3)^2} = \sqrt{\frac{1}{4} + \frac{25}{4} + 1} = \frac{\sqrt{30}}{2}$$

$$\left(x + \frac{1}{2}\right)^2 + \left(y - \frac{1}{2}\right)^2 + (z-2)^2 = \frac{30}{4} = \frac{15}{2}$$

26. $(x-4)^2 + [y-(-1)]^2 + (z-1)^2 = 5^2$

$$(x-4)^2 + (y+1)^2 + (z-1)^2 = 25$$

28. The distance from $(-2, 1, 1)$ to the xy-coordinate plane is the radius $r = 1$. Therefore, we have the following.

$$[x-(-2)]^2 + (y-1)^2 + (z-1)^2 = 1^2$$

$$(x+2)^2 + (y-1)^2 + (z-1)^2 = 1$$

30. $x^2 + y^2 + z^2 - 5x = 0$

$\left(x - \frac{5}{2}\right)^2 + (y-0)^2 + (z-0)^2 = \frac{25}{4}$

Center: $\left(\frac{5}{2}, 0, 0\right)$

Radius: $\frac{5}{2}$

32. $x^2 + y^2 + z^2 - 4y + 6z + 4 = 0$

$(x-0)^2 + (y-2)^2 + (z+3)^2 = 9$

Center: $(0, 2, -3)$

Radius: 3

34. $x^2 + y^2 + z^2 = 36$

Center: $(0, 0, 0)$

Radius: 6

36. To find the xy-trace, we let $z = 0$.

$$x^2 + y^2 + (0)^2 - 6x - 10y + 6(0) + 30 = 0$$

$$(x-3)^2 + (y-5)^2 = 4$$

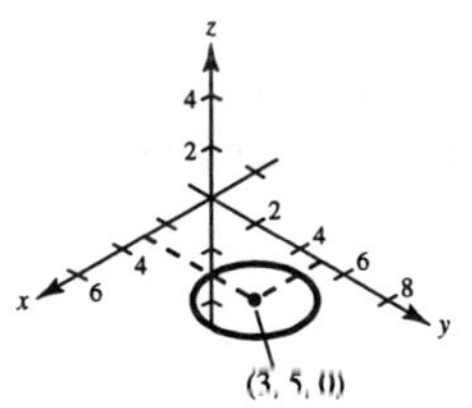

38. To find the yz-trace, we let $x = 0$.

$$(0)^2 + y^2 + z^2 - 6(0) - 10y + 6z + 30 = 0$$

$$(y-5)^2 + (z+3)^2 = 4$$

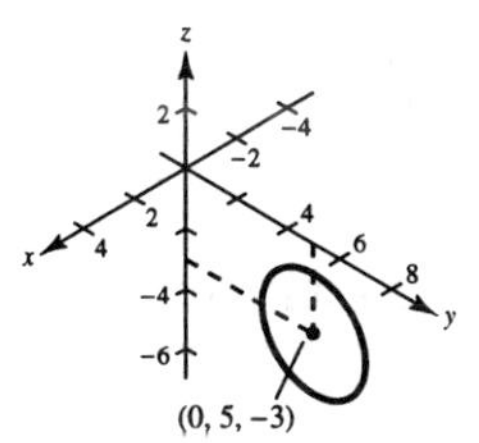

40. (a) When $x = 5$, we have $y^2 + z^2 = 144$.

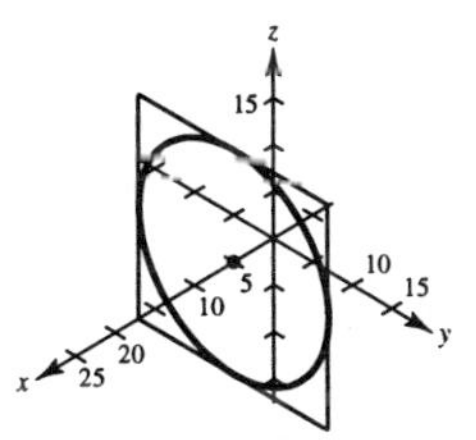

(b) When $y = 12$, we have $x^2 + z^2 = 25$.

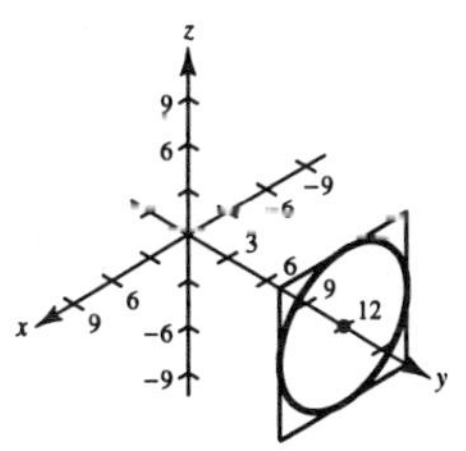

42. (a) When $x = 4$, we have $(y - 0)^2 + (z - 3)^2 = 9$.

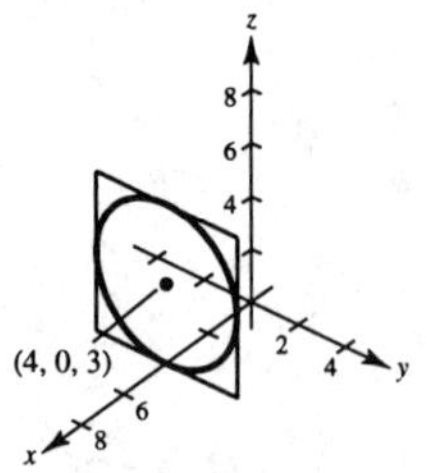

(b) When $z = 3$, we have $(x - 4)^2 + (y - 0)^2 = 9$.

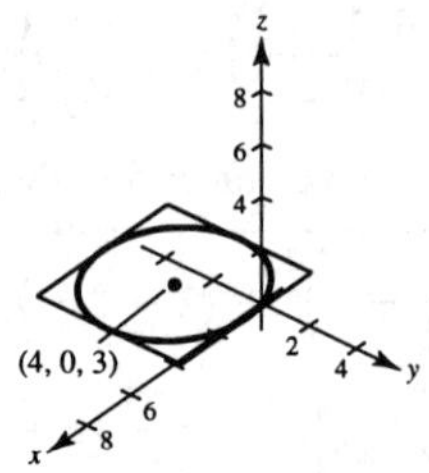

44. $x^2 + y^2 + z^2 = 3963^2$

Lines of longitude would be traces in planes passing through the z-axis. Each trace is a circle. Lines of latitude would be traces in planes parallel to the equator. They are circles.

Section 7.2 Surfaces in Space

2. To find the x-intercept, let $y = 0$ and $z = 0$.

$3x = 6 \Rightarrow x = 2$

To find the y-intercept, let $x = 0$ and $z = 0$.

$6y = 6 \Rightarrow y = 1$

To find the z-intercept, let $x = 0$ and $y = 0$.

$2z = 6 \Rightarrow z = 3$

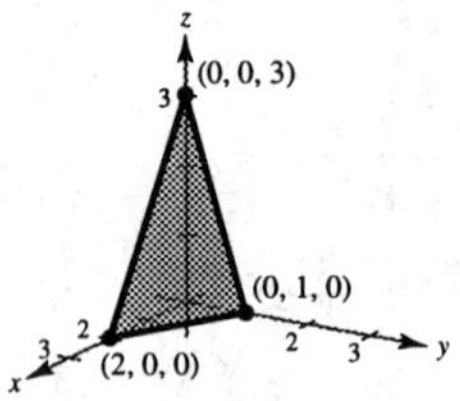

4. To find the x-intercept, let $y = 0$ and $z = 0$.

$x = 3$

To find the y-intercept, let $x = 0$ and $z = 0$.

$y = 3$

To find the z-intercept, let $x = 0$ and $y = 0$.

$z = 3$

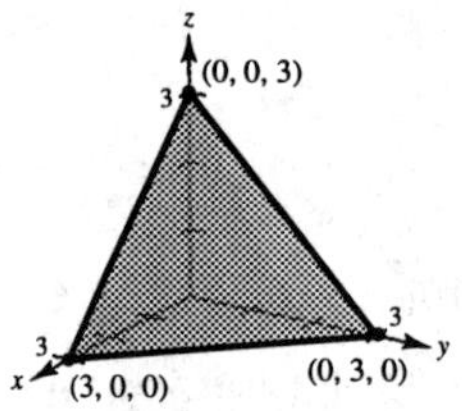

6. To find the x-intercept, let $y = 0$ and $z = 0$.

$2x = 4 \Rightarrow x = 2$

To find the y-intercept, let $x = 0$ and $z = 0$.

$-y = 4 \Rightarrow y = -4$

To find the z-intercept, let $x = 0$ and $y = 0$.

$z = 4$

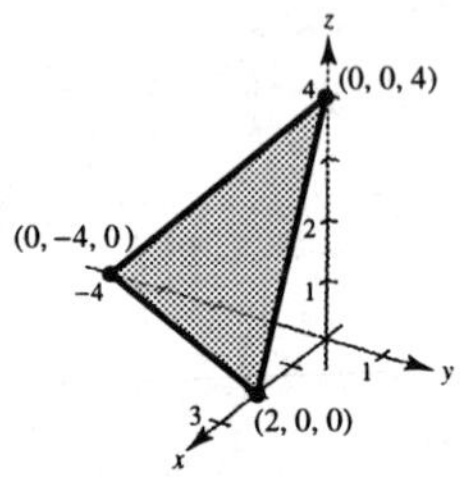

8. Since the coefficients of x and z are zero, the only intercept is the y-intercept. The plane is parallel to the xz-plane.

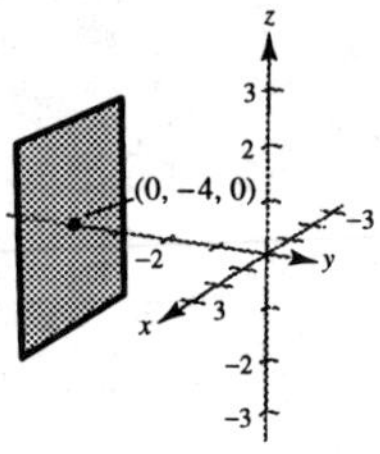

10. The x-intercept is 8, the y-intercept is 4, and the plane is parallel to the z-axis.

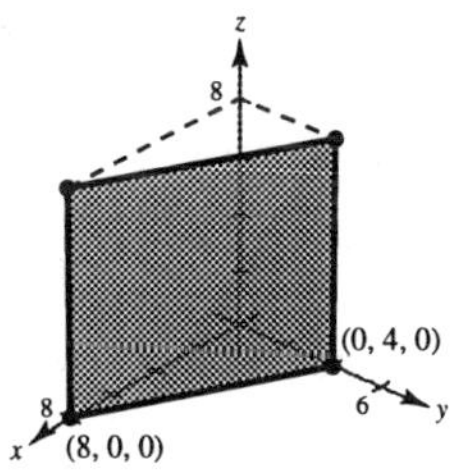

12. The x-intercept is 3, the z-intercept is -1, and the plane is parallel to the y-axis.

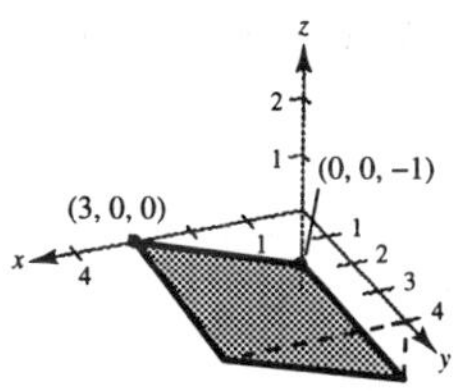

14. For the first plane, $a_1 = 3$, $b_1 = 1$, and $c_1 = -4$, and for the second plane, $a_2 = -9$, $b_2 = -3$, and $c_2 = 12$. Therefore, we have

$$a_2 = -3a_1, \quad b_2 = -3b_1, \quad c_2 = -3c_1.$$

The planes are parallel.

16. For the first plane, $a_1 = 1$, $b_1 = 0$, and $c_1 = 0$, and for the second plane, $a_2 = 0$, $b_2 = 1$, and $c_2 = 0$. Therefore, we have

$$a_1a_2 + b_1b_2 + c_1c_2 = (1)(0) + (0)(1) + (0)(0) = 0.$$

The planes are perpendicular.

18. For the first plane, $a_1 = 1$, $b_1 = 3$, and $c_1 = 1$, and for the second plane, $a_2 = 1$, $b_2 = 0$, and $c_2 = -5$. The planes are not parallel since

$$a_1 = a_2 \quad \text{and} \quad b_1 \neq b_2$$

and the planes are not perpendicular since

$$\begin{aligned} a_1a_2 + b_1b_2 + c_1c_2 &= (1)(1) + (3)(0) + (1)(-5) \\ &= -4 \neq 0. \end{aligned}$$

20. For the first plane, $a_1 = 2$, $b_1 = 0$, and $c_1 = -1$, and for the second plane, $a_2 = 4$, $b_2 = 1$, and $c_2 = 8$. Therefore, we have

$$a_1a_2 + b_1b_2 + c_1c_2 = (2)(4) + (0)(1) + (-1)(8) = 0.$$

The planes are perpendicular.

22. $D = \dfrac{|ax_0 + by_0 + cz_0 + d|}{\sqrt{a^2 + b^2 + c^2}} = \dfrac{|(3)(1) + (-1)(5) + 2(-4) - 6|}{\sqrt{(3)^2 + (-1)^2 + (2)^2}} = \dfrac{16}{\sqrt{14}} = \dfrac{8\sqrt{14}}{7} \approx 4.276$

24. $D = \dfrac{|ax_0 + by_0 + cz_0 + d|}{\sqrt{a^2 + b^2 + c^2}} = \dfrac{|(1)(10) + (-3)(0) + 4(0) - 6|}{\sqrt{(1)^2 + (-3)^2 + (4)^2}} = \dfrac{4}{\sqrt{26}} = \dfrac{2\sqrt{26}}{13} \approx 0.784$

26. $D = \dfrac{|ax_0 + by_0 + cz_0 + d|}{\sqrt{a^2 + b^2 + c^2}} = \dfrac{|3(2) + 3(-1) + 2(0) + (-6)|}{\sqrt{3^2 + 3^2 + 2^2}} = \dfrac{3}{\sqrt{40}} = \dfrac{3\sqrt{10}}{20}$

28. The graph of $-\dfrac{x^2}{\frac{4}{15}} + \dfrac{y^2}{1} - \dfrac{z^2}{\frac{4}{15}} = 1$ is a hyperboloid of two sheets that matches graph (e).

30. The graph of $\dfrac{x^2}{\frac{1}{4}} - y^2 + \dfrac{z^2}{\frac{1}{9}} = 0$ is an elliptic cone that matches graph (g).

32. The graph of $\dfrac{x^2}{3} - \dfrac{y^2}{4} = z$ is a hyperbolic paraboloid that matches graph (b).

34. The graph of $x^2 + y^2 + z^2 = 9$ is a sphere that matches graph (h).

36. $y = x^2 + z^2$

Trace in xy-plane $(z = 0)$: $y = x^2$	Parabola
Trace in plane $y = 1$: $x^2 + z^2 = 1$	Circle
Trace in yz-plane $(x = 0)$: $y = z^2$	Parabola

38. $y^2 + z^2 - x^2 = 1$

Trace in xy-plane $(z = 0)$: $y^2 - x^2 = 1$	Hyperbola
Trace in xz-plane $(y = 0)$: $z^2 - x^2 = 1$	Hyperbola
Trace in yz-plane $(x = 0)$: $y^2 + z^2 = 1$	Circle

40. The graph is an ellipsoid.

42. $\dfrac{x^2}{8} + \dfrac{y^2}{18} - \dfrac{z^2}{9} = 1$

The graph is a hyperboloid of one sheet.

44. $z = \dfrac{x^2}{1/4} + y^2$

The graph is an elliptic paraboloid.

46. The graph is a hyperboloid of two sheets.

48. $x^2 + \dfrac{y^2}{4} - z^2 = 0$

The graph is an elliptic cone.

50. $y = \dfrac{x^2}{4} + \dfrac{z^2}{4}$

The graph is an elliptic paraboloid.

52. $\dfrac{x^2}{1/2} + \dfrac{y^2}{1/2} - z^2 = 0$

The graph is an elliptic cone.

54.

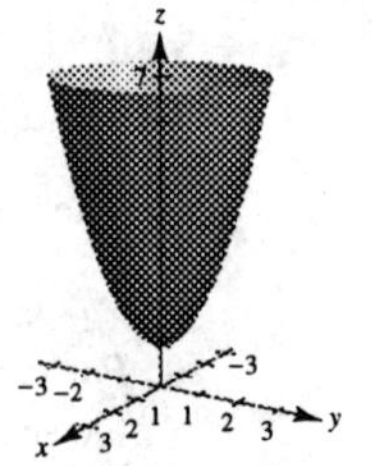

56.

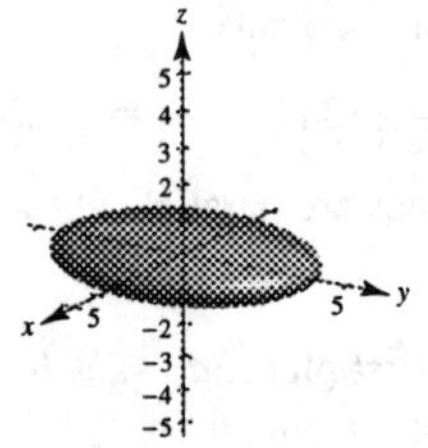

Section 7.3 Functions of Several Variables

2. (a) $f(0, 0) = 4$ (b) $f(0, 1) = 0$ (c) $f(2, 3) = -36$

(d) $f(1, y) = 3 - 4y^2$ (e) $f(x, 0) = 4 - x^2$ (f) $f(t, 1) = -t^2$

4. (a) $g(2, 3) = \ln 5$ (b) $g(5, 6) = \ln 11$ (c) $g(e, 0) = 1$

(d) $g(0, 1) = 0$ (e) $g(2, -3) = 0$ (f) $g(e, e) = \ln(2e) = 1 + \ln 2$

6. (a) $f(0, 5, 4) = \sqrt{5 + 4} = 3$

(b) $f(6, 8, -3) = \sqrt{11}$

8. (a) $F(0.09, 60) = 500\left(1 + \dfrac{0.09}{12}\right)^{60} \approx 782.84$

(b) $F(0.14, 240) = 500\left(1 + \dfrac{0.14}{12}\right)^{240} \approx 8090.14$

10. (a) $A(500, 0.10, 5) = 500e^{(0.10)(5)} \approx 824.36$

(b) $A(1500, 0.12, 20) = 1500e^{(0.12)(20)} \approx 16{,}534.76$

12. (a) $g(4, 1) = \displaystyle\int_4^1 \frac{1}{t}\,dt = \Big[\ln|t|\Big]_4^1 = -\ln 4$

(b) $g(6, 3) = \displaystyle\int_6^3 \frac{1}{t}\,dt = \Big[\ln|t|\Big]_6^3 = \ln 3 - \ln 6 = \ln\frac{1}{2} = -\ln 2$

14. (a) $f(x + \Delta x, y) = 3(x + \Delta x)y + y^2$

(b)
$$\begin{aligned}\frac{f(x, y + \Delta y) - f(x, y)}{\Delta y} &= \frac{[3x(y + \Delta y) + (y + \Delta y)^2] - (3xy + y^2)}{\Delta y}\\ &= \frac{3xy + 3x\Delta y + y^2 + 2y\Delta y + (\Delta y)^2 - 3xy - y^2}{\Delta y}\\ &= \frac{3x\Delta y + 2y\Delta y + (\Delta y)^2}{\Delta y} = 3x + 2y + \Delta y, \quad \Delta y \neq 0\end{aligned}$$

16. The domain is the set of all points in the xy-plane and the range is $[-1, \infty)$.

18. The domain is the half plane above the line $y = -x$ since $x + y > 0$, and the range is $(-\infty, \infty)$.

20. The domain is the set of all points outside and on the circle $x^2 + y^2 = 1$ since $x^2 + y^2 - 1 \geq 0$.

22. The domain is the set of all points in the xy-plane except those on the line $x = 1$.

24. The domain is the set of all points in the xy-plane except those on the line $y = x$.

26. Since $xy \geq 0$, the domain is the set of all points in the first and third quadrants, as well as the x- and y-axes.

28. The domain is the set of all points in the xy-plane except those on the y-axis since $x \neq 0$.

30. The contour map consists of curves $e^{1-x^2+y^2} = C$, or $1 - x^2 + y^2 = \ln C$, which are hyperbolas. Matches (d).

32. The contour map consists of curves $\ln|y - x^2| = C$, or $y - x^2 = e^C \implies y = x^2 + e^C$, which are parabolas. Matches (c).

34.

$c = 0,$	$0 = 6 - 2x - 3y,$	$2x + 3y = 6$
$c = 2,$	$2 = 6 - 2x - 3y,$	$2x + 3y = 4$
$c = 4,$	$4 = 6 - 2x - 3y,$	$2x + 3y = 2$
$c = 6,$	$6 = 6 - 2x - 3y,$	$2x + 3y = 0$
$c = 8,$	$8 = 6 - 2x - 3y,$	$2x + 3y = -2$
$c = 10,$	$10 = 6 - 2x - 3y,$	$2x + 3y = -4$

The level curves are parallel lines.

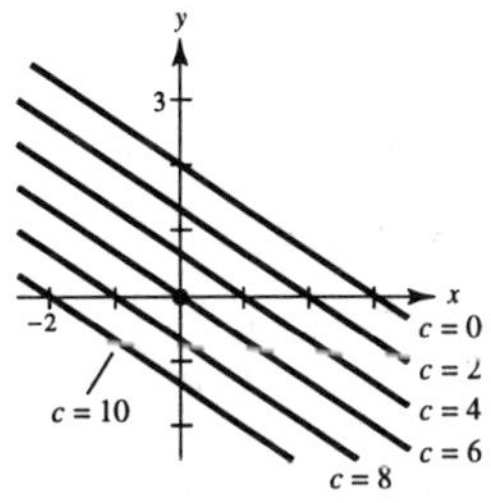

36.

$c = 0,$	$0 = x^2 + y^2$
$c = 2,$	$2 = x^2 + y^2$
$c = 4,$	$4 = x^2 + y^2$
$c = 6,$	$6 = x^2 + y^2$
$c = 8,$	$8 = x^2 + y^2$

The level curves are circles.

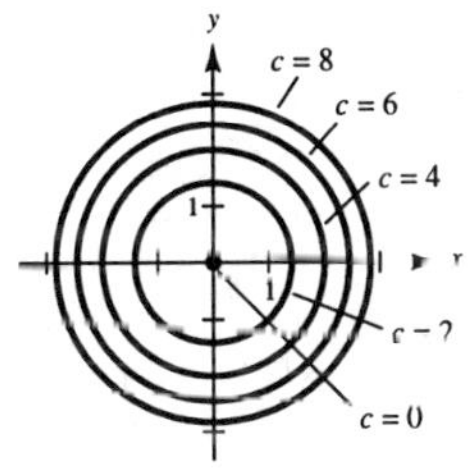

38.

$c = 1,$	$1 = e^{xy},$	$0 = xy$
$c = 2,$	$2 = e^{xy},$	$\ln 2 = xy$
$c = 3,$	$3 = e^{xy},$	$\ln 3 = xy$
$c = 4,$	$4 = e^{xy},$	$\ln 4 = xy$
$c = \frac{1}{2},$	$\frac{1}{2} = e^{xy},$	$-\ln 2 = xy$
$c = \frac{1}{3},$	$\frac{1}{3} = e^{xy},$	$-\ln 3 = xy$
$c = \frac{1}{4},$	$\frac{1}{4} = e^{xy},$	$-\ln 4 = xy$

The level curves are hyperbolas.

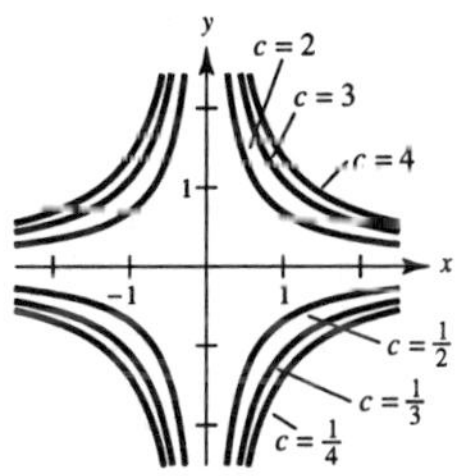

40. $c = -2, \quad \ln(x - y) = -2, \quad x - y = e^{-2}$

$c = -\frac{3}{2}, \quad \ln(x - y) = -\frac{3}{2}, \quad x - y = e^{-3/2}$

$c = -1, \quad \ln(x - y) = -1, \quad x - y = e^{-1}$

$c = -\frac{1}{2}, \quad \ln(x - y) = -\frac{1}{2}, \quad x - y = e^{-1/2}$

$c = 0, \quad \ln(x - y) = 0, \quad x - y = 1$

$c = \frac{1}{2}, \quad \ln(x - y) = \frac{1}{2}, \quad x - y = e^{1/2}$

$c = 1, \quad \ln(x - y) = 1, \quad x - y = e$

$c = \frac{3}{2}, \quad \ln(x - y) = \frac{3}{2}, \quad x - y = e^{3/2}$

$c = 2, \quad \ln(x - y) = 2, \quad x - y = e^2$

The level curves are lines.

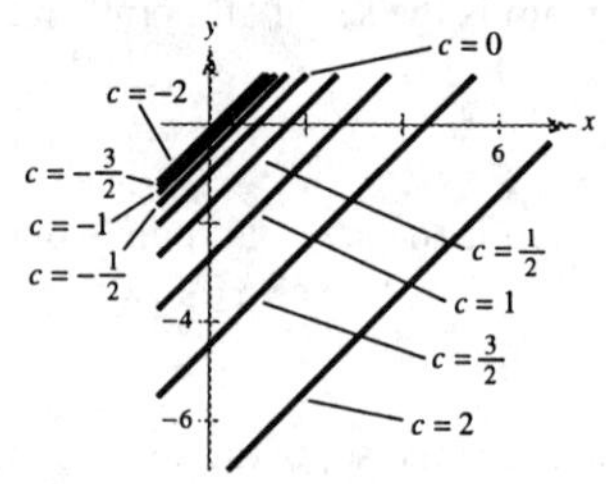

42. From Example 5, we have $f(x, y) = Cx^a y^{1-a}$. Thus,

$$\begin{aligned} f(2x, 2y) &= C(2x)^a(2y)^{1-a} \\ &= C2^a x^a 2^{1-a} y^{1-a} \\ &= 2Cx^a y^{1-a} \\ &= 2f(x, y). \end{aligned}$$

44. (a) $N(22, 12) = \left(\frac{22 - 4}{4}\right)^2(12) = 243$ board feet

(b) $N(30, 12) = \left(\frac{30 - 4}{4}\right)^2(12) = 507$ board feet

46. (a) $W(15, 10) = \frac{1}{5}$ hr $= 12$ min

(b) $W(12, 9) = \frac{1}{3}$ hr $= 20$ min

(c) $W(12, 6) = \frac{1}{6}$ hr $= 10$ min

(d) $W(4, 2) = \frac{1}{2}$ hr $= 30$ min

48. (a) The different colors represent various amplitudes.

(b) No, the level curves are uneven and sporadically spaced.

Section 7.4 Partial Derivatives

2. $f_x(x, y) = 2x$

$f_y(x, y) = -6y$

4. $f_x(x, y) = -\frac{1}{2}x^{-3/2} = -\frac{1}{2x^{3/2}}$

$f_y(x, y) = 6y^{1/2} = 6\sqrt{y}$

6. $z = x\sqrt{y} = xy^{1/2}$

$z_x = \sqrt{y}$

$z_y = \frac{1}{2}xy^{-1/2} = \frac{x}{2\sqrt{y}}$

8. $f_x(x, y) = \frac{(x^2 + y^2)(y) - (xy)(2x)}{(x^2 + y^2)^2} = \frac{y(y^2 - x^2)}{(x^2 + y^2)^2}$

$f_y(x, y) = \frac{(x^2 + y^2)(x) - (xy)(2y)}{(x^2 + y^2)^2} = \frac{x(x^2 - y^2)}{(x^2 + y^2)^2}$

10. $\frac{\partial z}{\partial x} = xe^{x+y} + e^{x+y} = (x + 1)e^{x+y}$

$\frac{\partial z}{\partial y} = xe^{x+y}$

12. $g_x(x, y) = \frac{1}{y}e^{x/y}$

$g_y(x, y) = -\frac{x}{y^2}e^{x/y}$

14. $g(x, y) = \ln \sqrt{x^2 + y^2} = \frac{1}{2}\ln(x^2 + y^2)$

$g_x(x, y) = \frac{1}{2}\left(\frac{2x}{x^2 + y^2}\right) = \frac{x}{x^2 + y^2}$

$g_y(x, y) = \frac{1}{2}\left(\frac{2y}{x^2 + y^2}\right) = \frac{y}{x^2 + y^2}$

16. $f_y(x, y) = 3x^2e^{x-y} - 3x^2ye^{x-y}$

$= 3x^2e^{x-y}(1 - y)$

18. $g_y(x, y) = 6xye^{y-x} + 3xy^2e^{y-x}$

$= 3xye^{y-x}(2 + y)$

20. Using the solution from Exercise 17, $g_x(-2, -2) = 36$.

22. $f_x = 2x - 3y \qquad f_x(1, -1) = 2 - 3(-1) = 5$

$f_y = -3x + 2y \qquad f_y(1, -1) = -3 + 2(-1) = -5$

24. $f_x = e^xy^2 \qquad f_x(0, 2) = e^0(2^2) = 4$

$f_y = 2e^xy \qquad f_y(0, 2) = 2e^0(2) = 4$

26. $f_x(x, y) = \frac{4y^3}{(x^2 + y^2)^{3/2}}, \qquad f_x(1, 0) = 0$

$f_y(x, y) = \frac{4x^3}{(x^2 + y^2)^{3/2}}, \qquad f_y(1, 0) = 4$

28. $z = \ln\sqrt{xy} = \frac{1}{2}(\ln x + \ln y)$

$\frac{\partial z}{\partial x} = \frac{1}{2x} \qquad z_x(-1, -1) = -\frac{1}{2}$

$\frac{\partial z}{\partial y} = \frac{1}{2y} \qquad z_y(-1, -1) = -\frac{1}{2}$

30. $w_x = \frac{y(y + z)}{(x + y + z)^2}, \qquad$ at $(1, 2, 0)$, $w_x = \frac{4}{9}$

$w_y = \frac{x(x + z)}{(x + y + z)^2}, \qquad$ at $(1, 2, 0)$, $w_y = \frac{1}{9}$

$w_z = \frac{-xy}{(x + y + z)^2}, \qquad$ at $(1, 2, 0)$, $w_z = -\frac{2}{9}$

32. $w_x = \frac{x}{(1 - x^2 - y^2 - z^2)^{3/2}}, \qquad w_x(0, 0, 0) = 0$

$w_y = \frac{y}{(1 - x^2 - y^2 - z^2)^{3/2}}, \qquad w_y(0, 0, 0) = 0$

$w_z = \frac{z}{(1 - x^2 - y^2 - z^2)^{3/2}}, \qquad w_z(0, 0, 0) = 0$

34. $w_x = ye^{z^2}, \qquad w_x(2, 1, 0) = 1$

$w_y = xe^{z^2}, \qquad w_y(2, 1, 0) = 2$

$w_z = 2xyze^{z^2}, \qquad w_z(2, 1, 0) = 0$

36. $f_x(x, y) = 9x^2 - 12y = 0 \Rightarrow y = \frac{3}{4}x^2$

$f_y(x, y) = -12x + 3y^2 = 0 \Rightarrow -12x + 3\left(\frac{3x^2}{4}\right)^2 = 0$

$-12x + \frac{27}{16}x^4 = 0$

$\frac{3}{16}x(9x^3 - 64) = 0$

$x = 0 \text{ or } x = \frac{4\sqrt[3]{3}}{3}$

$y = 0 \text{ or } y = \frac{4\sqrt[3]{9}}{3}$

Solutions: $(0, 0)$, $\left(\frac{4\sqrt[3]{3}}{3}, \frac{4\sqrt[3]{9}}{3}\right)$

38. $f_x(x, y) = \frac{2x}{x^2 + y^2 + 1} = 0 \Rightarrow x = 0$

$f_y(x, y) = \frac{2y}{x^2 + y^2 + 1} = 0 \Rightarrow y = 0$

Solution: $(0, 0)$

40. (a) $\frac{\partial z}{\partial x} = y, \qquad$ at $(1, 2, 2)$, $\qquad \frac{\partial z}{\partial x} = 2$

(b) $\frac{\partial z}{\partial y} = x, \qquad$ at $(1, 2, 2)$, $\qquad \frac{\partial z}{\partial y} = 1$

42. (a) $\dfrac{\partial z}{\partial x} = 2x,$ at $(2, 1, 8),$ $\dfrac{\partial z}{\partial x} = 4$

(b) $\dfrac{\partial z}{\partial y} = 8y,$ at $(2, 1, 8),$ $\dfrac{\partial z}{\partial y} = 8$

44. (a) $\dfrac{\partial z}{\partial x} = \dfrac{1}{y},$ at $(3, 1, 3),$ $\dfrac{\partial z}{\partial x} = 1$

(b) $\dfrac{\partial z}{\partial y} = -\dfrac{x}{y^2},$ at $(3, 1, 3),$ $\dfrac{\partial z}{\partial y} = -3$

46. (a) $\dfrac{\partial z}{\partial x} = 2x,$ at $(-2, 1, 3),$ $\dfrac{\partial z}{\partial x} = -4$

(b) $\dfrac{\partial z}{\partial y} = -2y,$ at $(-2, 1, 3),$ $\dfrac{\partial z}{\partial y} = -2$

48. $\dfrac{\partial z}{\partial x} = 4x^3 - 6xy^2$ $\qquad \dfrac{\partial^2 z}{\partial y \partial x} = -12xy$

$\dfrac{\partial z}{\partial y} = -6x^2y + 4y^3$ $\qquad \dfrac{\partial^2 z}{\partial x \partial y} = -12xy$

50. $\dfrac{\partial z}{\partial x} = \dfrac{x^2 + y^2}{2x^2y}$ $\qquad \dfrac{\partial^2 z}{\partial y \partial x} = \dfrac{y^2 - x^2}{2x^2y^2}$

$\dfrac{\partial z}{\partial y} = -\dfrac{x^2 + y^2}{2xy^2}$ $\qquad \dfrac{\partial^2 z}{\partial x \partial y} = \dfrac{y^2 - x^2}{2x^2y^2}$

52. The first partial derivatives are

$$\frac{\partial z}{\partial x} = 6x - y \quad \text{and} \quad \frac{\partial z}{\partial y} = -x + 6y^2$$

and the second partial derivatives are

$$\frac{\partial^2 z}{\partial x^2} = 6, \quad \frac{\partial^2 z}{\partial y \partial x} = -1 = \frac{\partial^2 z}{\partial x \partial y}, \quad \text{and} \quad \frac{\partial^2 z}{\partial y^2} = 12y.$$

54. The first partial derivatives are

$$\frac{\partial z}{\partial x} = -\frac{x}{\sqrt{9 - x^2 - y^2}} \quad \text{and} \quad \frac{\partial z}{\partial y} = -\frac{y}{\sqrt{9 - x^2 - y^2}}$$

and the second partial derivatives are

$$\frac{\partial^2 z}{\partial x^2} = \frac{y^2 - 9}{(9 - x^2 - y^2)^{3/2}}, \quad \frac{\partial^2 z}{\partial y \partial x} = -\frac{xy}{(9 - x^2 - y^2)^{3/2}} = \frac{\partial^2 z}{\partial x \partial y}, \quad \text{and} \quad \frac{\partial^2 z}{\partial y^2} = \frac{x^2 - 9}{(9 - x^2 - y^2)^{3/2}}.$$

56. The first partial derivatives are

$$\frac{\partial z}{\partial x} = \frac{y}{(x + y)^2} \quad \text{and} \quad \frac{\partial z}{\partial y} = \frac{-x}{(x + y)^2}$$

and the second partial derivatives are

$$\frac{\partial^2 z}{\partial x^2} = -\frac{2y}{(x + y)^3}, \quad \frac{\partial^2 z}{\partial y \partial x} = \frac{x - y}{(x + y)^3} = \frac{\partial^2 z}{\partial x \partial y},$$

$$\text{and} \quad \frac{\partial^2 z}{\partial y^2} = \frac{2x}{(x + y)^3}.$$

58. The first partial derivatives are

$$\frac{\partial z}{\partial x} = e^y + ye^x \quad \text{and} \quad \frac{\partial z}{\partial y} = xe^y + e^x$$

and the second partial derivatives are

$$\frac{\partial^2 z}{\partial x^2} = ye^x, \quad \frac{\partial^2 z}{\partial y \partial x} = e^y + e^x = \frac{\partial^2 z}{\partial x \partial y},$$

$$\text{and} \quad \frac{\partial^2 z}{\partial y^2} = xe^y.$$

60. $z_x = \dfrac{x}{\sqrt{x^2 + y^2}}$ $\qquad z_y = \dfrac{y}{\sqrt{x^2 + y^2}}$

$z_{xx} = \dfrac{y^2}{(x^2 + y^2)^{3/2}}$ $\qquad z_{yy} = \dfrac{x^2}{(x^2 + y^2)^{3/2}}$

$z_{xy} = z_{yx} = \dfrac{-xy}{(x^2 + y^2)^{3/2}}$

At $(0, 2),$ $\quad z_{xx} = \dfrac{2^2}{2^3} = \dfrac{1}{2}, \quad z_{yy} = 0, \quad z_{xy} = z_{yx} = 0.$

62. $f_x = 2xe^y$ $\qquad f_y = x^2e^y$

$f_{xx} = 2e^y$ $\qquad f_{xx}(-1, 0) = 2$

$f_{yy} = x^2e^y$ $\qquad f_{yy}(-1, 0) = 1$

$f_{xy} = f_{yx} = 2xe^y$ $\qquad f_{xy}(-1, 0) = f_{yx}(-1, 0) = -2$

64. $w_x = \dfrac{x}{\sqrt{x^2 + y^2 + z^2}}$

$w_y = \dfrac{y}{\sqrt{x^2 + y^2 + z^2}}$

$w_z = \dfrac{z}{\sqrt{x^2 + y^2 + z^2}}$

66. $w_x = \dfrac{x}{(1 - x^2 - y^2 - z^2)^{3/2}}$

$w_y = \dfrac{y}{(1 - x^2 - y^2 - z^2)^{3/2}}$

$w_z = \dfrac{z}{(1 - x^2 - y^2 - z^2)^{3/2}}$

68. (a) $\dfrac{\partial R}{\partial x_1} = 200 - 8x_1 - 8x_2;$ when $x_1 = 4$ and $x_2 = 12$, $\dfrac{\partial R}{\partial x_1} = 72.$

(b) $\dfrac{\partial R}{\partial x_2} = 200 - 8x_1 - 8x_2;$ when $x_1 = 4$ and $x_2 = 12$, $\dfrac{\partial R_1}{\partial x_2} = 72.$

70. (a) $\dfrac{\partial f}{\partial x} = 75x^{-0.25}y^{0.25} = 75\left(\dfrac{y}{x}\right)^{0.25}$, at (1000, 500), $\dfrac{\partial f}{\partial x} \approx 63.07$

(b) $\dfrac{\partial f}{\partial y} = 25x^{0.75}y^{-0.75} = 25\left(\dfrac{x}{y}\right)^{0.75}$, at (1000, 500), $\dfrac{\partial f}{\partial y} \approx 42.04$

72. $IQ_M = \dfrac{100}{C}$

$IQ_C = \dfrac{-100M}{C^2}$

$IQ_M(12, 10) = \dfrac{100}{10} = 10$

$IQ_C(12, 10) = \dfrac{-100(12)}{10^2} = -12$

74. (a) $\dfrac{\partial A}{\partial t} = 0.885 + 1.20h;$ at $t = 90°$, $h = 0.80$, $\dfrac{\partial A}{\partial t} = 1.845$

$\dfrac{\partial A}{\partial h} = -78.7 + 1.20t;$ at $t = 90°$, $h = 0.80$, $\dfrac{\partial A}{\partial h} = 29.3$

(b) The humidity has a greater effect since the coefficient of h is greater.

Section 7.5 Extrema of Functions of Two Variables

2. The first partial derivatives of f, $f_x(x, y) = 2(x + 1)$ and $f_y(x, y) = 2(y - 3)$, are zero at the point $(-1, 3)$. Moreover, since

$$f_{xx}(x, y) = 2, \quad f_{yy}(x, y) = 2, \quad \text{and} \quad f_{xy}(x, y) = 0,$$

it follows that

$$f_{xx}(-1, 3) > 0 \quad \text{and} \quad f_{xx}(-1, 3)f_{yy}(-1, 3) - [f_{xy}(-1, 3)]^2 = 4 > 0.$$

Thus, $(-1, 3, -4)$ is a relative minimum.

4. The first partial derivatives of f,

$$f_x(x, y) = \frac{2 - x}{\sqrt{25 - (x - 2)^2 - y^2}} \quad \text{and} \quad f_y(x, y) = \frac{-y}{\sqrt{25 - (x - 2)^2 - y^2}}$$

are zero at the point $(2, 0)$. Moreover, since

$$f_{xx}(x, y) = \frac{y^2 - 25}{[25 - (x - 2)^2 - y^2]^{3/2}}, \quad f_{yy}(x, y) = \frac{(x - 2)^2 - 25}{[25 - (x - 2)^2 - y^2]^{3/2}}, \quad \text{and}$$

$$f_{xy}(x, y) = \frac{(2 - x)y}{[25 - (x - 2)^2 - y^2]^{3/2}},$$

it follows that

$$f_{xx}(2, 0) = -\frac{1}{5} < 0 \quad \text{and} \quad f_{xx}(2, 0)f_{yy}(2, 0) - [f_{xy}(2, 0)]^2 = \frac{1}{25} > 0.$$

Thus, $(2, 0, 5)$ is a relative maximum.

6. The first partial derivatives of f, $f_x(x, y) = -2(x - 3)$ and $f_y(x, y) = -2(y + 2)$, are zero at the point $(3, -2)$. Moreover, since

$$f_{xx}(x, y) = -2, \quad f_{yy}(x, y) = -2, \quad \text{and} \quad f_{xy}(x, y) = 0,$$

it follows that

$$f_{xx}(3, -2) < 0 \quad \text{and} \quad f_{xx}(3, -2)f_{yy}(3, -2) - [f_{xy}(3, -2)]^2 = 4 > 0.$$

Thus, $(3, -2, 9)$ is a relative maximum.

8. The first partial derivatives of f, $f_x(x, y) = -2x + 8$ and $f_y(x, y) = -10x - 10$, are zero at the point $(4, -1)$. Moreover, since

$$f_{xx}(x, y) = -2, \quad f_{yy}(x, y) = -10, \quad \text{and} \quad f_{xy}(x, y) = 0,$$

it follows that

$$f_{xx}(4, -1) < 0 \quad \text{and} \quad f_{xx}(4, -1)f_{yy}(4, -1) - [f_{xy}(4, -1)]^2 = 20 > 0.$$

Thus, $(4, -1, 8)$ is a relative maximum.

10. The first partial derivatives of f, $f_x(x, y) = 2x + 6y = 2(x + 3y)$ and $f_y(x, y) = 6x + 20y - 4 = 2(3x + 10y - 2)$, are zero at the point $(-6, 2)$. Moreover, since

$$f_{xx}(x, y) = 2, \quad f_{yy}(x, y) = 20, \quad \text{and} \quad f_{xy}(x, y) = 6,$$

it follows that

$$f_{xx}(-6, 2) > 0 \quad \text{and} \quad f_{xx}(-6, 2)f_{yy}(-6, 2) - [f_{xy}(-6, 2)]^2 = 4 > 0.$$

Thus, $(-6, 2, 0)$ is a relative minimum.

12. The first partial derivatives of f, $f_x(x, y) = -6x + 3$ and $f_y(x, y) = -4y - 4$, are zero at $\left(\frac{1}{2}, -1\right)$. Moreover, since

$$f_{xx}(x, y) = -6, \quad f_{yy}(x, y) = -4, \quad \text{and} \quad f_{xy}(x, y) = 0,$$

it follows that $f_{xx}\left(\frac{1}{2}, -1\right) < 0$ and

$$f_{xx}\left(\tfrac{1}{2}, -1\right)f_{yy}\left(\tfrac{1}{2}, -1\right) - \left[f_{xy}\left(\tfrac{1}{2}, -1\right)\right]^2 = 24 > 0.$$

Thus, $\left(\frac{1}{2}, -1, \frac{31}{4}\right)$ is a relative maximum.

14. The first partial derivatives of f, $f_x(x, y) = 2x - 3y$ and $f_y(x, y) = -3x - 2y$, are zero at $(0, 0)$. Moreover, since

$$f_{xx}(x, y) = 2, \quad f_{yy}(x, y) = -2, \quad \text{and} \quad f_{xy}(x, y) = -3,$$

it follows that

$$f_{xx}(0, 0)f_{yy}(0, 0) - [f_{xy}(0, 0)]^2 = -13 < 0.$$

Thus, $(0, 0, 0)$ is a saddle point.

16. The first partial derivatives of f, $f_x(x, y) = 12 - y - 2x$ and $f_y(x, y) = 12 - x - 2y$, are zero at the point $(4, 4)$. Moreover, since

$$f_{xx}(x, y) = -2, \quad f_{yy}(x, y) = -2, \quad \text{and} \quad f_{xy}(x, y) = -1,$$

it follows that

$$f_{xx}(4, 4) < 0 \quad \text{and} \quad f_{xx}(4, 4)f_{yy}(4, 4) - [f_{xy}(4, 4)]^2 = 3 > 0.$$

Thus, $(4, 4, 48)$ is a relative maximum.

18. The first partial derivatives of f, $f_x(x, y) = -2xe^{-(x^2+y^2)}$ and $f_y(x, y) = -2ye^{-(x^2+y^2)}$, are zero at the point $(0, 0)$. Moreover, since

$$f_{xx}(x, y) = 2e^{-(x^2+y^2)}(2x^2 - 1), \quad f_{yy}(x, y) = 2e^{-(x^2+y^2)}(2y^2 - 1), \quad \text{and} \quad f_{xy}(x, y) = 4xye^{-(x^2+y^2)},$$

it follows that

$$f_{xx}(0, 0) = -2 < 0 \quad \text{and} \quad f_{xx}(0, 0)f_{yy}(0, 0) - [f_{xy}(0, 0)]^2 = 4 > 0.$$

Thus, $(0, 0, 1)$ is a relative maximum.

20. The first partial derivatives of f,

$$f_x(x, y) = \frac{4(x^2 - y^2 - 1)}{(x^2 + y^2 + 1)^2} \quad \text{and} \quad f_y(x, y) = \frac{8xy}{(x^2 + y^2 + 1)^2},$$

are zero at $(1, 0)$ and $(-1, 0)$. Moreover, since

$$f_{xx}(x, y) = \frac{8x(-x^2 + 3y^2 + 3)}{(x^2 + y^2 + 1)^3}, \quad f_{yy}(x, y) = \frac{8x(x^2 - 3y^2 + 1)}{(x^2 + y^2 + 1)^3}, \quad \text{and} \quad f_{xy}(x, y) = \frac{8y(-3x^2 + y^2 + 1)}{(x^2 + y^2 + 1)^3},$$

it follows that

$$f_{xx}(1, 0) = 2 > 0, \quad f_{xx}(-1, 0) = -2 < 0, \quad \text{and} \quad f_{xx}(\pm 1, 0)f_{yy}(\pm 1, 0) - [f_{xy}(\pm 1, 0)]^2 = 4 > 0.$$

Thus, $(1, 0, -2)$ is a relative minimum and $(-1, 0, 2)$ is a relative maximum.

22. Since $f_{xx}(x_0, y_0) < 0$ and $d = f_{xx}(x_0, y_0)f_{yy}(x_0, y_0) - [f_{xy}(x_0, y_0)]^2 = (-4)(-6) - (3)^2 = 15 > 0$, $f(x_0, y_0)$ is a relative maximum.

24. Since $f_{xx}(x_0, y_0) > 0$ and $d = f_{xx}(x_0, y_0)f_{yy}(x_0, y_0) - [f_{xy}(x_0, y_0)]^2 = (20)(8) - (9)^2 > 0$, $f(x_0, y_0)$ is a relative minimum.

26. The first partial derivatives of f,

$$f_x(x, y) = \frac{x}{\sqrt{x^2 + y^2}} \quad \text{and} \quad f_y(x, y) = \frac{y}{\sqrt{x^2 + y^2}},$$

are undefined at the point $(0, 0)$. Since

$$f_{xx}(x, y) = \frac{y^2}{(x^2 + y^2)^{3/2}}, \quad f_{yy}(x, y) = \frac{x^2}{(x^2 + y^2)^{3/2}}, \quad \text{and} \quad f_{xy}(x, y) = \frac{-xy}{(x^2 + y^2)^{3/2}},$$

it follows that $f_{xx}(0, 0)$ is undefined and the Second-Derivative Test fails. We note that $f(x, y) = \sqrt{x^2 + y^2}$ is nonnegative for all real numbers. Therefore, $(0, 0, 0)$ is a relative minimum.

28. The first partials of f are

$$f_x(x, y) = 3x^2 - 6x + 3 = 3(x^2 - 2x + 1) = 3(x - 1)^2$$

$$f_y(x, y) = 3y^2 + 12y + 12 = 3(y^2 + 4y + 4) = 3(y + 2)$$

The first partials are zero at $(1, -2)$.

Moreover,

$$f_{xx}(x, y) = 6x - 6, \quad f_{xx}(1, -2) = 0$$

$$f_{yy}(x, y) = 6y + 12, \quad f_{yy}(1, -2) = 0$$

$$f_{xy}(x, y) = 0$$

and hence the Second = Partials Test fails.
Note that $f(x, y) = (x - 1)^3 + (y + 2)^3$. By testing "nearby" points, we conclude that $(1, -2, 0)$ is a saddle point.

30. The first partials of f, $f_x(x, y) = \dfrac{4x}{3(x^2 + y^2)^{1/3}}$ and $f_y(x, y) = \dfrac{4y}{3(x^2 + y^2)^{1/3}}$ are zero at $(0, 0)$. Since

$$f_{xx}(x, y) = \frac{4}{9} \cdot \frac{x^2 + 3y^2}{(x^2 + y^2)^{4/3}}, \quad f_{yy} = \frac{4}{9} \cdot \frac{y^2 + 3x^2}{(x^2 + y^2)^{4/3}}, \quad \text{and} \quad f_{xy} = \frac{-8}{9} \cdot \frac{xy}{(x^2 + y^2)^{4/3}}$$

equal zero at $(0, 0)$, the Second-Partials Test fails. We note that $f(x, y) = (x^2 + y^2)^{2/3}$ is positive for all points $(x, y) \neq (0, 0)$, and hence $(0, 0, 0)$ is a relative minimum.

32. Critical points: Any points of form the $(0, y, z)$, $(x, -2, z)$, or $(x, y, 1)$. All are relative maxima.

34. True

36. True

38. The sum is given by

$$x + y + z = 32$$

$$z = 32 - x - y$$

and the product given by $P = xy^2z = xy^2(32 - x - y) = 32xy^2 - x^2y^2 - xy^3$ has first partials

$$P_x = 32y^2 - 2xy^2 - y^3 = y^2(32 - 2x - y)$$

$$P_y = 64xy - 2x^2y - 3xy^2 = xy(64 - 2x - 3y).$$

Setting these equal to zero produces the system

$$2x + y = 32$$

$$2x + 3y = 64.$$

Solving this system, we have $x = 8$, $y = 16$, and $z = 8$.

40. The sum is given by

$$x + y + z = 1$$

$$z = 1 - x - y$$

and the sum of the squares is given by $S = x^2 + y^2 + z^2 = x^2 + y^2 + (1 - x - y)^2$. The first partial derivatives of S are

$$S_x = 2x - 2(1 - x - y) = 4x + 2y - 2 \quad \text{and} \quad S_y = 2y - 2(1 - x - y) = 2x + 4y - 2.$$

Setting these equal to zero produces the system

$$2x + y = 1$$

$$x + 2y = 1.$$

Solving this system yields $x = y = z = \frac{1}{3}$.

42. The first partial derivatives of R are

$$R_{p_1} = 500 + 1.5p_2 - 3p_1$$

$$R_{p_2} = 800 + 1.5p_1 - 2p_2.$$

Setting these equal to zero produces the system

$$3p_1 - 1.5p_2 = 500$$

$$-1.5p_1 + 2p_2 = 800$$

which yields $p_1 \approx \$586.67$ and $p_2 = \$850.00$. The Second-Derivative Test verifies that these values maximize the total revenue.

44. The revenue function is given by

$$R = x_1p_1 + x_2p_2 = 1000p_1 - 4p_1^2 + 6p_1p_2 + 900p_2 - 3p_2^2$$

and the first partial derivatives of R are $R_{p_1} = 1000 - 8p_1 + 6p_2$ and $R_{p_2} = 6p_1 + 900 - 6p_2$. Setting these equal to zero produces the system

$$8p_1 - 6p_2 = 1000$$

$$-6p_1 + 6p_2 = 900$$

which yields $p_1 = \$950.00$ and $p_2 = \$1100.00$. The Second-Derivative Test verifies that these values maximize the total revenue.

46. Let x = length, y = width, z = height, and c = cost for constructing the sides. Then the total cost is given by $C = 1.5c(xy) + c(2xz + 2yz) = c(1.5xy + 2xz + 2yz)$. Let $k = C/c$, then

$$z = \frac{k - 1.5xy}{2x + 2y}.$$

The volume of the box is given by $V = xyz = xy\left(\frac{k - 1.5xy}{2x + 2y}\right)$.

The first partial derivatives of V are $V_x = \dfrac{y^2(2k - 3x^2 - 6xy)}{(2x + 2y)^2}$ and $V_y = \dfrac{x^2(2k - 3y^2 - 6xy)}{(2x + 2y)^2}$.

Setting these equal to zero produces the following system.

$$\left.\begin{aligned} 3x^2 + 6xy &= 2k \\ 3y^2 + 6xy &= 2k \end{aligned}\right\} \Rightarrow x = y$$

$$9x^2 = 2k$$

$$x = \frac{\sqrt{2k}}{3}$$

Thus, $x = \dfrac{\sqrt{2k}}{3}$, $y = \dfrac{\sqrt{2k}}{3}$, and $z = \dfrac{\sqrt{2k}}{4}$.

48. Let x = length, y = width, and z = height. The volume is $V = xyz$ which implies that $z = V/xy$. The surface area of the box is given by

$$A = 2xy + 2xz + 2yz = 2xy + 2x\left(\frac{V}{xy}\right) + 2y\left(\frac{V}{xy}\right) = 2xy + \frac{2V}{y} + \frac{2V}{x}.$$

The first partial derivatives of A are

$$A_x = 2y - \frac{2V}{x^2} \quad \text{and} \quad A_y = 2x - \frac{2V}{y^2}.$$

Setting these equal to zero produces the system

$$\left.\begin{aligned} 2(x^2y - V) &= 0 \\ 2(xy^2 - V) &= 0 \end{aligned}\right\} \Rightarrow x = y.$$

Solving this system, we have $x = y = z = \sqrt[3]{V}$, a cube.

50. $T = xW_1 + yW_2$

$= x(3 - 0.002x - 0.005y) + y(4.5 - 0.003x - 0.004y)$

$= 3x - 0.002x^2 + 4.5y - 0.004y^2 - 0.008xy$

$T_x = 3 - 0.004x - 0.008y$

$T_y = 4.5 - 0.008y - 0.008x$

Solving $T_x = T_y = 0$, we obtain $x = 375$ and $y = 187.5$. Hence, the lake should be stocked with approximately 375 smallmouth bass and 188 largemouth bass.

Section 7.6 Lagrange Multipliers

2. $F(x, y, \lambda) = xy - \lambda(2x + y - 4)$

$F_x = y - 2\lambda = 0, \qquad y = 2\lambda$

$F_y = x - \lambda = 0, \qquad x = \lambda$

$F_\lambda = -2x - y + 4 = 0, \qquad -4\lambda = -4$

Thus, $\lambda = 1$, $x = 1$, and $y = 2$, and $f(x, y)$ is maximum at $(1, 2)$. The maximum is $f(1, 2) = 2$.

4. $F(x, y, \lambda) = x^2 + y^2 - \lambda(-2x - 4y + 5)$

$F_x = 2x + 2\lambda = 0, \qquad x = -\lambda$

$F_y = 2y + 4\lambda = 0, \qquad y = -2\lambda$

$F_\lambda = 2x + 4y - 5 = 0, \qquad -10\lambda = 5$

Thus, $\lambda = -\frac{1}{2}$, $x = \frac{1}{2}$, and $y = 1$, and $f(x, y)$ is minimum at $\left(\frac{1}{2}, 1\right)$. The minimum is $f\left(\frac{1}{2}, 1\right) = \frac{5}{4}$.

6. $F(x, y, \lambda) = x^2 - y^2 - \lambda(x - 2y + 6)$

$F_x = 2x - \lambda = 0, \qquad x = \dfrac{1}{2}\lambda$

$F_y = -2y + 2\lambda = 0, \qquad y = \lambda$

$F_\lambda = -x + 2y - 6 = 0, \qquad \dfrac{3}{2}\lambda = 6$

Thus, $\lambda = 4$, $x = 2$, and $y = 4$, and $f(x, y)$ is minimum at $(2, 4)$. The minimum is $f(2, 4) = -12$.

8. $F(x, y, \lambda) = 3x + y + 10 - \lambda(x^2y - 6)$

$F_x = 3 - 2xy\lambda = 0$

$F_y = 1 - x^2\lambda = 0, \qquad \lambda = \dfrac{1}{x^2}$

$F_\lambda = -x^2y + 6 = 0, \qquad y = \dfrac{6}{x^2}$

For Fx, we can write

$$3 = 2x\left(\frac{6}{x^2}\right)\left(\frac{1}{x^2}\right)$$

$$3x^3 = 12$$

$$x = \sqrt[3]{4}.$$

Thus, $x = \sqrt[3]{4}$ and $y = \dfrac{3\sqrt[3]{4}}{2}$. The maximum is

$$f\left(\sqrt[3]{4}, \frac{3\sqrt[3]{4}}{2}\right) = \frac{9\sqrt[3]{4} + 20}{2} \approx 17.143.$$

10. Note: $f(x, y)$ has a minimum when $g(x, y) = x^2 + y^2$ is minimum.

$F(x, y, \lambda) = x^2 + y^2 - \lambda(2x + 4y - 15)$

$F_x = 2x - 2\lambda = 0, \qquad x = \lambda$

$F_y = 2y - 4\lambda = 0, \qquad y = 2\lambda$

$F_\lambda = -2x - 4y + 15 = 0, \qquad -10\lambda = -15$

Thus, $\lambda = \dfrac{3}{2}$, $x = \dfrac{3}{2}$, and $y = 3$, and $f(x, y)$ is minimum at $\left(\dfrac{3}{2}, 3\right)$. The minimum is $f\left(\dfrac{3}{2}, 3\right) = \dfrac{3\sqrt{5}}{2}$.

12. $F(x, y, \lambda) = 2x + y - \lambda(xy - 32)$

$$\left.\begin{aligned} F_x &= 2 - \lambda y = 0, \qquad & y &= 2/\lambda \\ F_y &= 1 - \lambda x = 0, \qquad & x &= 1/\lambda \end{aligned}\right\} \; y = 2x$$

$$F_\lambda = -xy + 32 = 0, \qquad -2x^2 = -32$$

Thus, $x = 4$ and $y = 8$, and $f(x, y)$ is minimum at $(4, 8)$. The minimum is $f(4, 8) = 16$.

14. $F(x, y, z, \lambda) = xyz - \lambda(x + y + z - 6)$

$$\left.\begin{aligned} F_x &= yz - \lambda = 0 \\ F_y &= xz - \lambda = 0 \\ F_z &= xy - \lambda = 0 \end{aligned}\right\} \; x = y = z$$

$$F_\lambda = -x - y - z + 6 = 0, \qquad -3x = -6$$

Thus, $x = y = z = 2$ and $f(x, y, z)$ is maximum at $(2, 2, 2)$. The maximum is $f(2, 2, 2) = 8$.

16. $F(x, y, \lambda) = x^2 - 8x + y^2 - 12y + 48 - \lambda(x + y - 8)$

$$\left.\begin{aligned} F_x &= 2x - 8 - \lambda = 0 \\ F_y &= 2y - 12 - \lambda = 0 \end{aligned}\right\} \; \begin{aligned} 2x - 8 &= 2y - 12 \\ y &= x + 2 \end{aligned}$$

$$F_\lambda = -x - y + 8 = 0, \qquad x + (x + 2) = 8$$

Thus, $x = 3$, and $y = 5$, and $f(x, y)$ is minimum at $(3, 5)$. The minimum is $f(3, 5) = -2$.

18. $F(x, y, z, \lambda) = x^2y^2z^2 - \lambda(x^2 + y^2 + z^2 - 1)$

$$\begin{aligned} F_x &= 2xy^2z^2 - 2\lambda x = 0, \qquad & y^2z^2 &= \lambda \\ F_y &= 2x^2yz^2 - 2\lambda y = 0, \qquad & x^2z^2 &= \lambda \\ F_z &= 2x^2y^2z - 2\lambda z = 0, \qquad & x^2y^2 &= \lambda \end{aligned} \Bigg\} \; x = y = z$$

$$F_\lambda = -(x^2 + y^2 + z^2 - 1) = 0,$$

$$3x^2 = 1 \implies x = y = z = \frac{1}{\sqrt{3}}.$$

Thus, the maximum is $f\left(\frac{1}{\sqrt{3}}, \frac{1}{\sqrt{3}}, \frac{1}{\sqrt{3}}\right) = \frac{1}{27}$.

20. $F(x, y, z, w, \lambda) = 2x^2 + y^2 + z^2 + 2w^2 - \lambda(x + y + 2z + 2w - 4)$

$$F_x = 4x - \lambda = 0, \qquad x = \frac{\lambda}{4}$$

$$F_y = 2y - \lambda = 0, \qquad y = \frac{\lambda}{2} = 2x$$

$$F_z = 2z - 2\lambda = 0, \qquad z = \lambda = 4x$$

$$F_w = 4w - 2\lambda = 0, \qquad w = \frac{\lambda}{2} = 2x$$

$$F_\lambda = -(x + y + 2z + 2w - 4) = 0$$

$$x + 2x + 2(4x) + 2(2x) = 4$$

$$15x = 4 \implies x = \frac{4}{15}$$

The maximum is $f\left(\frac{4}{15}, \frac{8}{15}, \frac{16}{15}, \frac{8}{15}\right) = \frac{32}{15}$.

22. $F(x, y, z, \lambda, \eta) = x^2 + y^2 + z^2 - \lambda(x + 2z - 4) - \eta(x + y - 8)$

$$F_x = 2x - \lambda - \eta = 0$$

$$F_y = 2y - \eta = 0, \qquad \eta = 2y$$

$$F_z = 2z - 2\lambda = 0, \qquad \lambda = z$$

$$F_\lambda = -x - 2z + 4 = 0$$

$$F_\eta = -x - y + 8 = 0$$

From F_x, we have $2x - 2y - z = 0$, and from F_λ and F_η, we have $x + 2x = 4$ and $x + y = 8$.

Solving this system yields $x = 4$, $y = 4$, and $z = 0$, and $f(x, y, z)$ is minimum at $(4, 4, 0)$. The minimum is $f(4, 4, 0) = 32$.

24. $F(x, y, z, \lambda, \eta) = xy + yz - \lambda(x + 2y - 6) - \eta(x - 3z)$

$$F_x = y - \lambda - \eta = 0, \qquad y = \lambda + \eta$$

$$F_y = x + z - 2\lambda = 0, \qquad \lambda = \frac{x + z}{2}$$

$$F_z = y + 3\eta = 0, \qquad \eta = -\frac{y}{3}$$

$$F_\lambda = -x - 2y + 6 = 0$$

$$F_\eta = -x + 3z = 0$$

From F_x, we have

$$y = \frac{x + z}{2} - \frac{y}{3}$$

$$3x - 8y + 3z = 0.$$

From F_λ, we have $x + 2y = 6$ and from F_η, we have $x - 3z = 0$. Solving this system yields $x = 3$, $y = \frac{3}{2}$, and $z = 1$, and $f(x, y, z)$ is maximum at $\left(3, \frac{3}{2}, 1\right)$. The maximum is $f\left(3, \frac{3}{2}, 1\right) = 6$.

26. Minimize: $f(x, y, z) = x^2 + y^2 + z^2$

Constraint: $x + y + z = 120$

$F(x, y, z, \lambda) = x^2 + y^2 + z^2 - \lambda(x + y + z - 120)$

$$\left.\begin{aligned} F_x &= 2x - \lambda = 0 \\ F_y &= 2y - \lambda = 0 \\ F_z &= 2z - \lambda = 0 \end{aligned}\right\} x = y = z$$

$$F_\lambda = -x - y - z + 120 = 0, \qquad 3x = 120$$

Thus, $x = y = z = 40$.

28. Minimize: $f(x, y, z) = x^2 + y^2 + z^2$

Constraint: $x + y + z = S$

$F(x, y, z, \lambda) = x^2 + y^2 + z^2 - \lambda(x + y + z - S)$

$$\left.\begin{aligned} F_x &= 2x - \lambda = 0 \\ F_y &= 2y - \lambda = 0 \\ F_z &= 2z - \lambda = 0 \end{aligned}\right\} x = y = z$$

$$F_\lambda = -x - y - z + S = 0, \qquad 3x = S$$

Thus, $x = y = z = S/3$.

30. $F(x, y, \lambda) = x^2 + (y - 10)^2 - \lambda[(x - 4)^2 + y^2 - 4]$

$$\left.\begin{aligned} F_x &= 2x - 2\lambda(x - 4) = 0, & \lambda &= \frac{x}{x - 4} \\ F_y &= 2(y - 10) - 2\lambda y = 0, & \lambda &= \frac{y - 10}{y} \end{aligned}\right\} 5x + 2y = 20$$

$$F_\lambda = -(x - 4)^2 - y^2 + 4 = 0, \qquad (x - 4)^2 + \left(\frac{20 - 5x}{2}\right)^2 = 4$$

Using the Quadratic Formula or a graphing utility gives us $x = \dfrac{232 \pm \sqrt{1856}}{58} = \dfrac{116 \pm 4\sqrt{29}}{29}$. ($x \approx 3.2572, 4.7428$)

Using $x = \dfrac{116 - 4\sqrt{29}}{29}$ (the other value of x results in a negative y-value) and $y = \dfrac{10\sqrt{29}}{29}$, we have

$$d = \sqrt{\left(\frac{116 - 4\sqrt{29}}{29}\right)^2 + \left(\frac{10\sqrt{29}}{29} - 10\right)^2} \approx 8.770.$$

32. $F(x, y, z, \lambda) = (x - 4)^2 + y^2 + z^2 - \lambda(x^2 + y^2 - z^2)$

$$F_x = 2(x - 4) - 2x\lambda = 0, \qquad 2x(1 - \lambda) = 8$$

$$\left.\begin{aligned} F_y &= 2y - 2y\lambda = 0 \\ F_z &= 2z + 2z\lambda = 0 \end{aligned}\right\} \quad \begin{aligned} 2y(1 - \lambda) &= 0 \\ 2z(1 + \lambda) &= 0 \end{aligned}$$

$$F_\lambda = -x^2 - y^2 + z^2 = 0, \qquad z = \sqrt{x^2 + y^2}$$

From F_y, we have $y = 0$ or $\lambda = 1$. From F_x, we know that $\lambda \neq 1$ (since $0 \neq 8$), thus, $y = 0$. From F_λ and F_z, we now have $x = z$ and $\lambda = -1$. Thus, $x = 2$, $y = 0$, and $z = 2$ and we have $d = \sqrt{(2 - 4)^2 + (0)^2 + (2)^2} = 2\sqrt{2} \approx 2.828$.

34. $F(x, y, z, \lambda) = xyz - \lambda(5xy + 2xz + 2yz - C)$

$$F_x = yz - (5\lambda y + 2\lambda z) = 0$$

$$F_y = xz - (5\lambda x + 2\lambda z) = 0$$

$$F_z = xy - (2\lambda x + 2\lambda y) = 0$$

$$F_\lambda = -(5xy + 2xz + 2yz - C) = 0$$

From the first two equations, $x = y$ or $z = 5\lambda$.

If $x = y$, then equation 3 implies $x = 4\lambda$. From equation 2, $z = 10\lambda = \frac{5}{2}x$ and the fourth equation gives $x^2 = \frac{C}{15}$.

Thus,

$$x = y = \sqrt{\frac{C}{15}}, \quad z = \frac{5}{2}\sqrt{\frac{C}{15}}.$$

(The case $z = 5\lambda$ results in $\lambda = 0$ or $z = 0$, which are impossible.)

36. $F(x_1, x_2, \lambda) = 0.25x_1^2 + 25x_1 + 0.05x_2^2 + 12x_2 - \lambda(x_1 + x_2 - 1000)$

$$F_{x_1} = 0.5x_1 + 25 - \lambda = 0, \qquad x_1 = 2\lambda - 50$$

$$F_{x_2} = 0.1x_2 + 12 - \lambda = 0, \qquad x_2 = 10\lambda - 120$$

$$F_\lambda = -(x_1 + x_2 - 1000) = 0$$

$$\begin{aligned} (2\lambda - 50) + (10\lambda - 120) &= 1000 \\ 12\lambda &= 1170 \\ \lambda &= 97.5 \end{aligned}$$

Hence, $x_1 = 145$ and $x_2 = 855$ units, respectively.

38. $F(p, q, r) = 2pq + 2pr + 2qr - \lambda(p + q + r - 1)$

$$\begin{aligned} F_p &= 2q + 2r - \lambda = 0, \\ F_q &= 2p + 2r - \lambda = 0, \\ F_r &= 2p + 2q - \lambda = 0, \end{aligned} \quad \left.\begin{aligned} \lambda &= 2q + 2r \\ \lambda &= 2p + 2r \\ \lambda &= 2p + 2q \end{aligned}\right\} p = q = r$$

$$F_\lambda = -(p + q + r - 1) = 0$$

$p + q + r = 3p = 1 \Rightarrow p = \frac{1}{3}$ and $f(p, q, r) = f\left(\frac{1}{3}, \frac{1}{3}, \frac{1}{3}\right) = \frac{2}{3}$.

40. (a) $F(x, y, \lambda) = 48x + 36y - \lambda(x^{0.6}y^{0.4} - 200)$

$$F_x = 48 - 0.6\lambda x^{-0.4}y^{0.4} = 0, \qquad 80 = \lambda\left(\frac{y}{x}\right)^{0.4}$$

$$F_y = 36 - 0.4\lambda x^{0.6}y^{-0.6} = 0, \qquad 90 = \lambda\left(\frac{x}{y}\right)^{0.6}$$

$$F_\lambda = -x^{0.6}y^{0.4} + 200 = 0$$

From F_x and F_y, we have $y/x = 8/9$ and

$$x^{0.6}\left(\frac{8}{9}x\right)^{0.4} = 200$$

$$x = 200\left(\frac{9}{8}\right)^{0.4} \approx 209.65$$

$$y = 200\left(\frac{8}{9}\right)^{0.6} \approx 186.35.$$

(b) $$\frac{f_x(x, y)}{f_y(x, y)} = \frac{48}{36}$$

$$\frac{60(y/x)^{0.4}}{40(x/y)^{0.6}} = \frac{48}{36}$$

$$\frac{y}{x} = \frac{48}{36} \cdot \frac{40}{60} = \frac{8}{9}$$

Thus, $y = (8/9)x$ and the conditions of part (a) are met.

42. (a) $F(x, y, \lambda) = 100x^{0.6}y^{0.4} - \lambda(48x + 36y - 100{,}000)$

$$F_\lambda = 48x + 36y - 100{,}000 = 0$$

From Exercise 40, we have $y = (8/9)x$. Thus, $x = 1250$ and $y = 10{,}000/9$ and

$$f\left(1250, \frac{10{,}000}{9}\right) \approx 119{,}247 \text{ units.}$$

(b) $$F_x = 60\left(\frac{y}{x}\right)^{0.4} - 48\lambda = 0$$

$$\lambda = \frac{60}{48}\left(\frac{10{,}000/9}{1250}\right)^{0.4} \approx 1.192$$

(c) $119{,}247 + (1.192)(25{,}000) \approx 149{,}047$ units

44. $F(x, y, z, \lambda) = x + 2y + 3z - \lambda(0.01x^2y^2z^2 - 0.13)$

$$F_x = 1 - 0.02\lambda xy^2z^2 = 0, \qquad 100 = 2\lambda xy^2z^2$$

$$F_y = 2 - 0.02\lambda x^2yz^2 = 0, \qquad 200 = 2\lambda x^2yz^2$$

$$F_z = 3 - 0.02\lambda x^2y^2z = 0, \qquad 300 = 2\lambda x^2y^2z$$

$$F_\lambda = -(0.01x^2y^2z^2 - 0.13) = 0, \qquad x^2y^2z^2 = 13$$

From equations 1 and 2, $x = 2y$. From equations 1 and 3, $x = 3z$. Hence, from equation 4,

$$x^2\left(\frac{x}{2}\right)^2\left(\frac{x}{3}\right)^2 = 13 \Rightarrow x^6 = 468 \Rightarrow x \approx 2.786$$

$y \approx 1.393$, $z \approx 0.929$.

46. Answers will vary.

48. $F(A, G, K) = 0.019A + 0.021G + 0.024K$ Maximize

$A + G + K = 20{,}000$ Constraint

$2000A - 2000K + G^2 = 0$ Constraint

From $F(A, G, K, \lambda, \mu) = 0.019A + 0.021G + 0.024K - \lambda(A + G + K - 20{,}000) - \mu(2000A - 2000K + G^2)$,

$$F_A = 0.019 - \lambda - 2000\mu = 0$$

$$F_K = 0.024 - \lambda + 2000\mu = 0$$

$$F_G = 0.021 - \lambda - 2G\mu = 0$$

Adding the first 2 equations, $0.043 = 2\lambda \Rightarrow \lambda = 0.0215$. Hence, $2000\mu = 0.019 - \lambda = 0.019 - 0.0215 = -0.0025$ which gives $\mu = -0.0025/2000$.

From the third equation,

$$0.021 - \lambda = 2G\mu$$

$$0.021 - 0.0215 = 2\left(\frac{-0.0025}{2000}\right)G \Rightarrow G = 200$$

$$F_\lambda = -A - G - K + 20{,}000 = 0 \Rightarrow A + K = 19{,}800$$

$$F_\mu = -2000A + 2000K - G^2 = 0 \Rightarrow -A + K = 20$$

Finally, $2K = 19{,}820 \Rightarrow K = 9910$ and $A = 9890$.

Invest \$9890 in Anheuser-Busch, \$200 in General Electric, and \$9910 in Kellogs.

Section 7.7 Least Squares Regression Analysis

2. (a) $\sum x_i = 0$

$\sum y_i = 4$

$\sum x_i y_i = 6$

$\sum x_i^2 = 20$

$$a = \frac{4(6) - (0)(4)}{4(20) - (0)^2} = \frac{3}{10}$$

$$b = \frac{1}{4}\left[4 - \frac{3}{10}(0)\right] = 1$$

The regression line is $y = \frac{3}{10}x + 1$.

(b) $\left(\frac{1}{10} - 0\right)^2 + \left(\frac{7}{10} - 1\right)^2 + \left(\frac{13}{10} - 1\right)^2 + \left(\frac{19}{10} - 2\right)^2 = \frac{1}{5}$

4. (a) $\sum x_i = 28$

$\sum y_i = 8$

$\sum x_i y_i = 37$

$\sum x_i^2 = 116$

$$a = \frac{8(37) - (28)(8)}{8(116) - (28)^2} = \frac{1}{2}$$

$$b = \frac{1}{8}\left[8 - \frac{1}{2}(28)\right] = -\frac{3}{4}$$

The regression line is $y = \frac{1}{2}x - \frac{3}{4}$.

(b) $\left(-\frac{1}{4} - 0\right)^2 + \left(\frac{1}{4} - 0\right)^2 + \left(\frac{3}{4} - 0\right)^2 + \left(\frac{3}{4} - 1\right)^2 + \left(\frac{5}{4} - 1\right)^2 + \left(\frac{5}{4} - 2\right)^2 + \left(\frac{7}{4} - 2\right)^2 + \left(\frac{9}{4} - 2\right)^2 = \frac{3}{2}$

6. $\sum x_i = 0$

$\sum y_i = 8$

$\sum x_i y_i = 32$

$\sum x_i^2 = 40$

Since $a = \frac{4}{5}$ and $b = 2$, the regression line is $y = \frac{4}{5}x + 2$.

8. $\sum x_i = 0$

$\sum y_i = 12$

$\sum x_i y_i = 14$

$\sum x_i^2 = 34$

Since $a = \frac{7}{17}$ and $b = 3$, the regression line is $y = \frac{7}{17}x + 3$.

10. $\sum x_i = 0$

$\sum y_i = 28$

$\sum x_i y_i = -94$

$\sum x_i^2 = 208$

Since $a = -\frac{47}{104}$ and $b = \frac{28}{5}$, the regression line is $y = -\frac{47}{104}x + \frac{28}{5}$.

12. $\sum x_i = 9$

$\sum y_i = 9$

$\sum x_i y_i = 39$

$\sum x_i^2 = 35$

Since $a = \frac{3}{2}$ and $b = -\frac{3}{2}$, the regression line is $y = \frac{3}{2}x - \frac{3}{2}$.

14. $\sum x_i = 36$

$\sum y_i = 19$

$\sum x_i y_i = 172$

$\sum x_i^2 = 322$

Since $a = \frac{29}{53}$ and $b = -\frac{37}{318}$, the regression line is $y = \frac{29}{53}x - \frac{37}{318}$.

16. The sum of the squared errors is as follows.

$$S = (-3a + b)^2 + (-a + b - 1)^2 + (a + b - 1)^2 + (3a + b - 2)^2$$

$$\frac{\partial S}{\partial a} = -6(-3a + b) - 2(-a + b - 1) + 2(a + b - 1) + 6(3a + b - 2)$$

$$\frac{\partial S}{\partial b} = 2(-3a + b) + 2(-a + b - 1) + 2(a + b - 1) + 2(3a + b - 2)$$

Setting the first partial derivatives equal to zero produces $40a = 12$ and $8b = 8$. Thus, $a = \frac{3}{10}$ and $b = 1$, and we have $y = \frac{3}{10}x + 1$.

18. The sum of the squared errors is as follows:

$$S = (-5a + b + 3)^2 + (-4a + b + 2)^2 + (-2a + b + 1)^2 + (-a + b - 1)^2$$

$$\frac{\partial S}{\partial a} = -10(-5a + b + 3) - 8(-4a + b + 2) - 4(-2a + b + 1) - 2(-a + b - 1)$$

$$= 92a - 24b - 48$$

$$\frac{\partial S}{\partial b} = 2(-5a + b + 3) + 2(-4a + b + 2) + 2(-2a + b + 1) + 2(-a + b - 1)$$

$$= -24a + 8b + 10.$$

Setting these partial derivatives equal to zero produces

$$92a - 24b = 48$$

$$-24a + 8b = -10.$$

Thus, $a = 0.9$ and $b = 1.45$ and $y = 0.9x + 1.45$.

20. $\sum x_i = 0$

$\sum y_i = 19$

$\sum x_i^2 = 40$

$\sum x_i^3 = 0$

$\sum x_i^4 = 544$

$\sum x_i y_i = -12$

$\sum x_i^2 y_i = 160$

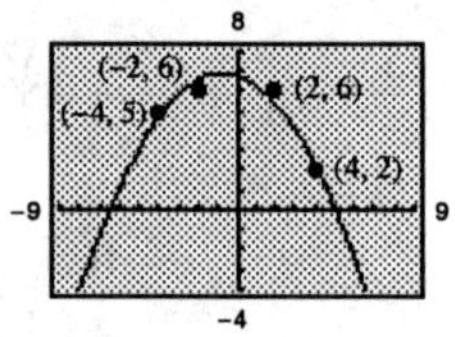

This produces the system

$$544a \quad + 40c = 160$$

$$40b \quad = -12$$

$$40a \quad + 4c = 19$$

which yields $a = -\frac{5}{24}$, $b = -\frac{3}{10}$, and $c = \frac{41}{6}$, and we have $y = -\frac{5}{24}x^2 - \frac{3}{10}x + \frac{41}{6}$.

22. $\sum x_i = 6$

$\sum y_i = 25$

$\sum x_i^2 = 14$

$\sum x_i^3 = 36$

$\sum x_i^4 = 98$

$\sum x_i y_i = 21$

$\sum x_i^2 y_i = 33$

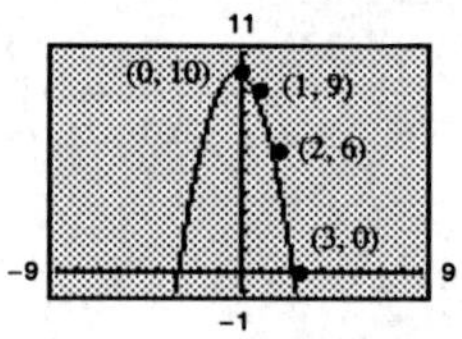

This produces the system

$$98a + 36b + 14c = 33$$

$$36a + 14b + 6c = 21$$

$$14a + 6b + 4c = 25$$

which yields $a = -\frac{5}{4}$, $b = \frac{9}{20}$, and $c = \frac{199}{20}$, and we have $y = -\frac{5}{4}x^2 + \frac{9}{20}x + \frac{199}{20}$.

24. Linear: $y = 1.0658x - 2.1974$

Quadratic: $y = -0.2184x^2 + 2.7901x - 3.0332$

26. Linear: $y = 4.9429x + 4.5667$

Quadratic: $y = 0.2286x^2 + 3.3429x + 6.7$

28. (25, 82), (30, 75), (35, 67), (40, 55)

(a) $\sum x_i = 130$

$\sum y_i = 279$

$\sum x_i y_i = 8845$

$\sum x_i^2 = 4350$

Thus, $a = -\frac{89}{50}$, $b = \frac{638}{5}$, and $y = -\frac{89}{50}x + \frac{638}{5}$.

(b) When $x = 32.95$, $y \approx 69$.

(c) When $y = 83$, $x \approx 25$.

30. (1.35, 118.5), (2.67, 135.2), (3.93, 167.3), (5.14, 197.6), (7.43, 204.7)

$\sum x_i = 20.52$

$\sum y_i = 823.3$

$\sum x_i y_i = 3715.033$

$\sum x_i^2 = 106.0208$

Thus, $a \approx 15.4177$ and $b \approx 101.3857$ and $y = 15.4177x + 101.3857$.

32. (a) $(-2, 0.9), (-1, 1.1)$; Thus, $y = 0.2x + 1.3$. In 1875, $x = -0.5$ and we have $y = 1.2$ billion.

(b) $(-1, 1.1), (0, 1.5), (1, 2.4), (1.8, 5.2), (1.92, 5.8)$

$y = 0.8211x^2 + 0.7704x + 1.1609$

For $x = 2$, $y \approx 5.99$ billion.

34. (a) Using a graphing utility, the least squares regression quadratic is

$y = 712.03t^2 - 5505.46t + 20{,}150.96 \quad (t = 0 \text{ is } 1970)$

(b) When $t = 30(2000)$, $y \approx 495{,}814$ thousand shares.

36. Linear: $y = 7.0424x - 16.98575$

Quadratic: $y = 0.7061x^2 - 4.1735x + 11.6029$

38. Linear: $y = 47.6372x - 45.0073$

Quadratic: $y = 5.3778x^2 + 6.0965x - 2.0411$

40. Negative correlation

$r = -0.9652$

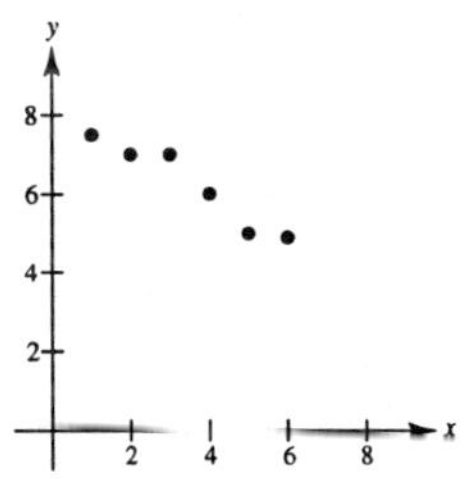

42. Positive correlation

$r = 0.9276$

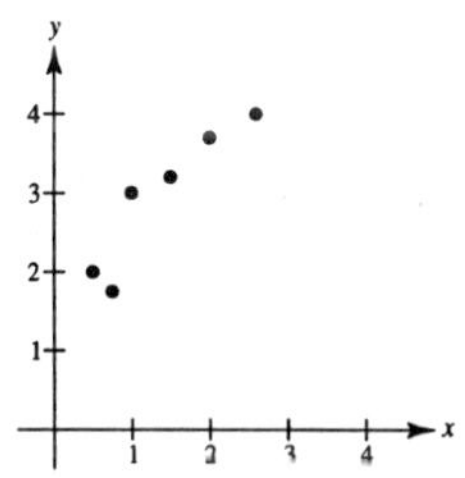

44. True, the slope is negative.

46. Yes, the correlation is poor.

Section 7.8 Double Integrals and Area in the Plane

2. $\displaystyle\int_x^{x^2} \frac{y}{x}\,dy = \left[\frac{y^2}{2x}\right]_x^{x^2} = \frac{x^3}{2} - \frac{x}{2} = \frac{x}{2}(x^2 - 1)$

4. $\displaystyle\int_0^{e^y} y\,dx = \left[yx\right]_0^{e^y} = ye^y$

6. $$\int_{x^2}^{\sqrt{x}} (x^2 + y^2)dy = \left[x^2y + \frac{1}{3}y^3\right]_{x^2}^{\sqrt{x}}$$
$$= x^{5/2} + \frac{1}{3}x^{3/2} - x^4 - \frac{1}{3}x^6$$
$$= \frac{x^{3/2}}{3}(3x + 1 - 3x^{5/2} - x^{9/2})$$

8. $$\int_{-\sqrt{1-y^2}}^{\sqrt{1-y^2}} (x^2 + y^2)dx = \left[\frac{1}{3}x^3 + xy^2\right]_{-\sqrt{1-y^2}}^{\sqrt{1-y^2}}$$
$$= \frac{2}{3}(1 - y^2)^{3/2} + 2y^2\sqrt{1 - y^2}$$
$$= \frac{2}{3}\sqrt{1 - y^2}\,[(1 - y^2) + 3y^2]$$
$$= \frac{2}{3}\sqrt{1 - y^2}\,(1 + 2y^2)$$

10. $$\int_y^3 \frac{xy}{\sqrt{x^2 + 1}}\,dx = y\sqrt{x^2 + 1}\Big]_y^3$$
$$= \sqrt{10}y - y\sqrt{y^2 + 1}$$
$$= y\left(\sqrt{10} - \sqrt{y^2 + 1}\right)$$

12. $$\int_0^2\int_0^2 (6 - x^2)dy\,dx = \int_0^2\left[(6 - x^2)y\right]_0^2 dx$$
$$= 2\int_0^2 (6 - x^2)dx$$
$$= 2\left[6x - \frac{1}{3}x^3\right]_0^2$$
$$= \frac{56}{3}$$

14. $$\int_0^1\int_0^x \sqrt{1 - x^2}\,dy\,dx = \int_0^1\left[\sqrt{1 - x^2}\,y\right]_0^x dx$$
$$= -\frac{1}{2}\int_0^1 \sqrt{1 - x^2}\,(-2x)dx$$
$$= -\frac{1}{3}(1 - x^2)^{3/2}\Big]_0^1$$
$$= \frac{1}{3}$$

16. $$\int_0^2\int_{3y^2-6y}^{2y-y^2} 3y\,dx\,dy = \int_0^2\left[3xy\right]_{3y^2-6y}^{2y-y^2} dy$$
$$= \int_0^2 (24y^2 - 12y^3)dy$$
$$= \left[8y^3 - 3y^4\right]_0^2$$
$$= 16$$

18. $$\int_0^1\int_y^{2y} (1 + 2x^2 + 2y^2)dx\,dy = \int_0^1\left[x + \frac{2x^3}{3} + 2xy^2\right]_y^{2y} dy$$
$$= \int_0^1\left(y + \frac{20}{3}y^3\right)dy$$
$$= \left[\frac{y^2}{2} + \frac{5y^4}{3}\right]_0^1$$
$$= \frac{13}{6}$$

20. $$\int_0^4\int_0^x \frac{2}{(x + 1)(y + 1)}\,dy\,dx = \int_0^4\left[\frac{2}{x + 1}\ln|y + 1|\right]_0^x dx$$
$$= \int_0^4 \frac{2}{x + 1}\ln|x + 1|dx$$
$$= \left[(\ln|x + 1|)^2\right]_0^4$$
$$= (\ln 5)^2$$

22. $$\int_0^a\int_0^{a-x} (x^2 + y^2)dy\,dx = \int_0^a\left[x^2y + \frac{1}{3}y^3\right]_0^{a-x} dx$$
$$= \int_0^a\left[x^2(a - x) + \frac{1}{3}(a - x)^3\right]dx = \left[\frac{ax^3}{3} - \frac{x^4}{4} - \frac{(a - x)^4}{12}\right]_0^a = \frac{a^4}{6}$$

24. Since (for fixed y) $\lim\limits_{b\to\infty}\left[-\frac{1}{2}ye^{-(x^2+y^2)}\right]_0^b = \frac{1}{2}ye^{-y^2}$, we have the following.

$$\int_0^\infty\int_0^\infty xye^{-(x^2+y^2)}\,dx\,dy = \int_0^\infty \frac{1}{2}ye^{-y^2}\,dy = \lim_{b\to\infty}\left[-\frac{1}{4}e^{-y^2}\right]_0^b = \frac{1}{4}$$

26. $\displaystyle\int_1^2\int_2^4 dx\,dy = \int_1^2\Big[x\Big]_2^4 dy = \int_1^2 2\,dy = 2y\Big]_1^2 = 2$

$\displaystyle\int_2^4\int_1^2 dy\,dx = \int_2^4\Big[y\Big]_1^2 dx = \int_2^4 dx = x\Big]_2^4 = 2$

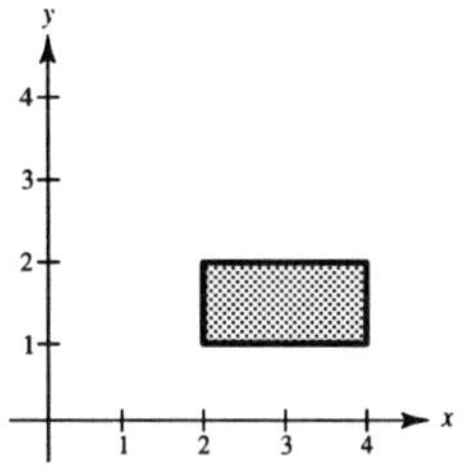

28. $\displaystyle\int_0^4\int_0^{\sqrt{x}} dy\,dx = \int_0^4 \sqrt{x}\,dx = \frac{2}{3}x^{3/2}\Big]_0^4 = \frac{16}{3}$

$\displaystyle\int_0^2\int_{y^2}^4 dx\,dy = \int_0^2 (4-y^2)\,dy = \left(4y - \frac{y^3}{3}\right)\Big]_0^2 = \frac{16}{3}$

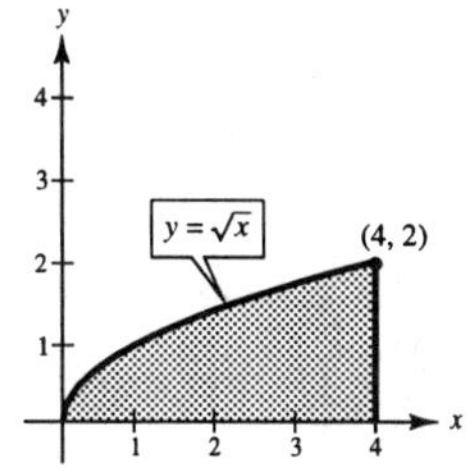

30. $\displaystyle\int_0^4\int_{\sqrt{x}}^2 dy\,dx = \int_0^4 (2-\sqrt{x})\,dx = \left(2x - \frac{2}{3}x^{3/2}\right)\Big]_0^4 = \frac{8}{3}$

$\displaystyle\int_0^2\int_0^{y^2} dx\,dy = \int_0^2 y^2\,dy = \frac{y^3}{3}\Big]_0^2 = \frac{8}{3}$

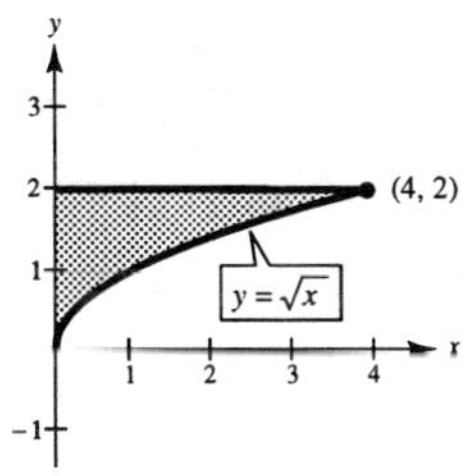

32. $\displaystyle\int_{-2}^2\int_0^{4-y^2} dx\,dy = \int_{-2}^2 (4-y^2)\,dy$

$\displaystyle= \left(4y - \frac{y^3}{3}\right)\Big]_{-2}^2 = \frac{32}{3}$

$\displaystyle\int_0^4\int_{-\sqrt{4-x}}^{\sqrt{4-x}} dy\,dx = \int_0^4 2\sqrt{4-x}\,dx$

$\displaystyle= -\frac{4}{3}(4-x)^{3/2}\Big]_0^4 = \frac{32}{3}$

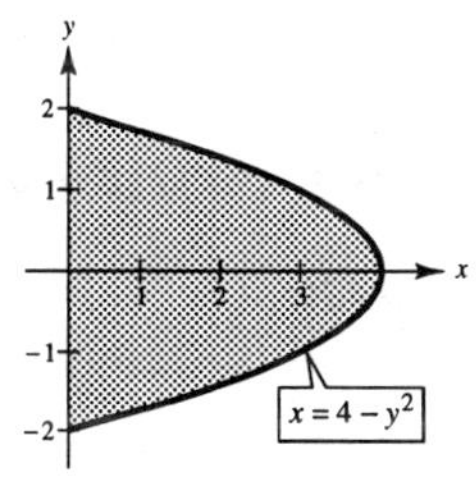

34. $\displaystyle\int_0^2\int_x^2 e^{-y^2}\,dy\,dx = \int_0^2\int_0^y e^{-y^2}\,dx\,dy$

$\displaystyle= \int_0^2 ye^{-y^2}dy$

$\displaystyle= -\frac{1}{2}e^{-y^2}\Big]_0^2$

$\displaystyle= \frac{1}{2}(1 - e^{-4}) \approx 0.491$

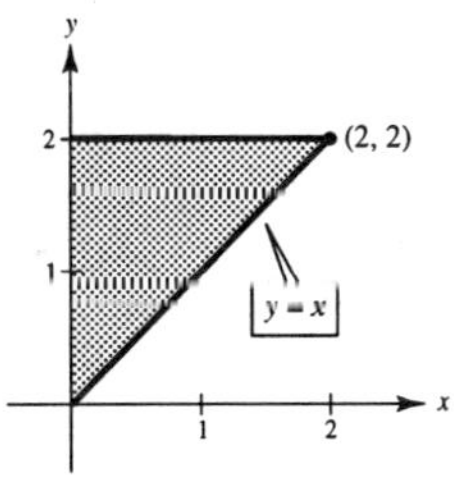

36. $\displaystyle A = \int_1^3\int_1^3 dy\,dx = \int_1^3 2\,dx = 2x\Big]_1^3 = 4$

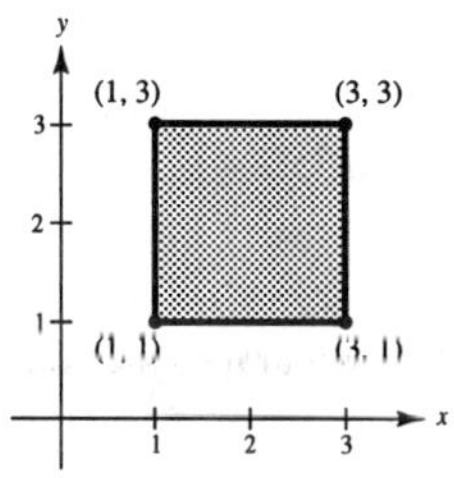

38. $A = \int_{-2}^{1}\int_{x+2}^{4-x^2} dy\,dx$

$= \int_{-2}^{1} [(4 - x^2) - (x + 2)]dx$

$= \int_{-2}^{1} (2 - x - x^2)\,dx$

$= \left[2x - \frac{x^2}{2} - \frac{x^3}{3}\right]_{-2}^{1}$

$= \left(2 - \frac{1}{2} - \frac{1}{3}\right) - \left(-4 - 2 + \frac{8}{3}\right)$

$= \frac{9}{2}$

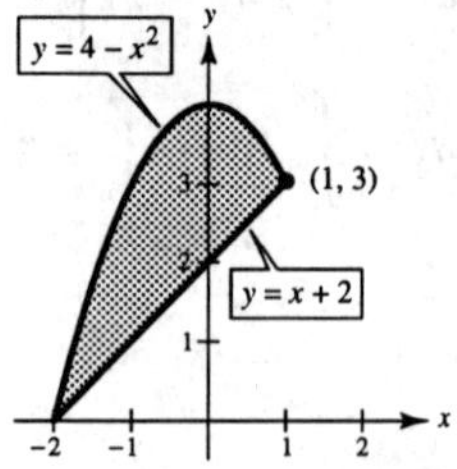

40. $A = \int_{2}^{5}\int_{0}^{1/\sqrt{x-1}} dy\,dx$

$= \int_{2}^{5} \frac{1}{\sqrt{x-1}}\,dx$

$= \left[2\sqrt{x-1}\right]_{2}^{5} = 2$

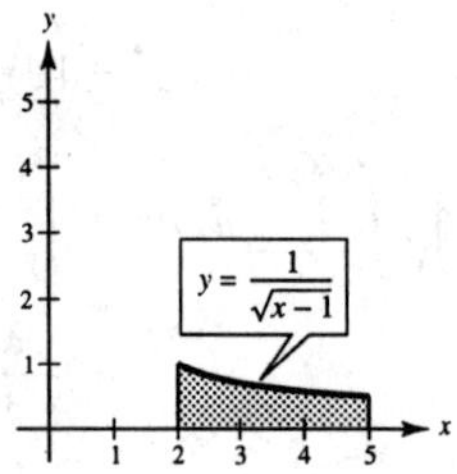

42. $A = \int_{0}^{1}\int_{x^{3/2}}^{x} dy\,dx$

$= \int_{0}^{1} (x - x^{3/2})\,dx$

$= \left[\frac{x^2}{2} - \frac{2}{5}x^{5/2}\right]_{0}^{1}$

$= \frac{1}{10}$

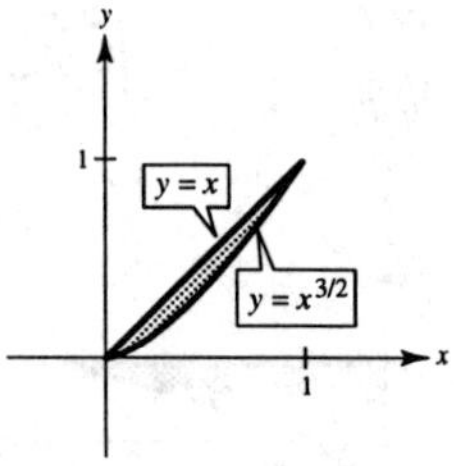

44. The point of intersection of the two graphs is given by equating $y = 9/x$ and $y = x$, which yields $x = y = 3$.

$A = \int_{0}^{3}\int_{0}^{x} dy\,dx + \int_{3}^{9}\int_{0}^{9/x} dy\,dx$

$= \int_{0}^{3} x\,dx + \int_{3}^{9} \frac{9}{x}\,dx$

$= \left[\frac{x^2}{2}\right]_{0}^{3} + \left[9\ln|x|\right]_{3}^{9}$

$= \frac{9}{2} + 9(\ln 9 - \ln 3)$

$= \frac{9}{2} + 9\ln 3 = \frac{9}{2}(1 + 2\ln 3)$

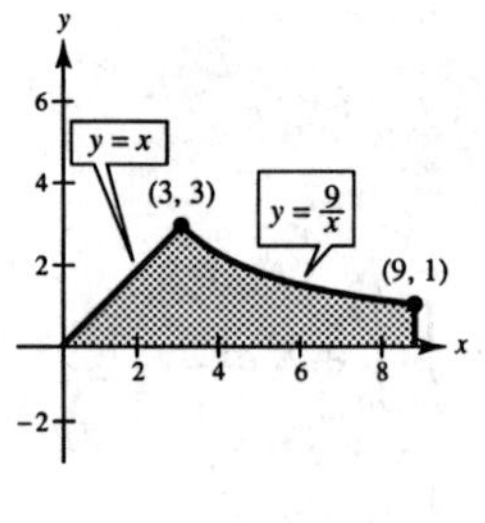

46. The points of intersection of the two graphs are found by equating $y = x^2 + 2x + 1$ and $y = 3(x + 1)$, which yields $(-1, 0)$ and $(2, 9)$.

$A = \int_{-1}^{2}\int_{x^2+2x+1}^{3x+3} dy\,dx$

$= \int_{-1}^{2} (-x^2 + x + 2)\,dx$

$= \left[-\frac{x^3}{3} + \frac{x^2}{2} + 2x\right]_{-1}^{2} = \frac{9}{2}$

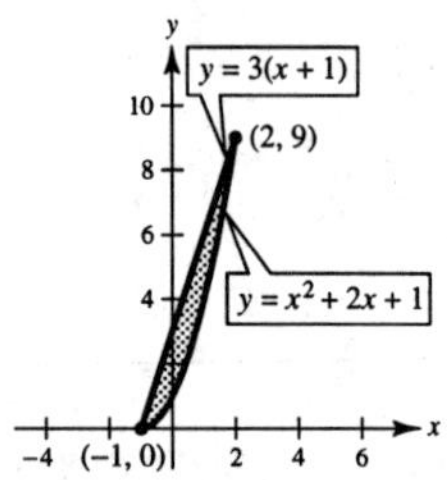

48. $\displaystyle\int_0^3\int_0^{x^2}\sqrt{x}\,\sqrt{1+x}\,dy\,dx \approx 24.3082$

50. $\displaystyle\int_1^2\int_y^{2y}\ln(x+y)\,dx\,dy = 2\ln 2 + \frac{3}{2}\ln\frac{27}{4} - \frac{9}{4}$

≈ 2.0006

52. $\displaystyle\int_0^4\int_0^y \frac{2}{(x+1)(y+1)}\,dx\,dy = (\ln 5)^2 = 2.5903$

54. True, the region is a rectangle.

Section 7.9 Applications of Double Integrals

2. $\displaystyle\int_{-a}^{a}\int_{-\sqrt{a^2-x^2}}^{\sqrt{a^2-x^2}} dy\,dx = \int_{-a}^{a}\Big[y\Big]_{-\sqrt{a^2-x^2}}^{\sqrt{a^2-x^2}} dx$

$\displaystyle= \int_{-a}^{a} 2\sqrt{a^2-x^2}\,dx$

$= \pi a^2$ (area of circle)

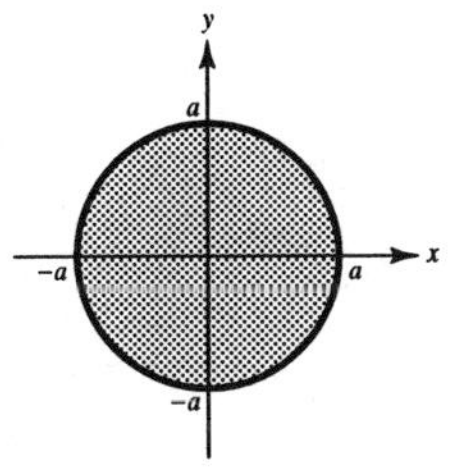

4. $\displaystyle\int_0^6\int_{y/2}^{3}(x+y)\,dx\,dy = \int_0^6\left[\frac{x^2}{2}+xy\right]_{y/2}^{3} dy$

$\displaystyle= \int_0^6\left(\frac{9}{2}+3y-\frac{5y^2}{8}\right)dy$

$\displaystyle= \left[\frac{9}{2}y+\frac{3y^2}{2}-\frac{5y^3}{24}\right]_0^6 = 36$

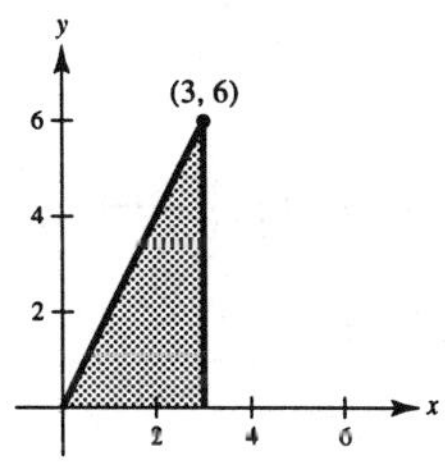

6. $\displaystyle\int_0^2\int_0^{4-x^2} xy^2\,dy\,dx = \int_0^2\left[\frac{xy^3}{3}\right]_0^{4-x^2} dx$

$\displaystyle= \int_0^2 \frac{x}{3}(4-x^2)^3\,dx$

$\displaystyle= -\frac{1}{24}(4-x^2)^4\Big]_0^2 = \frac{32}{3}$

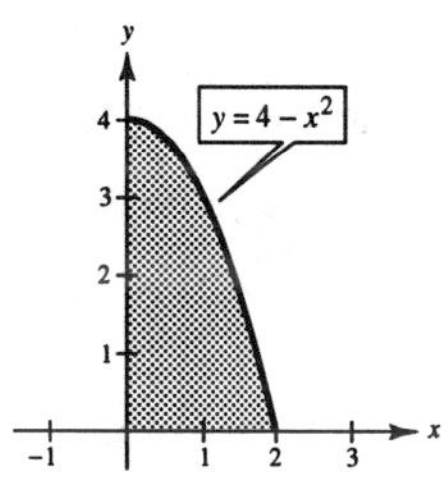

8. $\displaystyle\int_0^2\int_x^{2x}\frac{y}{x^2+y^2}\,dy\,dx = \int_0^2\int_{y/2}^{y}\frac{y}{x^2+y^2}\,dx\,dy + \int_2^4\int_{y/2}^{2}\frac{y}{x^2+y^2}\,dx\,dy$

$\displaystyle\int_0^2\int_x^{2x}\frac{y}{x^2+y^2}\,dy\,dx = \int_0^2\left[\frac{1}{2}\ln(x^2+y^2)\right]_x^{2x} dx$

$\displaystyle= \frac{1}{2}\int_0^2[\ln(5x^2)-\ln(2x^2)]\,dx = \frac{1}{2}\int_0^2 \ln\frac{5}{2}\,dx = \left(\frac{1}{2}\ln\frac{5}{2}\right)x\Big]_0^2 = \ln\frac{5}{2}$

10. $\displaystyle\int_{-5}^{5}\int_0^{\sqrt{25-x^2}} x\,dy\,dx = \int_0^5\int_{-\sqrt{25-y^2}}^{\sqrt{25-y^2}} x\,dx\,dy$

$\displaystyle\int_0^5\int_{-\sqrt{25-y^2}}^{\sqrt{25-y^2}} x\,dx\,dy = \int_0^5\left[\frac{x^2}{2}\right]_{-\sqrt{25-y^2}}^{\sqrt{25-y^2}} dy = 0$

12. $\displaystyle\int_0^{\ln 10}\int_{e^x}^{10}\frac{1}{\ln y}\,dy\,dx = \int_1^{10}\int_0^{\ln y}\frac{1}{\ln y}\,dx\,dy$

$\displaystyle= \int_1^{10}\left[\frac{x}{\ln y}\right]_0^{\ln y} dy = \int_1^{10} dy = 9$

14. $V = \int_0^4 \int_0^2 (6 - 2y)\,dy\,dx$

$= \int_0^4 \Big[6y - y^2\Big]_0^2\,dx$

$= \int_0^4 8\,dx = 32$

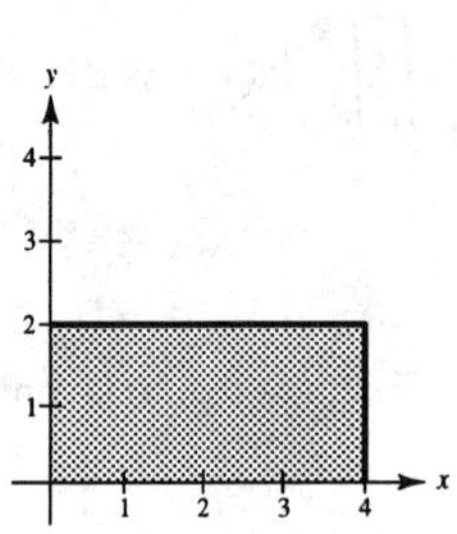

16. $V = \int_0^2 \int_0^x 6\,dy\,dx$

$= \int_0^2 6x\,dx$

$= 3x^2\Big]_0^2 = 12$

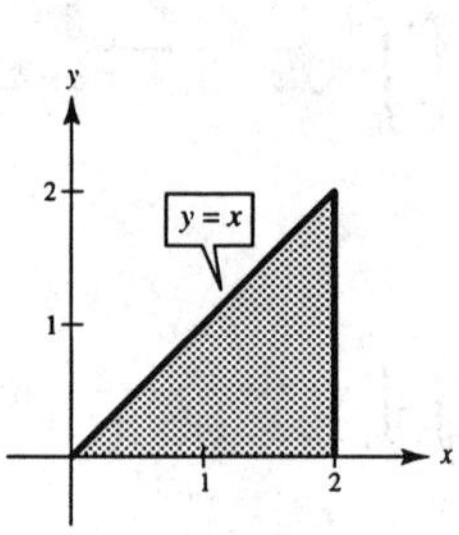

18. $V = \int_0^1 \int_0^{1-x} (1 - x - y)\,dy\,dx$

$= \int_0^1 \left[(1 - x)y - \frac{y^2}{2}\right]_0^{1-x} dx$

$= \int_0^1 \frac{1}{2}(1 - x)^2\,dx$

$= -\frac{1}{6}(1 - x)^3\Big]_0^1 = \frac{1}{6}$

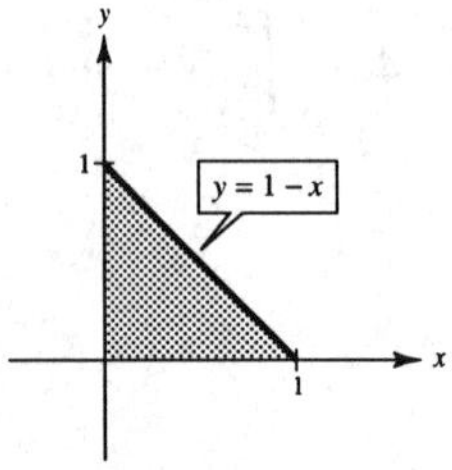

20. $V = \int_0^2 \int_0^y (4 - y^2)\,dx\,dy$

$= \int_0^2 y(4 - y^2)\,dy$

$= \left[2y^2 - \frac{y^4}{4}\right]_0^2 = 4$

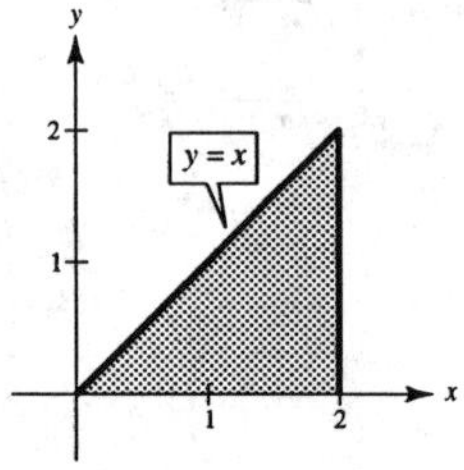

22. $\int_0^\infty \int_0^\infty \frac{1}{(x + 1)^2(y + 1)^2}\,dx\,dy = \int_0^\infty \lim_{b\infty} \left[\frac{-1}{(x + 1)(y + 1)^2}\right]_0^b dy$

$= \int_0^\infty \frac{1}{(y + 1)^2}\,dy$

$= \lim_{b\to\infty} \left[-\frac{1}{y + 1}\right]_0^b = 1$

24. $V = \int_0^\infty \int_0^\infty e^{-(x+y)/2}\,dy\,dx$

$= \int_0^\infty \lim_{b\to\infty} \Big[-2e^{-(x+y)/2}\Big]_0^b dx$

$= \int_0^\infty 2e^{-x/2}\,dx$

$= \lim_{b\to\infty} \Big[-4e^{-x/2}\Big]_0^b = 4$

26. $V = \int_0^4 \int_0^x x\,dy\,dx = \int_0^4 x^2\,dx = \frac{x^3}{3}\Big]_0^4 = \frac{64}{3}$

28. $V = \int_0^2 \int_0^{\sqrt{4-x^2}} (x + y)\,dy\,dx$

$= \int_0^2 \left[xy + \frac{y^2}{2}\right]_0^{\sqrt{4-x^2}} dx = \int_0^2 \left[x\sqrt{4 - x^2} + \frac{1}{2}(4 - x^2)\right] dx = \left[-\frac{1}{3}(4 - x^2)^{3/2} + 2x - \frac{x^3}{6}\right]_0^2 = \frac{16}{3}$

30. Average $= \frac{1}{8}\int_0^4 \int_0^2 xy\,dy\,dx = \frac{1}{8}\int_0^4 \left[\frac{xy^2}{2}\right]_0^2 dx = \frac{1}{8}\int_0^4 2x\,dx = \frac{x^2}{8}\Big]_0^4 = 2$

32. Average $= 2\int_0^1 \int_0^y e^{x+y}\,dx\,dy$

$= 2\int_0^1 \Big[e^{x+y}\Big]_0^y dy = 2\int_0^1 (e^{2y} - e^y)\,dy = 2\left[\frac{1}{2}e^{2y} - e^y\right]_0^1 = \Big[e^{2y} - 2e^y\Big]_0^1 = e^2 - 2e + 1 = (e - 1)^2$

34. $\text{Average} = \dfrac{1}{50}\displaystyle\int_{45}^{50}\int_{40}^{50}(192x_1 + 576x_2 - x_1^2 - 5x_2^2 - 2x_1x_2 - 5000)\,dx_1\,dx_2$

$= \dfrac{1}{50}\displaystyle\int_{45}^{50}\left[96x_1^2 + 576x_1x_2 - \frac{x_1^3}{3} - 5x_1x_2^2 - x_1^2x_2 - 5000x_1\right]_{40}^{50} dx_2$

$= \dfrac{1}{50}\displaystyle\int_{45}^{50}\left(\frac{482{,}000}{3} + 4860x_2 - 50x_2^2\right)dx_2$

$= \dfrac{1}{50}\left[\dfrac{482{,}000x}{3} + 2430x_2^2 - \dfrac{50x_2^3}{3}\right]_{45}^{50} = \$13{,}400$

Review Exercises for Chapter 7

2.

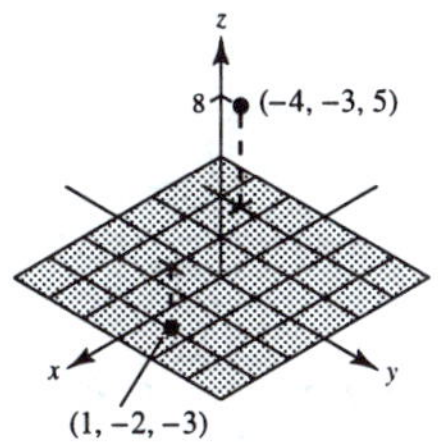

4. $d = \sqrt{[1-(-4)]^2 + (3-1)^2 + (7-5)^2}$

$= \sqrt{25 + 4 + 4}$

$= \sqrt{33}$

6. $\text{Midpoint} = \left(\dfrac{5-1}{2}, \dfrac{0-2}{2}, \dfrac{7+9}{2}\right) = (2, -1, 8)$

8. $(x-4)^2 + [y-(-5)]^2 + (z-3)^2 = 10^2$

$(x-4)^2 + (y+5)^2 + (z-3)^2 = 100$

10. $\text{Center} = \left(\dfrac{-2+4}{2}, \dfrac{5-3}{2}, \dfrac{1+3}{2}\right) = (1, 1, 2)$

$\text{Radius} = \sqrt{(1-4)^2 + [1-(-3)]^2 + (2-3)^2} = \sqrt{9 + 16 + 1} = \sqrt{26}$

$\text{Circle} = (x-1)^2 + (y-1)^2 + (z-2)^2 = 26$

12. $x^2 + y^2x + 4y + 4 + z^2 - 10z + 25 = 7 + 4 + 25$

$x^2 + (y+2)^2 + (z-5)^2 = 36 = 6^2$

Center: $(0, -2, 5)$ Radius: 6

14. Let $z = 0$: $(x-1)^2 + (y+3)^2 + (0-6)^2 = 72$

$(x-1)^2 + (y+3)^2 = 36$

$= 6^2$

Circle of radius 6

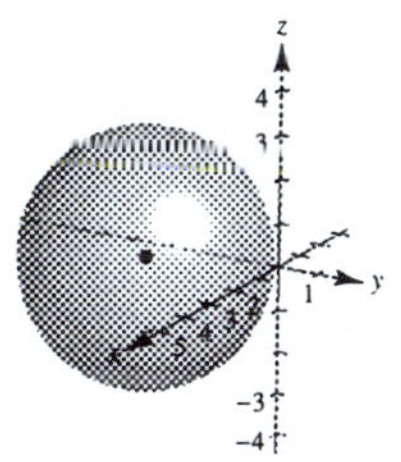

16. y-intercept: $(0, 2, 0)$

z-intercept: $(0, 0, 4)$

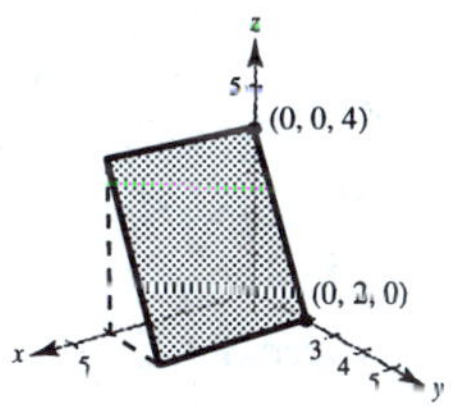

18. x-intercept: $(2, 0, 0)$

y-intercept: $(0, -8, 0)$

z-intercept: $(0, 0, 4)$

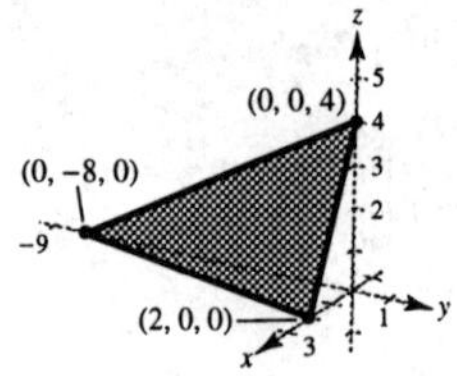

20. The graph is an elliptic cone whose standard equation is

$$\frac{x^2}{1/16} + \frac{y^2}{1/16} - \frac{z^2}{1/9} = 0.$$

22. The graph is a hyperboloid of one sheet.

24. The graph is a hyperboloid of one sheet whose standard equation is

$$\frac{y^2}{4} + \frac{z^2}{4} - x^2 = 1.$$

26. The graph is the plane $9x + 3y - z = 5$.

28. $f(x, y) = \dfrac{x^2}{y}$

(a) $f(6, 9) = \dfrac{6^2}{9} = 4$

(b) $f(8, 4) = \dfrac{8^2}{4} = 16$

(c) $f(t, 2) = \dfrac{t^2}{2}$

(d) $f(r, r) = \dfrac{r^2}{r} = r, \quad r \neq 0$

30. The domain is the set of all points except those on the line $y = -x$. The range is all real numbers except 0.

32. $z = \sqrt{9 - x^2 - y^2}$

$C = 0$: $\sqrt{9 - x^2 - y^2} = 0, \quad x^2 + y^2 = 9$

$C = 1$: $\sqrt{9 - x^2 - y^2} = 1, \quad x^2 + y^2 = 8$

$C = 2$: $\sqrt{9 - x^2 - y^2} = 2, \quad x^2 + y^2 = 5$

$C = 3$: $\sqrt{9 - x^2 - y^2} = 3, \quad x^2 + y^2 = 0$ (point)

The level curves are circles (except $C = 3$).

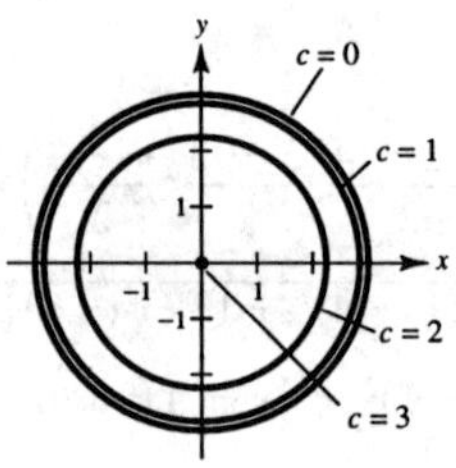

34. $z = 2e^{xy}$

$C = 1$: $2e^{xy} = 1, \quad xy = \ln\frac{1}{2}, \quad y = \dfrac{-\ln 2}{x}$

$C = 2$: $2e^{xy} = 2, \quad xy = \ln 1 = 0,$ axes

$C = 3$: $2e^{xy} = 3, \quad xy = \ln\left(\frac{3}{2}\right), \quad y = \dfrac{\ln(3/2)}{x}$

$C = 4$: $2e^{xy} = 4, \quad y = \dfrac{\ln 2}{x}$

$C = 5$: $2e^{xy} = 5, \quad y = \dfrac{\ln(5/2)}{x}$

The level curves are hyperbolas, except for $C = 2$, which consists of the x- and y- axes.

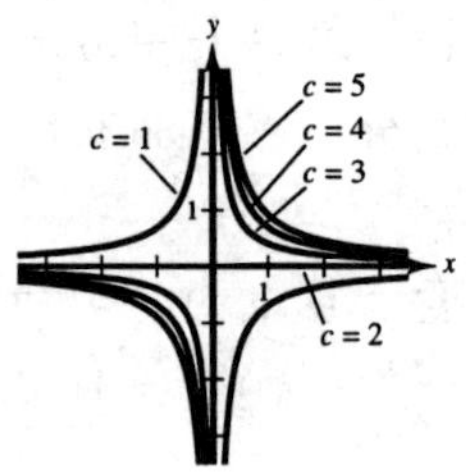

36. (a) The level curves represent lines of equal population density, and separate the four colors.

(b) No

(c) You could use more colors to represent more population densities.

38. (a)

Year	1991	1992	1993	1994	1995	1996
Model	5.4	7.0	8.8	10.7	13.0	14.7

(b) Because $\frac{\partial z}{\partial y} > \frac{\partial z}{\partial x}$, y has more influence.

(c) $f(x, 25) = 0.083x + 4.89$

For fixed assets $y = 25$ billion, the formula gives the shareholders equity as a function of net sales x.

40. $F = \frac{v^2}{gl} = \frac{(2)^2}{(3)(0.75)} \approx 1.778$ seconds

42. $f_x(x, y) = 10x + 4y$

$f_y(x, y) = 4x - 6y$

44. $z = (xy + 2x + 4y)^2$

$z_x = 2(xy + 2x + 4y)(y + 2)$

$z_y = 2(xy + 2x + 4y)(x + 4)$

46. $f_x(x, y) = \frac{1}{2}\left(\frac{2}{2x + 3y}\right) = \frac{1}{2x + 3y}$

$f_y(x, y) = \frac{1}{2}\left(\frac{3}{2x + 3y}\right) = \frac{3}{2(2x + 3y)}$

48. $f_x(x, y) = 2xe^{-2y}$

$f_y(x, y) = -2x^2e^{-2y}$

50. $w = xyz + 2xy - 9xz + 4yz - y^2z + 4z^2$

$w_x = yz + 2y - 9z$

$w_y = xz + 2x + 4z - 2yz$

$w_z = xy - 9x + 4y - y^2 + 8z$

52. $z = 4x^2 - y^2, \quad (2, 4, 0)$

(a) $z_x = 8x$; At $(2, 4, 0)$, $z_x = 16$.

(b) $z_y = -2y$; At $(2, 4, 0)$, $z_y = -8$.

54. $z = x^2 - y^2, \quad (5, -4, 9)$

(a) $z_x = 2x$; At $(5, -4, 9)$, $z_x = 10$.

(b) $z_y = -2y$; At $(5, -4, 9)$, $z_y = 8$.

56. $f(x, y) = \frac{y}{x + y}$

$f_x = \frac{-y}{(x + y)^2} \qquad f_y = \frac{x}{(x + y)^2}$

$f_{xx} = \frac{2y}{(x + y)^3} \qquad f_{yy} = \frac{-2x}{(x + y)^3} \qquad f_{xy} = f_{yx} = \frac{y - x}{(x + y)^3}$

58. $f(x, y) = x^2e^{-y^2}$

$f_x = 2xe^{-y^2} \qquad f_y = -2x^2ye^{-y^2}$

$f_{xx} = 2e^{-y^2} \qquad f_{yy} = 2x^2(2y^2 - 1)e^{-y^2}$

$f_{xy} = f_{yx} = -4xye^{-y^2}$

60. $R_{x_1} = 15 - \frac{1}{5}x_1 - \frac{1}{100}x_2 \qquad R_{x_1}(50, 40) = 4.6$

$R_{x_2} = 16 - \frac{1}{5}x_2 - \frac{1}{100}x_1 \qquad R_{x_2}(50, 40) = 7.5$

62. $D(x, y) = x^2 + 2y^2 - 18x - 24y + 2xy + 120$

$D(5, 2.5) = 32.5$

$D(7.5, 8) = 101.84$

64. The first partial derivatives of f, $f_x(x, y) = 3x^2 - 3y$ and $f_y(x, y) = -3x + 2y$, are zero at $(0, 0)$ and $\left(\frac{3}{2}, \frac{9}{4}\right)$. The second partial derivatives are

$$f_{xx}(x, y) = 6x$$
$$f_{yy}(x, y) = 2$$
$$f_{xy}(x, y) = -3.$$

The point $(0, 0, 0)$ is a saddle point since $f_{xx}(0, 0)f_{yy}(0, 0) - [f_{xy}(0, 0)]^2 = -9 < 0$. The point $\left(\frac{3}{2}, \frac{9}{4}, -\frac{27}{16}\right)$ is a relative minimum since

$$f_{xx}\left(\tfrac{3}{2}, \tfrac{9}{4}\right) = 9 > 0 \quad \text{and} \quad f_{xx}\left(\tfrac{3}{2}, \tfrac{9}{4}\right)f_{yy}\left(\tfrac{3}{2}, \tfrac{9}{4}\right) - \left[f_{xy}\left(\tfrac{3}{2}, \tfrac{9}{4}\right)\right]^2 = 9 > 0.$$

66. The first partial derivatives of f, $f_x = 1 - e^x$ and $f_y = 2y$, are zero at $(0, 0)$. The second partials are

$$f_{xx} = -e^x, \quad f_{yy} = 2, \quad f_{xy} = 0.$$

Since $d = f_{xx}f_{yy} - (f_{xy})^2 < 0$, $(0, 0, -1)$ is a saddle point. No relative extrema.

68. The first partial derivatives of f, $f_x = y - 2$ and $f_y = 2y + x + 3$, are zero at $(-7, 2)$. The second partials are

$$f_{xx} = 0, \quad f_{yy} = 2, \quad f_{xy} = 1.$$

At $(-7, 2)$, $d < 0 \Rightarrow$ saddle point. No relative extrema.

70. The first partial derivatives of f, $f_x = -2x$ and $f_y = 2y$, are zero at $(0, 0)$. The second partials are

$$f_{xx} = -2, \quad f_{yy} = 2, \quad f_{xy} = 0.$$

Since $d < 0$, $(0, 0)$ is a saddle point. No relative extrema.

72. $P = 10(x_1 + x_2) - C_1 - C_2$

$$= 10(x_1 + x_2) - \left(0.03x_1^2 + 4x_1 + 300\right) - \left(0.05x_2^2 + 7x_2 + 175\right)$$

$$P_{x_1} = 6 - 0.06x_1 = 0 \Rightarrow x_1 = 100$$

$$P_{x_2} = 3 - 0.01x_2 = 0 \Rightarrow x_2 = 300$$

By the Second-Partials Test, $x_1 = 100$, $x_2 = 300$ is a maximum.

74. $F(x, y, \lambda) = x^2 + y^2 - \lambda(x + y - 4)$

$$\left.\begin{aligned} F_x(x, y, \lambda) &= 2x - \lambda = 0 \\ F_y(x, y, \lambda) &= 2y - \lambda = 0 \end{aligned}\right\} x = y$$

$$F_\lambda(x, y, \lambda) = -x - y + 4 = 0, \qquad x = y = 2$$

Thus, $x = y = 2$. This implies that the extremum occurs at $(2, 2, 8)$.

[**Note:** This is a relative minimum.]

76. $F(x, y, z, \lambda) = xz + yz - \lambda(x + y + z - 6)$

$$F_x = z - \lambda = 0$$

$$F_y = z - \lambda = 0$$

$$F_z = x + y - \lambda = 0$$

$$F_\lambda = -(x + y + z - 6) = 0$$

$$x + y = \lambda, \quad z = \lambda: \quad \lambda + \lambda - 6 = 0 \Rightarrow \lambda = 3, z = 3, x + y = 3$$

The extrema are at $(x, 3 - x, 3)$ for any x.

78. $F(x, y, z) = xyz - \lambda(x + y + z - 32) - \mu(x - y + z)$

$$\left.\begin{aligned} F_x &= yz - \lambda - \mu = 0 \\ F_y &= xz - \lambda - \mu = 0 \\ F_z &= xy - \lambda - \mu = 0 \end{aligned}\right\} \; xy = yz \Rightarrow x = z$$

$$\left.\begin{aligned} F_\lambda &= -(x + y + z - 32) = 0 \\ F_\mu &= -(x - y + z) = 0 \end{aligned}\right\} \; y = 16$$

$f(8, 16, 8) = 1024$, relative maximum

80. Minimize $C = 0.20x_1^2 + 10x_1 + 0.15x_2^2 + 12x_2$ subject to the constraint $x_1 + x_2 = 1500$.

$F(x_1, x_2, \lambda) = 0.20x_1^2 + 10x_1 + 0.15x_2^2 + 12x_2 - \lambda(x_1 + x_2 - 1500)$

$$\left.\begin{aligned} F_{x_1}(x_1, x_2, \lambda) &= 0.40x_1 + 10 - \lambda = 0 \\ F_{x_2}(x_1, x_2, \lambda) &= 0.30x_2 + 12 - \lambda = 0 \end{aligned}\right\} \; 4x_1 - 3x_2 = 20$$

$F_\lambda(x_1, x_2, \lambda) = -x_1 - x_2 + 1500 = 0, \qquad x_2 = 1500 - x_1$

Thus, $x_1 \approx 646$ units and $x_2 \approx 854$ units.

82. (a) $\sum x_i = -2$

$\sum y_i = 0$

$\sum x_i^2 = 18$

$\sum x_i y_i = 8$

$y = \dfrac{20}{43}x + \dfrac{8}{43}$

$a = \dfrac{5(8) - (-2)(0)}{5(18) - (-2)^2} = \dfrac{20}{43}$

$b = \dfrac{1}{5}\left(0 - \dfrac{20}{43}(-2)\right) = \dfrac{8}{43}$

(b) $\left(\frac{20}{43}(-3) + \frac{8}{43} + 1\right)^2 + \left(\frac{20}{43}(-2) + \frac{8}{43} + 1\right)^2 + \left(\frac{20}{43}(0) + \frac{8}{43} - 0\right)^2 + \left(\frac{20}{43}(1) + \frac{8}{43} - 1\right)^2 + \left(\frac{20}{43}(2) + \frac{8}{43} - 1\right)^2$

≈ 0.279

84. (a) $y = 1.76x - 44.03$

(b) For each 1 point increase, about 1.76 million women enter the workforce.

86. $\sum x_i = 14$

$\sum y_i = 30$

$\sum x_i^2 = 54$

$\sum x_i^3 = 224$

$\sum x_i^4 = 978$

$\sum x_i y_i = 55$

$\sum x_i^2 y_i = 163$

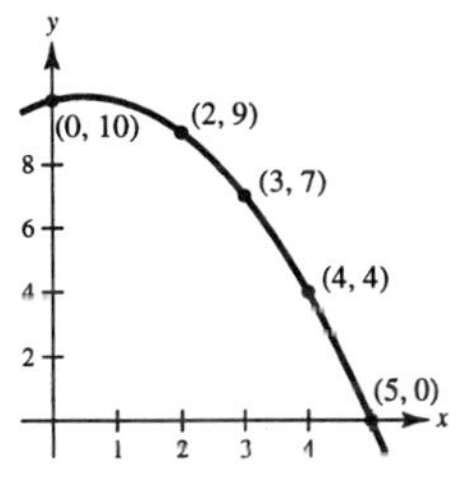

The system

$$\begin{aligned} 978a + 224b + 54c &= 163 \\ 224a + 54b + 14c &= 55 \\ 54a + 14b + 5c &= 30 \end{aligned}$$

has solutions $a = -\frac{1}{2}$, $b = \frac{1}{2}$, and $c = 10$. Therefore, the least squares quadratic is $y = -\frac{1}{2}x^2 + \frac{1}{2}x + 10$.

88. $$\int_{-2}^{2}\int_{0}^{4}(x - y^2)\,dx\,dy = \int_{-2}^{2}\left(\frac{x^2}{2} - xy^2\right)\bigg]_0^4 dy$$
$$= \int_{-2}^{2}(8 - 4y^2)\,dy$$
$$= 8y - \frac{4}{3}y^3\bigg]_{-2}^{2} = \frac{32}{3}$$

90. $$\int_{0}^{4}\int_{0}^{\sqrt{16-x^2}} 2x\,dy\,dx = \int_{0}^{4} 2xy\bigg]_0^{\sqrt{16-x^2}} dx$$
$$= \int_{0}^{4} 2x\sqrt{16 - x^2}\,dx$$
$$= \frac{2}{3}(16 - x^2)^{3/2}\bigg]_0^4 = \frac{128}{3}$$

92. $$A = \int_{1}^{4}\int_{0}^{4/x} dy\,dx = \int_{0}^{1}\int_{1}^{4} dx\,dy + \int_{1}^{4}\int_{1}^{4/y} dx\,dy$$
$$\int_{1}^{4}\int_{0}^{4/x} dy\,dx = \int_{1}^{4}\frac{4}{x}\,dx = 4\ln|x|\bigg]_1^4 = 4\ln 4$$

94. $$A = \int_{-1}^{2}\int_{x^2-2x-2}^{-x} dy\,dx = \int_{-1}^{2}[-x - (x^2 - 2x - 2)]\,dx$$
$$= \left[\frac{-x^3}{3} + \frac{x^2}{2} + 2x\right]_{-1}^{2} = \left(\frac{-8}{3} + 2 + 4\right) - \left(\frac{1}{3} + \frac{1}{2} - 2\right) = \frac{9}{2}$$

96. $$V = \int_{0}^{3}\int_{0}^{x}(x + y)\,dy\,dx = \int_{0}^{3}\left[xy + \frac{y^2}{2}\right]_0^x dx = \int_{0}^{3}\frac{3x^2}{2}\,dx = \frac{x^3}{2}\bigg]_0^3 = \frac{27}{2}$$

98. $$\text{Average} = \frac{\int_{0}^{5280}\int_{0}^{3960} 2.5x^{3/2}y^{3/4}\,dy\,dx}{\text{area}}$$
$$\approx 1.094 \times 10^8 \text{ dollars per square foot}$$

CHAPTER 8
Trigonometric Functions

CHAPTER 8
Trigonometric Functions

Section 8.1 Radian Measure of Angles

2. (a) Positive: $-120° + 360° = 240°$

Negative: $-120° - 360° = -480°$

(b) Positive: $390° - 360° = 30°$

Negative: $390° - 2(360°) = -330°$

4. (a) Positive: $-420° + 2(360°) = 300°$

Negative: $-420° + 360° = -60°$

(b) Positive: $230° + 360° = 590°$

Negative: $230° - 360° = 130°$

6. (a) Positive: $(11\pi/6) + 2\pi = 23\pi/6$

Negative: $(11\pi/6) - 2\pi = -\pi/6$

(b) Positive: $-(7\pi/6) + 2\pi = 5\pi/6$

Negative: $-(7\pi/6) - 2\pi = -19\pi/6$

8. (a) Positive: $(8\pi/9) + 2\pi = 26\pi/9$

Negative: $(8\pi/9) - 2\pi = -10\pi/9$

(b) Positive: $(8\pi/45) + 2\pi = 98\pi/45$

Negative: $(8\pi/45) - 2\pi = -82\pi/45$

10. $150°\left(\frac{\pi \text{ radians}}{180°}\right) = \frac{5\pi}{6}$ radians

12. $120°\left(\frac{\pi \text{ radians}}{180°}\right) = \frac{2\pi}{3}$ radians

14. $-240°\left(\frac{\pi \text{ radians}}{180°}\right) = -\frac{4\pi}{3}$ radians

16. $\frac{7\pi}{6}\left(\frac{180°}{\pi}\right) = 210°$

18. $\frac{-7\pi}{4}\left(\frac{180°}{\pi}\right) = -315°$

20. $\frac{43\pi}{60}\left(\frac{180°}{\pi}\right) = 129°$

22. $\frac{\pi}{9}\left(\frac{180°}{\pi}\right) = 20°$

24. $-\frac{7\pi}{12}\left(\frac{180°}{\pi}\right) = -105°$

26. The angle is $\theta = 90° - 45° = 45°$.

Since the triangle is an isosceles right triangle, we have the following.

$$a^2 + a^2 = 288^2$$
$$2a^2 = 288^2$$
$$a^2 = 41{,}472$$
$$a = 144\sqrt{2}$$

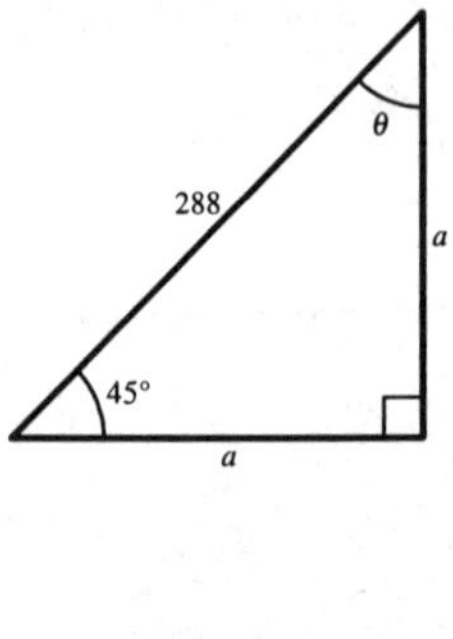

28. Since the triangle is isosceles, we have $\theta = 60°$. Thus, we note that the triangle is equilateral and

$$s = 2\left(\frac{4}{\sqrt{3}}\right) = \frac{8}{\sqrt{3}}.$$

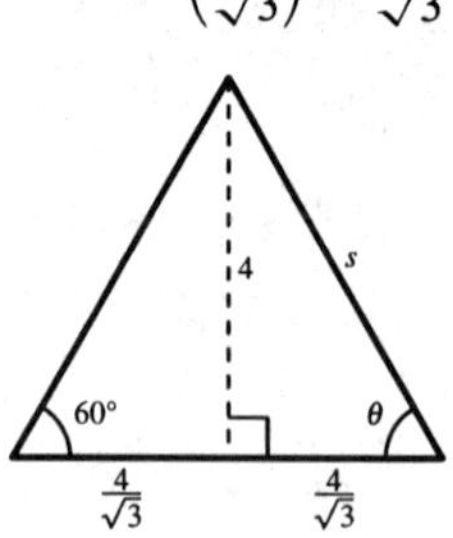

30. By similar triangles, we have $\frac{h}{1} = \frac{2}{3}$.

Thus, $h = \frac{2}{3}$.

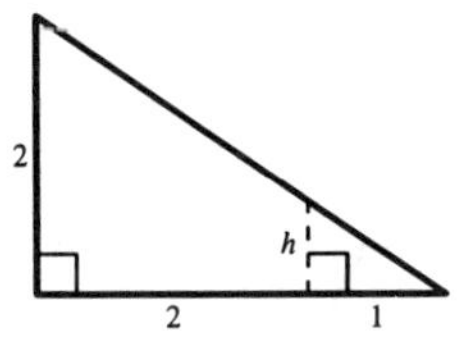

32. The triangle is equilateral and isosceles, and the equal angles are 60°. Hence, $a = 2.5$.

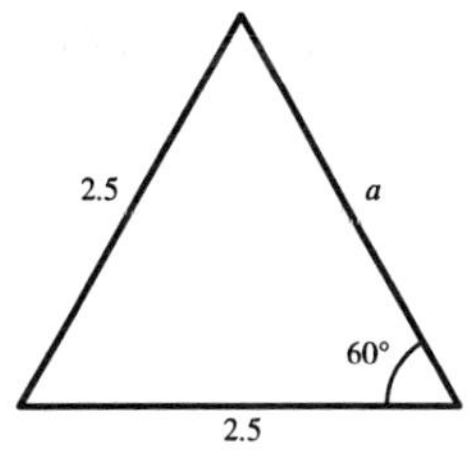

34.
$$h^2 + 4^2 = 8^2$$
$$h^2 = 48$$
$$h = 4\sqrt{3}$$
$$A = \frac{1}{2}bh = \frac{1}{2}(8)(4\sqrt{3}) = 16\sqrt{3}$$

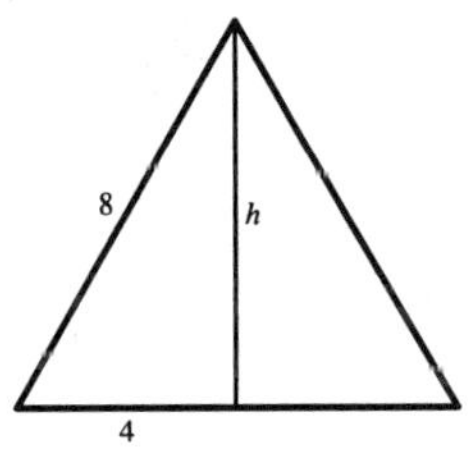

36.
$$h^2 + 6^2 = 12^2$$
$$h^2 = 108$$
$$h = 6\sqrt{3}$$
$$A = \frac{1}{2}bh = \frac{1}{2}(12)(6\sqrt{3}) = 36\sqrt{3}$$

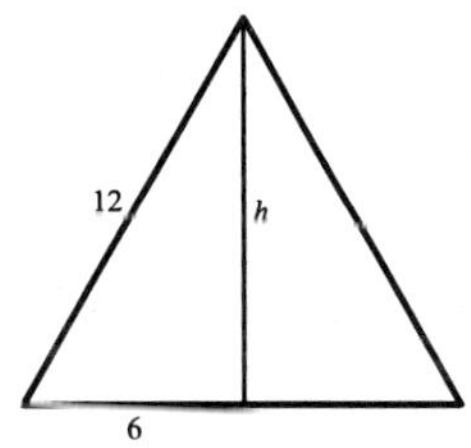

38. True, $-35° + 360° = 325°$

40. True, π radians $= 180°$

42. By the Pythagorean Theorem, we have

$$c^2 = 125^2 + 200^2$$
$$c = \sqrt{55{,}625} = 25\sqrt{89} \approx 235.85 \text{ feet.}$$

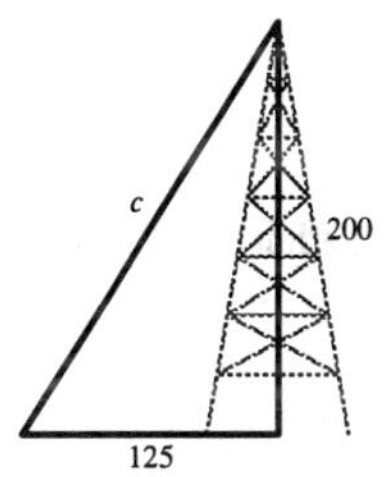

44. $s = r\theta = \frac{7}{2}\left(\frac{5\pi}{6}\right) = \frac{35\pi}{12} \approx 9.163$ inches

46. (a) $80°\left(\frac{\pi \text{ radians}}{180°}\right) = \frac{4\pi}{9}$ radians

(b) $s = r\theta = \left(\frac{5}{2}\right)\left(\frac{4\pi}{9}\right) = \frac{10\pi}{9} \approx 3.491$ feet

48. (360 rev/min)[$2\pi(1.68)$ in./rev]

≈ 3800.07 in./min

Section 8.2 The Trigonometric Functions

2. Since $x = 8$ and $y = -15$, it follows that $r = \sqrt{8^2 + (-15)^2} = 17$.

Therefore, we have the following.

$\sin\theta = -\frac{15}{17}$ $\quad\csc\theta = -\frac{17}{15}$

$\cos\theta = \frac{8}{17}$ $\quad\sec\theta = \frac{17}{8}$

$\tan\theta = -\frac{15}{8}$ $\quad\cot\theta = -\frac{8}{15}$

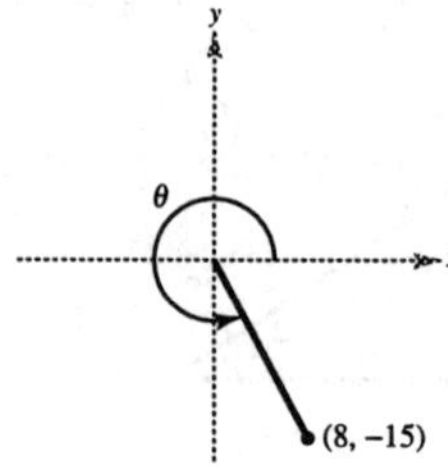

4. Since $x = 1$ and $y = -1$, it follows that $r = \sqrt{1^2 + (-1)^2} = \sqrt{2}$.

Therefore, we have the following.

$\sin\theta = -\frac{\sqrt{2}}{2}$ $\quad\csc\theta = -\sqrt{2}$

$\cos\theta = \frac{\sqrt{2}}{2}$ $\quad\sec\theta = \sqrt{2}$

$\tan\theta = -1$ $\quad\cot\theta = -1$

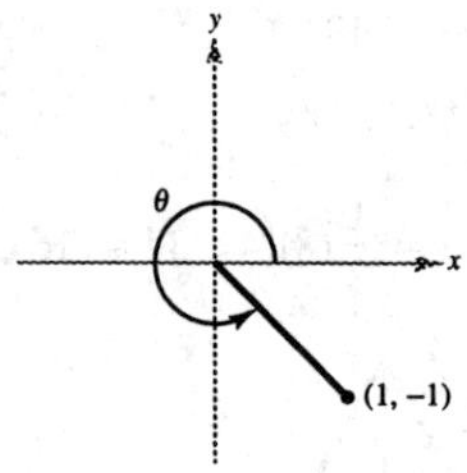

6. Since $x = -2$ and $y = -2$, it follows that $r = \sqrt{(-2)^2 + (-2)^2} = 2\sqrt{2}$.

Therefore, we have the following.

$\sin\theta = -\frac{\sqrt{2}}{2}$ $\quad\csc\theta = -\sqrt{2}$

$\cos\theta = -\frac{\sqrt{2}}{2}$ $\quad\sec\theta = -\sqrt{2}$

$\tan\theta = 1$ $\quad\cot\theta = 1$

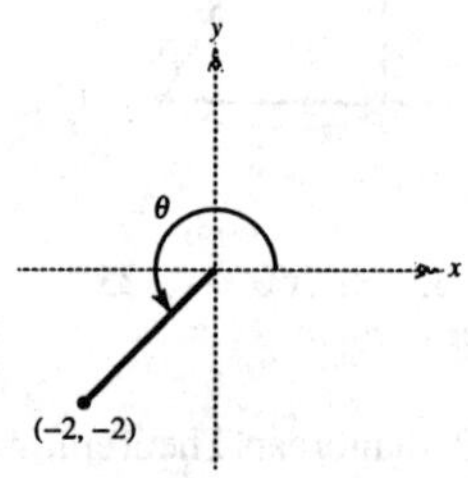

8. Since $y = 1$ and $r = 3$, the length of the adjacent side is $x = \sqrt{3^2 - 1^2} = 2\sqrt{2}$.

Therefore, we have the following.

$$\tan\theta = \frac{y}{x} = \frac{1}{2\sqrt{2}} = \frac{\sqrt{2}}{4}$$

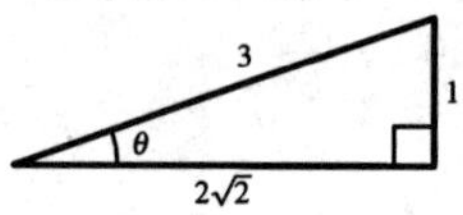

10. Since $x = 5$ and $r = 13$, the length of the opposite side is $y = \sqrt{13^2 - 5^2} = 12$.

Therefore, we have the following.

$$\cot\theta = \frac{x}{y} = \frac{5}{12}$$

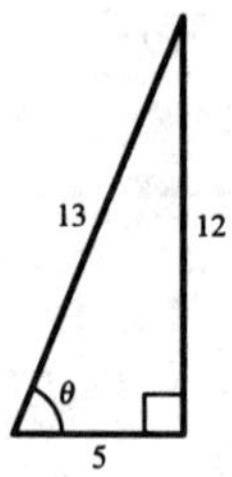

12. Since $x = 2$ and $y = 1$, the length of the hypotenuse is $r = \sqrt{2^2 + 1^2} = \sqrt{5}$.

Therefore, we have the following.

$$\sin\theta = \frac{y}{r} = \frac{1}{\sqrt{5}} = \frac{\sqrt{5}}{5}$$

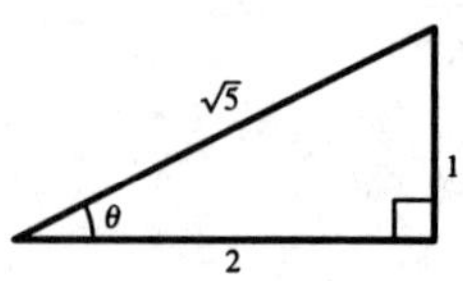

14. Since $x = 5$ and $y = 1$, the length of the hypotenuse is $r = \sqrt{5^2 + 1^2} = \sqrt{26}$.

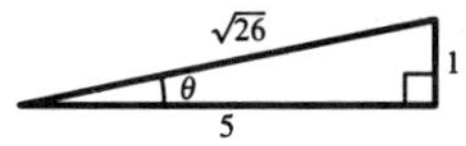

Therefore,

$$\sin\theta = \frac{\sqrt{26}}{26} \qquad \csc\theta = \sqrt{26}$$

$$\cos\theta = \frac{5\sqrt{26}}{26} \qquad \sec\theta = \frac{\sqrt{26}}{5}$$

$$\tan\theta = \frac{1}{5} \qquad \cot\theta = 5.$$

16. Since $x = 5$ and $r = 7$, the length of the opposite side is $y = \sqrt{7^2 - 5^2} = \sqrt{24} = 2\sqrt{6}$.

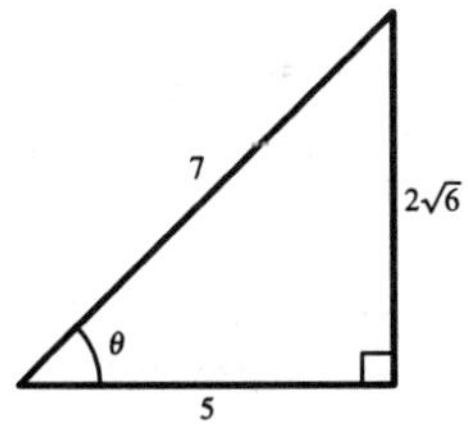

Therefore,

$$\sin\theta = \frac{2\sqrt{6}}{7} \qquad \csc\theta = \frac{7\sqrt{6}}{12}$$

$$\cos\theta = \frac{5}{7} \qquad \sec\theta = \frac{7}{5}$$

$$\tan\theta = \frac{2\sqrt{6}}{5} \qquad \cot\theta = \frac{5\sqrt{6}}{12}.$$

18. Since $\csc\theta = 4.25 = \frac{17}{4}$, we have $r = 17$ and $y = 4$. The length of the adjacent side is $x = \sqrt{17^2 - 4^2} = \sqrt{273}$.

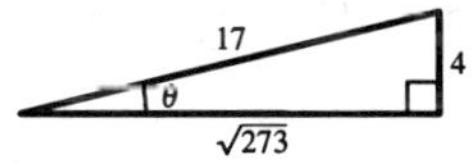

Therefore,

$$\sin\theta = \frac{4}{17} \qquad \csc\theta = \frac{17}{4}$$

$$\cos\theta = \frac{\sqrt{273}}{17} \qquad \sec\theta = \frac{17\sqrt{273}}{273}$$

$$\tan\theta = \frac{4\sqrt{273}}{273} \qquad \cot\theta = \frac{\sqrt{273}}{4}.$$

20. Since the sine is positive and the cosine is negative, θ must lie in Quadrant II.

22. Since the cotangent is negative and the cosine is positive, θ must lie in Quadrant IV.

24. Since the cosine is positive and the tangent is negative, θ must lie in Quadrant IV.

26. (a) $\sin\left(\frac{\pi}{4}\right) = \frac{\sqrt{2}}{2}$

$\cos\left(\frac{\pi}{4}\right) = \frac{\sqrt{2}}{2}$

$\tan\left(\frac{\pi}{4}\right) = 1$

(b) $\sin\left(\frac{5\pi}{4}\right) = -\frac{\sqrt{2}}{2}$

$\cos\left(\frac{5\pi}{4}\right) = -\frac{\sqrt{2}}{2}$

$\tan\left(\frac{5\pi}{4}\right) = 1$

28. (a) $\sin\left(-\frac{\pi}{2}\right) = -1$

$\cos\left(-\frac{\pi}{2}\right) = 0$

$\tan\left(-\frac{\pi}{2}\right)$ is undefined.

(b) $\sin\left(\frac{\pi}{2}\right) = 1$

$\cos\left(\frac{\pi}{2}\right) = 0$

$\tan\left(\frac{\pi}{2}\right)$ is undefined.

30. (a) $\sin 300° = -\frac{\sqrt{3}}{2}$

$\cos 300° = \frac{1}{2}$

$\tan 300° = -\sqrt{3}$

(b) $\sin 330° = -\frac{1}{2}$

$\cos 330° = \frac{\sqrt{3}}{2}$

$\tan 330° = -\frac{\sqrt{3}}{3}$

32. (a) $\sin\left(\frac{10\pi}{3}\right) = -\frac{\sqrt{3}}{2}$

$\cos\left(\frac{10\pi}{3}\right) = -\frac{1}{2}$

$\tan\left(\frac{10\pi}{3}\right) = \sqrt{3}$

(b) $\sin\left(\frac{17\pi}{3}\right) = -\frac{\sqrt{3}}{2}$

$\cos\left(\frac{17\pi}{3}\right) = \frac{1}{2}$

$\tan\left(\frac{17\pi}{3}\right) = -\sqrt{3}$

34. (a) $\sec 225° = \frac{1}{\cos 225°} \approx -1.4142$

(b) $\sec 135° = \frac{1}{\cos 135°} \approx -1.4142$

36. (a) $\cot 1.35 = \frac{1}{\tan 1.35} \approx 0.2245$

(b) $\tan 1.35 \approx 4.4552$

38. (a) $\tan 240° \approx 1.7321$

(b) $\cot 210° = \frac{1}{\tan 210°} \approx 1.7321$

40. (a) $\sin(-0.65) \approx -0.6052$

(b) $\sin 5.63 \approx -0.6077$

42. (a) $\theta = \frac{\pi}{4}$ or $\theta = \frac{7\pi}{4}$

(b) $\theta = \frac{3\pi}{4}$ or $\theta = \frac{5\pi}{4}$

44. (a) $\theta = \frac{\pi}{3}$ or $\theta = \frac{5\pi}{3}$

(b) $\theta = \frac{2\pi}{3}$ or $\theta = \frac{4\pi}{3}$

46. (a) $\theta = \frac{\pi}{3}$ or $\theta = \frac{2\pi}{3}$

(b) $\theta = \frac{4\pi}{3}$ or $\theta = \frac{5\pi}{3}$

48. Solving for $\tan\theta$ produces

$\tan\theta = \pm\sqrt{3}$

$\theta = \frac{\pi}{3}, \frac{2\pi}{3}, \frac{4\pi}{3}, \frac{5\pi}{3}$

50. $2\cos^2\theta - \cos\theta - 1 = 0$

$(2\cos\theta + 1)(\cos\theta - 1) = 0$

$\cos\theta = -\frac{1}{2}$ or $\cos\theta = 1$

$\theta = \frac{2\pi}{3}, \frac{4\pi}{3}$ $\quad\theta = 0, 2\pi$

52. $\sec\theta\csc\theta - 2\csc\theta = 0$

$\csc\theta(\sec\theta - 2) = 0$

$\csc\theta = 0$ or $\sec\theta = 2$

No solution $\quad\theta = \frac{\pi}{3}, \frac{5\pi}{3}$

54. Using the identity $\cos 2\theta = 2\cos^2\theta - 1$ produces the following.

$2\cos^2\theta + 3\cos\theta + 1 = 0$

$(2\cos\theta + 1)(\cos\theta + 1) = 0$

$\cos\theta = -\frac{1}{2}$ or $\cos\theta = -1$

$\theta = \frac{2\pi}{3}, \frac{4\pi}{3}$ $\quad\theta = \pi$

56. Using the identity $\cos\theta = 2\cos^2\frac{\theta}{2} - 1$ produces the following.

$$\cos\frac{\theta}{2} - 2\cos^2\frac{\theta}{2} + 1 = 1$$

$$0 = 2\cos^2\frac{\theta}{2} - \cos\frac{\theta}{2}$$

$$0 = \cos\frac{\theta}{2}\left(2\cos\frac{\theta}{2} - 1\right)$$

$$\cos\frac{\theta}{2} = 0 \quad \text{or} \quad \cos\frac{\theta}{2} = \frac{1}{2}$$

$$\frac{\theta}{2} = \frac{\pi}{2} \qquad \frac{\theta}{2} = \frac{\pi}{3}$$

$$\theta = \pi \qquad \theta = \frac{2\pi}{3}$$

58. Since $\cos 60° = \frac{1}{2} = \frac{x}{10}$ it follows that $x = 5$.

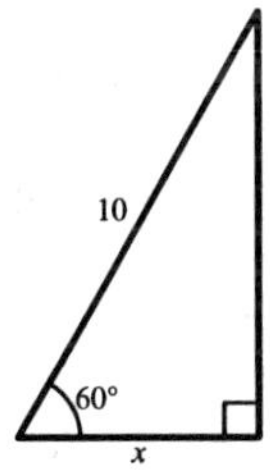

60. Since $\sin 45° = \frac{1}{\sqrt{2}} = \frac{30}{r}$ it follows that

$r = 30\sqrt{2} \approx 42.43.$

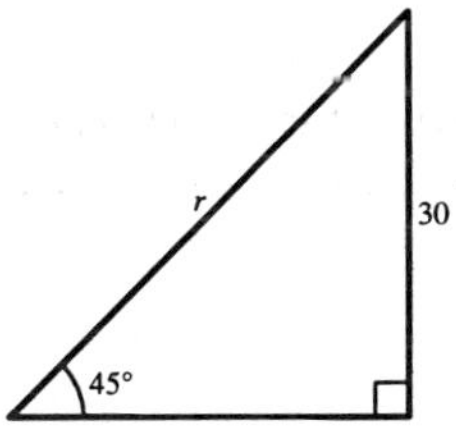

62. Since $\tan 20° = \frac{30}{x} \approx 0.36397$ it follows that

$$x = \frac{30}{0.36397} \approx 82.42$$

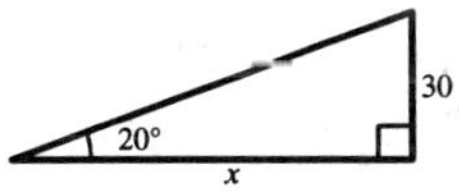

64. Since $\tan 50° = \frac{w}{100}$ it follows that

$w = 100\tan 50° \approx 119.18$ ft.

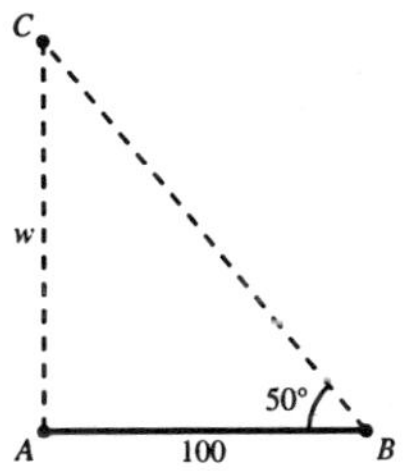

66. Since $\sin\theta = \frac{3.5}{17.5}$ it follows that

$$\theta = \sin^{-1}\left(\frac{3.5}{17.5}\right) = \sin^{-1}0.2$$

$\theta \approx 11.5°.$

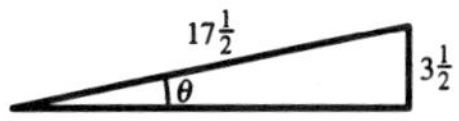

68. (a) $S(2) \approx 27.708$ (b) $S(14) \approx 33.012$ (c) $S(9) \approx 22.778$ (d) $S(21) \approx 28.082$

70.

x	0	2	4	6	8	10
$f(x)$	5.5	2.4271	−1.9271	−2.9271	−0.5729	0.5

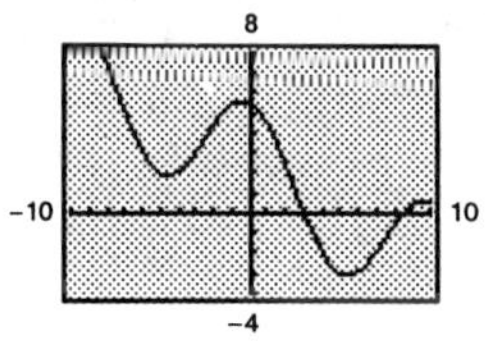

Section 8.3 Graphs of Trigonometric Functions

2. Period: $\frac{2\pi}{3}$

Amplitude: 3

4. Period: $\frac{2\pi}{1/3} = 6\pi$

Amplitude: 2

6. Period: $\frac{2\pi}{\pi/2} = 4$

Amplitude: $\frac{5}{2}$

8. Period: $\frac{2\pi}{2/3} = 3\pi$

Amplitude: 1

10. Period: $\frac{2\pi}{8} = \frac{\pi}{4}$

Amplitude: $\frac{1}{3}$

12. Period: $\frac{2\pi}{1/4} = 8\pi$

Amplitude: 5

14. Period: $\frac{2\pi}{\pi/10} = 20$

Amplitude: $\frac{2}{3}$

16. Period: $\frac{\pi}{2\pi} = \frac{1}{2}$

18. Period: $\frac{2\pi}{4} = \frac{\pi}{2}$

20. Period: $\frac{\pi}{2\pi/3} = \frac{3}{2}$

22. The graph of this function has a period of π and matches graph (e).

24. The graph of this function has a period of 2π and matches graph (a).

26. The graph of this function has a period of 2π and matches graph (d).

28. Period: 6π

Amplitude: 4

x-intercepts: $(0, 0)$, $(3\pi, 0)$, $(6\pi, 0)$, $(9\pi, 0)$, $(12\pi, 0)$

Maxima: $\left(\frac{3\pi}{2}, 4\right)$, $\left(\frac{15\pi}{2}, 4\right)$

Minima: $\left(\frac{9\pi}{2}, -4\right)$, $\left(\frac{21\pi}{2}, -4\right)$

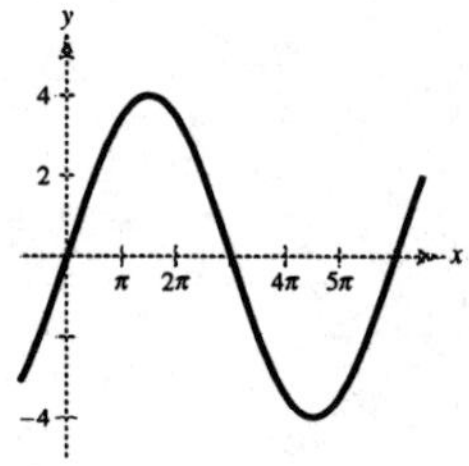

30. Period: 3π

Amplitude: $\frac{3}{2}$

x-intercepts: $\left(\frac{3\pi}{4}, 0\right)$, $\left(\frac{9\pi}{4}, 0\right)$, $\left(\frac{15\pi}{4}, 0\right)$, $\left(\frac{21\pi}{4}, 0\right)$

Maxima: $\left(0, \frac{3}{2}\right)$, $\left(3\pi, \frac{3}{2}\right)$, $\left(6\pi, \frac{3}{2}\right)$

Minima: $\left(\frac{3\pi}{2}, -\frac{3}{2}\right)$, $\left(\frac{9\pi}{2}, -\frac{3}{2}\right)$

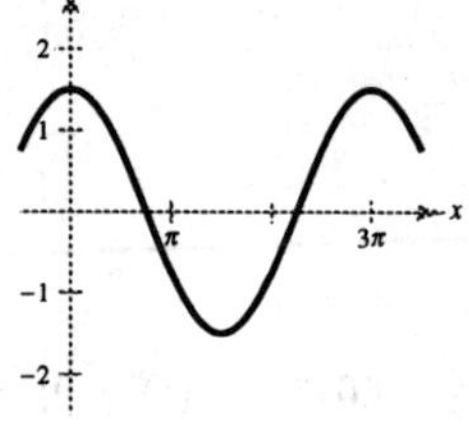

32. Period: $\frac{\pi}{2}$

Amplitude: 3

x-intercepts: $\left(\frac{\pi}{8}, 0\right), \left(\frac{3\pi}{8}, 0\right), \left(\frac{5\pi}{8}, 0\right), \left(\frac{7\pi}{8}, 0\right)$

Maxima: $\left(\frac{\pi}{4}, 3\right), \left(\frac{3\pi}{4}, 3\right)$

Minima: $(0, -3), \left(\frac{\pi}{2}, -3\right), (\pi, -3)$

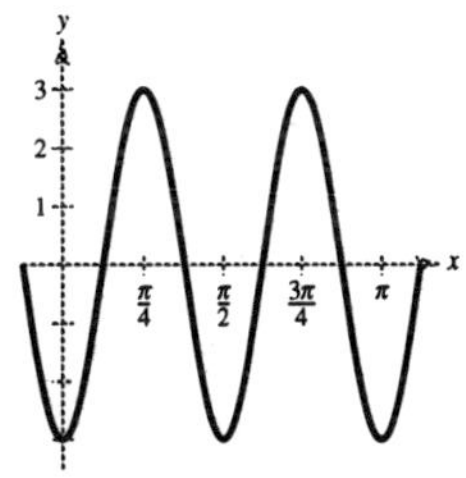

34. Period: 8

Amplitude: $\frac{3}{2}$

x-intercepts: $(0, 0), (4, 0), (8, 0), (12, 0), (16, 0)$

Maxima: $\left(2, \frac{3}{2}\right), \left(10, \frac{3}{2}\right)$

Minima: $\left(6, -\frac{3}{2}\right), \left(14, -\frac{3}{2}\right)$

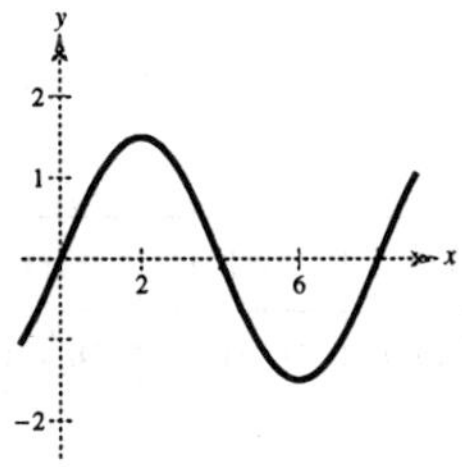

36. Period: 12

Amplitude: 10

x-intercepts: $(3, 0), (9, 0), (15, 0), (21, 0)$

Maxima: $(0, 10), (12, 10), (24, 10)$

Minima: $(6, -10), (18, -10)$

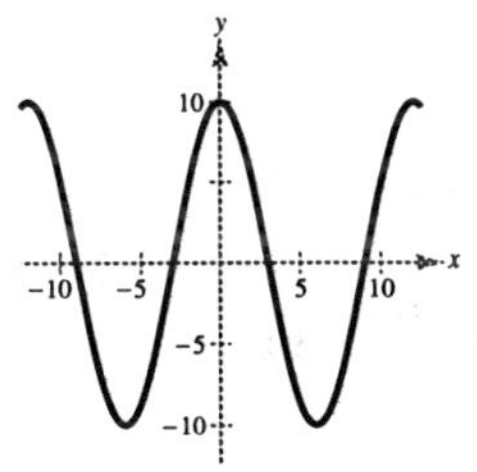

38. Period: π

x-intercepts: $\left(\frac{\pi}{2}, 0\right), \left(\frac{3\pi}{2}, 0\right)$

Asymptotes: $x = 0,\ x = \pi,\ x = 2\pi$

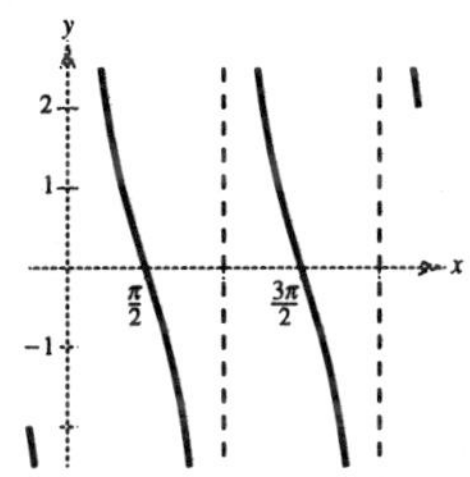

40. Period: 1

x-intercepts: $(0, 0), (1, 0)$

Asymptotes: $x = -\frac{1}{2},\ x = \frac{1}{2},\ x = \frac{3}{2}$

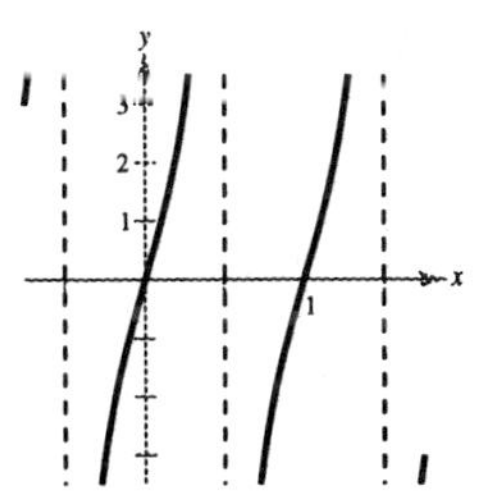

42. Period: 6π

Asymptotes: $x = 0,\ x = 3\pi,\ x = 6\pi$

Maximum: $\left(\frac{9\pi}{2}, -1\right)$

Minimum: $\left(\frac{3\pi}{2}, 1\right)$

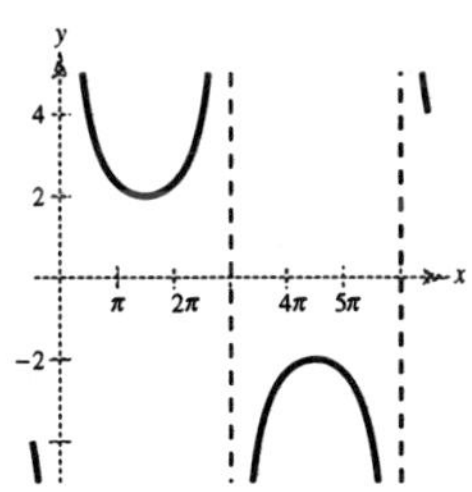

44. Period: 2

Asymptotes: $x = -\frac{1}{2}, x = \frac{1}{2}, x = \frac{3}{2}, x = \frac{5}{2}$

Maximum: $(1, -1)$

Minima: $(0, 1), (2, 1)$

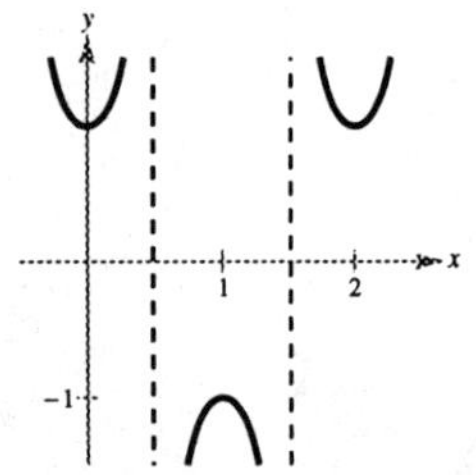

46. Period: π

Asymptotes: $x = -\frac{\pi}{2}, x = \frac{\pi}{2}, x = \frac{3\pi}{2}$

x-intercepts: $(0, 0), (\pi, 0)$

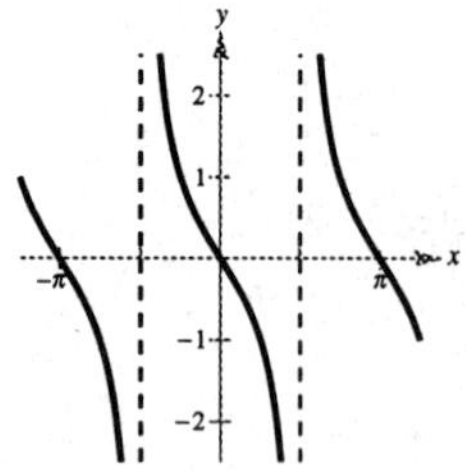

48.

x	−0.1	−0.01	−0.001	0.001	0.01	0.1
$f(x)$	0.6723	0.6667	0.6667	0.6667	0.6667	0.6723

From this table, we estimate that $\lim_{x\to 0} \frac{\sin 2x}{\sin 3x} = \frac{2}{3}$.

50.

x	−0.1	−0.01	−0.001	0.001	0.01	0.1
$f(x)$	−0.1499	−0.0150	−0.0015	0.0015	0.0150	0.1499

From this table, we estimate that $\lim_{x\to 0} \frac{3(1 - \cos x)}{x} = 0$.

52. False, the period is $\frac{\pi}{4/3} = \frac{3\pi}{4}$.

54. False, $\tan\left(\frac{5\pi/4}{2}\right) \neq 1$.

56. (a) Period $= \frac{2\pi}{\pi/2} = 4$ seconds

(b) Cycles per minute $= \frac{60}{4} = 15$

(c) See accompanying graph.

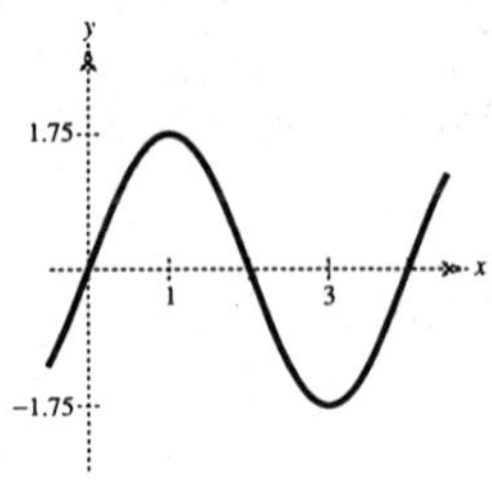

58. (a) Period $= \frac{2\pi}{5\pi/3} = \frac{6}{5}$ seconds

(b) Heartbeats per minute $= \frac{60}{6/5} = 50$

(c) See accompanying graph.

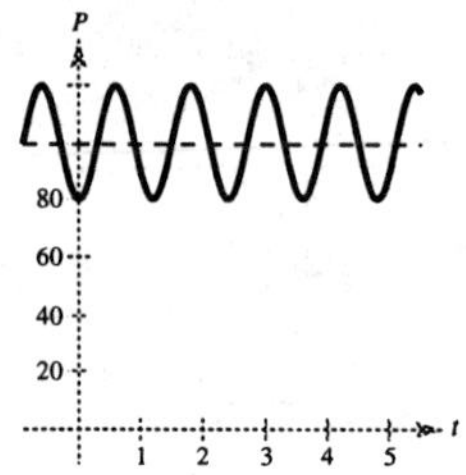

60. (Graphs of) $P = 5700 + 1200 \sin \dfrac{2\pi t}{24}$

$$p = 9800 + 2750 \cos \frac{2\pi t}{24}$$

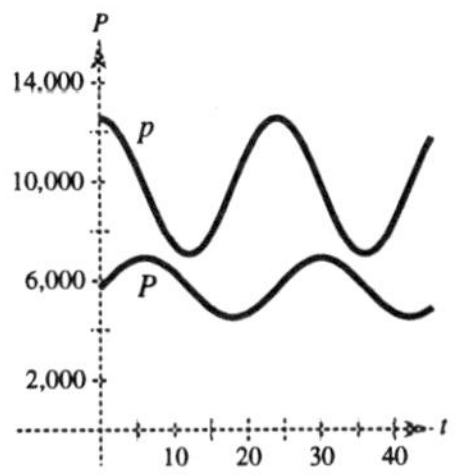

62. The graph has a period of 12, an amplitude of 43.75, and oscillates about the line $y = 74.50$, as shown in the accompanying graph.

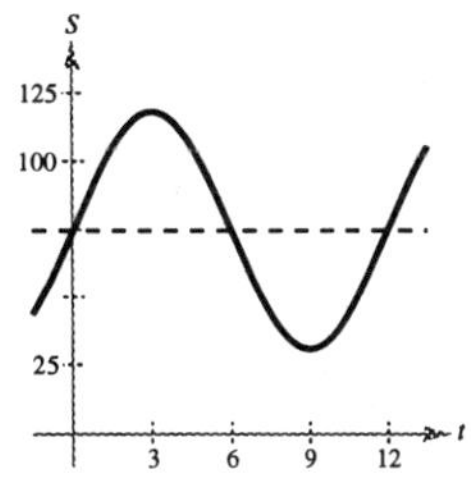

64. Use the procedure shown in Example 6 to calculate your three energy levels on December 25, 2000.

66.

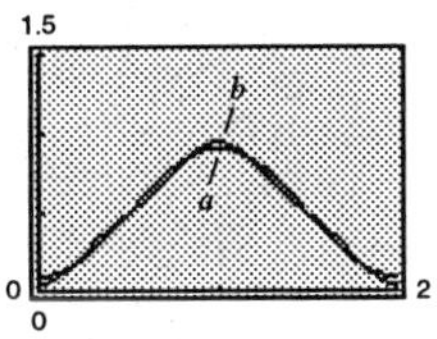

68. $\lim\limits_{x \to 0} \dfrac{\sin 5x}{\sin 2x} = \dfrac{5}{2}$

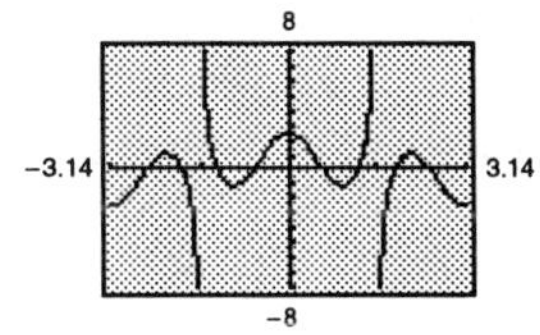

70. $H(t) = 54.33 - 20.38 \cos \dfrac{\pi t}{6} - 15.69 \sin\left(\dfrac{\pi t}{6}\right)$

$L(t) = 39.36 - 15.70 \cos \dfrac{\pi t}{6} - 14.16 \sin \dfrac{\pi t}{6}$

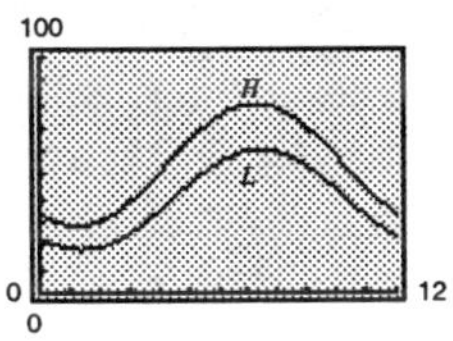

(a) By graphing the function $f(t) = H(t) - L(t)$, we see that the greatest difference occurs at $t = 6.6$ (June 18), and the smallest difference at $t = 0.6$ (January 18).

(b) $H(t)$ is a maximum for $t = 7.25$ (July 7), 16 days later.

72. (a) A

(b) B

(c) B

(d) Reciprocals of each other

Section 8.4 Derivatives of Trigonometric Functions

2. $y' = \cos x$

4. $g'(t) = -\pi \sin t$

6. $f'(x) = \cos x - \sin x$

8. $f'(x) = (x + 1)(-\sin x) + (\cos x)(1)$

$= \cos x - (x + 1)\sin x$

10. $f'(x) = \dfrac{x \cos x - \sin x}{x^2}$

12. $y' = \dfrac{\sec x \tan x + \sec^2 x}{\sec x + \tan x}$

$= \dfrac{\sec x(\tan x + \sec x)}{\sec x + \tan x} = \sec x$

14. $y' = e^{-x}\cos x - e^{-x}\sin x$

$= e^{-x}(\cos x - \sin x)$

16. $y' = 2\csc x(-\csc x\cot x) = -2\csc^2 x\cot x$

18. $y = \frac{1}{2}\csc 2x$

$y' = \frac{1}{2}(2)(-\csc 2x\cot 2x) = -\csc 2x\cot 2x$

20. $y' = x^2\left(-\frac{1}{x^2}\cos\frac{1}{x}\right) + 2x\sin\frac{1}{x}$

$= 2x\sin\frac{1}{x} - \cos\frac{1}{x}$

22. $y' = e^x\sec^2 e^x$

24. $y = \frac{1}{4}\sin^2 2x = \frac{1}{4}(\sin 2x)^2$

$y' = \frac{1}{4}[2\sin 2x(2\cos 2x)]$

$= \sin 2x\cos 2x = \frac{1}{2}\sin 4x$

26. $y' = \frac{1}{2} + \frac{1}{2}\cos 2x$

$= \frac{1}{2}(1 + 2\cos^2 x - 1) = \cos^2 x$

28. $y' = -\left(\frac{-\sin x}{\cos x}\right) = \tan x$

30. $y' = 1 - \csc^2 x = -\cot^2 x$

32. $y' = \sec^7 x\tan x - \sec^5 x\tan x$

$= \sec^5 x\tan x(\sec^2 x - 1)$

$= \sec^5 x\tan^3 x$

34. $y = \frac{1}{2}(x\tan x - \sec x)$

$y' = \frac{1}{2}(x\sec^2 x + \tan x - \sec x\tan x)$

36. $y = \sec x, \quad \left(\frac{\pi}{3}, 2\right)$

$y' = \sec x\tan x, \quad y'\left(\frac{\pi}{3}\right) = 2\sqrt{3}$

$y - 2 = 2\sqrt{3}\left(x - \frac{\pi}{3}\right)$

$y = 2\sqrt{3}x + 2 - 2\sqrt{3}\left(\frac{\pi}{3}\right)$

38. $y = \csc^2 x, \quad \left(\frac{\pi}{2}, 1\right)$

$y' = -2\csc^2 x\cot x, \quad y'\left(\frac{\pi}{2}\right) = 0$

$y - 1 = 0\left(x - \frac{\pi}{2}\right)$

$y = 1$

40. $y = \sin x\cos x, \quad \left(\frac{3\pi}{2}, 0\right)$

$y' = -\sin^2 x + \cos^2 x, \quad y'\left(\frac{3\pi}{2}\right) = -1$

$y - 0 = -1\left(x - \frac{3\pi}{2}\right)$

$y = -x + \frac{3\pi}{2}$

42. $y = \sqrt{\sin x} = (\sin x)^{1/2}, \quad \left(\frac{\pi}{6}, \frac{\sqrt{2}}{2}\right)$

$$y' = \frac{1}{2}(\sin x)^{-1/2}\cos x, \quad y'\left(\frac{\pi}{6}\right) = \frac{1}{2}\left(\frac{2}{\sqrt{2}}\right)\left(\frac{\sqrt{3}}{2}\right) = \frac{\sqrt{6}}{4}$$

$$y - \frac{\sqrt{2}}{2} = \frac{\sqrt{6}}{4}\left(x - \frac{\pi}{6}\right)$$

$$y = \frac{\sqrt{6}}{4}x + \frac{\sqrt{2}}{2} - \frac{\sqrt{6}}{24}\pi$$

44. $\tan(x + y) = x$

$$\left(1 + \frac{dy}{dx}\right)\sec^2(x + y) = 1$$

$$\frac{dy}{dx} = \cos^2(x + y) - 1 = -\sin^2(x + y)$$

At $(0, 0)$, we have $\frac{dy}{dx} = 0$.

46. Since

$$y' = \frac{x \sin x - (10 - \cos x)(1)}{x^2} = \frac{x \sin x + \cos x - 10}{x^2}$$

it follows that

$$xy' + y = \frac{x \sin x + \cos x - 10}{x} + \frac{10 - \cos x}{x} = \frac{x \sin x}{x} = \sin x.$$

48. Since

$$y' = e^x\left(-\sqrt{2}\sin\sqrt{2}x + \sqrt{2}\cos\sqrt{2}x\right) + e^x\left(\cos\sqrt{2}x + \sin\sqrt{2}x\right)$$
$$= e^x\left[\left(1 - \sqrt{2}\right)\sin\sqrt{2}x + \left(1 + \sqrt{2}\right)\cos\sqrt{2}x\right]$$

and

$$y'' = e^x\left[\left(1 - \sqrt{2}\right)\sqrt{2}\cos\sqrt{2}x - \left(1 + \sqrt{2}\right)\sqrt{2}\sin\sqrt{2}x\right] + e^x\left[\left(1 - \sqrt{2}\right)\sin\sqrt{2}x + \left(1 + \sqrt{2}\right)\cos\sqrt{2}x\right]$$
$$= e^x\left[\left(2\sqrt{2} - 1\right)\cos\sqrt{2}x + \left(-2\sqrt{2} - 1\right)\sin\sqrt{2}x\right]$$

it follows that $y'' - 2y' + 3y = 0$.

50. Since $y' = \frac{5}{2}\cos\frac{5x}{2}$, the slope of the tangent line at $(0, 0)$ is $\frac{5}{2}$. There are two complete cycles in the interval $[0, 2\pi]$.

52. Since $y' = \frac{3}{2}\cos\frac{3x}{2}$, the slope of the tangent line at $(0, 0)$ is $\frac{3}{2}$. There is one complete cycle in the interval $[0, 2\pi]$

54. Since $y' = \frac{1}{2}\cos\frac{x}{2}$, the slope of the tangent line at $(0, 0)$ is $\frac{1}{2}$. There are no complete cycles in the interval $[0, 2\pi]$.

56. The first derivative is zero when

$$2\cos x - 2\sin 2x = 0$$
$$2(\cos x - 2\sin x\cos x) = 0$$
$$2\cos x(1 - 2\sin x) = 0$$
$$\cos x = 0 \text{ or } \sin x = \frac{1}{2}.$$

Critical numbers: $x = \frac{\pi}{6}, \frac{\pi}{2}, \frac{5\pi}{6}, \frac{3\pi}{2}$

Relative maxima: $\left(\frac{\pi}{6}, \frac{3}{2}\right), \left(\frac{5\pi}{6}, \frac{3}{2}\right)$

Relative minima: $\left(\frac{\pi}{2}, 1\right), \left(\frac{3\pi}{2}, -3\right)$

58. The first derivative is zero when

$$e^{-x}(\cos x - \sin x) = 0$$
$$\cos x = \sin x$$
$$1 = \tan x.$$

Critical numbers: $x = \frac{\pi}{4}, \frac{5\pi}{4}$

Relative maximum: $\left(\frac{\pi}{4}, \frac{\sqrt{2}}{2}e^{-\pi/4}\right)$

Relative minimum: $\left(\frac{5\pi}{4}, -\frac{\sqrt{2}}{2}e^{-5\pi/4}\right)$

60. $y = \sec \frac{x}{2}$

$$y' = \frac{1}{2}\sec\frac{x}{2}\tan\frac{x}{2} = 0$$
$$\tan\frac{x}{2} = 0$$

No relative extrema on $(0, 2\pi)$.

62. $T'(t) = 21\left(\frac{2\pi}{365}\right)\sin\frac{2\pi(t - 32)}{365} = 0 \Rightarrow t = 32, 214.5$

(a) Warmest day: $t = 214.5$ (August 2 & 3)

(b) Coldest day: $t = 32$ (Feb. 1)

64. $D'(t) = 1.9\left(\frac{6}{\pi}\right)\sin\left(\frac{\pi(t + 0.2)}{6}\right) = 0$

$\pi(t + 0.2)/6 = k\pi,$ k is an integer

$$t + 0.2 = 6k$$
$$t = 6k - 0.2$$

$t = 5.8$ $(k = 1)$ (June)

$D(5.8) = 14.1$ hours

66. $D = 1.5\cos\frac{\pi t}{6} + 3.5,\quad 0 \le t \le 24$

(a) $D'(t) = -1.5\left(\frac{\pi}{6}\right)\sin\frac{\pi t}{6} = -\frac{\pi}{4}\sin\frac{\pi t}{6}$

(b) $D'(4) = -\frac{\pi}{4}\sin\frac{2\pi}{3} \approx -0.68$

$D'(20) = -\frac{\pi}{4}\sin\frac{10\pi}{3} \approx 0.68$

(c) D greatest at $t = 0$ and $t = 12$

D least at $t = 6$ and $t = 18$

(d) Greatest depth: 5 feet

Least depth: 2 feet

68. (a) $f(x) = \frac{x}{2} + \cos\frac{x}{2},\quad (0, 4\pi)$

$$f'(x) = \frac{1}{2} - \frac{1}{2}\sin\frac{x}{2}$$

(b) $f'(x) = 0$ when $x = \pi$.

(c)

Interval	$(0, \pi)$	$(\pi, 4\pi)$
Sign of f'	+	+
Conclusion	f is increasing	f is increasing

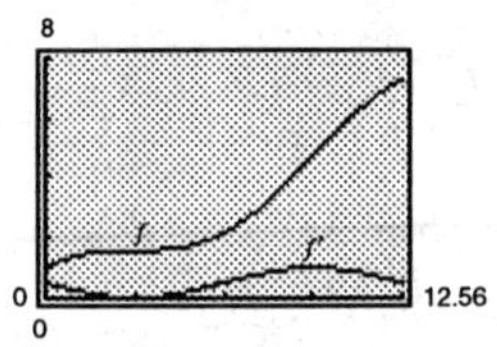

70. (a) $f(x) = x\sin x,\quad (0, \pi)$

$f'(x) = x\cos x + \sin x$

(b) $f'(x) = 0$ when $x = -\tan x$ and $x \approx 2.029$.

(c)

Interval	$(0, 2.029)$	$(2.029, \pi)$
Sign of f'	+	−
Conclusion	f is increasing	f is decreasing

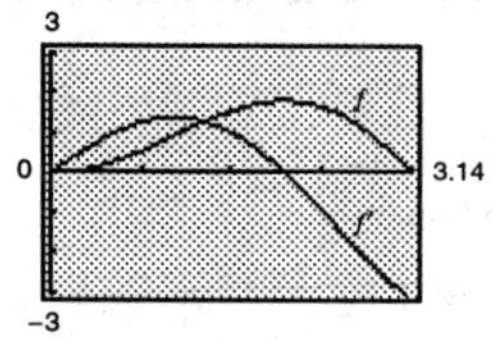

72. (a) $f(x) = 4e^{-0.5x} \sin \pi x, \quad (0, 4)$

$f'(x) = 2e^{-0.5x}(2\pi \cos \pi x - \sin \pi x)$

(b) $f'(x) = 0$ when $2\pi = \tan \pi x, \quad x \approx 0.45,\ 1.45,\ 2.45,$ and 3.45.

(c)

Interval	(0, 0.45)	(0.45, 1.45)	(1.45, 2.45)	(2.45, 3.45)	(3.45, 4)
Sign of f'	+	−	+	−	+
Conclusion	Increasing	Decreasing	Increasing	Decreasing	Increasing

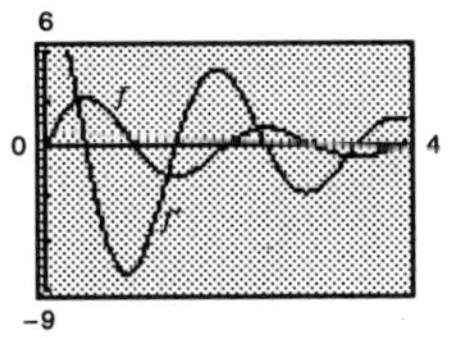

74. $f(x) = \dfrac{x^2 - 2}{\sin x} - 5x$

Relative maxima: $(4.01, -38.5)$, $(0.70, -5.8)$

Relative minimum: $(1.95, -7.8)$

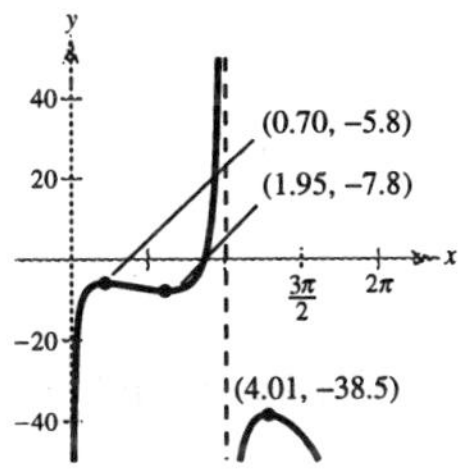

76. $f(x) = \ln x \sin x$

Relative minima: $(0.35, -0.36)$, $(4.84, -1.56)$

Relative maximum: $(2.13, 0.64)$

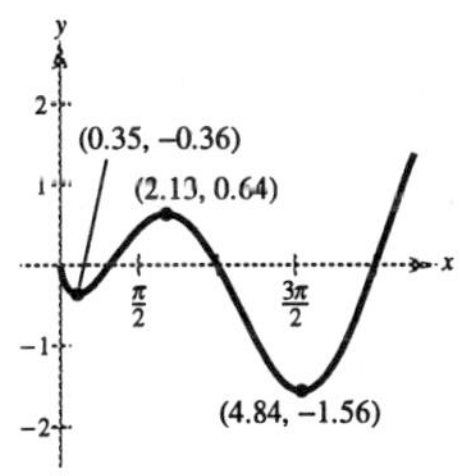

78. $f(x) = \sin\sqrt{x}$

Relative maximum: $(2.47, 1)$

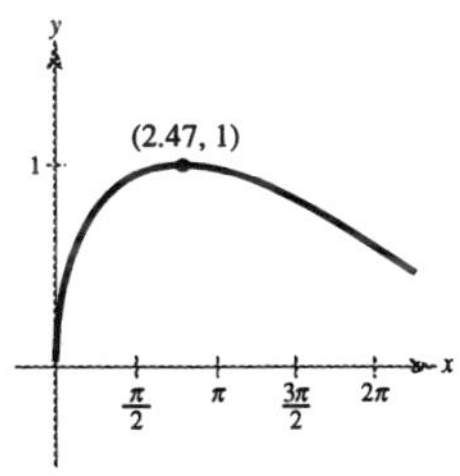

Section 8.5 Integrals of Trigonometric Functions

2. $\displaystyle\int (t^2 - \sin t)\,dt = \frac{t^3}{3} + \cos t + C$

4. $\displaystyle\int (\theta^2 + \sec^2 \theta)\,d\theta = \frac{\theta^3}{3} + \tan \theta + C$

6. $\displaystyle\int (\sec y \tan y - \sec^2 y)\,dy = \sec y - \tan y + C$

8. $\displaystyle\int \cos 6x\,dx = \frac{1}{6}\int 6 \cos 6x\,dx = \frac{1}{6}\sin 6x + C$

10. $\displaystyle\int x \sin x^2\,dx = \frac{1}{2}\int 2x \sin x^2\,dx = -\frac{1}{2}\cos x^2 + C$

12. $\displaystyle\int \csc^2 \frac{x}{2}\,dx = 2\int \frac{1}{2}\csc^2 \frac{x}{2}\,dx = -2 \cot \frac{x}{2} + C$

14. $\displaystyle\int \csc 2x \cot 2x\,dx = \frac{1}{2}\int 2 \csc 2x \cot 2x\,dx$

$\displaystyle = -\frac{1}{2}\csc 2x + C$

16. $\displaystyle\int \sqrt{\cot x}\,\csc^2 x\,dx = -\int (\cot x)^{1/2}(-\csc^2 x)\,dx$

$\displaystyle = -\frac{2}{3}(\cot x)^{3/2} + C$

18. $\int \tan 5x\,dx = \frac{1}{5}\int 5\tan 5x\,dx$

$= -\frac{1}{5}\ln|\cos 5x| + C$

20. $\int \sec\frac{x}{2}\,dx = 2\int \frac{1}{2}\sec\frac{x}{2}\,dx$

$= 2\ln\left|\sec\frac{x}{2} + \tan\frac{x}{2}\right| + C$

22. $\int \frac{\sin x}{\cos^2 x}\,dx = \int \frac{1}{\cos x}\cdot\frac{\sin x}{\cos x}\,dx$

$= \int \sec x\tan x\,dx = \sec x + C$

24. $\int \frac{\cos t}{1+\sin t}\,dt = \ln|1+\sin t| + C$

26. $\int \frac{\sin\sqrt{x}}{\sqrt{x}}\,dx = 2\int (\sin\sqrt{x})\left(\frac{1}{2\sqrt{x}}\right)dx$

$= -2\cos\sqrt{x} + C$

28. $\int \frac{1-\cos\theta}{\theta - \sin\theta}\,d\theta = \ln|\theta - \sin\theta| + C$

30. $\int e^{\sin x}\cos x\,dx = e^{\sin x} + C$

32. $\int e^{\sec x}\sec x\tan x\,dx = e^{\sec x} + C$

34. $\int (\csc 2\theta - \cot 2\theta)^2\,d\theta = \int (\csc^2 2\theta - 2\csc 2\theta\cot 2\theta + \cot^2 2\theta)d\theta$

$= \int (2\csc^2 2\theta - 2\csc 2\theta\cot 2\theta - 1)d\theta$

$= -\cot 2\theta + \csc 2\theta - \theta + C = \csc 2\theta - \cot 2\theta - \theta + C$

36. Using integration by parts, we let $u = x$ and $dv = \sin x\,dx$. Then $du = dx$ and $v = -\cos x$.

$\int x\sin x\,dx = -x\cos x + \int \cos x\,dx$

$= -x\cos x + \sin x + C$

38. Using integration by parts, we let $u = \theta$ and $dv = \sec\theta\tan\theta\,d\theta$. Then $du = d\theta$ and $v = \sec\theta$.

$\int \theta\sec\theta\tan\theta\,d\theta = \theta\sec\theta - \int \sec\theta\,d\theta$

$= \theta\sec\theta - \ln|\sec\theta + \tan\theta| + C$

40. $\int_0^{\pi/2} \sin 2x\,dx = -\frac{1}{2}\cos 2x\Big]_0^{\pi/2}$

$= \frac{1}{2} + \frac{1}{2}$

$= 1$

42. $\int_0^{\pi/2} (x + \cos x)dx = \left[\frac{x^2}{2} + \sin x\right]_0^{\pi/2}$

$= \frac{\pi^2}{8} + 1$

$= \frac{\pi^2 + 8}{8} \approx 2.2337$

44. $\int_0^{\pi/8} \sin 2x\cos 2x\,dx = \frac{1}{2}\int_0^{\pi/8} \sin 2x(2\cos 2x)dx$

$= \left[\frac{1}{2}\cdot\frac{\sin^2 2x}{2}\right]_0^{\pi/8} = \frac{1}{8}$

46. $\int_0^{\pi/4} \sec x\tan x\,dx = \sec x\Big]_0^{\pi/4}\ \sqrt{2} - 1$

≈ 0.4142

48. Area $= \int_0^{\pi/4} \tan x\, dx$

$= -\ln |\cos x| \Big]_0^{\pi/4}$

$= -\ln\left(\frac{1}{\sqrt{2}}\right)$

$= \ln \sqrt{2} \approx 0.3466$ square unit

50. Area $= \int_0^{\pi} (2 \sin x + \sin 2x)dx$

$= \left[-2 \cos x - \frac{1}{2} \cos 2x\right]_0^{\pi}$

$= 4$ square units

52. From Exercise 36, we have

Area $= \int_0^{\pi} x \sin x\, dx$

$= \left[-x \cos x + \sin x\right]_0^{\pi}$

$= \pi$ square units

54. $V = \pi \int_{\pi/6}^{5\pi/6} \csc^2 x\, dx$

$= -\pi \cot x \Big]_{\pi/6}^{5\pi/6}$

$= 2\sqrt{3}\pi$ cubic units

56. (a) $\frac{1}{8}\left[f(0) + 2f\left(\frac{1}{4}\right) + 2f\left(\frac{1}{2}\right) + 2f\left(\frac{3}{4}\right) + f(1)\right] \approx 0.8958$

(b) $\frac{1}{12}\left[f(0) + 4f\left(\frac{1}{4}\right) + 2f\left(\frac{1}{2}\right) + 4f\left(\frac{3}{4}\right) + f(1)\right] \approx 0.9045$

58. $\frac{1}{b-a}\int_a^b \left(74.50 + 43.75 \sin \frac{\pi t}{6}\right)dt = \frac{1}{b-a}\left[74.50t - \frac{262.5}{\pi} \cos \frac{\pi t}{6}\right]_a^b$

(a) $\frac{1}{3-0}\left[74.50t - \frac{262.5}{\pi} \cos \frac{\pi t}{6}\right]_0^3 \approx 102.352$ thousand units $= 102{,}352$ units

(b) $\frac{1}{6-3}\left[74.50t - \frac{262.5}{\pi} \cos \frac{\pi t}{6}\right]_0^6 \approx 102.352$ thousand units $= 102{,}352$ units

(c) $\frac{1}{12-0}\left[74.50t - \frac{262.5}{\pi} \cos \frac{\pi t}{6}\right]_0^{12} = 74.5$ thousand units $= 74{,}500$ units

60. (a) $P'(t) = (-826.32)(0.069) \sin (0.069t + 2.61)$

$= -57.016 \sin(0.069t + 2.61) > 0$ on $[8, 16]$

(b) $\int_8^{16} P(t)dt \sim 601.8$ million dollars

(c) $\frac{601.8}{8} \approx 75.2$ million dollars

62. The volume is given by the following.

$$V = \int_0^3 0.9 \sin \frac{\pi t}{3}\, dt = -\frac{3}{\pi}(0.9) \cos \frac{\pi t}{3}\Big]_0^3 = -\frac{2.7}{\pi}(-1 - 1) \approx 1.7189 \text{ liters}$$

64. $\frac{1}{200-0}\int_0^{200} 100{,}000\left[1 + \sin \frac{2\pi(t-60)}{365}\right]dt = 500\left[t - \frac{365}{2\pi} \cos \frac{2\pi(t-60)}{365}\right]_0^{200} \approx 136{,}495$ lb

66. $\int_0^{\pi/2} \cos \sqrt{x}\, dx \approx 1.0057$

68. $\int_0^2 (4 + x + \sin \pi x)dx = 10$

Section 8.6 L'Hôpital's Rule

2. Yes, $\dfrac{\infty}{\infty}$

4. No, $\dfrac{0}{1} \to 0$

6. Yes, $\dfrac{\infty}{\infty}$

8.

x	2.9	2.99	2.999	3	3.001	3.01	3.1
$f(x)$	1.169	1.1669	1.16669	Undefined	1.16664	1.1664	1.1639

$$\lim_{x\to 3} \frac{x^2 + x - 12}{x^2 - 9} = \frac{7}{6}$$

10.

x	−0.1	−0.01	−0.001	0	0.001	0.01	0.1
$f(x)$	0.033	0.0033	0.00033	Undefined	−0.00033	−0.0033	−0.033

$$\lim_{x\to 0} \frac{\cos 2x - 1}{6x} = 0$$

12. $\displaystyle\lim_{x\to\infty} \frac{e^x}{x^2} = \infty$

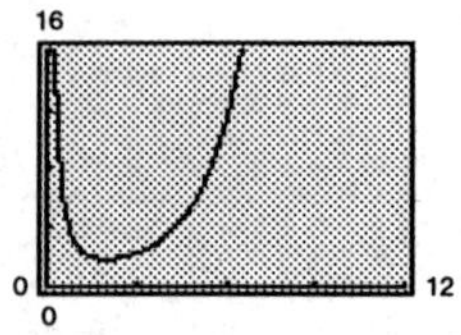

14. $\displaystyle\lim_{x\to 1} \frac{x^2 + 3x - 4}{x^2 + 2x - 3} = \frac{5}{4}$

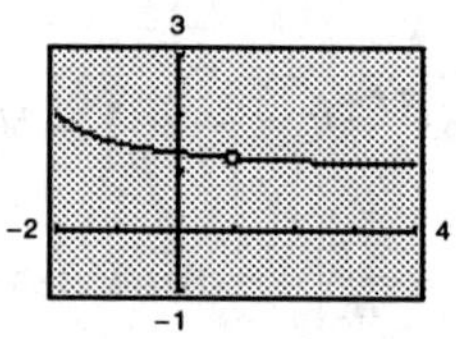

16. $\displaystyle\lim_{x\to 3} \frac{x^2 + x - 12}{x^2 - 9} = \lim_{x\to 3} \frac{2x + 1}{2x} = \frac{7}{6}$ (see Exercise 8)

18. $\displaystyle\lim_{x\to 0} \frac{\cos 2x - 1}{6x} = \lim_{x\to 0} \frac{-2\sin 2x}{6} = 0$ (see Exercise 10)

20. $\displaystyle\lim_{x\to\infty} \frac{e^x}{x^2} = \lim_{x\to\infty} \frac{e^x}{2x} = \lim_{x\to\infty} \frac{e^x}{2} = \infty$ (see Exercise 12)

22. $\displaystyle\lim_{x\to 1} \frac{x^2 + 3x - 4}{x^2 + 2x - 3} = \lim_{x\to 1} \frac{2x + 3}{2x + 2} = \frac{5}{4}$

24. $\displaystyle\lim_{x\to 0} \frac{2x - 1 + e^{-x}}{3x} = \lim_{x\to 0} \frac{2 - e^{-x}}{3} = \frac{1}{3}$

26. $\displaystyle\lim_{x\to\infty} \frac{3x}{e^x} = \lim_{x\to\infty} \frac{3}{e^x} = 0$

28. $\displaystyle\lim_{x\to\infty} \frac{3x^2 + 5}{2x^2 - 11} = \lim_{x\to\infty} \frac{6x}{4x} = \frac{3}{2}$

30. $\displaystyle\lim_{x\to 0} \frac{e^{3x} - 1}{2x} = \lim_{x\to 0} \frac{3e^{3x}}{2} = \frac{3}{2}$

32. $\displaystyle\lim_{x\to 2} \frac{\ln(x - 1)}{x - 2} = \lim_{x\to 2} \frac{1/(x - 1)}{1} = 1$

34. $\displaystyle\lim_{x\to \pi/2} \frac{2\cos x}{3\cos x} = \lim_{x\to \pi/2} \frac{-2\sin x}{-3\sin x} = \frac{2}{3}$

36. $\displaystyle\lim_{x\to \pi} \frac{\cos(x/2)}{x - \pi} = \lim_{x\to \pi} \frac{-\frac{1}{2}\sin(x/2)}{1} = -\frac{1}{2}$

38. $\displaystyle\lim_{x\to 0} \frac{\sqrt{4 - x^2} - 2}{x} = \lim_{x\to 0} \frac{\frac{1}{2}(4 - x^2)^{-1/2}(-2x)}{1}$

$$= \lim_{x\to 0} \frac{-x}{\sqrt{4 - x^2}}$$

$$= 0$$

40. $\lim_{x\to\infty} \frac{x^2}{e^x \ln x} = \lim_{x\to\infty} \frac{2x}{e^x \ln x + e^x/x}$

$= \lim_{x\to\infty} \frac{2}{(e^x \ln x + e^x/x) + (e^x/x - e^x/x^2)} = 0$

42. $\lim_{x\to\infty} \frac{3x^3 - 7x + 5}{4x^3 + 2x - 11} = \lim_{x\to\infty} \frac{9x^2 - 7}{12x^2 + 2} = \lim_{x\to\infty} \frac{18x}{24x} = \frac{3}{4}$

44. $\lim_{x\to 0} \frac{2x^3 - x^2 - 3x}{x^4 + 2x^3 - 9x^2 - 18x} = \lim_{x\to 0} \frac{6x^2 - 2x - 3}{4x^3 + 6x^2 - 18x - 18}$

$= \frac{1}{6}$

46. $\lim_{x\to -1} \frac{\ln(x+2)}{x+2} = \frac{0}{1} = 0$

48. $\lim_{x\to 0} \frac{\sin \pi x}{x} = \lim_{x\to 0} \frac{\pi \cos \pi x}{1} = \pi$

50. $\lim_{x\to\infty} \frac{x^3}{e^{2x}} = \lim_{x\to\infty} \frac{3x^2}{2e^{2x}} = \lim_{x\to\infty} \frac{6x}{4e^{2x}} = \lim_{x\to\infty} \frac{6}{8e^{2x}} = 0$

52. $\lim_{x\to\infty} \frac{(\ln x)^2}{x^4} = \lim_{x\to\infty} \frac{2(\ln x)\frac{1}{x}}{4x^3} = \lim_{x\to\infty} \frac{2 \ln x}{4x^4}$

$= \lim_{x\to\infty} \frac{\frac{2}{x}}{16x^3} = 0$

54. $\lim_{x\to\infty} \frac{x^m}{e^{nx}} = 0$ (The x terms eventually disappear.)

(m, n positive integers)

56.

x	1	2	4	8	12	20	30
$\frac{e^x}{x^6}$	2.718	0.115	0.013	0.011	0.055	7.58	14,659

$\lim_{x\to\infty} \frac{e^x}{x^6} = \infty$

58. The limit of the numerator is not 0.

60. The limit of the denominator is not 0.

62. (a)

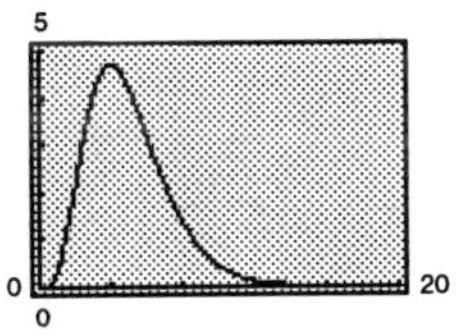

(b) $\lim_{x\to\infty} \frac{x^4}{e^x} = 0$

64. (a)

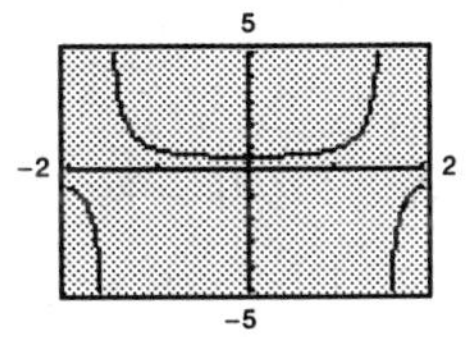

(b) $\lim_{x\to 0} \frac{3 \tan x}{5x} = \frac{3}{5}$

66. False, $\lim_{x\to\infty} \frac{x}{1-x} = -1$

68. False. For example, $\lim_{x\to\infty} \frac{x+1}{x} = 1$, but $x + 1 \neq x$.

70. $\lim_{k\to 0^+} \frac{x^k - 1}{k} = -\infty$

The graph of f approaches the graph of $y = \ln x$.

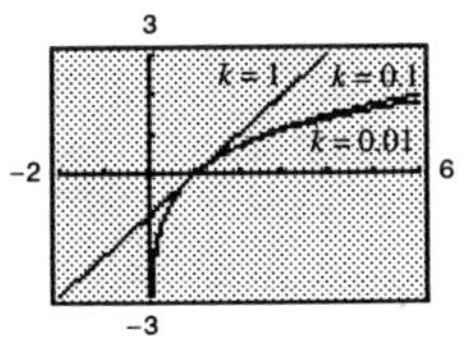

72. Answers will vary.

Review Exercises for Chapter 8

2. Positive coterminal angle:

$$\frac{2\pi}{9} + 2\pi = \frac{20\pi}{9}$$

Negative coterminal angle:

$$\frac{2\pi}{9} - 2\pi = -\frac{16\pi}{9}$$

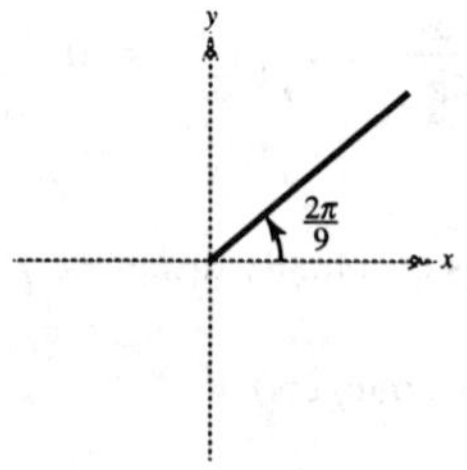

4. Positive coterminal angle:

$$-405° + 2(360°) = 315°$$

Negative coterminal angle:

$$-405° + 360° = -45°$$

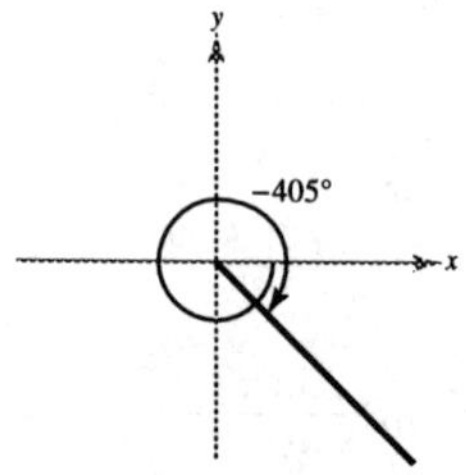

6. $-75°\left(\frac{\pi \text{ radians}}{180°}\right) = \frac{-5}{12}\pi$ radians

8. $810°\left(\frac{\pi \text{ radians}}{180°}\right) = \frac{9}{2}\pi$ radians

10. $\frac{9\pi}{10}\left(\frac{180°}{\pi}\right) = 162°$

12. $\frac{-21\pi}{9}\left(\frac{180°}{\pi}\right) = -420°$

14. $C = \sqrt{1+3} = 2; \quad \theta = 90° - 60° = 30°$

16. $\frac{5}{5} = \frac{b}{3} \Rightarrow b = 3; \quad \theta = 45°$

18. $C = \sqrt{50^2 + 75^2} = \sqrt{8125} \approx 90.14$ feet

20. The reference angle is $\frac{\pi}{5}$.

22. The reference angle is $2(360°) - 640° = 80°$.

24. $\sin 240° = \frac{-\sqrt{3}}{2}$

26. $\csc \frac{\pi}{4} = \sqrt{2}$

28. $\cot\left(\frac{-5\pi}{6}\right) = \sqrt{3}$

30. $\cos\left(\frac{-4\pi}{3}\right) = -\frac{1}{2}$

32. $\sec \frac{12\pi}{5} \approx 3.2361$

34. $\cos 105° \approx -0.2588$

36. $\tan 30° = \frac{y}{30} \Rightarrow y = 30 \tan 30° \approx 17.32$

38. $\sin 45° = \frac{100}{x} \Rightarrow x = \frac{100}{\sin 45°} = 100\sqrt{2} \approx 141.42$

40. $2\sin^2 x = 1$

$$\sin x = \pm\frac{\sqrt{2}}{2}$$

$$x = \frac{\pi}{4} + \frac{k\pi}{2}$$

42. $$\cos^3 x = \cos x$$
$$\cos x(\cos^2 x - 1) = 0$$
$$\cos x(-\sin^2 x) = 0$$
$$\cos x = 0 \quad \text{or} \quad \sin x = 0$$
$$x = \frac{\pi}{2} + k\pi \qquad x = k\pi$$

Combining, $x = \dfrac{\pi}{2} + \dfrac{k\pi}{2}$.

44. $$2\sec^2 x + \tan^2 x - 3 = 0$$
$$2\sec^2 x + (1 - \sec^2 x) - 3 = 0$$
$$\sec^2 x = 2$$
$$\sec x = \pm\sqrt{2}$$
$$x = \frac{\pi}{4} + \frac{k\pi}{2}$$

46. $76° - 32° = 44°$

$90° - 76° = 14°$

$d = 35{,}000 \tan 58° - 35{,}000 \tan 14° \approx 47{,}285$ feet

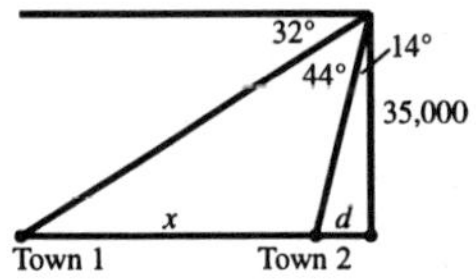

48. $y = \sin 2\pi x$

Period: $\dfrac{2\pi}{2\pi} = 1$

Amplitude: 1

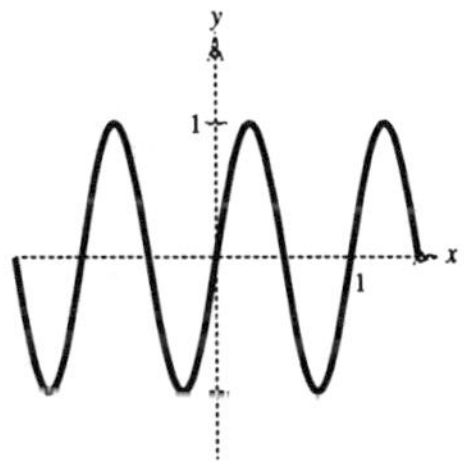

50. $y = \cot \dfrac{x}{2}$

Period: $\dfrac{\pi}{\frac{1}{2}} = 2\pi$

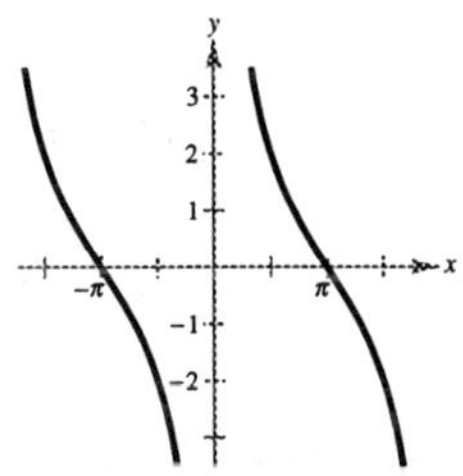

52. $y = 8\cos\left(-\dfrac{x}{4}\right)$

Period: 8π

Amplitude: 8

x-intercepts: $(2\pi, 0), (6\pi, 0), \ldots$

Maximum: $(0, 8), (8\pi, 8)$

Minimum: $(4\pi, -8)$

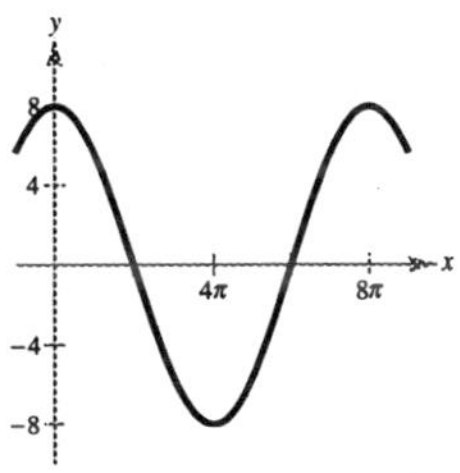

54. $y = 3\csc 2x$

Period: $\dfrac{2\pi}{2} = \pi$

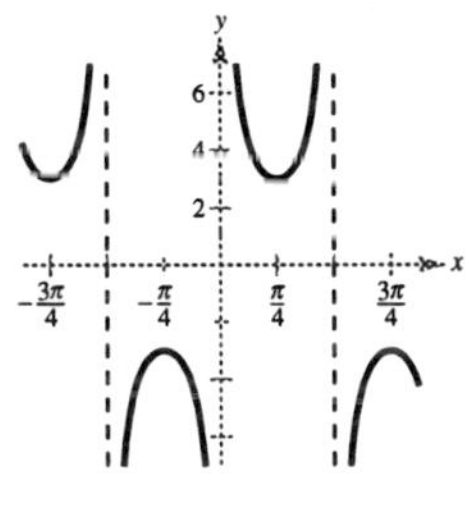

56. $S = 25 + \dfrac{2}{365}t + 20\sin\dfrac{2\pi t}{365}$

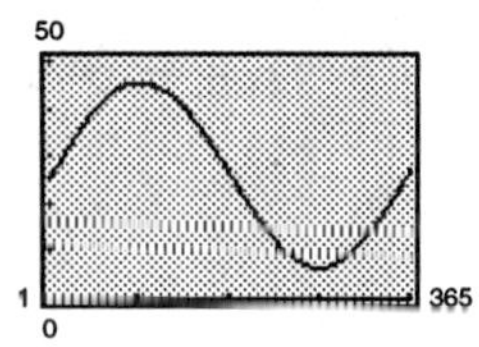

58. $y' = 4\sec^2(4x - \pi)$

60. $y' = -3\csc 3x \cot 3x - 3\csc^2 3x$
$$= -3\csc 3x(\cot 3x + \csc 3x)$$

62. $y' = \dfrac{(x-1)[-\sin(x-1)] - \cos(x-1)}{(x-1)^2}$
$$= -\frac{\sin(x-1) + \cos(x-1)}{x-1}$$

64. $y' = -x\sin x + \cos x - \cos x = -x\sin x$

66. $y' = 2 \sec 2x(2 \sec 2x \tan 2x) = 4 \sec^2 2x \tan 2x$

68. $y' = \frac{1}{2}(2 \cos 2x)e^{\sin 2x} = \cos 2xe^{\sin 2x}$

70. $y = -x \cos x, \quad (0, 0)$

$y' = x \sin x - \cos x, \quad y'(0) = -1$

$y - 0 = -1(x - 0)$

$y = -x$

72. $y = \dfrac{2x}{\cos x}, \quad (0, 0)$

$$y' = \frac{(\cos x)(2) - 2x(-\sin x)}{\cos^2 x}, \quad y'(0) = 2$$

$y - 0 = 2(x - 0)$

$y = 2x$

74. $y = x^2 \cot x, \quad \left(\dfrac{\pi}{4}, \dfrac{\pi^2}{16}\right)$

$$y' = 2x \cot x - x^2 \csc^2 x, \quad y'\left(\frac{\pi}{4}\right) = \frac{\pi}{2} - \frac{\pi^2}{8}$$

$$y - \frac{\pi^2}{16} = \left(\frac{\pi}{2} - \frac{\pi^2}{8}\right)\left(x - \frac{\pi}{4}\right) = \frac{\pi}{8}(4 - \pi)\left(x - \frac{\pi}{4}\right)$$

$$y = \frac{\pi}{8}(4 - \pi)x + \frac{\pi^2}{16} - \frac{\pi^2}{32}(4 - \pi)$$

$$= \frac{\pi}{8}(4 - \pi)x - \frac{\pi^2}{16} + \frac{\pi^3}{32}$$

76. Relative maxima: $\left(\dfrac{\pi}{4}, \dfrac{1}{2}\right), \left(\dfrac{5\pi}{4}, \dfrac{1}{2}\right)$

Relative minima: $\left(\dfrac{3\pi}{4}, -\dfrac{1}{2}\right), \left(\dfrac{7\pi}{4}, -\dfrac{1}{2}\right)$

$\left(\textit{Note:}\ f(x) = \sin x \cos x = \dfrac{1}{2} \sin 2x\right)$

78. Relative minimum: $(\pi, -1)$

80. (a) Maximum daily sales on 45.50 hundred units occurs on the 92nd day of the year.

(b) Minimum daily sales on 6.50 hundred units occurs on the 273rd day of the year.

82. $\displaystyle\int \csc 2x \cot 2x\, dx = \frac{1}{2}\int 2 \csc 2x \cot 2x\, dx = -\frac{1}{2} \csc 2x + C$

84. $\displaystyle\int x \cos x^2\, dx = \frac{1}{2}\int 2x \cos x^2\, dx = \frac{1}{2} \sin x^2 + C$

86. $\displaystyle\int_{-\pi/6}^{\pi/6} \sec^2 x\, dx = \tan x\Big]_{-\pi/6}^{\pi/6} = \frac{1}{\sqrt{3}} - \left(\frac{-1}{\sqrt{3}}\right) = \frac{2}{\sqrt{3}}$

88. $\displaystyle\int_{-\pi/2}^{\pi/2} (2x + \cos x)dx = \Big[x^2 + \sin x\Big]_{-\pi/2}^{\pi/2} = 2$

90. $\displaystyle\int_{\pi/4}^{\pi/2} \cot x\, dx = \ln|\sin x|\Big]_{\pi/4}^{\pi/2} = \ln 1 - \ln\left(\frac{\sqrt{2}}{2}\right)$

$$= \frac{1}{2}\ln 2 \approx 0.3466$$

92. $\displaystyle\int_0^{\pi/4} (2 \cos x + \cos 2x)dx = \left[2 \sin x + \frac{1}{2} \sin 2x\right]_0^{\pi/4}$

$$= \sqrt{2} + \frac{1}{2} \approx 1.9142$$

94. (a) Average $= \frac{1}{12-0}\int_0^{12}\left[15 + 6\sin\frac{\pi(t-8)}{6}\right]dt$

$$= \frac{1}{12}\left[15t - \frac{36}{\pi}\cos\frac{\pi(t-8)}{6}\right]_0^{12}$$

$$= \frac{1}{12}\left(180 - \frac{36}{\pi}\cos\frac{2\pi}{3}\right) - \left[\frac{-36}{\pi}\cos\left(\frac{-4\pi}{3}\right)\right]$$

$$= \frac{1}{12}(180) = 15 \text{ million}$$

(b) Average $= \frac{1}{12-6}\int_6^{12}\left[15 + 6\sin\frac{\pi(t-8)}{6}\right]dt$

$$\approx \frac{1}{6}(101.46) \approx 16.91 \text{ million}$$

96. $R(t) = -2357.41\sin(0.0519t + 4.28) - 1822.11$

(a) 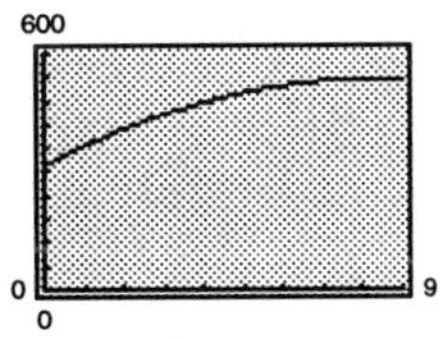

(b) $R'(t) = (-2357.41)(0.0519)\cos(0.0519t + 4.28)$

$$= -122.3496\cos(0.0519t + 4.28)$$

1990: $R'(3) = 33.4$ dollars/year

1995: $R'(8) = 2.1$ dollars/year

(c) $\int_0^9 R(t)dt \approx 4211.1$ dollars

98. $\lim_{x\to 2}\frac{x^2 + 2x + 11}{x^3 + 4} = \frac{19}{12}$

100. $\lim_{x\to 2}\frac{4x^2 + 13x - 42}{3x^2 - 23x + 34} = \lim_{x\to 2}\frac{8x + 13}{6x - 23} = \frac{29}{-11} = -\frac{29}{11}$

102. $\lim_{x\to 0}\frac{\cos^2 x - 1}{\sin x} = \lim_{x\to 0}\frac{-\sin^2 x}{\sin x} = 0$

104. $\lim_{x\to\infty}\frac{e^{5x}}{x^2} = \lim_{x\to\infty}\frac{5e^{5x}}{2x} = \lim_{x\to\infty}\frac{25e^{5x}}{2} = \infty$

106. $f(x)$ grows faster than $g(x)$.

CHAPTER 9
Probability and Calculus

CHAPTER 9
Probability and Calculus

Section 9.1 Discrete Probability

2. (a) $S = \{HH, HT, T1, T2, T3, T4, T5, T6\}$

(b) $A = \{T4, T5, T6\}$

(c) $B = \{HH\}$

4. (a) $S = \{III, IIO, IIU, IOI, IUI, OII, UII, IOO, IOU, IUO, IUU, OIO, OIU, UIO, UIU, OOI, OUI, UOI, UUI, OOO, OOU, OUO, UOO, OUU, UOU, UUO, UUU\}$

(b) $A = \{III, IIO, IIU, IOI, IUI, OII, UII\}$

(c) $B = \{III, IIO, IIU, IOI, IUI, OII, UII, IOU, IUO, IUU, OIU, UIO, UIU, OUI, UOI, UUI, OUU, UOU, UUO, UUU\}$

6. (a) $P\text{ (college)} = \frac{12}{72} + \frac{9}{72} = \frac{21}{72} = \frac{7}{24}$

(b) $P\text{ (not college)} = 1 - \frac{7}{24} - \frac{17}{24}$

(c) $P\text{ (girl, not college)} = \frac{32}{72} = \frac{4}{9}$

8. $P\text{ (black face card)} = \frac{6}{52} = \frac{3}{26}$

10.

Random variable	0	1	2	3	4
Frequency	1	4	6	4	1

12.

Random variable	0	1	2	3	4
Frequency	1	4	6	4	1

14. (a) $P(x \le 2) = \frac{8}{20} + \frac{6}{20} + \frac{3}{20} = \frac{17}{20}$

(b) $P(x > 2) = 1 - P(x \le 2) = 1 - \frac{17}{20} = \frac{3}{20}$

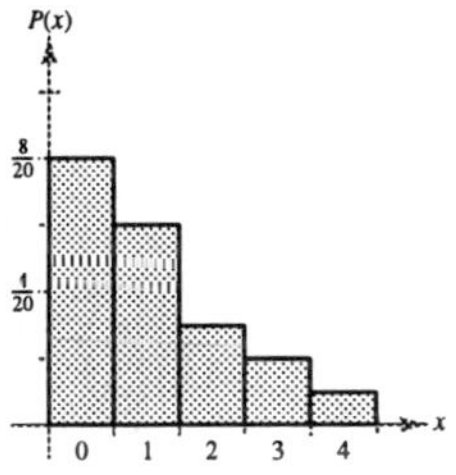

16. (a) $P(1 \le x \le 2) = P(1) + P(2)$

$= 0.189 + 0.441 = 0.630$

(b) $P(x < 2) = P(0) + P(1)$

$= 0.027 + 0.189 = 0.216$

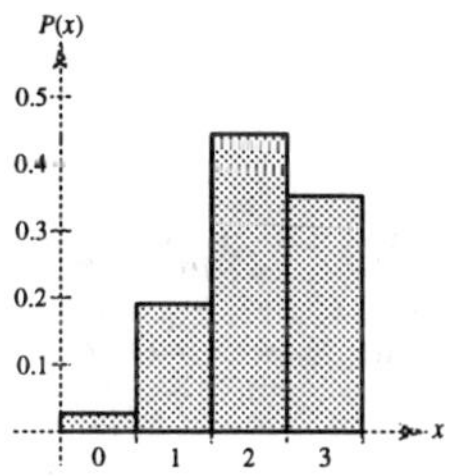

18. (a) $S = \{(1, 1), (1, 2), (1, 3), (1, 4), (1, 5), (1, 6), (2, 1), (2, 2), (2, 3), (2, 4), (2, 5), (2, 6),$
$(3, 1), (3, 2), (3, 3), (3, 4), (3, 5), (3, 6), (4, 1), (4, 2), (4, 3), (4, 4), (4, 5), (4, 6),$
$(5, 1), (5, 2), (5, 3), (5, 4), (5, 5), (5, 6), (6, 1), (6, 2), (6, 3), (6, 4), (6, 5), (6, 6)\}$

(b)

x	2	3	4	5	6	7	8	9	10	11	12
$P(x)$	$\frac{1}{36}$	$\frac{2}{36}$	$\frac{3}{36}$	$\frac{4}{36}$	$\frac{5}{36}$	$\frac{6}{36}$	$\frac{5}{36}$	$\frac{4}{36}$	$\frac{3}{36}$	$\frac{2}{36}$	$\frac{1}{36}$

(c)

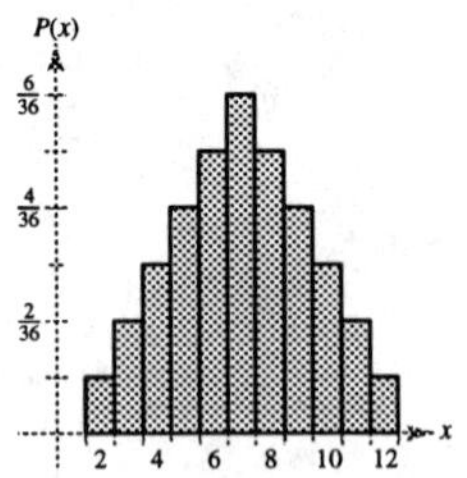

(d) $P(5 \le x \le 9) = \frac{4}{36} + \frac{5}{36} + \frac{6}{36} + \frac{5}{36} + \frac{4}{36} = \frac{24}{36} = \frac{2}{3}$

20. Answers will vary.

22. $E(x) = 1\left(\frac{4}{10}\right) + 2\left(\frac{2}{10}\right) + 3\left(\frac{2}{10}\right) + 4\left(\frac{1}{10}\right) + 5\left(\frac{1}{10}\right) = \frac{23}{10} = 2.3$

$V(x) = \left(1 - \frac{23}{10}\right)^2\left(\frac{4}{10}\right) + \left(2 - \frac{23}{10}\right)^2\left(\frac{2}{10}\right) + \left(3 - \frac{23}{10}\right)^2\left(\frac{2}{10}\right) + \left(4 - \frac{23}{10}\right)^2\left(\frac{1}{10}\right) + \left(5 - \frac{23}{10}\right)^2\left(\frac{1}{10}\right)$

$= \frac{181}{100} = 1.81$

$\sigma = \sqrt{\frac{181}{100}} = \frac{\sqrt{181}}{10} \approx 1.345$

24. $E(x) = -5000(0.008) - 2500(0.052) + 300(0.940) = \$112$

$V(x) = (-5000 - 112)^2(0.008) + (-2500 - 112)^2(0.052) + (300 - 112)^2(0.940) = 597{,}056$

$\sigma = \sqrt{597{,}056} \approx 772.694$

26.

x	0	1	2	3	4
$P(x)$	$\frac{1}{16}$	$\frac{4}{16}$	$\frac{6}{16}$	$\frac{4}{16}$	$\frac{1}{16}$

$E(x) = 0\left(\frac{1}{16}\right) + 1\left(\frac{4}{16}\right) + 2\left(\frac{6}{16}\right) + 3\left(\frac{4}{16}\right) + 4\left(\frac{1}{16}\right) = 2$

$V(x) = (0 - 2)^2\left(\frac{1}{16}\right) + (1 - 2)^2\left(\frac{4}{16}\right) + (2 - 2)^2\left(\frac{6}{16}\right) + (3 - 2)^2\left(\frac{4}{16}\right) + (4 - 2)^2\left(\frac{1}{16}\right) = 1$

28. $E(x) = 30(0.10) + 40(0.20) + 50(0.50) + 60(0.15) + 80(0.05) = 49$ thousand dollars

$V(x) = (30 - 49)^2(0.10) + (40 - 49)^2(0.20) + (50 - 49)^2(0.50) + (60 - 49)^2(0.15) + (80 - 49)^2(0.05) = 119$

$\sigma = \sqrt{119} \approx 10.909$

30. $E(x) = 0(0.936) + 1000(0.040) + 5000(0.020) + 10{,}000(0.004) = \$180$

32.

x	2950	400	20	0
$P(x)$	$\frac{1}{3000}$	$\frac{1}{3000}$	$\frac{25}{3000}$	$\frac{2973}{3000}$

$$E(x) = 2950\left(\tfrac{1}{3000}\right) + 400\left(\tfrac{1}{3000}\right) + 20\left(\tfrac{25}{3000}\right) + 0\left(\tfrac{2973}{3000}\right)$$
$$\approx \$1.28$$

Net gain $= 1.28 - 2 = -\$0.72$

ALTERNATE METHOD:

x	2948	398	18	-2
$P(x)$	$\frac{1}{3000}$	$\frac{1}{3000}$	$\frac{25}{3000}$	$\frac{2973}{3000}$

$$E(x) = 2948\left(\tfrac{1}{3000}\right) + 398\left(\tfrac{1}{3000}\right) + 18\left(\tfrac{25}{3000}\right) + (-2)\left(\tfrac{2973}{3000}\right)$$
$$\approx -\$0.72$$

34. City 1: Expected value $= 0.4(20) + 0.6(-4) = 5.6$

City 2: Expected value $= 0.25(50) + 0.75(-9) = 5.75$

The company should open the store in city 2.

36. (a)

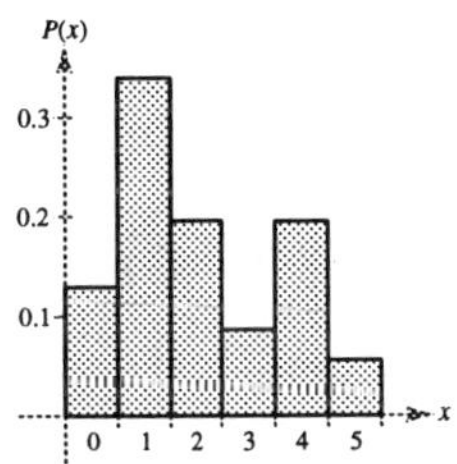

(b) $E(x) = 0(0.129) + 1(0.34) + 2(0.196) + 3(0.086) + 4(0.195) + 5(0.055)$

$= 2.045$

$V(x) = 2.215$

$\sigma = 1.488$

38. First option: Expected value $= 10{,}000 + 0.05(10{,}000) = \$10{,}500$

Second option: $0.5(7000) + 0.5(13{,}000) + 500 = \$10{,}500$

Same!

No, the choice would not change.

Section 9.2 Continuous Random Variables

2. $f(x) = \dfrac{1}{5} > 0$ on $[0, 5]$

$$\int_0^5 \frac{1}{5}\,dx = \frac{1}{5}x\bigg]_0^5 = 1$$

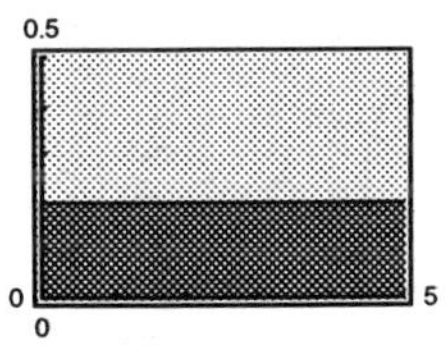

4. $f(x) = \dfrac{x}{18} \ge 0$ on $[0, 6]$

$$\int_0^6 \frac{x}{18}\,dx = \frac{x^2}{36}\bigg]_0^6 = 1$$

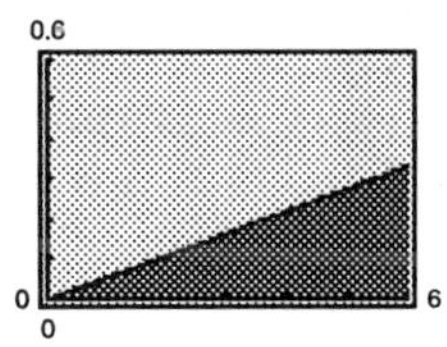

6. $f(x) = \dfrac{x(6-x)}{36} \ge 0$ on $[0, 6]$

$$\int_0^6 \frac{x(6-x)}{36}\,dx = \frac{1}{36}\int_0^6 (6x - x^2)\,dx$$

$$= \frac{1}{36}\left(3x^2 - \frac{x^3}{3}\right)\Bigg]_0^6 = 1$$

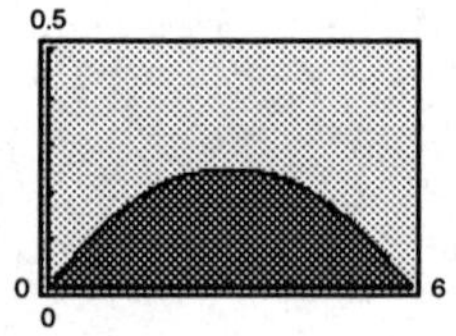

8. $f(x) = \dfrac{1}{6}e^{-x/6} > 0$ on $[0, \infty)$

$$\int_0^\infty \frac{1}{6}e^{-x/6}\,dx = \lim_{b\to\infty}\left[-e^{-x/6}\right]_0^b = 1$$

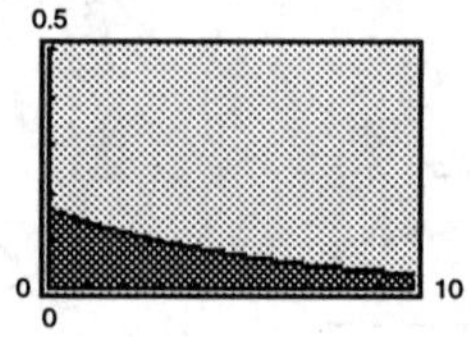

10. $f(x) = 12x^2(1-x) \ge 0$ on $[0, 1]$

$$\int_0^1 (12x^2 - 12x^3)\,dx = \left[4x^3 - 3x^4\right]_0^1 = 1$$

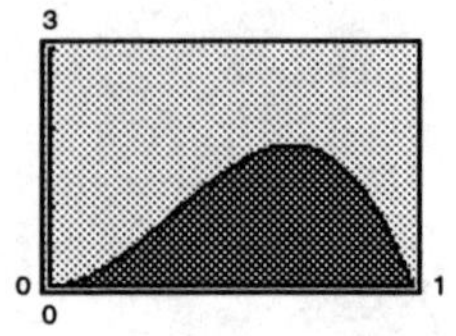

12. $f(x) = \dfrac{2}{9}x(3-x) \ge 0$ on $[0, 3]$

$$\int_0^3 \left(\frac{2}{3}x - \frac{2}{9}x^2\right)dx = \left[\frac{x^2}{3} - \frac{2x^3}{27}\right]_0^3 = 1$$

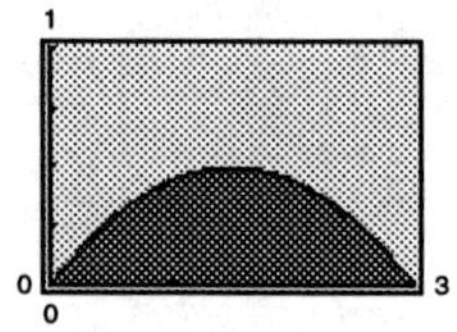

14. $f(x) = \dfrac{1}{4} > 0$ on $[8, 12]$

$$\int_8^{12} \frac{1}{4}\,dx = \frac{1}{4}x\Bigg]_8^{12} = 1$$

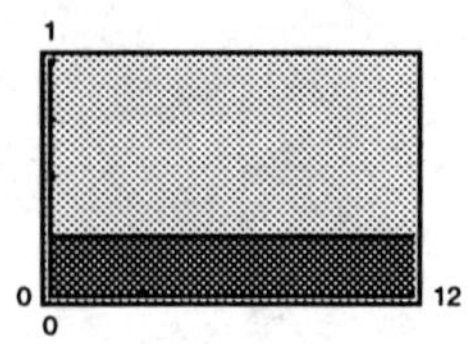

16. $\displaystyle\int_0^4 kx^3\,dx = \frac{kx^4}{4}\Bigg]_0^4 = 64k = 1 \Rightarrow k = \frac{1}{64}$

18. $\displaystyle\int_0^1 k\sqrt{x}(1-x)\,dx = k\int_0^1 (x^{1/2} - x^{3/2})\,dx$

$$= k\left[\frac{2}{3}x^{3/2} - \frac{2}{5}x^{5/2}\right]_0^1$$

$$= \frac{4k}{15} = 1 \Rightarrow k = \frac{15}{4}$$

20. $\displaystyle\int_a^b \frac{k}{b-a}\,dx = \frac{kx}{b-a}\Bigg]_a^b$

$$= k = 1 \Rightarrow k = 1$$

22. $\displaystyle\int_a^b \frac{x}{50}\,dx = \left[\frac{x^2}{100}\right]_a^b = \frac{b^2 - a^2}{100}$

(a) $P(0 < x < 6) = \dfrac{36 - 0}{100} = \dfrac{9}{25}$

(b) $P(4 < x < 6) = \dfrac{36 - 16}{100} = \dfrac{1}{5}$

(c) $P(8 < x < 10) = \dfrac{100 - 64}{100} = \dfrac{9}{25}$

(d) $P(x \ge 2) = P(2 \le x \le 10) = \dfrac{100 - 4}{100} = \dfrac{24}{25}$

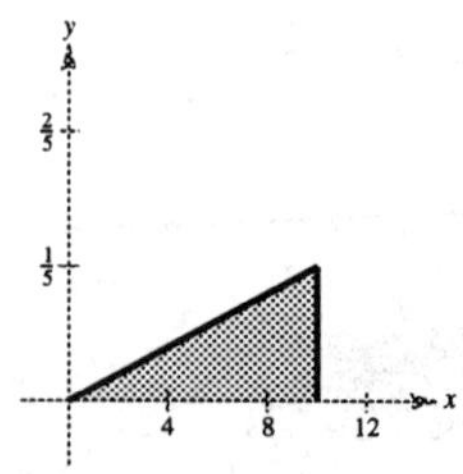

24. $\displaystyle\int_a^b \frac{5}{4(x+1)^2}\,dx = -\frac{5}{4(x+1)}\bigg]_a^b = -\frac{5}{4}\left(\frac{1}{b+1} - \frac{1}{a+1}\right)$

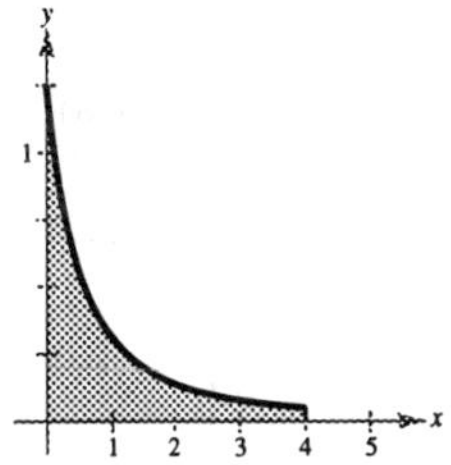

(a) $P(0 < x < 2) = -\frac{5}{4}\left(\frac{1}{3} - 1\right) = \frac{5}{6}$

(b) $P(2 < x < 4) = -\frac{5}{4}\left(\frac{1}{5} - \frac{1}{3}\right) = \frac{1}{6}$

(c) $P(1 < x < 3) = -\frac{5}{4}\left(\frac{1}{4} - \frac{1}{2}\right) = \frac{5}{16}$

(d) $P(x \le 3) = P(0 \le x \le 3) = -\frac{5}{4}\left(\frac{1}{4} - 1\right) = \frac{15}{16}$

26. $\displaystyle\int_a^b \frac{3}{256}(16 - t^2)dt = \frac{3}{256}\left[16t - \frac{t^3}{3}\right]_a^b = \frac{t}{256}\left[48 - t^2\right]_a^b = \frac{48(b-a) + (a^3 - b^3)}{256}$

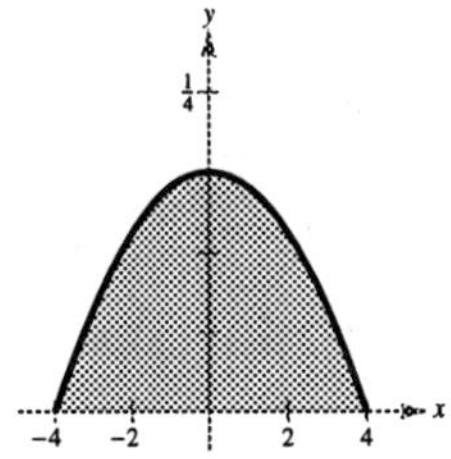

(a) $P(t < -2) = P(-4 < t < -2) = \dfrac{48(2) + (-64 + 8)}{256} = \dfrac{5}{32}$

(b) $P(t > 2) = P(2 < t < 4) = \dfrac{48(2) + (8 - 64)}{256} = \dfrac{5}{32}$

(c) $P(-1 < t < 1) = \dfrac{48(2) + (-1 - 1)}{256} = \dfrac{47}{128}$

(d) $P(t > -2) = 1 - P(t < -2) = 1 - \dfrac{5}{32} = \dfrac{27}{32}$

28. Using substitution, let $u = 9 - t$ and then $t = 9 - u$ and $dt = -du$.

$$\int \frac{5}{324} t\sqrt{9-t}\,dt = \frac{5}{324}\int (9-u)u^{1/2}(-du) = \frac{5}{324}\int (u^{3/2} - 9u^{1/2})du = \frac{5}{324}\left(\frac{2}{5}u^{5/2} - 6u^{3/2}\right)$$

$$\int_a^b \frac{5}{324} t\sqrt{9-t}\,dt = \frac{5}{324}\left[\frac{2}{5}(9-t)^{5/2} - 6(9-t)^{3/2}\right]_a^b$$

(a) $P(t < 3) = \displaystyle\int_0^3 f(t)dt = 1 - \frac{1}{3}\sqrt{6} \approx 0.1835$

(b) $P(4 < t < 8) = \displaystyle\int_4^8 f(t)dt = \frac{25}{81}\sqrt{5} - \frac{7}{81} \approx 0.6037$

30. $\displaystyle\int_a^b \frac{1}{5}e^{-t/5}\,dt = -e^{-t/5}\bigg]_a^b = e^{-a/5} - e^{-b/5}$

(a) $P(0 < t < 6) = 1 - e^{-6/5} \approx 0.699$

(b) $P(2 < t < 6) = e^{-2/5} - e^{-6/5} \approx 0.369$

(c) $P(t > 8) = e^{-8/5} - \displaystyle\lim_{b\to\infty} e^{-b/5}$

$= e^{-8/5} \approx 0.202$

32. $\displaystyle\int_0^x \frac{1}{3.5}e^{-t/3.5}\,dt = 1 - 0.90$

$$-e^{-t/3.5}\bigg]_0^x = 0.10$$

$$1 - e^{-x/3.5} = 0.10$$

$$e^{-x/3.5} = 0.90$$

$$-\frac{x}{3.5} = \ln 0.90$$

$$x = -3.5 \ln 0.90$$

$x \approx 0.369$ year

34. $\displaystyle\int_a^b \frac{1}{36}xe^{-x/16}\,dx = \frac{-1}{6}e^{-x/6}(x+6)\Big]_a^b$

$$P(0 < x < b) = 0.90$$

$$-\frac{1}{6}e^{-x/6}(x+6)\Big]_0^b = 0.90$$

$$-\frac{1}{6}\Big[e^{-b/6}(b+6) - 6\Big] = 0.90$$

$$-\frac{1}{6}(b+6)e^{-b/6} = -0.10$$

$$(b+6)e^{-b/6} = 0.60$$

Solving for b with a graphing utility or using a computer, $b \approx 23.34$ tons.

36. Using substitition, let $u = 1 - x$, then $x = 1 - u$ and $dx = -du$.

$$\int_a^b \frac{1155}{32}x^3(1-x)^{3/2}\,dx = -\frac{1}{16}(1-x)^{5/2}(105x^3 + 70x^2 + 40x + 16)\Big]_a^b$$

(a) $P(0 < x < 0.25) \approx 0.025$

(b) $P(0.50 < x < 1.00) \approx 0.736$

Section 9.3 Expected Value and Variance

2. (a) $\displaystyle\mu = \int_a^b xf(x)dx = \int_0^4 x\left(\frac{1}{4}\right)dx = \frac{x^2}{8}\Big]_0^4 = 2$

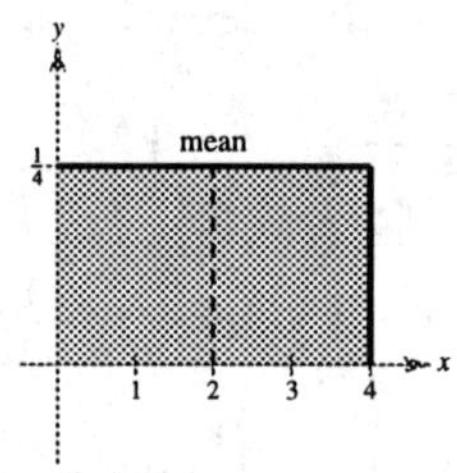

(b) $\displaystyle\sigma^2 = \int_a^b x^2 f(x)dx - \mu^2$

$$= \int_0^4 x^2\left(\frac{1}{4}\right)dx - (2)^2$$

$$= \frac{x^3}{12}\Big]_0^4 - 4 = \frac{16}{3} - \frac{12}{3} = \frac{4}{3}$$

(c) $\displaystyle\sigma = \frac{2}{\sqrt{3}}$

4. (a) $\displaystyle\mu = \int_a^b xf(x)dx = \int_1^4 x\left(\frac{4}{3x^2}\right)dx = \frac{4}{3}\int_1^4 \frac{1}{x}\,dx = \frac{4}{3}\ln|x|\Big]_1^4 = \frac{4}{3}\ln 4 \approx 1.848$

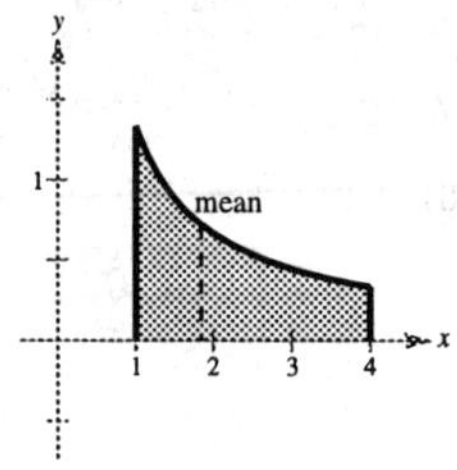

(b) $\displaystyle\sigma^2 = \int_a^b x^2 f(x)dx - \mu^2$

$$= \int_1^4 x^2\left(\frac{4}{3x^2}\right)dx - \left[\frac{4}{3}\ln 4\right]^2$$

$$= \frac{4}{3}x\Big]_1^4 - \frac{16}{9}(\ln 4)^2$$

$$= 4 - \frac{16}{9}(\ln 4)^2 \approx 0.583$$

(c) $\sigma = \sqrt{0.583} \approx 0.76$

6. (a) $\mu = \int_a^b xf(x)dx$

$= \int_0^4 x\left(\frac{3}{16}\sqrt{4-x}\right)dx$

$= \frac{3}{16}\int_0^4 x\sqrt{4-x}\,dx$

$= -\frac{1}{40}(4-x)^{3/2}(3x+8)\Big]_0^4 = \frac{8}{5} = 1.6$

(b) $\sigma^2 = \int_a^b x^2 f(x)dx - \mu^2$

$= \int_0^4 x^2\left(\frac{3}{16}\sqrt{4-x}\right)dx - \left(\frac{8}{5}\right)^2$

$= \frac{3}{16}\int_0^4 x^2\sqrt{4-x}\,dx - \frac{64}{25}$

$= -\frac{1}{280}(4-x)^{3/2}(15x^2+48x+128)\Big]_0^4 - \frac{64}{25}$

$= \frac{128}{35} - \frac{64}{25} = \frac{192}{175} \approx 1.097$

(c) $\sigma \approx 1.047$

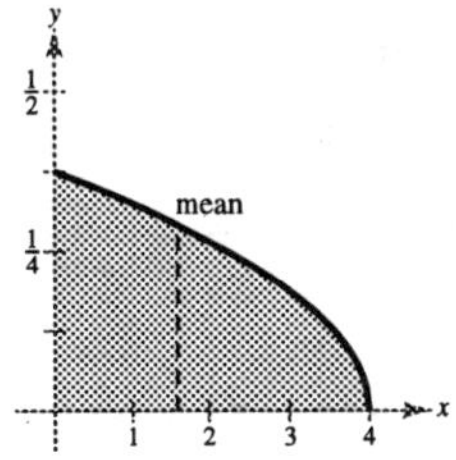

8. $\mu = \int_a^b xf(x)dx = \int_0^4 x\left[\frac{3}{32}x(4-x)\right]dx = \frac{3}{32}\int_0^4 (4x^2 - x^3)dx = \frac{3}{32}\left(\frac{4x^3}{3} - \frac{x^4}{4}\right)\Big]_0^4 = 2$

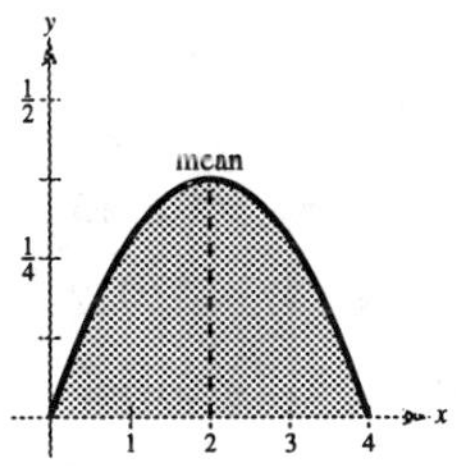

10. $\mu = \int_a^b xf(x)dx = \int_0^9 x\left(\frac{1}{18}\sqrt{9-x}\right)dx = \frac{1}{18}\int_0^9 x\sqrt{9-x}\,dx = -\frac{1}{45}(9-x)^{3/2}(x+6)\Big]_0^9 = \frac{18}{5} = 3.6$

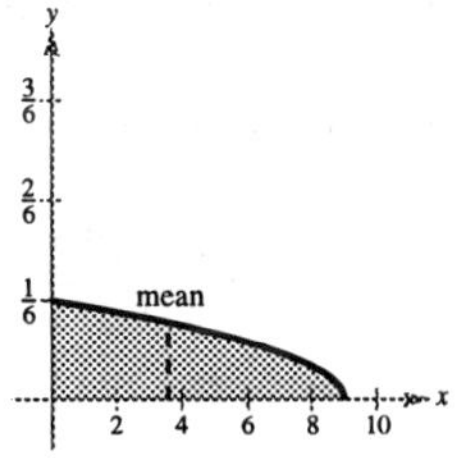

12. $P(0 < x < m) = \int_0^m \frac{2}{5}e^{-2t/5}\,dt = -e^{-2t/5}\Big]_0^m = 1 - e^{-2m/5} = \frac{1}{2}$

$e^{-2m/5} = \frac{1}{2}$

$-\frac{2m}{5} = \ln\frac{1}{2}$

$m = -\frac{5}{2}\ln\frac{1}{2} = \frac{5}{2}\ln 2 \approx 1.733$

14. $f(x) = \frac{1}{9}$, $[0, 9]$ is a uniform probability density function.

Mean: $\dfrac{0+9}{2} = \dfrac{9}{2}$

Variance: $\dfrac{(9-0)^2}{12} = \dfrac{27}{4}$

Standard deviation: $\dfrac{9-0}{\sqrt{12}} = \dfrac{3\sqrt{3}}{2}$

16. $f(x) = \frac{5}{3}e^{-5x/3}$, $[0, \infty)$ is an exponential probability density function.

Mean: $\dfrac{1}{5/3} = \dfrac{3}{5}$

Variance: $\dfrac{1}{(5/3)^2} = \dfrac{9}{25}$

Standard deviation: $\dfrac{1}{5/3} = \dfrac{3}{5}$

18. $f(x) = \dfrac{1}{6\sqrt{2\pi}}e^{-(x-30)^2/72}$, $(-\infty, \infty)$
is a normal probability density function.

Mean: $\mu = 30$

Variance: $\sigma^2 = 36$

Standard deviation: $\sigma = 6$

20. Mean $= 0$

Standard deviation $= 1$

$P(-1.21 \le x \le 1.21) \approx 0.7737$

22. Mean $= \frac{4}{3}$

Standard deviation $= \frac{4}{3}$

$P(x \ge 0.27) \approx 0.8167$

24. Mean $= 2$

Standard deviation $= 1.5$

$P(-2.5 \le x \le 2.5) \approx 0.6292$

26. $\mu = 75$, $\sigma = 12$

(a) $P(x > 69) = 0.6915$

(b) $P(x < 99) = 0.9772$

(c) $P(x < 57) = 0.0668$

(d) $P(63 < x < 78) = 0.4401$

28. $f(x) = \frac{1}{5}$, $[0, 5]$ where $t = 0$ represents 10:00 A.M. and $t = 5$ represents 10:05 A.M.

(a) Mean: $\mu = \frac{5}{2}$ or 2.5 minutes after 10:00 A.M.

Standard deviation: $\sigma = \dfrac{5}{\sqrt{12}} = \dfrac{5\sqrt{3}}{6} \approx 1.443$

(b) $P(3 < t < 5) = \displaystyle\int_3^5 \frac{1}{5}\,dt = \frac{1}{5}t\Big]_3^5 = \frac{2}{5} = 0.40$

30. $\mu = \dfrac{1}{a} = 2 \Rightarrow a = \dfrac{1}{2}$

(a) $f(t) = ae^{-at} = \dfrac{1}{2}e^{-t/2}$

(b) $P(t > 3) = \displaystyle\int_3^\infty \frac{1}{2}e^{-t/2}\,dt = -e^{-t/2}\Big]_3^\infty = 0 + e^{-3/2} \approx 0.223$

32. $\mu = \dfrac{1}{a} = 3.5 \Rightarrow a = \dfrac{2}{7}$

(a) $f(t) = ae^{-at} = \dfrac{2}{7}e^{-2t/7}$

(b) $\mu = \sigma = \dfrac{1}{a} = \dfrac{7}{2}$

$P(\mu - \sigma < t < \mu + \sigma) = P(0 < t < 7) = \displaystyle\int_0^7 \frac{2}{7}e^{-2t/7}\,dt = -e^{-2t/7}\Big]_0^7 = 1 - e^{-2} \approx 0.865$

34. (a) $\dfrac{x-\mu}{\sigma} = \dfrac{340-400}{24} = -2.5$

(b) $P(x > 340) = P(z > -2.5) = P(-2.5 < z < 0) + 0.5000 = 0.4938 + 0.5000 = 0.9938$

36. (a) $\mu = \displaystyle\int_2^{10} \frac{3}{256}x(x-2)(10-x)dx = \frac{3}{256}\int_2^{10}(-x^3+12x^2-20x)dx = \frac{3}{256}\left[-\frac{x^4}{4}+4x^3-10x^2\right]_2^{10} = 6$

$$\sigma^2 = \int_2^{10}\frac{3}{256}x^2(x-2)(10-x)dx - (6)^2$$

$$= \frac{3}{256}\int_2^{10}(-x^4+12x^3-20x^2)dx - 36 = \frac{3}{256}\left[-\frac{x^5}{5}+3x^4-\frac{20x^3}{3}\right]_2^{10} = \frac{196}{5} - 36 = \frac{16}{5}$$

$$\sigma = \sqrt{\frac{16}{5}} = \frac{4}{\sqrt{5}} \approx 1.789$$

(b) $\displaystyle\int_2^m \frac{3}{256}(x-2)(10-x)dx = \frac{3}{256}\int_2^m(-x^2+12x-20)dx = \frac{3}{256}\left[-\frac{x^3}{3}+6x^2-20x\right]_2^m$

$$= \frac{3}{256}\left(-\frac{m^3}{3}+6m^2-20m\right) - \frac{3}{256}\left(-\frac{56}{3}\right) = \frac{1}{2}$$

Therefore, $-m^3 + 18m^2 - 60m + 56 = 128$

$$0 = m^3 - 18m^2 + 60m + 72.$$

By the Rational Zero Theorem or a graphing utility, we have $m = 6$.

(c) $P\left(6 - \dfrac{4}{\sqrt{5}} < x < 6 + \dfrac{4}{\sqrt{5}}\right) \approx P(4.2111 < x < 7.7889)$

$$= \int_{4.2111}^{7.7889}\frac{3}{256}(x-2)(10-x)dx = \frac{3}{256}\left[-\frac{x^3}{3}+6x^2-20x\right]_{4.2111}^{7.7889} \approx 0.626$$

38. $\mu = \displaystyle\int_0^1 \frac{1155}{32}x^4(1-x)^{3/2}dx = \frac{1155}{32}\int_0^1 x^4(1-x)^{3/2}dx$

Let $u = \sqrt{1-x}$, then $x = 1 - u^2$ and $dx = -2u\,du$.

$$\frac{1155}{32}\int_0^1 x^4(1-x)^{3/2}dx = \frac{1155}{32}\int_1^0(1-u^2)^4u^3(-2u)du = -\frac{1155}{16}\int_1^0(u^4-4u^6+6u^8-4u^{10}+u^{12})du$$

$$= -\frac{1155}{16}\left[\frac{u^5}{5}-\frac{4u^7}{7}+\frac{2u^9}{3}-\frac{4u^{11}}{11}+\frac{u^{13}}{13}\right]_1^0 \approx 0.615$$

40. (a) $P(t > 4) = \displaystyle\int_4^6 \frac{1}{2\sqrt{t-2}}dt \approx 0.5859$

(b) $\mu = \displaystyle\int_3^6 t\frac{1}{2\sqrt{t-2}}dt \approx 4.333$

42. Mean = 12.5 = Median

44. Mean $= \displaystyle\int_0^3 x\left(\frac{4}{3}-\frac{2}{3}x\right)dx = 0$

Median $= \displaystyle\int_0^m\left(\frac{4}{3}-\frac{2}{3}x\right)dx = \frac{1}{2}$

$$\left[\frac{4}{3}x - \frac{x^2}{3}\right]_0^m = \frac{1}{2}$$

$$\frac{4}{3}m - \frac{m^2}{3} = \frac{1}{2} \Rightarrow m \approx 0.4189$$

46. Mean $= \dfrac{3}{2}$

Median $= \dfrac{3}{2}\ln 2 \approx 1.0397$

48. $f(x) = \dfrac{2x+1}{10}, \quad 1 \le x \le 2$

$$\text{Expected value} = \int_1^2 x\left(\frac{2x+1}{10}\right) dx$$
$$= \frac{1}{10}\left[\frac{2x^3}{3} + \frac{x^2}{2}\right]_1^2$$
$$= \frac{37}{60}$$

50. (a) $\mu = 5.3, \ \sigma = 0.8$

(b) We want to find x_0 such that $P(x < x_0) = 0.1$. By trial and error, or using a graphing utility, $x \approx 4.275$ years.

52. $u = 4.5, \ \sigma = 0.5$

$$P(4 < x < 5) = \int_4^5 \frac{1}{0.5\sqrt{2\pi}} e^{-(x-4.5)^2/2(0.5)^2}\, dx$$
$$= \frac{2}{\sqrt{2\pi}} \int_4^5 e^{-2(x-4.5)^2}\, dx$$
$$\approx 0.6826$$

54. $f(x) = \dfrac{1}{16\sqrt{2\pi}} e^{-(x-266)^2/2(16)^2}$

$$= \frac{1}{40.106} e^{-(x-266)^2/512}$$

(a)

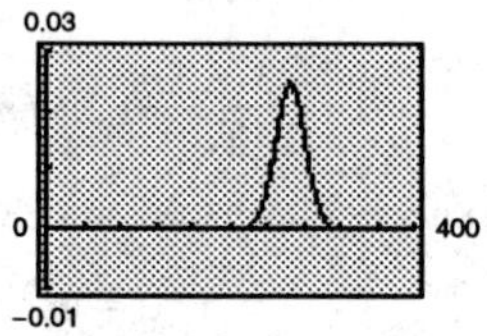

(b) $P(240 \le x \le 280) = \displaystyle\int_{240}^{280} f(x)dx \approx 0.757$ or 76%

Review Exercises for Chapter 9

2. The sample space consists of the letters
$S = \{c, a, l, u, s\}$

4. If the prizes are numbered 1, 2, 3, 4, and 5,
$S = \{12, 13, 14, 15, 23, 24, 25, 34, 35, 45\}$

6. $S = \{$SSSS, SSSU, SSUS, SUSS, USSS, SSUU, SUSU, SUUS, USUS, UUSS, USSU, SUUU, USUU, UUSU, UUUS, UUUU$\}$

8.

x	0	1	2	3	4
$n(x)$	1	4	6	4	1

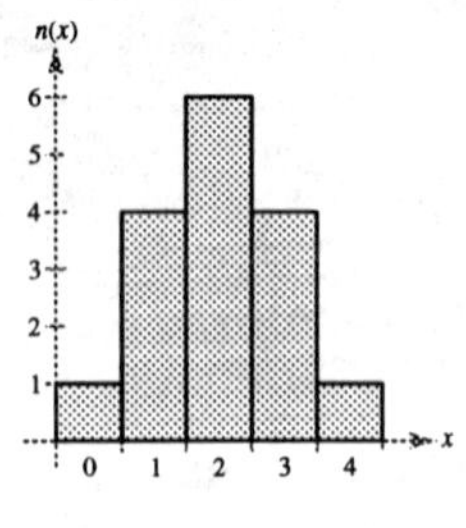

10. (a) $P(x < 0) = \frac{1}{11} + \frac{2}{11} = \frac{3}{11}$

(b) $P(x > 1) = \frac{3}{11} + \frac{1}{11} = \frac{4}{11}$

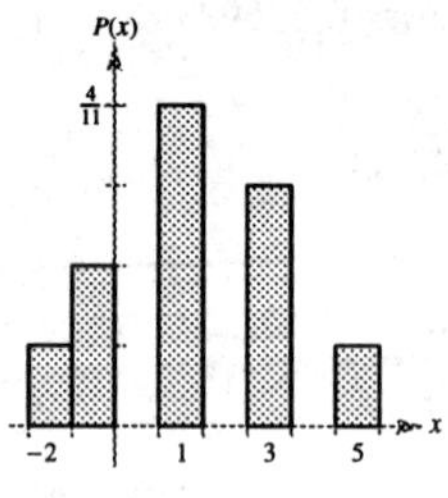

12. (a) P (face card) $= \dfrac{12}{52} = \dfrac{3}{13}$

(b) P (not a face card) $= 1 - \dfrac{3}{13} = \dfrac{10}{13}$

(c) P (black card that is not a face card) $= \dfrac{5}{13}$

(d) P (6 or less) $= \dfrac{24}{52} = \dfrac{6}{13}$ (ace = 1)

14. Mean $= \dfrac{(0.08)(650) + (0.23)(375) + (0.16)(800)}{3}$

$$= \frac{266.25}{3} = \$88.75$$

16.

x	\$1997	\$997	\$17	$-$\$3
$P(x)$	$\frac{1}{2000}$	$\frac{1}{2000}$	$\frac{50}{2000}$	$\frac{1948}{2000}$

$E(x) = 1997\left(\frac{1}{2000}\right) + 997\left(\frac{1}{2000}\right) + 17\left(\frac{50}{2000}\right) + (-3)\left(\frac{1948}{2000}\right) = -\$1.00$

OR

x	\$2000	\$1000	\$20	\$0
$P(x)$	$\frac{1}{2000}$	$\frac{1}{2000}$	$\frac{50}{2000}$	$\frac{1948}{2000}$

$E(x) = 2000\left(\frac{1}{2000}\right) + 1000\left(\frac{1}{2000}\right) + 20\left(\frac{50}{2000}\right) + 0\left(\frac{1948}{2000}\right) = \$2$

Expected net gain: \$2.00 − \$3.00 = −\$1.00

18. Following the pattern in Exercise 17, or using a graphing utility, you obtain

$E(x) \approx 52.547$

$V(x) \approx 51.56$

$\sigma \approx 7.18$

20. $E(x) = 0(0.12) + 1(0.31) + 2(0.43) + 3(0.12) + 4(0.02)$

$= 1.61$

$V(x) = (0 - 1.61)^2(0.12) + (1 - 1.61)^2(0.31) + (2 - 1.61)^2(0.43) + (3 - 1.61)^2(0.12) + (4 - 1.61)^2(0.02)$

$= 0.8379$

$\sigma = \sqrt{V(x)} \approx 0.9154$

22.

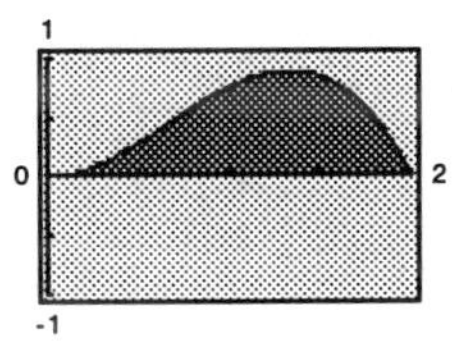

$$\int_0^2 \frac{3}{4}x^2(2 - x)\,dx = \frac{3}{4}\int_0^2 (2x^2 - x^3)\,dx = \frac{3}{4}\left[\frac{2x^3}{3} - \frac{x^4}{4}\right]_0^2 = 1$$

24.

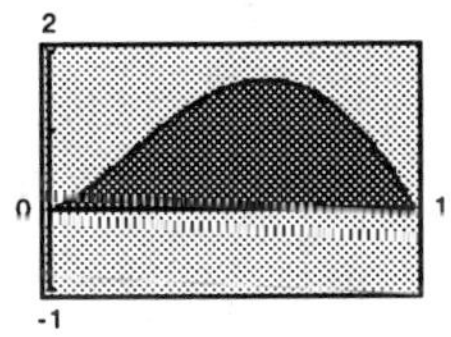

$$\int_0^1 8.75x^{3/2}(1 - x)\,dx = 8.75\int_0^1 (x^{3/2} - x^{5/2})\,dx = 8.75\left[\frac{2}{5}x^{5/2} - \frac{2}{7}x^{7/2}\right]_0^1 = 1$$

26. $$P(-1 < x < 2) = \int_{-1}^2 \frac{1}{36}(9 - x^2)\,dx = \frac{1}{36}\left[9x - \frac{x^3}{3}\right]_{-1}^2 = \frac{2}{3}$$

28. $$P(4 < x < 9) = \int_4^9 \frac{3}{128}\sqrt{x}\,dx = \frac{3}{128}\left(\frac{2}{3}x^{3/2}\right)\Big]_4^9 = \frac{19}{64}$$

30. $P(t < 4) = \int_0^4 \frac{5}{324} t\sqrt{9-t}\,dt$

Using substitution ($u = 9 - t, t = 9 - u, dt = -du$), or a graphing utility, you obtain

$P(t < 4) \approx 0.3099.$

32. $P(x \geq 9) = 1 - P(x \leq 9)$

$$= 1 - \int_0^9 \frac{1}{64} xe^{-x/8}\,dx$$

$$= 1 - \left[-\frac{1}{8}e^{-x/8}(x+8)\right]_0^9$$

$$= 1 + \frac{1}{8}\left[e^{-x/8}(x+8)\right]_0^9$$

$$= \frac{17}{8}e^{-9/8} \approx 0.690$$

34. $\text{Mean} = \int_0^8 x\left(\frac{8-x}{32}\right)dx$

$$= \frac{1}{32}\left[4x^2 - \frac{x^3}{3}\right]_0^8 = \frac{8}{3}$$

36. $\text{Mean} = \frac{10}{3}$

38. $\text{Mean} = \int_0^4 x\left(\frac{3}{16}\right)\sqrt{x}\,dx = 2.4$

$\text{Variance} = \int_0^4 (x - 2.4)^2 \frac{3}{16}\sqrt{x}\,dx \approx 1.097$

Standard deviation ≈ 1.047

40. Mean $= 1.25$

Variance $= 1.5625$

Standard deviation $= 1.25$

42. $\int_0^m 12x^2(1-x)dx = 4x^3 - 3x^4\Big]_0^m = 4m^3 - 3m^4 = \frac{1}{2} \Rightarrow m \approx 0.6143$

44. $\int_0^m \frac{5}{6}e^{-5x/6}\,dx = 1 - e^{-5m/6}$

$$= \frac{1}{2} \Rightarrow m \approx 0.8318$$

46. $P(t > 400) = \int_{400}^{\infty} f(t)dt$

$$= 1 - \int_0^{400} f(t)dt \approx 0.3189$$

48. $\mu = 14.50,\ \sigma = 1.40$

$f(x) = \frac{1}{\sigma\sqrt{2\pi}}e^{-(x-\mu)^2/2\sigma^2}$

(a) $P(13 < x < 15) = \int_{13}^{15} f(x)\,dx$

≈ 0.4975 or 49.75%

(b) $P(x > m) = 0.10$

By trial and error, $m \approx \$16.29$.

50. $f(x) = \frac{1}{25}xe^{-x/5},\ (0, \infty)$

(a) $P(x < 5) \approx 0.1813$

(b) $P(5 < x < 10) \approx 0.1484$

(c) $P(x > 10) \approx 0.6703$

CHAPTER 10
Series and Taylor Polynomials

CHAPTER 10
Series and Taylor Polynomials

Section 10.1 Sequences

Solutions to Even-Numbered Exercises

2. $\frac{1}{2}, \frac{2}{3}, \frac{3}{4}, \frac{4}{5}, \frac{5}{6}$

4. $0, \frac{1}{6}, \frac{2}{11}, \frac{3}{18}, \frac{4}{27}$

6. $5, \frac{19}{4}, \frac{43}{9}, \frac{77}{16}, \frac{121}{25}$

8. $3, 6, 9, 12, 15$

10. This sequence diverges since $\lim_{n\to\infty} \frac{n}{2} = \infty$.

12. This sequence converges since $\lim_{n\to\infty} \frac{1}{n^{3/2}} = 0$.

14. This sequence converges since

$$\lim_{n\to\infty} \frac{\sqrt{n}}{\sqrt{n+1}} = \lim_{n\to\infty} \sqrt{\frac{n}{n+1}} = \sqrt{1} = 1.$$

16. This sequence diverges since $\lim_{n\to\infty} [1 + (-1)^n]$ does not exist.

18. This sequence diverges since

$$\lim_{n\to\infty} \frac{n!}{n} = \lim_{n\to\infty} (n-1)! = \infty.$$

20. This sequence converges since

$$\lim_{n\to\infty} \frac{n}{\sqrt{n^2+1}} = \lim_{n\to\infty} \sqrt{\frac{n^2}{n^2+1}} = \sqrt{1} = 1.$$

22. This sequence converges since $\lim_{n\to\infty} (0.5)^n = 0$.

24. This sequence converges since

$$\lim_{n\to\infty} \frac{(n-2)!}{n!} = \lim_{n\to\infty} \frac{1}{n(n-1)} = 0.$$

26. This sequence converges since

$$\lim_{n\to\infty} \left(\frac{n-1}{n} - \frac{n}{n-1}\right) = \lim_{n\to\infty} \frac{1-2n}{n^2-n} = 0.$$

28. The terms of the sequence seem to approach 1.

Hence, $\lim_{n\to\infty} a_n = \lim_{n\to\infty} \frac{n}{n+2} = 1$.

30. $a_n = 4n - 1$

32. $a_n = \frac{1}{n^2}$

34. $a_n = \frac{n+1}{2n-1}$

36. $a_n = \frac{2^{n-1}}{3^n}$

38. $a_n = 1 + \frac{1}{2^n}$

40. $a_n = (-1)^{n+1}(2n)$

42. $a_n = \frac{x^{n-1}}{(n-1)!}$

44. Since $a_n = 3 + \frac{n}{2}$, we have $a_5 = \frac{11}{2}$ and $a_6 = 6$.

46. Since $a_n = \frac{3}{4}n - \frac{1}{4}$, we have $a_5 = \frac{14}{4} = \frac{7}{2}$ and $a_6 = \frac{17}{4}$.

48. Since $a_n = 5(2)^{n-1}$, we have $a_5 = 80$ and $a_6 = 160$.

50. Since $a_n = 9\left(\frac{2}{3}\right)^{n-1}$, the next two terms are $\frac{16}{9}$ and $\frac{32}{27}$.

52. Since $a_n = 108 - 8n$, the sequence is arithmetic.

54. Since $a_n = 378\left(-\dfrac{1}{3}\right)^{n-1}$, the sequence is geometric.

56. One example is $a_n = 100 - \dfrac{1}{n}$.

58. $A_n = 5000(1.10)^n$

$A_1 = \$5500.00$ $\quad A_6 \approx \$8857.81$

$A_2 = \$6050.00$ $\quad A_7 \approx \$9743.59$

$A_3 = \$6655.00$ $\quad A_8 \approx \$10{,}717.94$

$A_4 = \$7320.50$ $\quad A_9 \approx \$11{,}789.74$

$A_5 = \$8052.55$ $\quad A_{10} \approx \$12{,}968.71$

60. $S_6 = 130 + 70 + 40 = 240$

$S_7 = 240 + 130 + 70 = 440$

$S_8 = 440 + 240 + 130 = 810$

$S_9 = 810 + 440 + 240 = 1490$

$S_{10} = 1490 + 810 + 440 = 2740$

62. $a_n = 4.232e^{0.0458n} - 3.453$

$a_0 = 4.232e^{0.0458(0)} - 3.453 = 0.779$ (1980)

$a_1 = 0.977$ $\quad a_9 = 2.938$

$a_2 = 1.185$ $\quad a_{10} = 3.237$

$a_3 = 1.402$ $\quad a_{11} = 3.551$

$a_4 = 1.630$ $\quad a_{12} = 3.879$

$a_5 = 1.868$ $\quad a_{13} = 4.223$

$a_6 = 2.117$ $\quad a_{14} = 4.583$

$a_7 = 2.379$ $\quad a_{15} = 4.959$

$a_8 = 2.652$ $\quad a_{16} = 5.353$ (1996)

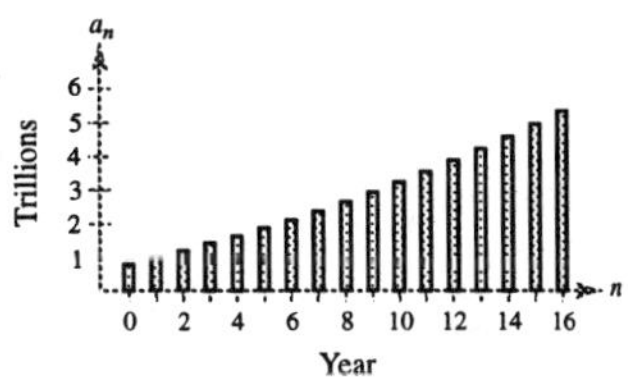

64. (a) $S_1 = \dfrac{1(1+1)}{2} = 1$

$S_2 = \dfrac{2(2+1)}{2} = 3 = 1 + 2$

$S_3 = \dfrac{3(3+1)}{2} = 6 = 1 + 2 + 3$

$S_4 = \dfrac{4(4+1)}{2} = 10 = 1 + 2 + 3 + 4$

$S_5 = \dfrac{5(5+1)}{2} = 15 = 1 + 2 + 3 + 4 + 5$

(b) $S_{50} = \dfrac{50(50+1)}{2} = 1275$

66. (a) $h_n = 12\left(\frac{2}{3}\right)^n$

(b) This sequence converges since $\lim_{n\to\infty} 12\left(\frac{2}{3}\right)^n = 0$.

68. $P_n = 13{,}000(1.055)^n$

$P_1 = \$13{,}715$

$P_2 = \$14{,}469.33$

$P_3 = \$15{,}265.14$

$P_4 = \$16{,}104.72$

$P_5 = \$16{,}990.48$

70. $A_1 = 16$

$A_2 = 16 + 16.10 = 16(1 + 0.10)$

$A_3 = 16(1 + 0.10 + 0.20) = 1.6(10 + 1 + 2)$

$A_4 = 1.6(10 + 1 + 2 + 3)$

$\vdots$

$A_{100} = 1.6(10 + 1 + 2 + \cdots + 99)$

From Exercise 64, $1 + 2 + \cdots + 99 = \dfrac{99(100)}{2} = 4950.$

Hence, $A_{100} = 1.6(10 + 4950) = \$7936.00.$

72. (a) $a_n = 0.55996n^2 + 5.32511n + 14.34978$

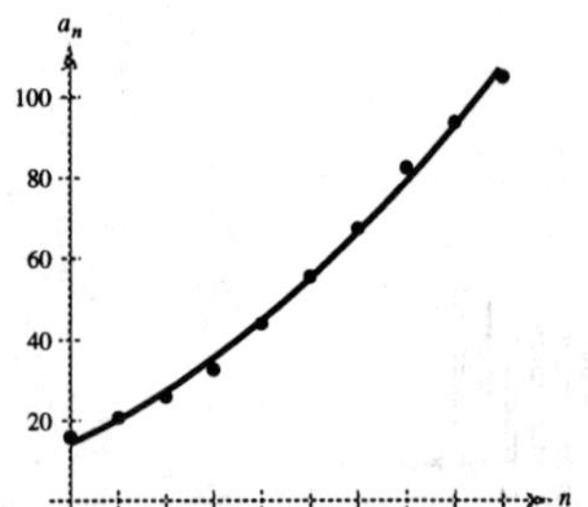

(b) For 2005, $n = 18$ and $a_{18} \approx \$291.6$ billion

Section 10.2 Series and Convergence

2. $S_1 = 3$

$S_2 = -\dfrac{3}{2} = -1.5$

$S_3 = \dfrac{21}{4} = 5.25$

$S_4 = -\dfrac{39}{8} = -4.875$

$S_5 = \dfrac{165}{16} = 10.3125$

4. $S_1 = 1$

$S_2 = \dfrac{1}{2} = 0.5$

$S_3 = \dfrac{4}{6} = \dfrac{2}{3} \approx 0.6667$

$S_4 = \dfrac{15}{24} = \dfrac{5}{8} = 0.625$

$S_5 = \dfrac{76}{120} = \dfrac{19}{30} \approx 0.6333$

6. This series diverges by the nth-Term Test since

$$\lim_{n\to\infty} \frac{n}{2n + 3} = \frac{1}{2} \neq 0.$$

8. This series diverges by the nth-Term Test since

$$\lim_{n\to\infty} \frac{n}{\sqrt{n^2 + 1}} = 1 \neq 0.$$

10. This series diverges by the Test for Convergence of a Geometric Series since $|r| = \left|-\frac{3}{2}\right| > 1.$

12. This series diverges by the Test for Convergence of a Geometric Series since $|r| = |-1.03| > 1.$

14. This series converges by the Test for Convergence of a Geometric Series since $|r| = \left|-\frac{1}{2}\right| < 1.$

16. This series converges by the Test for Convergence of a Geometric Series since $|r| = |-0.6| < 1.$

18. Since $a = 2$ and $r = \frac{2}{3}$, we have

$$S = \frac{2}{1 - (2/3)} = 6.$$

20. Since $a = 2$ and $r = -\frac{2}{3}$, we have

$$S = \frac{2}{1 + (2/3)} = \frac{6}{5}.$$

22. Since $a = 4$ and $r = \frac{1}{4}$, we have

$$S = \frac{4}{1-(1/4)} = \frac{16}{3}.$$

24. Since $a = 8$ and $r = \frac{3}{4}$, we have

$$S = \frac{8}{1-(3/4)} = 32.$$

26. Since $a = 4$ and $r = -\frac{1}{2}$, we have

$$S = \frac{4}{1+(1/2)} = \frac{8}{3}.$$

28. $$\sum_{n=0}^{\infty}[(0.7)^n + (0.9)^n] = \sum_{n=0}^{\infty}(0.7)^n + \sum_{n=0}^{\infty}(0.9)^n = \frac{1}{1-0.7} + \frac{1}{1-0.9} = \frac{10}{3} + 10 = \frac{40}{3}$$

30. $$\sum_{n=0}^{\infty}[(0.4)^n - (0.8)^n] = \sum_{n=0}^{\infty}(0.4)^n - \sum_{n=0}^{\infty}(0.8)^n = \frac{1}{1-0.4} - \frac{1}{1-0.8} = \frac{5}{3} - 5 = -\frac{10}{3}$$

32. This series converges by the Test for Convergence of a Geometric Series since

$$\sum_{n=0}^{\infty}\frac{4}{2^n} = \sum_{n=0}^{\infty}4\left(\frac{1}{2}\right)^n \text{ and } r = \frac{1}{2} < 1.$$

34. This series diverges by the nth-Term Test since $\lim_{n\to\infty}\frac{n+1}{2n-1} = \frac{1}{2} \neq 0$.

36. This series converges by the Test for Convergence of a Geometric Series since $\sum_{n=0}^{\infty}\frac{1}{4^n} = \sum_{n=0}^{\infty}\left(\frac{1}{4}\right)^n$ and $r = \frac{1}{4} < 1$.

38. This series diverges by the Test for Convergence of a Geometric Series since $\sum_{n=1}^{\infty}\frac{2^n}{100} = \sum_{n=1}^{\infty}(0.01)2^n$ and $r = 2 > 1$.

40. This series diverges by the nth-Term Test since $\lim_{n\to\infty} n! = \infty$.

42. $$0.23\overline{23} = \sum_{n=0}^{\infty}0.23(0.01)^n = \frac{0.23}{1-0.01} = \frac{0.23}{0.99} = \frac{23}{99}$$

44. $$0.21\overline{21} = \sum_{n=0}^{\infty}0.21(0.01)^n = \frac{0.21}{1-0.01} = \frac{0.21}{0.99} = \frac{7}{33}$$

46. (a) $$\sum_{i=0}^{n-1}8000(0.75)^i = \frac{8000[1-(0.75)^{(n-1)+1}]}{1-0.75} = 32{,}000(1-0.75^n)$$

(b) $$\sum_{i=1}^{\infty}8000(0.75)^i = \frac{8000}{1-0.75} = 32{,}000$$

48. $$T = 1 + 2\sum_{n=1}^{\infty}(0.8)^n$$

$$= 1 + 2\left[-1 + \sum_{n=0}^{\infty}(0.8)^n\right] = 1 + 2\left[\frac{1}{1-0.8}\right] = -1 + 2\left(\frac{1}{0.2}\right)$$

$$= -1 + 10 = 9 \text{ seconds}$$

50. $A = \sum_{n=1}^{N} P\left(1 + \frac{r}{12}\right)^n = -P + \sum_{n=0}^{N} P\left(1 + \frac{r}{12}\right)^n = -P + \frac{P\left[1 - \left(1 + \frac{r}{12}\right)^{N+1}\right]}{1 - \left(1 + \frac{r}{12}\right)}$

$= -P - \frac{12}{r}P\left[1 - \left(1 + \frac{r}{12}\right)^{N+1}\right] = P\left[-1 - \frac{12}{r} + \frac{12}{r}\left(1 + \frac{r}{12}\right)^{N+1}\right]$

$= P\left[-\left(1 + \frac{12}{r}\right) + \frac{12}{r}\left(1 + \frac{r}{12}\right)\left(1 + \frac{r}{12}\right)^N\right] = P\left[-\left(1 + \frac{12}{r}\right) + \left(\frac{12}{r} + 1\right)\left(1 + \frac{r}{12}\right)^N\right]$

$= P\left[-1 + \left(1 + \frac{r}{12}\right)^N\right]\left(1 + \frac{12}{r}\right) = P\left[\left(1 + \frac{r}{12}\right)^N - 1\right]\left(1 + \frac{12}{r}\right)$

52. $A_1 = \left(\frac{1}{4}\right)A$

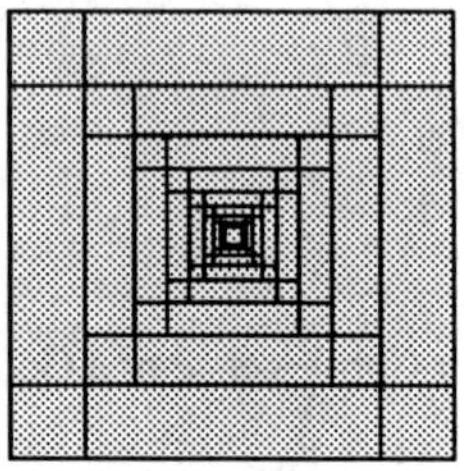

$A_2 = \left(\frac{1}{4}\right)A_1 = \left(\frac{1}{4}\right)^2 A$

$A_3 = \left(\frac{1}{4}\right)A_2 = \left(\frac{1}{4}\right)^3 A$

$\vdots$

$A_n = \left(\frac{1}{4}\right)^n A$

The shaded area is given by

$\sum_{n=1}^{\infty} A\left(\frac{1}{4}\right)^n = \sum_{n=0}^{\infty} A\left(\frac{1}{4}\right)^n - A = \frac{A}{1 - (1/4)} - A = \frac{4A}{3} - A = \frac{A}{3}$

which is $\frac{1}{3}$ the area of the square.

54. $\sum_{n=1}^{\infty} \left(\frac{1}{2}\right)^n = \sum_{n=0}^{\infty} \left(\frac{1}{2}\right)\left(\frac{1}{2}\right)^n = \frac{1/2}{1 - (1/2)} = 1$

56. Total $= \sum_{n=0}^{9} 942.6\, e^{0.13n} = 942.6 \sum_{n=0}^{9} (e^{0.13})^n \approx 942.6(19.23) = \$18{,}123.6$ million

58. $\sum_{n=0}^{\infty} 5.5(0.85)^n = 5.5 + 5.5(0.85) + 5.5(0.85)^2 + \ldots$

$= 5.5\frac{1}{1 - 0.85} \approx 36.67$ revolutions

60. $\sum_{n=0}^{\infty} 80\left(\frac{1}{8}\right)^n = 80 + 80\left(\frac{1}{8}\right) + 80\left(\frac{1}{8}\right)^2 + \ldots$

$= 80\frac{1}{1 - 1/8} \approx 91.43$ sq. inches

62. $\sum_{n=1}^{\infty} 2n^3\left(\frac{1}{5}\right)^n = \frac{115}{64} \approx 1.7969$

64. $\sum_{n=1}^{\infty} n\left(\frac{4}{11}\right)^n = \frac{44}{49} \approx 0.8980$

66. $\sum_{n=1}^{\infty} \ln 2\left(\frac{1}{8}\right)^{2n} = \frac{1}{63}\ln 2 \approx 0.0110$

Section 10.3 p-Series and the Ratio Test

2. $\sum_{n=1}^{\infty} \frac{1}{\sqrt{n}}$ is a p-series with $p = \frac{1}{2}$.

4. $\sum_{n=1}^{\infty} \frac{1}{n^{3/4}}$ is a p-series with $p = \frac{3}{4}$.

6. $\sum_{n=1}^{\infty} \frac{1}{n+1}$ is not a p-series.

8. This series diverges since $p = \frac{1}{3} < 1$.

10. This series converges since $p = \frac{4}{3} > 1$.

12. This series converges since $p = \pi > 1$.

14. $1 + \frac{1}{4} + \frac{1}{9} + \frac{1}{16} + \frac{1}{25} + \cdots = \sum_{n=1}^{\infty} \frac{1}{n^2}$

Therefore, this series converges since $p = 2 > 1$.

16. $1 + \frac{1}{2} + \frac{1}{3} + \frac{1}{4} + \frac{1}{5} + \cdots = \sum_{n=1}^{\infty} \frac{1}{n}$

Therefore, this series diverges since $p = 1$.

18. Since $a_n = n(2/3)^n$, we have

$$\lim_{n\to\infty} \left|\frac{a_{n+1}}{a_n}\right| = \lim_{n\to\infty} \left|\frac{(n+1)2^{n+1}}{3^{n+1}} \cdot \frac{3^n}{n2^n}\right| = \lim_{n\to\infty} \frac{2(n+1)}{3n} = \frac{2}{3} < 1$$

and the series converges.

20. Since $a_n = n(3/2)^n$, we have

$$\lim_{n\to\infty} \left|\frac{a_{n+1}}{a_n}\right| = \lim_{n\to\infty} \left|\frac{(n+1)3^{n+1}}{2^{n+1}} \cdot \frac{2^n}{n3^n}\right| = \lim_{n\to\infty} \frac{3(n+1)}{2n} = \frac{3}{2} > 1$$

and the series diverges.

22. Since $a_n = n^2/2^n$, we have

$$\lim_{n\to\infty} \left|\frac{a_{n+1}}{a_n}\right| = \lim_{n\to\infty} \left|\frac{(n+1)^2}{2^{n+1}} \cdot \frac{2^n}{n^2}\right| = \lim_{n\to\infty} \frac{n^2+2n+1}{2n^2} = \frac{1}{2} < 1$$

and the series converges.

24. Since $a_n = (-1)^n e^{-n}$, we have

$$\lim_{n\to\infty} \left|\frac{a_{n+1}}{a_n}\right| = \lim_{n\to\infty} \left|\frac{(-1)^{n+1}}{e^{n+1}} \cdot \frac{e^n}{(-1)^n}\right| = \lim_{n\to\infty} \frac{1}{e} = \frac{1}{e} < 1$$

and the series converges.

26. Since $a_n = 4^n/n!$, we have

$$\lim_{n\to\infty} \left|\frac{a_{n+1}}{a_n}\right| = \lim_{n\to\infty} \left|\frac{4^{n+1}}{(n+1)!} \cdot \frac{n!}{4^n}\right| = \lim_{n\to\infty} \frac{4}{n+1} = 0$$

and the series converges.

28. Since $a_n = \frac{3^n}{n+1}$, we have

$$\lim_{n\to\infty} \left|\frac{a_{n+1}}{a_n}\right| = \lim_{n\to\infty} \left|\frac{3^{n+1}}{(n+1)+1} \cdot \frac{n+1}{3^n}\right| = \lim_{n\to\infty} 3\left(\frac{n+1}{n+2}\right) = 3 > 1$$

and the series diverges.

30. Since $a_n = \frac{2n!}{n^5}$, we have

$$\lim_{n\to\infty} \left|\frac{a_{n+1}}{a_n}\right| = \lim_{n\to\infty} \frac{2(n+1)!}{(n+1)^5} \cdot \frac{n^5}{2n!} = \lim_{n\to\infty} (n+1)\left(\frac{n}{n+1}\right)^5 = \infty$$

and the series diverges.

32. $\sum_{n=1}^{\infty} \frac{1}{n^4} \approx 1 + \frac{1}{16} + \frac{1}{81} + \frac{1}{256} \approx 1.0788$

The error is less than $\frac{1}{(p-1)N^{p-1}} = \frac{1}{(4-1)4^{4-1}} = \frac{1}{192}$.

34. $\sum_{n=1}^{\infty} \frac{1}{n^{10}} \approx 1 + \frac{1}{1024} + \frac{1}{59{,}049} \approx 1.00099$

The error is less than $\frac{1}{(p-1)N^{p-1}} = \frac{1}{(10-1)3^{10-1}} = \frac{1}{177{,}147} \approx 5.645 \times 10^{-6}$.

36. $\lim_{n\to\infty} \left|\frac{a_{n+1}}{a_n}\right| = \lim_{n\to\infty} \frac{1/(n+1)^3}{1/n^3} = \lim_{n\to\infty} \left(\frac{n}{n+1}\right)^3 = 1,$

and the ratio test is inconclusive. (The series is a convergent p-series.)

38. $\sum_{n=1}^{\infty} \frac{2}{n} = \frac{2}{1} + \frac{2}{2} + \frac{2}{3} + \cdots$

$S_1 = 2, S_2 = 3, \ldots$ Matches (d), diverges

40. $\sum_{n=1}^{\infty} \frac{2}{n^2} = 2 + \frac{1}{2} + \frac{2}{9} + \cdots$ Matches (c), converges

42. $\sum_{n=1}^{\infty} \frac{10}{3\sqrt[3]{n^2}} = \frac{10}{3}\sum_{n=1}^{\infty} \frac{1}{n^{2/3}}$

This series is a divergent p-series since $p = \frac{2}{3} < 1$.

44. This series converges by the Geometric Series Test since $r = \frac{5}{6} < 1$.

46. This series diverges by the nth-Term Test since $\lim_{n\to\infty} \ln n = \infty$.

48. This series converges by the Ratio Test since

$$a_n = \frac{n3^n}{n!} = \frac{3^n}{(n-1)!} \text{ and } \lim_{n\to\infty} \left|\frac{a_{n+1}}{a_n}\right| = \lim_{n\to\infty} \left|\frac{3^{n+1}}{n!} \cdot \frac{(n-1)!}{3^n}\right| = \lim_{n\to\infty} \frac{3}{n} = 0.$$

50. This series converges by the Ratio Test since $a_n = n(0.4)^n$ and

$$\lim_{n\to\infty} \left|\frac{a_{n+1}}{a_n}\right| = \lim_{n\to\infty} \left|\frac{(n+1)(0.4)^{n+1}}{n(0.4)^n}\right| = \lim_{n\to\infty} \frac{0.4(n+1)}{n} = 0.4 < 1.$$

52. This series is a divergent p-series since $p = 0.95 < 1$.

54. $\sum_{n=1}^{\infty} \frac{1}{(2n)^2} \approx \frac{1}{4} + \frac{1}{16} + \frac{1}{36} + \frac{1}{64} + \frac{1}{100} + \cdots + \frac{1}{10{,}000} \approx 0.4063, \quad \frac{\pi^2}{24} \approx 0.41123$

56. Answers will vary.

Section 10.4 Power Series and Taylor's Theorem

2. $\sum_{n=1}^{\infty} \frac{(-1)^n(x-2)^n}{3^n} = -\frac{(x-2)}{3} + \frac{(x-2)^2}{3^2} - \frac{(x-2)^3}{3^3} + \frac{(x-2)^4}{3^4} - \frac{(x-2)^5}{3^5} + \cdots$

4. $\sum_{n=0}^{\infty} \frac{(-1)^n nx^n}{n!} = 0 - x + x^2 - \frac{x^3}{2} + \frac{x^4}{6} - \cdots$

6. $R = k$ **8.** $R = 1$ **10.** $R = \infty$ **12.** $R = 1$

14. $R = 0$ **16.** $R = 3$ **18.** $R = c$ **20.** $R = 1$

22. $R = \infty$ **24.** $R = 1$

26.
$$\begin{aligned} f(x) &= e^{-x} & f(0) &= 1 \\ f'(x) &= -e^{-x} & f'(0) &= -1 \\ f''(x) &= e^{-x} & f''(0) &= 1 \\ f'''(x) &= -e^{-x} & f'''(0) &= -1 \\ & & &\vdots \\ & & f^{(n)}(0) &= (-1)^n \end{aligned}$$

The power series for f is

$$e^{-x} = f(0) + f'(0)x + \frac{f''(0)x^2}{2!} + \cdots = 1 - x + \frac{x^2}{2!} - \frac{x^3}{3!} + \cdots + \frac{(-1)^n x^n}{n!} + \cdots = \sum_{n=0}^{\infty} \frac{(-1)^n x^n}{n!}$$

and the radius of convergence is $R = \infty$.

28.
$$\begin{aligned} f(x) &= e^{-2x} & f(0) &= 1 \\ f'(x) &= -2e^{-2x} & f'(0) &= -2 \\ f''(x) &= 4e^{-2x} & f''(0) &= 4 \\ f'''(x) &= -8e^{-2x} & f'''(0) &= -8 \\ & & &\vdots \\ & & f^{(n)}(0) &= (-1)^n 2^n \end{aligned}$$

The power series for f is

$$e^{-2x} = f(0) + f'(0)x + \frac{f''(0)x^2}{2!} + \cdots = 1 - 2x + \frac{4x^2}{2!} - \frac{8x^3}{3!} + \cdots = \sum_{n=0}^{\infty} \frac{(-2x)^n}{n!}$$

and the radius of convergence is $R = \infty$.

30.
$$\begin{aligned} f(x) &= (2-x)^{-1} & f(0) &= \frac{1}{2} \\ f'(x) &= (2-x)^{-2} & f'(0) &= \frac{1}{4} \\ f''(x) &= 2(2-x)^{-3} & f''(0) &= \frac{2}{8} \\ f'''(x) &= 6(2-x)^{-4} & f'''(0) &= \frac{6}{16} \\ & & &\vdots \\ & & f^{(n)}(0) &= \frac{n!}{2^{n+1}} \end{aligned}$$

The power series for f is

$$\frac{1}{2-x} = f(0) + f'(0)x + \frac{f''(0)x^2}{2!} + \cdots = \frac{1}{2} + \frac{1}{4}x + \frac{2x^2}{8 \cdot 2!} + \frac{6x^3}{16 \cdot 3!} + \cdots = \frac{1}{2} + \frac{x}{2^2} + \frac{x^2}{2^3} + \frac{x^3}{2^4} + \cdots = \sum_{n=0}^{\infty} \frac{x^n}{2^{n+1}}$$

and the radius of convergence is $R = 2$.

32. $f(x) = x^{1/2}$ $\qquad f(4) = 2$

$f'(x) = \frac{1}{2}x^{-1/2} \qquad f'(4) = \frac{1}{2 \cdot 2} = \frac{1}{2^2}$

$f''(x) = -\frac{1}{4}x^{-3/2} \qquad f''(4) = -\frac{1}{4 \cdot 2^3} = -\frac{1}{2^5}$

$f'''(x) = \frac{3}{8}x^{-5/2} \qquad f'''(4) = \frac{3}{8 \cdot 2^5} = \frac{3}{2^8}$

$f^{(4)}(x) = -\frac{3 \cdot 5}{16}x^{-7/2} \qquad f^{(4)}(4) = -\frac{3 \cdot 5}{16 \cdot 2^7} = -\frac{3 \cdot 5}{2^{11}}$

The power series for f is

$$\sqrt{x} = f(4) + f'(4)(x-4) + \frac{f''(4)(x-4)^2}{2!} + \cdots = 2 + \frac{x-4}{2^2} - \frac{(x-4)^2}{2^5 \cdot 2!} + \frac{3(x-4)^3}{2^8 \cdot 3!} - \frac{3 \cdot 5(x-4)^4}{2^{11} \cdot 4!} + \cdots$$

$$= 2 + \frac{x-4}{4} - \sum_{n=2}^{\infty} \frac{(-1)^n 1 \cdot 3 \cdot 5 \cdots (2n-3)(x-4)^n}{2^{3n-1}n!}$$

and the radius of convergence is $R = 4$.

34. $f(x) = (1+x)^{1/2} \qquad f(0) = 1$

$f'(x) = \frac{1}{2}(1+x)^{-1/2} \qquad f'(0) = \frac{1}{2}$

$f''(x) = -\frac{1}{4}(1+x)^{-3/2} \qquad f''(0) = -\frac{1}{4} = -\frac{1}{2^2}$

$f'''(x) = \frac{3}{8}(1+x)^{-5/2} \qquad f'''(0) = \frac{3}{8} = \frac{1 \cdot 3}{2^3}$

$f^{(4)}(x) = -\frac{15}{16}(1+x)^{-7/2} \qquad f^{(4)}(0) = -\frac{15}{16} = -\frac{1 \cdot 3 \cdot 5}{2^4}$

The power series for f is

$$\sqrt{1+x} = f(0) + f'(0)x + \frac{f''(0)x^2}{2!} + \cdots$$

$$= 1 + \frac{1}{2}x - \frac{x^2}{2^2 \cdot 2!} + \frac{3x^3}{2^3 \cdot 3!} - \frac{3 \cdot 5x^4}{2^4 \cdot 4!} + \cdots = 1 + \frac{x}{2} - \sum_{n=2}^{\infty} \frac{(-1)^n 1 \cdot 3 \cdot 5 \cdots (2n-3)x^n}{2^n n!}$$

and the radius of convergence is $R = 1$.

36. $f(x) = (1+x)^{1/3} \qquad f(0) = 1$

$f'(x) = \frac{1}{3}(1+x)^{-2/3} \qquad f'(0) = \frac{1}{3}$

$f''(x) = -\frac{2}{9}(1+x)^{-5/3} \qquad f''(0) = -\frac{2}{9} = -\frac{2}{3^2}$

$f'''(x) = \frac{10}{27}(1+x)^{-8/3} \qquad f'''(0) = \frac{10}{27} = \frac{2 \cdot 5}{3^3}$

$f^{(4)}(x) = -\frac{80}{81}(1+x)^{-11/3} \qquad f^{(4)}(0) = -\frac{80}{81} = -\frac{2 \cdot 5 \cdot 8}{3^4}$

The power series for f is

$$\sqrt[3]{1+x} = f(0) + f'(0)x + \frac{f''(0)x^2}{2!} + \frac{f'''(0)x^3}{3!} + \cdots$$

$$= 1 + \frac{1}{3}x - \frac{2x^2}{3^2 \cdot 2!} + \frac{2 \cdot 5x^3}{3^3 3!} - \frac{2 \cdot 5 \cdot 8x^4}{3^4 4!} + \cdots = 1 + \frac{x}{3} - \sum_{n=2}^{\infty} \frac{(-1)^n 2 \cdot 5 \cdot 8 \cdots (3n-4)x^n}{3^n n!}$$

and the radius of convergence is $R = 1$.

38. (a) $f(x) = \sum_{n=1}^{\infty} \frac{x^n}{n5^n}$

$R = 5$

(b) $f'(x) = \sum_{n=1}^{\infty} \frac{x^{n-1}}{5^n}$

$R = 5$

(c) $f''(x) = \sum_{n=2}^{\infty} \frac{(n-1)x^{n-2}}{5^n}$

$R = 5$

(d) $\int f(x)\,dx = C + \sum_{n=1}^{\infty} \frac{x^{n+1}}{(n+1)n5^n}$

$R = 5$

40. (a) $f(x) = \sum_{n=0}^{\infty} \frac{(-1)^{n+1}(x-1)^n}{n}$

$R = 1$

(b) $f'(x) = \sum_{n=1}^{\infty} (-1)^{n+1}(x-1)^{n-1}$

$R = 1$

(c) $f''(x) = \sum_{n=2}^{\infty} (-1)^{n+1}(n-1)x^{n-2}$

$R = 1$

(d) $\int f(x)\,dx = C + \sum_{n=0}^{\infty} \frac{(-1)^{n+1}(x-1)^{n+1}}{n(n+1)}$

$R = 1$

42. Since the power series for e^{x^3} is $e^{x^3} = \sum_{n=0}^{\infty} \frac{x^{3n}}{n!}$ it follows that $e^{-x^3} = \sum_{n=0}^{\infty} \frac{(-x^3)^n}{n!} = \sum_{n=0}^{\infty} \frac{(-1)^n x^{3n}}{n!}$.

44. Since $e^x = \sum_{n=0}^{\infty} \frac{x^n}{n!}$ and $e^{-x} = \sum_{n=0}^{\infty} \frac{(-1)^n x^n}{n!}$ we have

$$f(x) = \frac{1}{2}(e^x + e^{-x}) = \frac{1}{2}\left[\sum_{n=0}^{\infty} \frac{x^n}{n!} + \sum_{n=0}^{\infty} \frac{(-1)^n x^n}{n!}\right] = \frac{1}{2}\sum_{n=0}^{\infty} \frac{x^n}{n!}[1 + (-1)^n]$$

$$= \frac{1}{2}\left[2 + 0 + \frac{2x^2}{2!} + 0 + \frac{2x^4}{4!} + 0 + \frac{2x^6}{6!} + \cdots\right] = 1 + \frac{x^2}{2!} + \frac{x^4}{4!} + \frac{x^6}{6!} + \cdots = \sum_{n=0}^{\infty} \frac{x^{2n}}{(2n)!}.$$

46. Since the power series for $\frac{1}{1+x}$ is $\sum_{n=0}^{\infty} (-1)^n x^n$, it follows that the power series for $\frac{2x}{1+x}$ is

$$(2x)\sum_{n=0}^{\infty} (-1)^n x^n = \sum_{n=0}^{\infty} (-1)^n 2x^{n+1}.$$

48. Since $\frac{1}{1+x} = \sum_{n=0}^{\infty} (-1)^n x^n$ we have $\ln(1+x) = \int \frac{1}{1+x}\,dx = \sum_{n=0}^{\infty} \frac{(-1)^n x^{n+1}}{n+1}$.

50. Since $\frac{e^x + e^{-x}}{2} = \sum_{n=0}^{\infty} \frac{x^{2n}}{(2n)!}$ we have

$$\frac{e^x - e^{-x}}{2} = \frac{d}{dx}\left[\frac{e^x + e^{-x}}{2}\right] = \sum_{n=1}^{\infty} \frac{2nx^{2n-1}}{(2n)!} = \sum_{n=1}^{\infty} \frac{x^{2n-1}}{(2n-1)!} = \sum_{n=0}^{\infty} \frac{x^{2n+1}}{(2n+1)!}.$$

52. $e^x = \sum_{n=0}^{\infty} \frac{x^n}{n!} = 1 + x + \frac{x^2}{2!} + \frac{x^3}{3!} + \frac{x^4}{4!} + \cdots$

$$\frac{d}{dx}[e^x] = 0 + 1 + \frac{2x}{2!} + \frac{3x^2}{3!} + \frac{4x^3}{4!} + \cdots = 1 + x + \frac{x^2}{2!} + \frac{x^3}{3!} + \cdots = e^x$$

54. $e^x = 1 + x + \frac{x^2}{2} + \frac{x^3}{3!} + \frac{x^4}{4!} + \frac{x^5}{5!} + \cdots$

$$e^{-1} = 1 - 1 + \frac{1}{2} - \frac{1}{3!} + \frac{1}{4!} - \frac{1}{5!} + \frac{1}{6!} - \frac{1}{7!} + \cdots = \frac{1}{2} - \frac{1}{6} + \frac{1}{24} - \frac{1}{120} + \frac{1}{720} - \frac{1}{5040} + \cdots \approx 0.3679$$

56. $f(x) = \sum_{n=1}^{\infty} \frac{(-1)^{n+1}(x-1)^n}{n}, \quad 0 < x \leq 2$

$$f(1.5) = \sum_{n=1}^{\infty} \frac{(-1)^{n+1}(0.5)^n}{n} = \sum_{n=1}^{\infty} \frac{(-1)^{n+1}}{2^n n} \approx \frac{1}{2} - \frac{1}{2^2 \cdot 2} + \frac{1}{2^3 \cdot 3} - \frac{1}{2^4 \cdot 4} + \cdots - \frac{1}{2^{50} \cdot 50} \approx 0.4055$$

58. $f(1.95) = \sum_{n=1}^{\infty} \frac{(-1)^{n+1}(1.95-1)^n}{n} = \sum_{n=1}^{\infty} \frac{(-1)^n(0.95)^n}{n} \approx 0.6678$

Section 10.5 Taylor Polynomials

2. (a) $S_1(x) = x$

(b) $S_2(x) = x - \frac{x^2}{2}$

(c) $S_3(x) = x - \frac{x^2}{2} + \frac{x^3}{3}$

(d) $S_4(x) = x - \frac{x^2}{2} + \frac{x^3}{3} - \frac{x^4}{4}$

4. (a) $S_1(x) = 1 - 2x$

(b) $S_2(x) = 1 - 2x + 3x^2$

(c) $S_3(x) = 1 - 2x + 3x^2 - 4x^3$

(d) $S_4(x) = 1 - 2x + 3x^2 - 4x^3 + 5x^4$

6. $f(x) = \frac{4}{x+1}$

(a) $S_1(x) = 4 - 4x$

(b) $S_2(x) = 4 - 4x + 4x^2$

(c) $S_3(x) = 4 - 4x + 4x^2 - 4x^3$

(d) $S_4(x) = 4 - 4x + 4x^2 - 4x^3 + 4x^4$

8.

x	0	0.25	0.50	0.75
$\ln(x^2 + 1)$	0.00000	0.06062	0.22314	0.44629
x^2	0.00000	0.06250	0.25000	0.56250
$x^2 - \frac{x^4}{2}$	0.00000	0.06055	0.21875	0.40430
$x^2 - \frac{x^4}{2} + \frac{x^6}{3}$	0.00000	0.06063	0.22396	0.46362
$x^2 - \frac{x^4}{2} + \frac{x^6}{3} - \frac{x^8}{4}$	0.00000	0.06062	0.22298	0.43859

10. (a) $S_2(x) = 1 - x^2$

(b) $S_4(x) = 1 - x^2 + \frac{x^4}{2}$

(c) $S_6(x) = 1 - x^2 + \frac{x^4}{2} - \frac{x^6}{6}$

(d) $S_8(x) = 1 - x^2 + \frac{x^4}{2} - \frac{x^6}{6} + \frac{x^8}{24}$

12. $S_4(x) = x + x^2 + \frac{x^3}{2} + \frac{x^4}{6}$

14. $y = \frac{1}{8}x^4 - \frac{1}{2}x^2 + 1$ is a fourth-degree polynomial through $(0, 1)$; matches (c).

16. $y = e^{-1/2}\left[\frac{1}{3}(x-1)^3 - (x-1) + 1\right]$ is cubic; matches (b).

18. $S_6(x) = x^2 - x^3 + \frac{x^4}{2} - \frac{x^5}{6} + \frac{x^6}{24}$

$$f\left(\frac{1}{4}\right) \approx \frac{1}{16} - \frac{1}{64} + \frac{1}{512} - \frac{1}{6144} + \frac{1}{98{,}304}$$

$$\approx 0.0487$$

20. (See Section 10.4, Exercise #32.)

$$S_6(x) = 2 + \frac{x-4}{4} - \frac{(x-4)^2}{2^5 \cdot 2!} + \frac{3(x-4)^3}{2^8 \cdot 3!} - \frac{3 \cdot 5(x-4)^4}{2^{11} \cdot 4!} + \frac{3 \cdot 5 \cdot 7(x-4)^5}{2^{14} \cdot 5!} - \frac{3 \cdot 5 \cdot 7 \cdot 9(x-4)^6}{2^{17}6!}$$

$$f(5) = 2 + \frac{1}{4} - \frac{1}{2^5(2)} + \frac{3}{2^8(6)} - \frac{15}{2^{11}(24)} + \frac{105}{2^{14}(120)} - \frac{945}{2^{17}(720)} \approx 2.2361$$

22. $S_6(x) = x^2 - \frac{x^4}{2} + \frac{x^6}{3}$

$$\int_{-1/4}^{1/4} \ln(x^2+1)\,dx \approx \int_{-1/4}^{1/4} \left(x^2 - \frac{x^4}{2} + \frac{x^6}{3}\right)dx = \left[\frac{x^3}{3} - \frac{x^5}{10} + \frac{x^7}{21}\right]_{-1/4}^{1/4} \approx 0.0102$$

24. $S_6(x) = 1 - \frac{1}{3}x^2 + \frac{2}{9}x^4 - \frac{14}{81}x^6$

$$\int_0^{1/2} \frac{1}{\sqrt[3]{1+x^2}}\,dx \approx \int_0^{1/2} \left(1 - \frac{x^2}{3} + \frac{2x^4}{9} - \frac{14x^6}{81}\right)dx = \left[x - \frac{x^3}{9} + \frac{2x^5}{45} - \frac{2x^7}{81}\right]_0^{1/2} \approx 0.4873$$

26. The $(n+1)$ derivative of $f(x) = \frac{1}{x}$ is $f^{(n+1)}(x) = \frac{(-1)^{n+1}(n+1)!}{x^{n+2}}$. The maximum value of $|f^{(n+1)}(x)|$ in the interval $\left[1, \frac{3}{2}\right]$ is $|f^{(n+1)}(1)| = (n+1)!$. Therefore, the nth remainder is bounded by

$$|R_n| \le \left|\frac{(n+1)!}{(n+1)!}(x-1)^{n+1}\right|$$

$$|R_n| \le \frac{1}{2^{n+1}}.$$

When $n = 9$, $1/2^{10} \approx 0.00098$. Thus, $n = 9$ will approximate $1/x$ with an error less than 0.001 in $[1, 3/2]$.

28. $|R_5| \le \frac{f^{(6)}(z)}{6!}(x-1)^6 = \frac{(x-1)^6}{z^7}$

In the interval $[1, 3/2]$, $R_5 \le \frac{(1/2)^6}{(1)^7} = \frac{1}{64} \approx 0.0156.$

30. (a) $\sum_{n=0}^{\infty} P(n) = \sum_{n=0}^{\infty} \frac{1}{2}\left(\frac{2}{3}\right)^{n+1} = \sum_{n=0}^{\infty} \frac{1}{3}\left(\frac{2}{3}\right)^n = \frac{1/3}{1-(2/3)} = 1$

(b) Expected value $= \sum_{n=0}^{\infty} n\,P(n) = \sum_{n=0}^{\infty} \frac{n}{3}\left(\frac{2}{3}\right)^n = \frac{1}{3}\sum_{n=0}^{\infty} n\left(\frac{2}{3}\right)^n = \left[\frac{1}{3} \cdot \frac{1}{[1-(2/3)]^2}\right]\left(\frac{2}{3}\right) = 2$

(See Example 5 in the text.)

(c) Expected daily profit $= (\$10)(2) = \$20$

Section 10.6 Newton's Method

2. $x_2 = x_1 - \dfrac{f(x_1)}{f'(x_1)} = 1 - \dfrac{1}{6} = \dfrac{5}{6} \approx 0.833$

4. $f'(x) = 5x^4 + 1$

n	x_n	$f(x_n)$	$f'(x_n)$	$\dfrac{f(x_n)}{f'(x_n)}$	$x_n - \dfrac{f(x_n)}{f'(x_n)}$
1	0.5000	−0.4687	1.3125	−0.3571	0.8571
2	0.8571	0.3198	3.6983	0.0865	0.7707
3	0.7707	0.0426	2.7640	0.0154	0.7553
4	0.7553	0.0011	2.6272	0.0004	0.7549

Approximation: $x \approx 0.755$

6. $f'(x) = \dfrac{1}{x} + \dfrac{1}{x^2} = \dfrac{x+1}{x^2}$

n	x_n	$f(x_n)$	$f'(x_n)$	$\dfrac{f(x_n)}{f'(x_n)}$	$x_n - \dfrac{f(x_n)}{f'(x_n)}$
1	1.5000	−0.2612	1.1111	−0.2351	1.7351
2	1.7351	−0.0253	0.9085	−0.0278	1.7629
3	1.7629	0.0003	0.8890	−0.0003	1.7632

Approximation: $x \approx 1.763$

8. $f'(x) = e^{3x}(8 - 3x)$

n	x_n	$f(x_n)$	$f'(x_n)$	$\dfrac{f(x_n)}{f'(x_n)}$	$x_n - \dfrac{f(x_n)}{f'(x_n)}$
1	−0.5000	−0.2190	2.1197	−0.1033	0.3967
2	−0.3967	0.0333	2.7958	0.0119	−0.4086
3	−0.4086	0.00053	2.7080	0.00019	−0.4088
4	−0.4088	0.136×10^{-6}	2.7066	0.50×10^{-7}	−0.4088

Approximation: $x \approx -0.4088$

10. $f'(x) = 4x^3 + 3x^2$

n	x_n	$f(x_n)$	$f'(x_n)$	$\dfrac{f(x_n)}{f'(x_n)}$	$x_n - \dfrac{f(x_n)}{f'(x_n)}$
1	1	1	7	$\frac{1}{7}$	$\frac{6}{7}$
2	$\frac{6}{7}$	0.1695	4.7230	0.0359	0.82125
3	0.82125	0.00879	4.2390	0.00207	0.81918
4	0.81918	0.000028	4.2120	0.66×10^{-5}	0.81918

Approximation: $x \approx 0.81918$

—CONTINUED—

10. —CONTINUED—

n	x_n	$f(x_n)$	$f'(x_n)$	$\frac{f(x_n)}{f'(x_n)}$	$x_n - \frac{f(x_n)}{f'(x_n)}$
1	-1	-1	-1	1	-2
2	-2	7	-20	-0.35	-1.65
3	-1.65	1.9199	-9.801	-0.1959	-1.454
4	-1.454	0.3962	-5.955	-0.665	-1.3876
5	-1.3876	0.3545	-4.910	-0.00722	-1.3803
6	-1.3803	0.00038	-4.804	-0.00008	-1.3803

Approximation: $x \approx -1.3803$

12. $f'(x) = 3x^2 - 2x$

n	x_n	$f(x_n)$	$f'(x_n)$	$\frac{f(x_n)}{f'(x_n)}$	$x_n - \frac{f(x_n)}{f'(x_n)}$
1	-1	1	5	0.2	-1.2
2	-1.2	-0.168	6.72	-0.025	-1.175
3	-1.175	-0.00286	6.492	-0.00044	-1.1746
4	-1.1746	-0.88×10^{-6}	6.488	-0.1356×10^{-6}	-1.1746

Approximation: $x \approx -1.1746$

14. Let $2x + 1 = \sqrt{x+4}$ and define $h(x) = 2x + 1 - \sqrt{x+4}$, then $h'(x) = 2 - \frac{1}{2\sqrt{x+4}}$.

n	x_n	$h(x_n)$	$h'(x_n)$	$\frac{h(x_n)}{h'(x_n)}$	$x_n - \frac{h(x_n)}{h'(x_n)}$
1	0.6000	0.0552	1.7669	0.0312	0.5688
2	0.5688	0.0001	1.7661	0.0001	0.5687
3	0.5687	-0.0001	1.7661	-0.0001	0.5688

Approximation: $x \approx 0.569$

16. Let $x = e^{-x}$ and define $h(x) = x - e^{-x}$, then $h'(x) = 1 + e^{-x}$.

n	x_n	$h(x_n)$	$h'(x_n)$	$\frac{h(x_n)}{h'(x_n)}$	$x_n - \frac{h(x_n)}{h'(x_n)}$
1	0.5000	-0.1065	1.6065	-0.0663	0.5663
2	0.5663	-0.0013	1.5676	-0.0008	0.5671
3	0.5671	-0.0001	1.5672	-0.0001	0.5672

Approximation: $x \approx 0.567$

18. There are three real zeros.

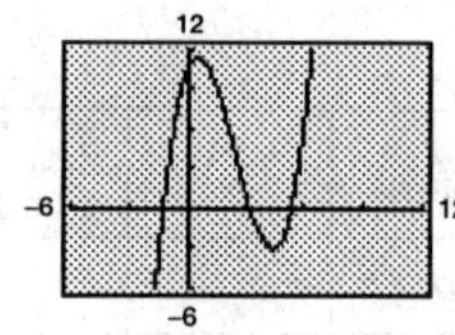

$$f(x) = \frac{1}{2}x^3 - \frac{11}{3}x^2 + \frac{10}{3}x + \frac{32}{3} = \frac{3x^3 - 22x^2 + 20x + 64}{6}$$

$$f'(x) = \frac{3}{2}x^2 - \frac{22}{3}x + \frac{10}{3} = \frac{9x^2 - 44x + 20}{6}$$

$$x_{n+1} = x_n - \frac{3x_n^3 - 22x_n^2 + 20x_n + 64}{9x_n^2 - 44x_n + 20} = \frac{6x_n^3 - 22x_n^2 - 64}{9x_n^2 - 44x_n + 20}$$

i	1	2	3	4
x_i	-1	-1.2603	-1.2363	-1.2361

i	1	2	3	4
x_i	3	3.2258	3.2360	3.2361

i	1	2	3	4	5
x_i	5	5.4400	5.3398	5.3334	5.3333

Zeros: $x \approx -1.236,\ 3.236,\ 5.333$

20. $f(x) = \frac{4}{11}x^4 + \frac{3}{7}x^2 - 2, \quad x \approx \pm 1.3523$

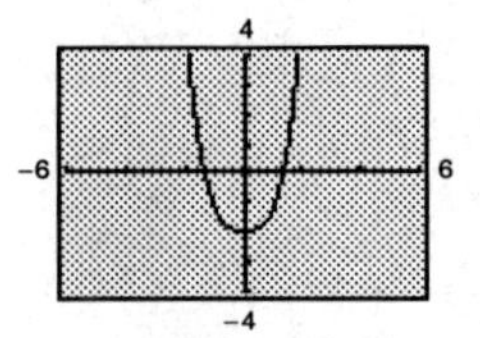

22. $f(x) = 0.16e^{x^2} - 4.1x^3 + 6.8, \quad x \approx 1.2238,\ 2.3947$

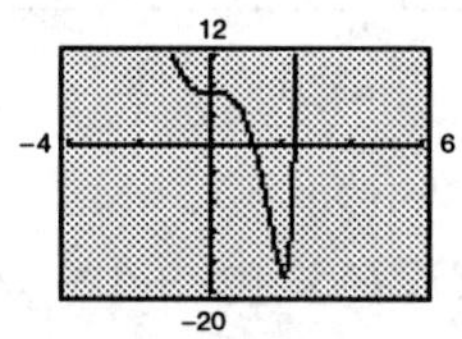

24. $f(x) = x + \sin(x + 1)$

$x \approx -0.4890$

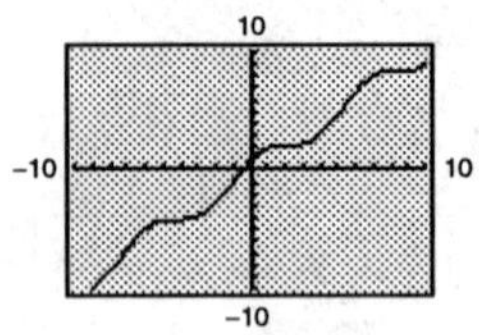

26. $f(x) = \sin \pi x + x - 1$

$x \approx 0.2635,\ 1.0,\ 1.7365$

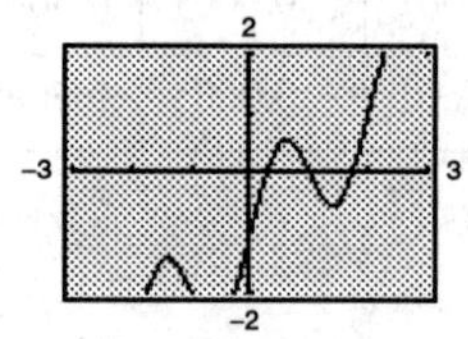

28. $y = 4x^3 - 12x^2 + 12x - 3$

$y' = 12x^2 - 24x + 12$

Newton's Method fails because $f'(x_2) = 0$.

n	x_n	$f(x_n)$	$f'(x_n)$	$\frac{f(x_n)}{f'(x_n)}$	$x_n - \frac{f(x_n)}{f'(x_n)}$
1	$\frac{3}{2}$	$\frac{3}{2}$	3	$\frac{1}{2}$	1
2	1	1	0	—	—

30. $y = x^3 - 2x - 2$

$y' = 3x^2 - 2$

Newton's Method fails because it fails to converge.

n	x_n	$f(x_n)$	$f'(x_n)$	$\frac{f(x_n)}{f'(x_n)}$	$x_n - \frac{f(x_n)}{f'(x_n)}$
1	0	-2	-2	1	-1
2	-1	-1	1	-1	0
3	0	-2	-2	1	-1
4	-1	-1	1	-1	0

32. Let $f(x) = x^n - a$, then $f'(x) = nx^{n-1}$.

$$x_{i+1} = x_i - \frac{x_i^n - a}{nx_i^{n-1}}$$

$$= \frac{nx_i^n - x_i^n + a}{nx_i^{n-1}} = \frac{(n-1)x_i^n + a}{nx_i^{n-1}}$$

34. Let $f(x) = x^2 - 5$, then $f'(x) = 2x$.

$$x_{i+1} = \frac{x_i^2 + 5}{2x_i}$$

i	1	2	3	4
x_i	2	2.25	2.23611	2.23610

$\sqrt{5} \approx 2.236$

36. Let $f(x) = x^3 - 15$, then $f'(x) = 3x^2$.

$$x_{i+1} = \frac{2x_i^3 + 15}{3x_i^2}$$

i	1	2	3	4
x_i	2.5	2.4667	2.4662	2.4662

$\sqrt[3]{15} \approx 2.466$

38. $x_{i+1} = x_i(2 - 11x_i)$

i	1	2	3	4
x_i	0.1	0.09	0.0909	0.0909

$\frac{1}{11} \approx 0.091$

40. Average cost: $\overline{C} = \dfrac{C}{x} = 0.0001x^2 + 0.02x + 0.04 + \dfrac{800}{x}$

$$\overline{C}' - 0.0002x + 0.02 - \frac{800}{x^2} = \frac{0.0002x^3 + 0.02x^2 - 800}{x^2} = 0$$

Let $f(x) = 0.0002x^3 + 0.02x^2 - 800$, then $f'(x) = 0.0006x^2 + 0.04x$.
Since $f(130) = -22.6$ and $f(132) = 8.4736$, the solution is in the interval (130, 132).

n	x_n	$f(x_n)$	$f'(x_n)$	$\frac{f(x_n)}{f'(x_n)}$	$x_n - \frac{f(x_n)}{f'(x_n)}$
1	131	−7.1618	15.5366	−0.4610	131.4610
2	131.4610	0.0215	15.6276	0.0014	131.4596

$x \approx 131$ units

42. $C = 100\left(\dfrac{200}{x^2} + \dfrac{x}{x+30}\right)$

$$C' = 100\left[-\frac{400}{x^3} + \frac{30}{(x+30)^2}\right] = 1000\left[\frac{-40(x+30)^2 + 3x^3}{x^3(x+30)^2}\right] = 1000\left[\frac{3x^3 - 40x^2 - 2400x - 36{,}000}{x^3(x+30)^2}\right] = 0$$

Let $f(x) = 3x^3 - 40x^2 - 2400x - 36{,}000$, then $f'(x) = 9x^2 - 80x - 2400$.
Since $f(39) = -12{,}483$ and $f(41) = 9449$, the solution is in the interval (39, 41).

n	x_n	$f(x_n)$	$f'(x_n)$	$\frac{f(x_n)}{f'(x_n)}$	$x_n - \frac{f(x_n)}{f'(x_n)}$
1	40	−4000	8800	−0.4545	40.4545
2	40.4545	65.9841	9092.7391	0.0073	40.4472

$x \approx 40$ units

44. $P(t) = A(t) - C(t) = V(t)e^{-0.10t} - \int_0^t 1000e^{-0.10u}\,du = 100{,}000e^{0.9\sqrt{t}}e^{-0.10t} + \Big[10{,}000e^{-0.10u}\Big]_0^t$

$= 100{,}000e^{0.9\sqrt{t}-0.10t} + 10{,}000(e^{-0.10t} - 1) = 10{,}000(10e^{0.9\sqrt{t}-0.10t} + e^{-0.10t} - 1)$

$$P'(t) = 10{,}000\left[10\left(\frac{0.45}{\sqrt{t}} - 0.10\right)e^{0.9\sqrt{t}-0.10t} - 0.10e^{-0.10t}\right] = 10{,}000e^{-0.10t}\left[\left(\frac{4.5}{\sqrt{t}} - 1\right)e^{0.9\sqrt{t}} - 0.10\right] = 0$$

Let $f(t) = (4.5 - \sqrt{t})e^{0.9\sqrt{t}} - 0.10\sqrt{t}$, then

$$f'(t) = \left(\frac{3.05}{2\sqrt{t}} - 0.45\right)e^{0.9\sqrt{t}} - \frac{0.05}{\sqrt{t}}.$$

Since $f(20) \approx 1.1125$ and $f(21) \approx -5.5636$, the solution is in the interval $(20, 21)$.

n	x_n	$f(x_n)$	$f'(x_n)$	$\frac{f(x_n)}{f'(x_n)}$	$x_n - \frac{f(x_n)}{f'(x_n)}$
1	20.5	−2.0824	−6.6715	0.3121	20.1879
2	20.1879	−0.0554	−6.3194	0.0088	20.1791
3	20.1791	0.0001	−6.3096	0.0000	20.1791

The timber should be harvested in 20.179 years or in the first quarter of 2010.

Review Exercises for Chapter 10

2. $a_n = \dfrac{n-2}{n^2+2}$: $\dfrac{-1}{3}, 0, \dfrac{1}{11}, \dfrac{1}{9}, \dfrac{1}{9}$

4. $a_n = \dfrac{(-1)^n}{n^3}$: $-1, \dfrac{1}{8}, \dfrac{-1}{27}, \dfrac{1}{64}, \dfrac{-1}{125}$

6. This sequence converges since

$$\lim_{n\to\infty} \frac{1}{\sqrt{n}} = 0.$$

8. This sequence converges since

$$\lim_{n\to\infty} \frac{10}{e^n} = 0.$$

10. This sequence converges since

$$\lim_{n\to\infty} \frac{n}{\sqrt{n^2+1}} = 1.$$

12. This sequence converges since

$$\lim_{n\to\infty} \frac{(n-1)!}{(n+1)!} = \lim_{n\to\infty} \frac{1}{(n+1)n} = 0.$$

14. $a_n = \dfrac{n}{n^2+1}$, $n = 1, 2, 3, \ldots$

OR

$a_n = \dfrac{n+1}{(n+1)^2+1}$, $n = 0, 1, 2, \ldots$

16. $a_n = \left(\dfrac{2}{5}\right)^{n-1}$, $n = 1, 2, 3, \ldots$

OR

$a_n = \left(\dfrac{2}{5}\right)^n$, $n = 0, 1, 2, \ldots$

18. (a) $f(n) = 22 - n$, $n = 1, 2, \ldots, 7$

$(n = 1$ bottom row, $n = 7$ top row$)$

(b) $n = 1$: $f(1) = 22 - 1 = 21$ logs in bottom row

$n = 2$: $f(2) = 22 - 2 = 20$ logs in second row

⋮

$n = 7$: $f(7) = 22 - 7 = 15$ logs in top row

20. $S_1 = \frac{1}{2}$

$S_2 = \frac{1}{2} - \frac{1}{4} = \frac{1}{4}$

$S_3 = \frac{1}{2} - \frac{1}{4} + \frac{1}{6} = \frac{5}{12}$

$S_4 = \frac{1}{2} - \frac{1}{4} + \frac{1}{6} - \frac{1}{8} = \frac{7}{24}$

$S_5 = \frac{1}{2} - \frac{1}{4} + \frac{1}{6} - \frac{1}{8} + \frac{1}{10} = \frac{47}{120}$

22. $S_1 = 1$

$S_2 = 1 + \frac{1}{4} = \frac{5}{4}$

$S_3 = 1 + \frac{1}{4} + \frac{1}{9} = \frac{49}{36}$

$S_4 = 1 + \frac{1}{4} + \frac{1}{9} + \frac{1}{16} = \frac{205}{144}$

$S_5 = 1 + \frac{1}{4} + \frac{1}{9} + \frac{1}{16} + \frac{1}{25} = \frac{5269}{3600}$

24. This geometric series converges since

$$r = \frac{1}{3} < 1.$$

26. This series diverges by the nth-Term Test since

$$\lim_{n\to\infty} \frac{\sqrt{n^3}}{n} = \lim_{n\to\infty} \sqrt{n} = \infty \neq 0.$$

28. $\lim_{n\to\infty} \frac{n^3}{1 - n^3} = -1 \neq 0$, and the series diverges.

30. $\lim_{n\to\infty} 3\left(\frac{-4}{3}\right)^n \neq 0$, and the series diverges.

32. $S_n = \sum_{k=0}^{n} 3\left(\frac{1}{6}\right)^n = 3\frac{1 - (1/6)^{n+1}}{1 - 1/6} = \frac{18}{5}\left[1 - \left(\frac{1}{6}\right)^{n+1}\right]$

34. $S_n = \sum_{k=0}^{n} 2\left(\frac{1}{\sqrt{3}}\right)^n = 2\left[\frac{1 - (1/\sqrt{3})^{n+1}}{1 - 1/\sqrt{3}}\right] = \frac{2\sqrt{3}}{\sqrt{3} - 1}\left[1 - \left(\frac{1}{\sqrt{3}}\right)^{n+1}\right]$

36. $\sum_{n=0}^{\infty} 4\left(-\frac{1}{4}\right)^n = 4\frac{1}{1 - (-1/4)} = \frac{16}{5}$ (Geometric Series)

38. $\sum_{n=0}^{\infty} [(1.5)^n + (0.2)^n]$ diverges since $\lim_{n\to\infty} a_n \neq 0$.

40. 1 year: 8000

2 years: $8000 + 0.85(8000) = 8000(1 + 0.85)$

3 years: $8000 + 0.85[8000(1 + 0.85)] = 8000(1 + 0.85 + 0.85^2)$

⋮

n years: $8000[1 + 0.85 + 0.85^2 + \cdots + (0.85)^{n-1}] = 8000\left(\frac{1 - (0.85)^n}{1 - 0.85}\right) = \frac{160{,}000}{3}[1 - (0.85)^n]$

42. $\sum_{n=1}^{\infty} \frac{2}{n^{2/3}}$ diverges by the p-series test since $p = \frac{2}{3} < 1$.

44. This series converges by the p-series test since $p = e > 1$.

46. $a_n = 4 - \frac{1}{2}n$ Matches (c)

48. $a_n = 6\left(-\frac{2}{3}\right)^{n-1}$ Matches (b)

50. $\sum_{n=1}^{\infty} \frac{1}{n^3} \approx 1 + \frac{1}{2^3} + \frac{1}{3^3} \approx 1.0170$

Error $< \frac{1}{2(3)^2} = \frac{1}{18}$

52. $\sum_{n=1}^{\infty} \frac{1}{n^{12}} \approx 1 + \frac{1}{2^{12}} + \frac{1}{3^{12}} + \frac{1}{4^{12}} \approx 1.000246$

Error $< \frac{1}{11(4)^{11}} \approx 2.167 \times 10^{-8}$

54. This series diverges by the Ratio Test since

$$\lim_{n\to\infty} \left|\frac{(n + 1)!}{4^{n+1}} \cdot \frac{4^n}{n!}\right| = \lim_{n\to\infty} \frac{n + 1}{4} = \infty.$$

56. This series diverges by the Ratio Test since

$$\lim_{n\to\infty} \left|\frac{2^{n+1}}{n + 1} \cdot \frac{n}{2^n}\right| = \lim_{n\to\infty} \frac{2n}{n + 1} = 2 > 1.$$

58. This series converges by the Ratio Test since

$$\lim_{n\to\infty}\left|\frac{2(n+1)}{1-4^{(n+1)}}\cdot\frac{1-4^n}{2n}\right| = \lim_{n\to\infty}\frac{n+1}{n}\left(\frac{4^n-1}{4^{n+1}-1}\right) = \frac{1}{4} < 1.$$

60. $R = \frac{1}{2}$

62. $R = 2$

64. $f(x) = x^{-1/2}$ $\qquad f(1) = 1$

$f'(x) = -\frac{1}{2}x^{-3/2}$ $\qquad f'(1) = -\frac{1}{2}$

$f''(x) = \frac{3}{4}x^{-5/2}$ $\qquad f''(1) = \frac{3}{4} = \frac{1\cdot 3}{2^2}$

$f'''(x) = -\frac{15}{8}x^{-7/2}$ $\qquad f'''(1) = -\frac{15}{8} = -\frac{1\cdot 3\cdot 5}{2^3}$

The power series for f is

$$\frac{1}{\sqrt{x}} = 1 - \frac{1}{2}(x-1) + \frac{3}{2^2 2!}(x-1)^2 - \frac{3\cdot 5}{2^3 3!}(x-1)^3 + \cdots = 1 - \sum_{n=1}^{\infty}\frac{(-1)^{n+1}1\cdot 3\cdot 5\cdots(2n-1)(x-1)^n}{2^n n!}.$$

66. $f(x) = (1+x)^{1/4}$ $\qquad f(0) = 1$

$f'(x) = \frac{1}{4}(1+x)^{-3/4}$ $\qquad f'(0) = \frac{1}{4}$

$f''(x) = -\frac{3}{16}(1+x)^{-7/4}$ $\qquad f''(0) = -\frac{3}{16} = -\frac{3}{4^2}$

$f'''(x) = \frac{21}{64}(1+x)^{-11/4}$ $\qquad f'''(0) = \frac{21}{64} = \frac{3\cdot 7}{4^3}$

$f^{(4)}(x) = -\frac{231}{256}(1+x)^{-15/4}$ $\qquad f^{(4)}(0) = -\frac{231}{256} = -\frac{3\cdot 7\cdot 11}{4^4}$

The power series for f is

$$\sqrt[4]{1+x} = 1 + \frac{1}{4}x - \frac{3}{4^2 2!}x^2 + \frac{3\cdot 7}{4^3 3!}x^3 - \frac{3\cdot 7\cdot 11}{4^4 4!}x^4 + \cdots = 1 + \frac{x}{4} - \sum_{n=2}^{\infty}\frac{(-1)^n 3\cdot 7\cdot 11\cdots(4n-5)x^n}{4^n n!}.$$

68. $e^{2x+1} = e\cdot e^{2x} = e\left(1 + 2x + \frac{(2x)^2}{2!} + \cdots\right) = \sum_{n=0}^{\infty} e\frac{(2x)^n}{n!}$

70. $\frac{1}{x^3+1} = 1 - x^3 + x^6 - x^9 + \ldots = \sum_{n=0}^{\infty}(-1)^n x^{3n}$

72. $e^{x+1} \approx e + ex + \frac{1}{2}ex^2 + \frac{1}{6}ex^3 + \frac{1}{24}ex^4 + \frac{1}{120}ex^5 + \frac{1}{720}ex^6$

74. $\sqrt{x+2} \approx \sqrt{2} + \frac{1}{4}\sqrt{2}x - \frac{1}{32}\sqrt{2}x^2 + \frac{1}{128}\sqrt{2}x^3 - \frac{5}{2048}\sqrt{2}x^4 + \frac{7}{8192}\sqrt{2}x^5 - \frac{21}{65{,}536}\sqrt{2}x^6$

76. $f(x) = \frac{1}{\sqrt{x}},\quad c = 1,\quad f(1.15) \approx 0.9325051221$

78. $f(x) = e^{x-1},\quad c = 0,\quad f(1.75) \approx 2.112340045$

80. Error $= R_n = \dfrac{f^{(n+1)}(z)}{(n+1)!}(x-c)^{n+1}$

$f(x) = e^{-2x} \Longrightarrow f^{(6)}(x) = 64e^{-2x} \le 64$ on $[0, 1]$

$$R_n \le \frac{64}{6!}x^6 < \frac{4}{45}(1) = \frac{4}{45}$$

82. $$e^{-x^2} = 1 - x^2 + \frac{x^4}{2} - \frac{x^6}{6} + \cdots$$

$$\int_0^{0.5} e^{-x^2}\,dx \approx \left[x - \frac{x^3}{3} + \frac{x^5}{10} - \frac{x^7}{42}\right]_0^{0.5} \approx 0.4613$$

84. $$\sqrt{1+x} = 1 + \frac{1}{2}x - \frac{1}{8}x^2 + \frac{1}{16}x^3 - \frac{5}{128}x^4 + \frac{7}{256}x^5 - \frac{21}{1024}x^6 + \cdots$$

$$\int_0^{0.5} \sqrt{1+x}\,dx \approx 0.55807$$

86. Expected value $= \displaystyle\sum_{n=0}^{\infty} nP(n) = \sum_{n=0}^{\infty} \frac{1}{2}n\left(\frac{1}{3}\right)^{n+1}$

From Exercise 81, $\displaystyle\sum_{n=0}^{\infty} n\left(\frac{1}{3}\right)^{n+1} = \frac{1}{4}$

Hence, expected value $= \dfrac{1}{2}\left(\dfrac{1}{4}\right) = \dfrac{1}{8} = 0.125$ unit

Expected production cost $= (23.00)\left(\dfrac{1}{8}\right) = \$2.875$

88. $f'(x) = 3x^2 + 2$

Since $f(-1) = -2$ and $f(0) = 1$, the zero is in the interval $(-1, 0)$.

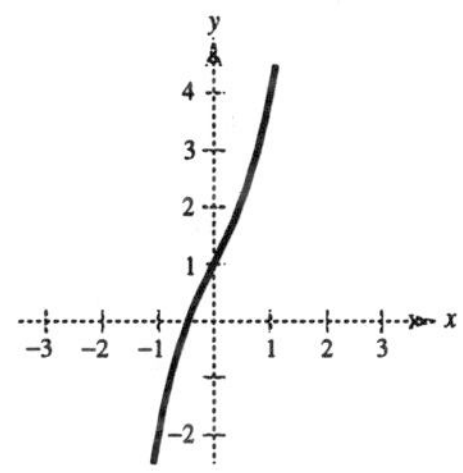

n	x_n	$f(x_n)$	$f'(x_n)$	$\dfrac{f(x_n)}{f'(x_n)}$	$x_n - \dfrac{f(x_n)}{f'(x_n)}$
1	-0.5000	-0.1250	2.7500	-0.0455	-0.4545
2	-0.4545	-0.0029	2.6197	-0.0011	-0.4534
3	-0.4534	-0.0000	2.6167	-0.0000	-0.4534

Approximation: $x \approx -0.453$

90. $f(x) = e^x - 3$

$f'(x) = e^x$

$$x_{n+1} = x_n - \frac{f(x_n)}{f'(x_n)}$$

$x_1 = 1$ (initial guess)

$x_2 = 1.1036$

$x_3 = 1.0986$

$x \approx 1.099$

92. Let $h(x) = f(x) - g(x) = 2 - x - x^5 - 2 = -x^5 - x$

$x = 0$ is the only point of intersection.

94. Let $h(x) = f(x) - g(x) = 2x^2 - 5e^{-x}$

$h'(x) = 4x + 5e^{-x}$

$x_1 = 1$ (initial guess)

$x_2 = 0.9725$

$x_3 = 0.9724$

$x \approx 0.972$

96. (a) $A_1 = 5000\left(1 + \frac{0.05}{4}\right)^1 \approx 5062.5$

$A_2 = 5125.78$

$A_3 = 5189.85$

$A_4 = 5254.73$

$A_5 = 5320.41$

$A_6 = 5386.92$

$A_7 = 5454.25$

$A_8 = 5522.43$

(b) $A_{40} = 8218.10$

APPENDICES

APPENDIX A
Alternate Introduction to the Fundamental Theorem of Calculus

Solutions to Even-Numbered Exercises

2. 3.685, 4.038

4. 0.760, 0.635

6. 1.166, 1.270

8. 3.892

10.

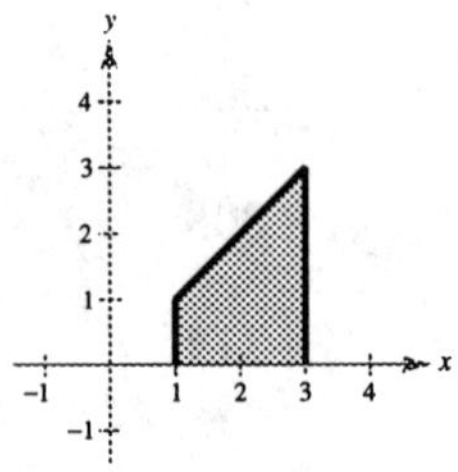

12. $\int_0^2 (4 - 2x)\, dx$

14. $\int_0^2 x^2\, dx$

16. $\int_{-1}^1 \frac{1}{x^2 + 1}\, dx$

18. 8a;

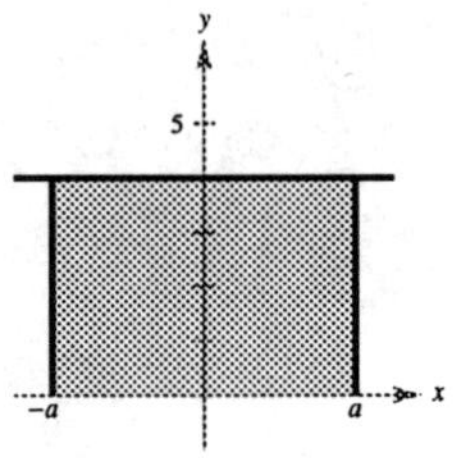

20. 4;

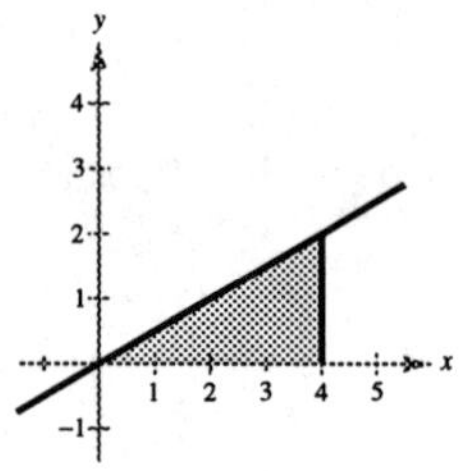

22. $\frac{25}{2}$;

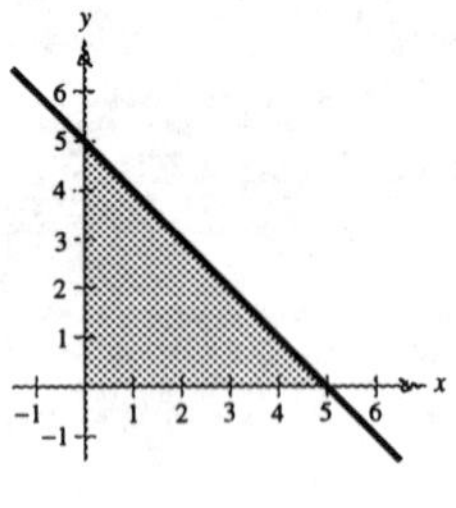

24. a^2;

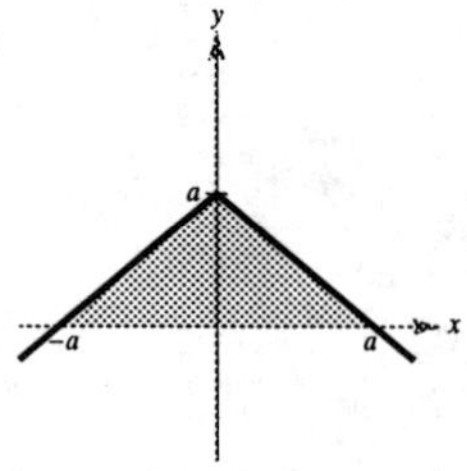

26. $\frac{1}{2}\pi r^2$;

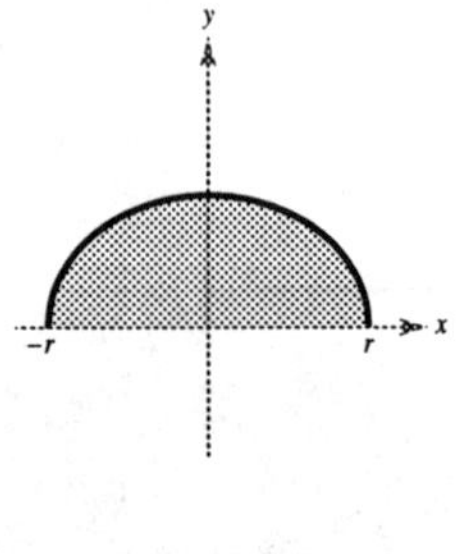

28. $\frac{3}{2}$

APPENDIX D
Differential Equations

Section D.1 Solutions of Differential Equations

Solutions to Even-Numbered Exercises

2. $y = 2x^3 - x + 1$

$y' = 6x^2 - 1$

4. $y = 3e^{x^2}$

$y' = 6xe^{x^2}$ and $y' - \frac{2}{x}y = 6xe^{x^2} - 2x(3e^{x^2}) = 0$

6. $y = 4x^2$

$y' = 8x$ and $y' - \frac{2}{x}y = 8x - \frac{2}{x}(4x^2) = 0$

8. $y = \frac{1}{x}$

$y' = \frac{-1}{x^2}$

$y'' = \frac{2}{x^3}$ and $xy'' + 2y' = x\left(\frac{2}{x^3}\right) + 2\left(\frac{-1}{x^2}\right) = 0$

10. $y = e^{x^3}$

$y' = 3x^2e^{x^3}$

$y'' = (9x^4 + 6x)e^{x^3}$ and $y'' - 3x^2y' - 6xy = (9x^4 + 6x)e^{x^3} - 3x^2(3x^2e^{x^3}) - 6xe^{x^3} = 0$

12. By differentiation, we have

$$\frac{dy}{dx} = \frac{1}{2}(4 - x^2)^{-1/2}(-2x) = -\frac{x}{\sqrt{4 - x^2}}.$$

14. By differentiation, we have

$$\frac{dy}{dx} = -4Ce^{-4x} = -4y.$$

16. Since $y' = -Ce^{-t}$, we have $y' + y - 10 = -Ce^{-t} + Ce^{-t} + 10 - 10 = 0$.

18. Since $y' = x\left(\frac{2x}{x^2}\right) + \ln x^2 + 3x^{1/2} + C = 2 + \ln x^2 + 3\sqrt{x} + C$, we have

$$y' - \frac{y}{x} = 2 + \ln x^2 + 3\sqrt{x} + C - \frac{1}{x}(x \ln x^2 + 2x^{3/2} + Cx) = 2 + \ln x^2 + 3\sqrt{x} + C - \ln x^2 - 2\sqrt{x} - C = 2 + \sqrt{x}.$$

20. Since $y' = C_2e^x$ and $y'' = C_2e^x$, we have $y'' - y' = C_2e^x - C_2e^x = 0$.

22. Since $y' = 4C_1e^{4x} - C_2e^{-x}$ and $y'' = 16C_1e^{4x} + C_2e^{-x}$, we have

$$y'' - 3y' - 4y = (16C_1e^{4x} + C_2e^{-x}) - 3(4C_1e^{4x} - C_2e^{-x}) - 4(C_1e^{4x} + C_2e^{-x}) = 0.$$

24. Since $y' = \frac{3}{5}x^2 - 1 + \frac{C}{2\sqrt{x}}$, we have $2xy' - y = 2x\left(\frac{3}{5}x^2 - 1 + \frac{C}{2\sqrt{x}}\right) - \left(\frac{1}{5}x^3 - x + C\sqrt{x}\right) = x^3 - x$.

26. Since $y' = x\left(\frac{1}{x}\right) + (\ln x + C)(1) = 1 + \ln x + C$, we have $x + y - xy' = x + x(\ln x + C) - x(1 + \ln x + C) = 0$.

28. Since $y' = (1 - 2x)Ce^{x-x^2} = (1 - 2x)y$, we have $y' + (2x - 1)y = (1 - 2x)y + (2x - 1)y = 0$.

30. By implicit differentiation, we have $2yy' + 2xy' + 2y - 2x = 0$ which implies that

$$y' = \frac{2x - 2y}{2x + 2y} = \frac{x - y}{x + y}.$$

Thus,

$$(x + y)y' - x + y = (x + y)\left(\frac{x - y}{x + y}\right) - x + y = 0.$$

32. By implicit differentiation, we have $2x - 2yy' = 0$, which implies that

$$y' = \frac{x}{y} \text{ and } y'' = \frac{y(1) - xy'}{y^2} = \frac{y - x(x/y)}{y^2} = \frac{y^2 - x^2}{y^3}.$$

Thus,

$$y^3y'' + x^2 - y^2 = y^3\left(\frac{y^2 - x^2}{y^3}\right) + x^2 - y^2 = 0.$$

34.

$$y = 5 \ln x$$

$$y' = \frac{5}{x}$$

$$y'' = -\frac{5}{x^2}$$

$$y''' = \frac{10}{x^3}$$

$$y^{(4)} = -\frac{30}{x^4}$$

$$y^{(4)} - 16y = -\frac{30}{x^4} - \frac{80}{x} \neq 0$$

Therefore, it is *not* a solution of the given differential equation.

36.

$$y = 4e^{2x}$$

$$y' = 8e^{2x}$$

$$y'' = 16e^{2x}$$

$$y''' = 32e^{2x}$$

$$y^{(4)} = 64e^{2x}$$

Therefore, we have

$$y^{(4)} - 16y = 64e^{2x} - 16(4e^{2x}) = 0.$$

38. $y = 4e^x + \frac{2}{9}xe^{-2x}$

$y' = 4e^x - \frac{4}{9}xe^{-2x} + \frac{2}{9}e^{-2x}$

$y'' = 4e^x + \frac{8}{9}xe^{-2x} - \frac{8}{9}e^{-2x}$

$y''' = 4e^x - \frac{16}{9}xe^{-2x} + \frac{24}{9}e^{-2x}$

Thus,

$$y''' - 3y' + 2y = \left(4e^x - \tfrac{16}{9}xe^{-2x} + \tfrac{24}{9}e^{-2x}\right) - 3\left(4e^x - \tfrac{4}{9}xe^{-2x} + \tfrac{2}{9}e^{-2x}\right) + 2\left(4e^x + \tfrac{2}{9}xe^{-2x}\right) = 2e^{-2x} \neq 0.$$

This is *not* a solution.

40. $y = x \ln x$

$y' = 1 + \ln x$

$$y'' = \frac{1}{x}$$

$$y''' = -\frac{1}{x^2}$$

Thus, $y''' - 3y' + 2y = -\dfrac{1}{x^2} - \dfrac{3}{x} + 2x \ln x \neq 0$. Therefore, it is *not* a solution of the given differential equation.

42. Since $4x + 6yy' = 0$, we have $y' = -2x/3y$ and it follows that $2x + 3yy' = 0$. To find the particular solution, we use the fact that $y = 2$ when $x = 1$. That is, $2(1)^2 + 3(2)^2 = C \Rightarrow C = 14$. Thus, the particular solution is $2x^2 + 3y^2 = 14$.

44. Since $y' = C_1 + 3C_2x^2$ and $y'' = 6C_2x$, it follows that $x^2y'' - 3xy' + 3y = 0$. To find the particular solution, we use the fact that $y = 0$ and $y' = 4$ when $x = 2$. That is,

$$\left.\begin{aligned} 0 &= 2C_1 + 8C_2 \\ 4 &= C_1 + 12C_2 \end{aligned}\right\} \Rightarrow C_1 = -2, \quad C_2 = \tfrac{1}{2}.$$

Thus, the particular solution is $y = -2x + \frac{1}{2}x^3$.

46. Since $y' = (1 - 2x)Ce^{x-x^2} = (1 - 2x)y$, it follows that $y' + (2x - 1)y = 0$. To find the particular solution, use the fact that $y = 2$ when $x = 1$. That is, $2 = Ce^0 \Rightarrow C = 2$. Thus, the particular solution is $y = 2e^{x-x^2}$.

48. Since

$$y' = \left(C_1 + C_2x + \tfrac{1}{12}x^4\right)(2e^{2x}) + \left(C_2 + \tfrac{1}{3}x^3\right)e^{2x} = \left(2C_1 + C_2 + 2C_2x + \tfrac{1}{3}x^3 + \tfrac{1}{6}x^4\right)e^{2x}$$

and

$$\begin{aligned} y'' &= \left(2C_1 + C_2 + 2C_2x + \tfrac{1}{3}x^3 + \tfrac{1}{6}x^4\right)(2e^{2x}) + \left(2C_2 + x^2 + \tfrac{2}{3}x^3\right)e^{2x} \\ &= \left(4C_1 + 4C_2 + 4C_2x + x^2 + \tfrac{4}{3}x^3 + \tfrac{1}{3}x^4\right)e^{2x} \end{aligned}$$

it follows that $y'' - 4y' + 4y = x^2e^{2x}$. To find the particular solution, we use the fact that $y = 2$ and $y' = 1$ when $x = 0$. That is,

$$\begin{aligned} 2 &= C_1e^0 \Rightarrow C_1 = 2 \\ 1 &= (2C_1 + C_2)e^0 \Rightarrow C_2 = -3. \end{aligned}$$

Thus, the particular solution is $y = \left(2 - 3x + \frac{1}{12}x^4\right)e^{2x}$.

50. When $C = 0$, the graph is a single point: $(0, 0)$.

When $C = 1$, the graph is a hyperbola: $4y^2 - x^2 = 1$.

When $C = -1$, the graph is a hyperbola: $x^2 - 4y^2 = 1$.

When $C = 4$, the graph is a hyperbola: $y^2 - \dfrac{x^2}{4} = 1$.

When $C = -4$, the graph is a hyperbola: $\dfrac{x^2}{4} - y^2 = 1$.

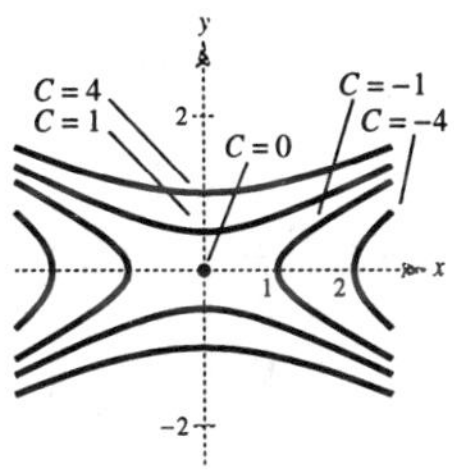

52. When $C = 0$, the graph is the x-axis.

When $C = 1$, the graph is the exponential curve $y = e^{-x}$.

When $C = -1$, the graph is the exponential curve $y = -e^{-x}$.

When $C = 2$, the graph is the exponential curve $y = 2e^{-x}$.

When $C =$ 2, the graph is the exponential curve $y = -2e^{-x}$.

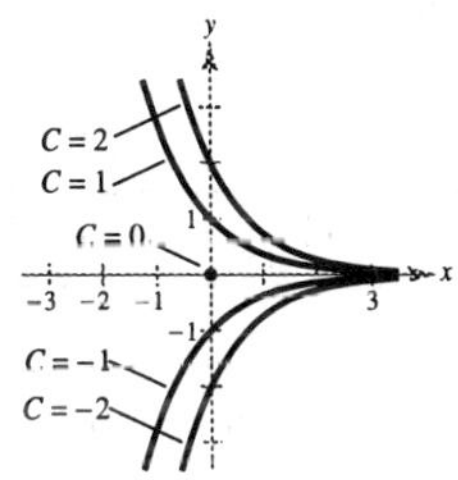

54. $y = \displaystyle\int \frac{1}{1 + x}\,dx = \ln|1 + x| + C$

56. By using partial fractions:

$$\frac{1}{x^2-1} = \frac{A}{x-1} + \frac{B}{x+1}$$

$$1 = A(x+1) + B(x-1)$$

When $x = 1$: $1 = 2A \Rightarrow A = \frac{1}{2}$.

When $x = -1$: $1 = -2B \Rightarrow B = -\frac{1}{2}$.

We have $y = \displaystyle\int \frac{1}{x^2-1}\,dx = \frac{1}{2}\int\left[\frac{1}{x-1} - \frac{1}{x+1}\right]dx = \frac{1}{2}[\ln|x-1| - \ln|x+1|] + C = \frac{1}{2}\ln\left|\frac{x-1}{x+1}\right| + C.$

58. Let $u = x$ and $dv = e^x\,dx$, then $du = dx$ and $v = e^x$. Integration by parts gives us the following.

$$y = \int xe^x\,dx = xe^x - \int e^x\,dx = xe^x - e^x + C = (x-1)e^x + C.$$

60. Since $y = 4$ when $x = 3$, we have $2(3)^2 - (4)^2 = C$ which implies that $C = 2$ and the particular solution is $2x^2 - y^2 = 2$.

62. Since $y = 1$ when $x = 2$, we have $(1)^2 = 2C(2)$ which implies that $C = \frac{1}{4}$ and the particular solution is

$$y^2 = 2\left(\tfrac{1}{4}\right)x$$

$$2y^2 = x.$$

64. Since $A = 1000$ when $t = 0$ and $A = 3320.12$ when $t = 10$, we have the following.

$$1000 = Ce^{k(0)} \Rightarrow C = 1000$$

$$3320.12 = 1000e^{k(10)}$$

$$3.32012 = e^{10k}$$

$$\ln 3.32012 = 10k$$

$$k = \frac{1}{10}\ln 3.32012 \approx 0.12$$

Thus, the particular solution is $A = 1000e^{0.12t}$.

66. $C = 50{,}000$ and $2000 = 50{,}000 - 50{,}000e^{-k} \Rightarrow \frac{1}{25} = 1 - e^{-k} \Rightarrow k = -\ln\left(\frac{24}{25}\right) \approx 0.04082$ and

$x = 50{,}000 - 50{,}000e^{-0.04082t}$

Years, t	2	4	6	8	10
Units, x	3920	7532	10,862	13,930	16,758

68. False. From Example 1, $y = e^x$ is a solution of $y'' - y = 0$, but $y = e^x + 1$ is not.

70. No, you need two more conditions ("initial conditions") to find C and k.

Section D.2 Separation of Variables

2. Yes, $\dfrac{dy}{dx} = \dfrac{x+1}{x}$

$$dy = \frac{x+1}{x}\,dx$$

4. No, the variables cannot be separated.

6. Yes, $x\dfrac{dy}{dx} = \dfrac{1}{y}$

$$y\,dy = \frac{1}{x}\,dx$$

8. $\dfrac{dy}{dx} = x^2y$

$$\int \frac{1}{y}\,dy = \int x^2\,dx$$

$$\ln|y| = \frac{x^3}{3} + C_1$$

$$e^{x^3/3 + C_1} = y$$

$$y = Ce^{x^3/3}$$

10. $(y + 1)\dfrac{dy}{dx} = 2x$

$$\int (y + 1)dy = \int 2x\,dx$$

$$\frac{(y + 1)^2}{2} = x^2 + C_1$$

$$C = 2x^2 - (y + 1)^2$$

12. $(1 + y)\dfrac{dy}{dx} - 4x = 0$

$$\int (1 + y)\,dy = \int 4x\,dx$$

$$\frac{(y + 1)^2}{2} = 2x^2 + C_1$$

$$C = 4x^2 - (y + 1)^2$$

14. $y' - y = 5$

$$\frac{dy}{dx} = y + 5$$

$$\int \frac{1}{y + 5}\,dy = \int dx$$

$$\ln|y + 5| = x + C_1$$

$$y + 5 = e^{x + C_1}$$

$$y = Ce^x - 5$$

16. $\dfrac{dy}{dx} = \sqrt{\dfrac{x}{y}}$

$$\int \sqrt{y}\,dy = \int \sqrt{x}\,dx$$

$$\frac{2}{3}y^{3/2} = \frac{2}{3}x^{3/2} + C_1$$

$$y^{3/2} = x^{3/2} + C$$

18. $\dfrac{dy}{dx} = \dfrac{x^2 + 2}{3y^2}$

$$\int 3y^2\,dy = \int (x^2 + 2)\,dx$$

$$y^3 = \frac{x^3}{3} + 2x + C_1$$

$$C = x^3 + 6x - 3y^3$$

20. $xy' = y$

$$x\frac{dy}{dx} = y$$

$$\int \frac{1}{y}\,dy = \int \frac{1}{x}\,dx$$

$$\ln|y| = \ln|x| + \ln|C|$$

$$\ln|y| = \ln|Cx|$$

$$y = Cx$$

22. $y' - y(x + 1) = 0$

$$\frac{dy}{dx} = y(x + 1)$$

$$\int \frac{1}{y}\,dy = \int (x + 1)\,dx$$

$$\ln|y| = \frac{(x + 1)^2}{2} + C_1$$

$$y = e^{[(x+1)^2/2] + C_1} = Ce^{(x+1)^2/2}$$

24. $e^x(y' + 1) = 1$

$$\left(\frac{dy}{dx} + 1\right) = e^{-x}$$

$$\frac{dy}{dx} = e^{-x} - 1$$

$$\int dy = \int (e^{-x} - 1)dx$$

$$y = -e^{-x} - x + C$$

26. $yy' - 2xe^x = 0$

$$y\frac{dy}{dx} = 2xe^x$$

$$\int y\,dy = \int 2xe^x\,dx$$

$$\frac{y^2}{2} = 2(x-1)e^x + C_1$$

$$y^2 = 4(x-1)e^x + C$$

28. $\sqrt{y}\frac{dy}{dx} = -\sqrt{x}$

$$\int \sqrt{y}\,dy = -\int \sqrt{x}\,dx$$

$$\frac{2}{3}y^{3/2} = -\frac{2}{3}x^{3/2} + C_1$$

$$x^{3/2} + y^{3/2} = C$$

When $x = 1$, $y = 4$. Therefore, $C = 9$ and the particular solution is $x^{3/2} + y^{3/2} = 9$.

30. $dP = 6P\,dt$

$$\int \frac{1}{P}\,dP = \int 6\,dt$$

$$\ln|P| = 6t + C_1$$

$$P = Ce^{6t}$$

When $t = 0$, $P = 5$. Therefore, $C = 5$ and the particular solution is $P = 5e^{6t}$.

32. $\frac{dT}{dt} = -k(T - 70)$

$$\int \frac{1}{T-70}\,dt = \int k\,dt$$

$$\ln|T - 70| = -kt + C_1$$

$$T - 70 = e^{-kt+C_1}$$

$$T = Ce^{-kt} + 70$$

When $t = 0$, $T = 140$. Therefore, $C = 70$ and $T = 70(1 + e^{-kt})$.

34. $\frac{dy}{dx} = \frac{2y}{3x}$

$$\int \frac{1}{2y}\,dy = \int \frac{1}{3x}\,dx$$

$$\frac{1}{2}\ln|y| = \frac{1}{3}\ln|x| + C_1$$

$$3\ln|y| = 2\ln|x| + C_2$$

$$\ln|y^3| = \ln x^2 + \ln C$$

$$y^3 = Cx^2$$

When $x = 8$, $y = 2$. Therefore, $C = \frac{1}{8}$ and the particular solution is $y^3 = \frac{1}{8}x^2$ or $8y^3 = x^2$.

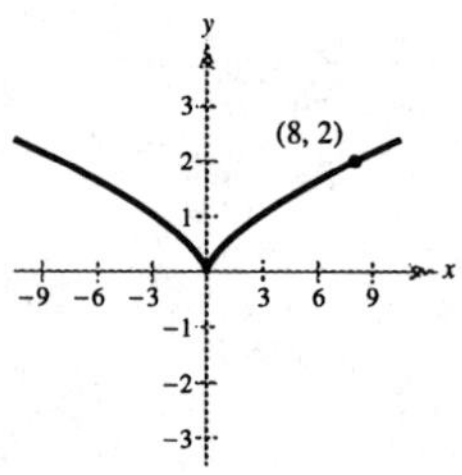

36. $\frac{dv}{dt} = 3.456 - 0.14v$

$$\int \frac{dv}{3.456 - 0.14v} = \int dt$$

$$-\frac{50}{7}\ln|3.456 - 0.14v| = t + C_1$$

$$3.456 - 0.14v = e^{-(7/50)(t+C_1)} = Ce^{-(7/50)t}$$

$$v = \frac{50}{7}(3.456 - Ce^{-0.14t})$$

When $t = 0$, $v = 0$. Therefore $C = 3.456$ and the solution is $v = \frac{50}{7}(3.456)(1 - Ce^{-0.14t}) = \frac{864}{35}(1 - Ce^{-0.14t})$.

For Exercise 38, use the following solution for Newton's Law of Cooling:

$$\int \frac{dT}{T - T_0} = \int k\,dt$$

$$\ln|T - T_0| = kt + C_1$$

$$T - T_0 = d^{kt+C_1}$$

$$T = Ce^{kt} + T_0$$

38. From the above, we have $T = Ce^{kt} + 70$. When $t = 0$, $T = 350$. Therefore, $C = 280$ and the equation for the temperature is $T = 280e^{kt} + 70$. When $t = \frac{3}{4}$, $T = 150$. Therefore, we have $150 = 280e^{3k/4} + 70$ which implies that $k = \frac{4}{3}\ln\frac{2}{7} \approx -1.6704$. Thus, the equation for the temperature is $T = 280e^{-1.6704t} + 70$. When $T = 80°$, we can solve for the time to obtain

$$t = \frac{\ln(1/28)}{-1.6704} \approx 1.9949 \text{ hours}$$

$$\approx 119.69 \text{ minutes.}$$

40. $\dfrac{dV}{dt} = kV^{2/3}$

$$\int V^{-2/3}\,dV = \int k\,dt$$

$$3V^{1/3} = kt + C_1$$

$$V^{1/3} = \frac{1}{3}kt + C$$

$$V = \left(\frac{1}{3}kt + C\right)^3$$

42. $\dfrac{dS}{dt} = \dfrac{kS}{t^2}$

$$\int \frac{1}{S}\,dS = \int kt^{-2}\,dt$$

$$\ln S = -\frac{k}{t} + C_1$$

$$S = Ce^{-k/t}$$

Since $S = 10$ (in thousands of units) when $t = 1$, we have $10 = Ce^{-k}$. Also, $\lim_{t\to\infty} Ce^{-k/t} = C = 50$. Thus,

$$10 = 50e^{-k}$$

$$\ln\frac{1}{5} = -k$$

$$k = \ln 5$$

$$S = 50e^{-(\ln 5)/t}.$$

44. $y = Cx^{-k}$

$x = 75{,}000$ when $y = 7.1$ and $x = 25{,}000$ when $y = 64.7$.

Thus,

$$7.1 = C(75{,}000)^{-k}$$

$$64.7 = C(25{,}000)^{-k}$$

Dividing eliminates C:

$$7.1/64.7 = \left(\frac{75{,}000}{25{,}000}\right)^{-k} = \left(\frac{1}{3}\right)^k$$

$$\ln(7.1/64.7) = k\ln\left(\frac{1}{3}\right) \Rightarrow k \approx 2.0113$$

Hence, $C = 7.1(75{,}000)^k \approx 45{,}351{,}451{,}618$ and $y = 45{,}351{,}451{,}618\,x^{(-2.0113)}$

(a) $x = 20{,}000 \Rightarrow y = 101.3$ million

(b) $x = 100{,}000 \Rightarrow y = 3.98$ million

Section D.3 First-Order Linear Differential Equations

2. $y' - 5(2x - y) = 0$

$$y' + 5y = 10x$$

4. $$xy' + y = x^3y$$

$$xy' + (1 - x^3)y = 0$$

$$y' + \frac{1 - x^3}{x}y = 0$$

6. $$x = x^2(y' + y)$$

$$x^2y' + x^2y = x$$

$$y' + y = \frac{1}{x}$$

8. For this linear differential equation, $y' + 5y = 15$, we have $P(x) = 5$ and $Q(x) = 15$. Therefore, the integrating factor is

$$u(x) = e^{\int 5\,dx} = e^{5x}$$

and the general solution is

$$y = \frac{1}{u(x)}\int Q(x)u(x)dx = e^{-5x}\int 15e^{5x}dx = e^{-5x}(3e^{5x} + C) = 3 + Ce^{-5x}.$$

10. For this linear differential equation, $y' + 3y = e^{-3x}$, we have $P(x) = 3$ and $Q(x) = e^{-3x}$. Therefore, the integrating factor is

$$u(x) = e^{\int 3\,dx} = e^{3x}$$

and the general solution is

$$y = \frac{1}{u(x)}\int Q(x)u(x)dx = e^{-3x}\int e^{-3x}e^{3x}\,dx = e^{-3x}(x + C).$$

12. For this linear differential equation,

$$y' + \frac{2}{x}y = 3x + 1$$

we have $P(x) = 2/x$ and $Q(x) = 3x + 1$. Therefore, the integrating factor is

$$u(x) = e^{\int 2/x\,dx} = e^{2\ln x} = e^{\ln x^2} = x^2$$

and the general solution is

$$y = \frac{1}{u(x)}\int Q(x)u(x)dx = \frac{1}{x^2}\int (3x + 1)(x^2)dx = \frac{1}{x^2}\left(\frac{3x^4}{4} + \frac{x^3}{3} + C\right) = \frac{3}{4}x^2 + \frac{1}{3}x + \frac{C}{x^2}.$$

14. For this linear differential equation, $y' + 5y = e^{5x}$, we have $P(x) = 5$ and $Q(x) = e^{5x}$. Therefore, the integrating factor is

$$u(x) = e^{\int 5dx} = e^{5x}$$

and the general solution is

$$y = \frac{1}{u(x)}\int Q(x)u(x)dx = e^{-5x}\int e^{5x}e^{5x}dx = e^{-5x}\left(\frac{1}{10}e^{10x} + C\right) = \frac{1}{10}e^{5x} + Ce^{-5x}.$$

16. For this linear differential equation,

$$y' + \frac{2}{x^3}y = \frac{1}{x^3}e^{1/x^2}$$

we have $P(x) = 2/x^3$ and $Q(x) = (1/x^3)e^{1/x^2}$. Therefore, the integrating factor is

$$u(x) = e^{\int (2/x^3)\,dx} = e^{-1/x^2}$$

and the general solution is

$$y = \frac{1}{u(x)}\int Q(x)u(x)dx = e^{1/x^2}\int \frac{1}{x^3}e^{1/x^2}e^{-1/x^2}\,dx = e^{1/x^2}\int \frac{1}{x^3}dx = e^{1/x^2}\left(-\frac{1}{2x^2} + C\right).$$

18. For this linear differential equation,

$$y' + \frac{1}{x}y = x\ln x$$

we have $P(x) = 1/x$ and $Q(x) = x\ln x$. Therefore, the integrating factor is

$$u(x) = e^{\int (1/x)\,dx} = e^{\ln x} = x$$

and the general solution is

$$y = \frac{1}{u(x)}\int Q(x)u(x)dx = \frac{1}{x}\int x^2\ln x\,dx = \frac{1}{x}\left(\frac{x^3}{3}\ln x - \frac{x^3}{9} + C\right) = \frac{x^2}{3}\ln x - \frac{x^2}{9} + \frac{C}{x}.$$

20. Separation of Variables:

$$\frac{dy}{dx} = 5 - 10y$$

$$\int \frac{1}{5 - 10y}\,dy = \int dx$$

$$-\frac{1}{10}\ln|5 - 10y| = x + C_1$$

$$\ln|5 - 10y| = -10x + C_2$$

$$5 - 10y = e^{-10x + C_2}$$

$$-10y = C_3e^{-10x} - 5$$

$$y = \frac{1}{10}(5 - C_3e^{-10x}) = \frac{1}{2} + Ce^{-10x}$$

First-Order Linear: $P(x) = 10, \quad Q(x) = 5$

$$u(x) = e^{\int 10\,dx} = e^{10x}$$

$$y = e^{-10x}\int 5e^{10x}\,dx = e^{-10x}\left(\frac{1}{2}e^{10x} + C\right) = \frac{1}{2} + Ce^{-10x}$$

22. Separation of Variables:

$$\frac{dy}{dx} = x(1 - 4y)$$

$$\int \frac{1}{1 - 4y}\,dy = \int x\,dx$$

$$-\frac{1}{4}\ln|1 - 4y| = \frac{x^2}{2} + C_1$$

$$\ln|1 - 4y| = -2x^2 + C_2$$

$$1 - 4y = e^{-2x^2 + C_2}$$

$$1 - 4y = C_3e^{-2x^2}$$

$$y = \frac{1}{4}(1 - C_3e^{-2x^2}) = \frac{1}{4} + Ce^{-2x^2}$$

—CONTINUED—

22. —CONTINUED—

First-Order Linear: $P(x) = 4x,\quad Q(x) = x$

$$u(x) = e^{\int 4x\,dx} = e^{2x^2}$$

$$y = e^{-2x^2}\int xe^{2x^2}\,dx = e^{-2x^2}\left(\frac{1}{4}e^{2x^2} + C\right) = \frac{1}{4} + Ce^{-2x^2}$$

24. Since $P(x) = 3x^2$ and $Q(x) = 3x^2$, the integrating factor is $u(x) = e^{\int 3x^2\,dx} = e^{x^3}$ and the general solution is

$$y = e^{-x^3}\int 3x^2e^{x^3}\,dx = e^{-x^3}(e^{x^3} + C) = 1 + Ce^{-x^3}.$$

Since $y = 6$ when $x = 0$, it follows that $C = 5$ and the particular solution is $y = 1 + 5e^{-x^3}$.

26. Since $P(x) = 2x - 1$ and $Q(x) = 0$, the integrating factor is $u(x) = e^{\int (2x-1)dx} = e^{x^2-x}$ and the general solution is

$$y = e^{x-x^2}\int (0)(e^{x^2-x})dx = Ce^{x-x^2}.$$

Since $y = 2$ when $x = 1$, it follows that $C = 2$ and the particular solution is $y = 2e^{x-x^2}$.

28. Since $P(x) = 2$ and $Q(x) = e^{-2x}$, the integrating factor is $u(x) = e^{\int 2\,dx} = e^{2x}$ and the general solution is

$$y = e^{-2x}\int e^{-2x}e^{2x}\,dx = e^{-2x}(x + C).$$

Since $y = 4$ when $x = 1$, it follows that $C = 4e^2 - 1$ and the particular solution is $y = e^{-2x}(x + 4e^2 - 1)$.

30. Since $P(x) = -4/x$ and $Q(x) = 10/x^2$, the integrating factor is

$$u(x) = e^{\int -(4/x)dx} = e^{-4\ln x} = x^{-4} = \frac{1}{x^4}$$

and the general solution is

$$y = x^4\int\left(\frac{10}{x^2}\right)\left(\frac{1}{x^4}\right)dx = x^4\left(-\frac{2}{x^5} + C\right) = -\frac{2}{x} + Cx^4.$$

Since $y = 10$ when $x = 1$, it follows that $C = 12$ and the particular solution is $y = -(2/x) + 12x^4$.

32. $$\frac{dS}{dt} = k_1(L - S) + k_2t$$

$S' + k_1S = k_1L + k_2t$

Since $P(t) = k_1$ and $Q(t) = k_1L + k_2t$, the integrating factor is $u(t) = e^{\int k_1\,dt} = e^{k_1t}$ and the general solution is

$$S = e^{-k_1t}\int (k_1L + k_2t)e^{k_1t}\,dt = e^{-k_1t}\left(Le^{k_1t} + \frac{k_2}{k_1}te^{k_1t} - \frac{k_2}{k_1^2}e^{k_1t} + C\right) = L + \frac{k_2}{k_1}t - \frac{k_2}{k_1^2} + Ce^{-k_1t}.$$

Since $S = 0$ when $t = 0$, it follows that $C = \dfrac{k_2}{k_1^2} - L$ and the particular solution is

$$S = \left(L - \frac{k_2}{k_1^2}\right) + \frac{k_2}{k_1}t + \left(\frac{k_2}{k_1^2} - L\right)e^{-k_1t} = \left(L - \frac{k_2}{k_1^2}\right)(1 - e^{-k_1t}) + \frac{k_2}{k_1}t.$$

34. Since $\eta = \dfrac{p/x}{dp/dx} = 1 - \dfrac{500}{3x}$, we have

$$\frac{p}{x} = \left(1 - \frac{500}{3x}\right)\frac{dp}{dx}$$

$$\frac{p}{x - (500/3)} = \frac{dp}{dx}$$

$$\int \frac{1}{x - (500/3)}\,dx = \int \frac{1}{p}\,dp$$

$$\ln\left|x - \frac{500}{3}\right| = \ln|p| + C_1$$

$$x - \frac{500}{3} = Cp.$$

Since $p = 2$ when $x = 100$, it follows that $C = -100/3$ and the particular solution is

$$x - \frac{500}{3} = -\frac{100}{3}p$$

$$\text{or } p = 5 - \frac{3x}{100}.$$

36.

$$D(t) = S(t)$$

$$4000 + 5p(t) - 4p'(t) = 2800 + 7p(t) + 2p'(t)$$

$$1200 - 2p(t) - 6p'(t) = 0$$

$$p'(t) + \frac{1}{3}p(t) = 200$$

$$P(t) = \frac{1}{3},\ Q(t) = 200$$

$$u(t) = e^{\int (1/3)\,dt} = e^{(1/3)t}$$

$$p(t) = e^{-(1/3)t}\int 200e^{(1/3)t}\,dt$$

$$= e^{-(1/3)t}(600e^{(1/3)t} + C)$$

$$= 600 + Ce^{-(1/3)t}$$

$$p(0) = 1000 \Longrightarrow C = 400$$

$$p(t) = 200(3 + 2e^{-(1/3)t})$$

38. $\dfrac{dv}{dt} + \dfrac{k}{m}v = -g$

$u(t) = e^{\int (k/m)\,dt} = e^{kt/m}$

$$v = \frac{1}{u(t)}\int Q(t)u(t)\,dt = \frac{1}{e^{kt/m}}\int -ge^{kt/m}\,dt = \frac{1}{e^{kt/m}}\left(\frac{-gm}{k}e^{kt/m} + C\right) = \frac{-gm}{k} + Ce^{-kt/m}$$

40. (a)

$$\frac{dy}{dt} = \frac{1-y}{4}$$

$$\frac{4}{1-y}\,dy = dt$$

$$-4\ln(1-y) = t + C_1$$

$$y(0) = 0 \Longrightarrow C_1 = 0$$

Hence, $\ln(1-y) = -\frac{1}{4}t$

$$1 - y = e^{-1/4\,t}$$

$$y = 1 - e^{-1/4t}$$

(b)

$$\frac{1}{2} = 1 - e^{-1/4\,t}$$

$$e^{-1/4\,t} = \frac{1}{2}$$

$$-\frac{1}{4}t = \ln\frac{1}{2} = -\ln 2$$

$$t = 4\ln 2 \approx 2.77 \text{ years}$$

(c) $y = 1 - e^{-1/4\,(4)} = 1 - e^{-1} \approx 0.632$ or 63.2%

Section D.4 Applications of Differential Equations

2. The general solution is $y = Ce^{kx}$. Since $y = 4$ when $x = 0$, it follows that $C = 4$. Thus, $y = 4e^{kx}$. Since $y = 6$ when $x = 1$, it follows that $6 = 4e^k$ which implies that

$$k = \ln\left(\frac{3}{2}\right) \approx 0.4055.$$

Thus, the particular solution is $y = 4e^{0.4055x}$.

4. The general solution is $y = Ce^{kx}$. Since $y = 60$ when $x = 0$, it follows that $C = 60$. Thus, $y = 60e^{kx}$. Since $y = 30$ when $x = 5$, it follows that $30 = 60e^{5k}$ which implies that

$$k = \frac{1}{5}\ln\frac{1}{2} \approx -0.1386.$$

Thus, the particular solution is $y = 60e^{-0.1386x}$.

6. The general solution is $y = Ce^{kx}$. Since $y = 4$ when $x = 1$ and $y = 1$ when $x = 2$, it follows that $4 = Ce^{k}$ and $1 = Ce^{2k}$. By equating C-values from these two equations, we have

$$4e^{-k} = e^{-2k}$$

$$4 = e^{-k} \Rightarrow k = -\ln 4 \approx -1.3863.$$

This implies that $C = 4e^{\ln 4} = 4(4) = 16$. Thus, the particular solution is $y = 16e^{-1.3863x}$.

8. Let $t = 0$ represent the year 1993. The general solution is $y = Ce^{kt}$ where $C = 400{,}000$ and $k = 0.015$. Thus, the particular solution is $y = 400{,}000e^{0.015t}$. In the year 2000, we have $t = 7$ and the population of the city is estimated to be $y = 400{,}000e^{0.015(7)} \approx 444{,}284$.

10. The general solution is $S = L(1 - e^{-kt})$.

(a) Since $L = 100$ and $S = 25$ when $t = 2$, we have the following.

$$25 = 100(1 - e^{-2k})$$

$$\frac{1}{4} = 1 - e^{-2k}$$

$$e^{-2k} = \frac{3}{4}$$

$$-2k = \ln\frac{3}{4}$$

$$k = -\frac{1}{2}\ln\frac{3}{4} \approx 0.1438$$

Thus, the particular solution is $y = 100(1 - e^{-0.1438t})$.

(b) Since $L = 500$ and $S = 50$ when $t = 1$, we have the following.

$$50 = 500(1 - e^{-k})$$

$$\frac{1}{10} = 1 - e^{-k}$$

$$e^{-k} = \frac{9}{10}$$

$$-k = \ln\frac{9}{10}$$

$$k = -\ln\frac{9}{10} = \ln\frac{10}{9} \approx 0.1054$$

Thus, the particular solution is $y = 500(1 - e^{-0.1054t})$.

12. The general solution is $y = Ce^{100kx}(100 - y)$. Since $y = 10$ when $x = 0$, it follows that $C = \frac{1}{9}$. Thus, $y = \frac{1}{9}e^{100kx}(100 - y)$. Since $y = 30$ when $x = 5$, it follows that

$$\frac{27}{7} = e^{500k}$$

$$100k = \frac{1}{5}\ln\frac{27}{7} \approx 0.269985 \approx 0.2700.$$

Thus, the particular solution is

$$y = \frac{1}{9}e^{0.2700x}(100 - y)$$

$$y(9 + e^{0.2700x}) = 100e^{0.2700x}$$

$$y = \frac{100e^{0.2700x}}{9 + e^{0.2700x}} = \frac{100}{1 + 9e^{-0.2700x}}.$$

14. The general solution is $y = Ce^{1000kx}(1000 - y)$. Since $y = 100$ when $x = 0$, it follows that $C = \frac{1}{9}$. Thus, $y = \frac{1}{9}e^{1000kx}(1000 - y)$. Since $y = 750$ when $x = 4$, it follows that

$$27 = e^{4000k}$$

$$1000k = \frac{1}{4}\ln 27 \approx 0.8240.$$

Thus, the particular solution is

$$y = \frac{1}{9}e^{0.8240x}(1000 - y)$$

$$y(9 + e^{0.8240x}) = 1000e^{0.8240x}$$

$$y = \frac{1000e^{0.8240x}}{9 + e^{0.8240x}} = \frac{1000}{1 + 9e^{-0.8240x}}.$$

16. The differential equation is given by the following.

$$\frac{dS}{dt} = kS(L - S)$$

$$\int \frac{dS}{S(L - S)} = \int k\,dt$$

$$\frac{1}{L}[\ln|S| - \ln|L - S|] = kt + C_1$$

$$\frac{S}{L - S} = Ce^{Lkt}$$

$$S = \frac{CLe^{Lkt}}{1 + Ce^{Lkt}} = \frac{CL}{C + e^{-Lkt}}$$

When $t = 0$, $S = 10$. Thus, $C = \dfrac{10}{L - 10}$.

When $t = 2$, $S = 50$. Thus, $k = \dfrac{1}{2L}\ln\left[\dfrac{5(L - 10)}{L - 50}\right]$. Therefore, $S = \dfrac{L}{1 + [(L - 10)/10]e^{-(1/2)\ln[5(L-10)/L-50]t}}$.

18. The general solution is $P = \dfrac{CL}{C + e^{-Lkn}}$.

(a) Since $L = 1$ and $P = 0.50$ when $n = 0$, and $P = 0.85$ when $n = 4$, we have the following.

$$0.50 = \frac{C}{C + 1} \Rightarrow C = 1$$

$$0.85 = \frac{1}{1 + e^{-4k}} \Rightarrow k = -\frac{1}{4}\ln\frac{3}{17} \approx 0.4337$$

Therefore, $P = \dfrac{1}{1 + e^{-0.4337n}}$.

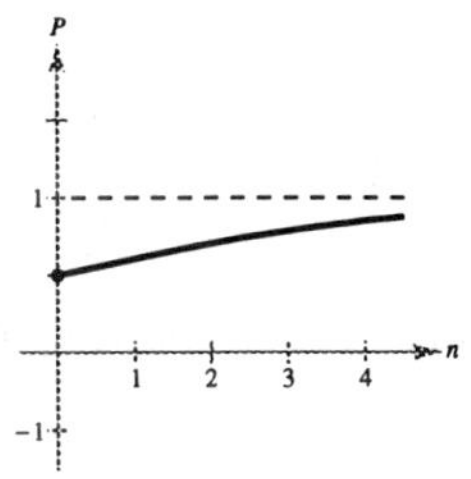

(b) Since $L = 0.80$ and $P = 0.25$ when $n = 0$, and $P = 0.60$ when $n = 10$, we have the following.

$$0.25 = \frac{0.80C}{C + 1} \Rightarrow C = \frac{5}{11}$$

$$0.60 = \frac{(5/11)(0.80)}{(5/11) + e^{-8k}} \Rightarrow k = -\frac{1}{8}\ln\frac{5}{33} \approx 0.2359$$

Therefore, $P = \dfrac{4}{5 + 11e^{-0.1887n}}$.

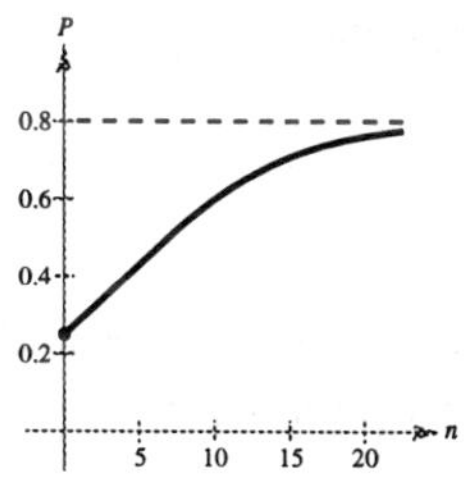

20. The general solution is $y = -1/(kt + C)$. Since $y = 75$ when $t = 0$, we have $C = -1/75$. Therefore,

$$y = -\frac{1}{kt - (1/75)} = \frac{75}{1 - 75kt}.$$

Since $y = 12$ when $t = 1$, we have

$$12 = \frac{75}{1 - 75k} \Rightarrow k = -\frac{7}{100}.$$

Thus, we have

$$y = \frac{75}{1 + 5.25t} = \frac{300}{4 + 21t}.$$

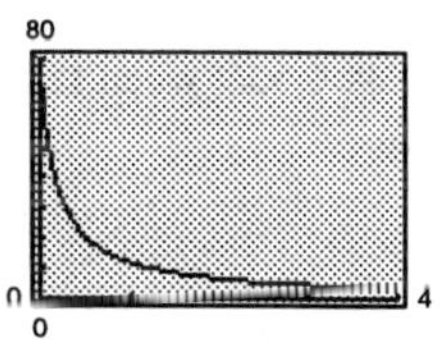

22. The general solution is $y = 5000e^{-Ce^{-kt}}$. Since $y = 500$ when $t = 0$, it follows that $500 = 5000e^{-C}$ which implies that $C = -\ln \frac{1}{10} = \ln 10$. Therefore, we have $y = 5000e^{(-\ln 10)e^{-kt}}$. Since $y = 625$ when $t = 1$, it follows that

$$625 = 5000e^{(-\ln 10)e^{-k}}$$

$$e^{-k} = \frac{\ln(1/8)}{\ln(1/10)}$$

$$k = -\ln\left(\frac{\ln(1/8)}{\ln(1/10)}\right) \approx 0.1019.$$

Thus, we have $y = 5000e^{(-2.3026)e^{(-0.1019)t}}$

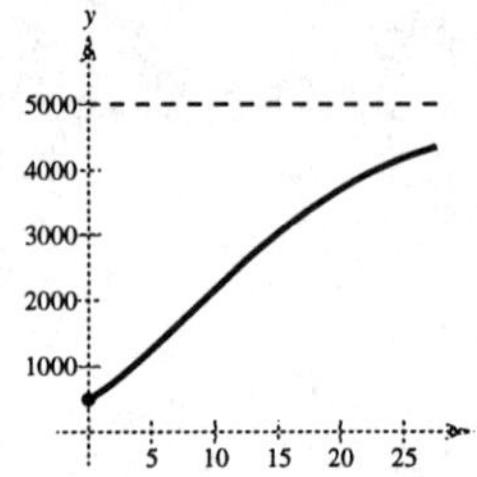

24. From Example 3, the general solution is

$$y = 400e^{-Ce^{-kt}}.$$

Since $y = 30$ when $t = 0$,

$$30 = 400e^{-C} \Longrightarrow C = \ln\left(\frac{40}{3}\right) \approx 2.5903.$$

Since $y = 90$ when $t = 1$,

$$90 = 400e^{-2.5903e^{-k}}$$

$$\frac{9}{40} = e^{-2.5903e^{-k}}$$

$$\ln\left(\frac{9}{40}\right) = -2.5903e^{-k}$$

$$k = -\ln\left(\frac{\ln(9/40)}{-2.5903}\right) \approx 0.5510.$$

Thus,

$$y = 400e^{-2.5903e^{-0.5519t}}.$$

Finally, when $t = 3$, $y \approx 244$ rabbits.

26. Following Example 4, the differential equation is

$$\frac{dy}{dt} = ky(1 - y)(2 - y)$$

and its general solution is

$$\frac{y(2 - y)}{(1 - y)^2} = Ce^{2kt}.$$

$$y = 0.4 \text{ when } t = 0 \Longrightarrow \frac{(0.4)(1.6)}{(0.6)^2} = \frac{16}{9} = C$$

$$y = 0.8 \text{ when } t = 5 \Longrightarrow \frac{(0.8)(1.2)}{(0.2)^2} = 24 = \frac{16}{9}e^{2k(5)}$$

$$\Longrightarrow \frac{27}{2} = e^{10k}$$

$$\Longrightarrow k = \frac{1}{10}\ln\left(\frac{27}{2}\right) \approx 0.2603$$

Hence, the particular solution is

$$\frac{y(2 - y)}{(1 - y)^2} = \frac{16}{9}e^{0.5205t}.$$

Using a symbolic algebra utility or graphing utility, we find that when $t = 8$, $y \approx 0.91$.

28. Since $Q' + \frac{1}{20}Q = \frac{5}{2}$ is a first-order linear differential equation with $P(x) = \frac{1}{20}$ and $R(x) = \frac{5}{2}$, we have the integrating factor $u(t) = e^{\int (1/20)dt} = e^{(1/20)t}$, and the general solution is

$$Q = e^{-0.05t}\int \frac{5}{2}e^{0.05t}\,dt = e^{-0.05t}(50e^{0.05t} + C) = 50 + Ce^{-0.05t}.$$

Since $Q = 0$ when $t = 0$, we have $C = -50$ and $Q = 50(1 - e^{-0.05t})$. Finally, when $t = 30$, we have $Q \approx 38.843$ lbs/gal.

30. The general solution is $y = Ce^{kt}$. Since $y = 0.60C$ when $t = 1$, we have $0.60C = Ce^{k} \Longrightarrow k = \ln 0.60 \approx -0.5108$. Thus, $y = Ce^{-0.5108t}$. When $y = 0.20C$, we have

$$0.20C = Ce^{-0.5108t}$$

$$\ln 0.20 = -0.5108t$$

$$t \approx 3.15 \text{ hours.}$$

32. $\displaystyle\int \frac{1}{kP+N}\,dp = \int dt$

$$\frac{1}{k}\ln|kP+N| = t + C_1$$

$$kP + N = C_2e^{kt}$$

$$P = Ce^{kt} - \frac{N}{k}$$

34. From Exercise 33, we have $A = \dfrac{P}{r}(e^{rt}-1)$.

(a) For $P = \$100{,}000$, $r = 12\%$, and $t = 5$ years,

$$A = \frac{100{,}000}{0.12}(e^{(0.12)(5)} - 1) \approx \$685{,}099.00.$$

(b) For $P = \$250{,}000$, $r = 15\%$, and $t = 10$ years,

$$A = \frac{250{,}000}{0.15}(e^{(0.15)(10)} - 1) \approx \$5{,}802{,}815.12.$$

36. For $A = \$800{,}000$, $P = \$75{,}000$, and $r = 13\%$,

$$800{,}000 = \frac{75{,}000}{0.13}(e^{0.13t} - 1)$$

$$\frac{104}{75} = e^{0.13t} - 1$$

$$0.13t = \ln\frac{179}{75}$$

$$t \approx 6.692 \text{ years.}$$

38. From Exercise 37, we have $C = C_0e^{-Rt/V}$.

(a) For $V = 2$, $R = 0.5$, and $C_0 = 0.6$, we have $C = 0.6e^{-0.25t}$.

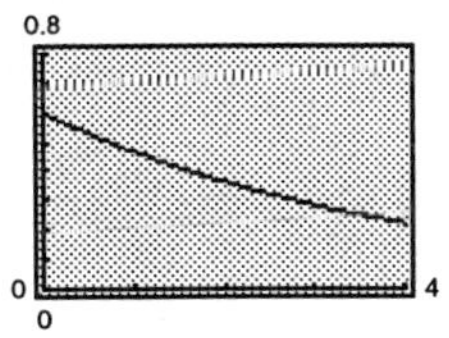

(b) For $V = 2$, $R = 1.5$, and $C_0 = 0.6$, we have $C = 0.6e^{-0.75t}$.

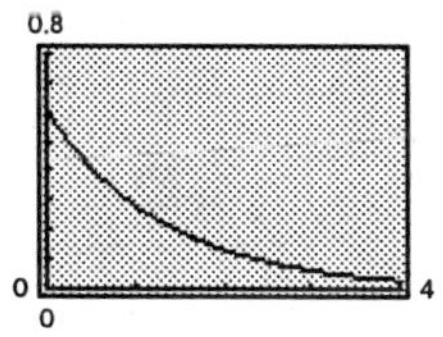

PART III

ANSWERS TO SPECIAL FEATURES

Answers to the

GROUP DISCUSSIONS

CHAPTER 1

Section 1.1

a. $(x, y) \rightarrow (-x, y)$
Reflection in the y-axis
Congruent figures

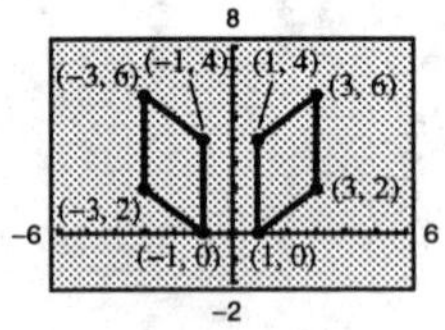

b. $(x, y) \rightarrow (x, -y)$
Reflection in the x-axis
Congruent figures

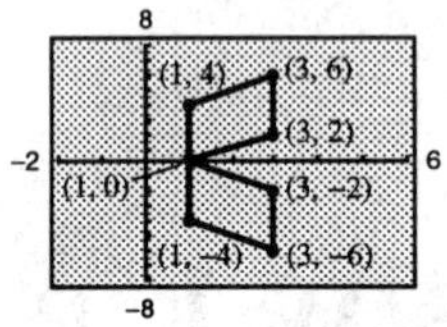

c. $(x, y) \rightarrow (-x, -y)$
Reflection through (0, 0) (or equivalently, rotation by 180°)
Congruent figures

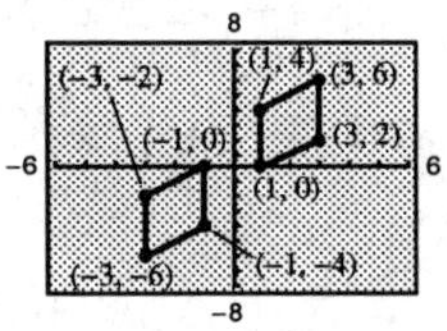

Section 1.2

Graphically—determine at what time the revenues are equal (i.e. where the graphs intersect) and then be sure that after this time the Dell graph is above the Sun graph. Numerically—compare the revenues in the table and determine when the Dell revenue exceeds the Sun revenue. Analytically—solve the revenue equations simultaneously and find the time when the revenues are the same. Using the revenue equations, plug in a time value slightly larger than the value already found and check that the Dell revenue is greater than the Sun revenue. $t = 5.48$ represents June 1995.

Section 1.3

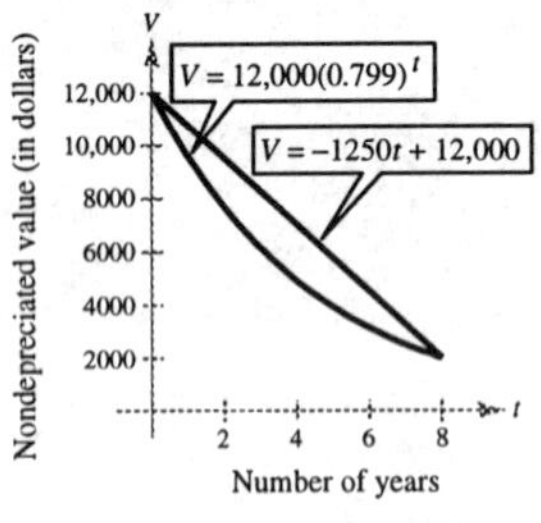

Linear is simpler but may not be as realistic.

Section 1.4

Option b gives a lower price ($14,650 versus $14,850) so this would be preferred by the customer. $f(g(x))$ represents option b; $g(f(x))$ represents option a.

Section 1.5

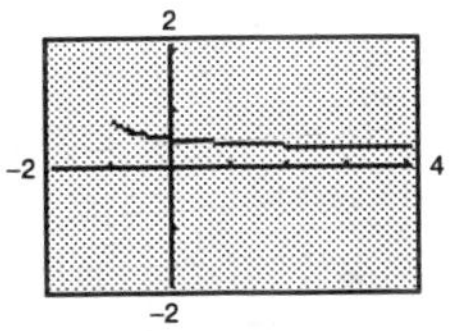

The analytic approach should be preferred to the others.

Section 1.6

The function $A = P\left(1 + \frac{r}{n}\right)^{nt}$ is continuous. The bank that uses the continuous formula will have the amount in the account that is closest to the correct amount; the one using the discontinuous function will be "rounding down" the amount in the account.

CHAPTER 2

Section 2.1

Conclusion can be verified analytically by taking the derivative of $f(x)$ and finding the tangent line slope and making sure that the point of tangency satisfies both the function and the tangent line equation.

Section 2.2

a. millions of people per year

b. feet per second

c. dollars per unit

Section 2.3

If less than 24,400 are sold the profit is increasing; if more than 24,400 are sold, the profit is decreasing. If exactly 24,400 are sold the profit is maximized. The owner should always try to sell 24,400 hamburgers and charge \$1.78/hamburger.

Section 2.4

Derivative of a product (quotient) is not the product (quotient) of the derivatives. To convince the class, simply multiply (or divide) first and then differentiate to show that the product (quotient) rule IS correct.

Section 2.5

a. simple	b. simple	c. general
d. simple	e. general	f. general

Section 2.6

earth/moon = 5.9

Section 2.7

$$\frac{dp}{dx} = \frac{(0.000003x^2 + 0.01)p^2}{-3} = \frac{1}{dx/dp}$$

dp/dx represents the rate of change in the price of a product as the demand changes.

Section 2.8

Revenue is increasing over the 10 week period. The company can continue to drop the price for 240 more weeks before the revenue starts to decline. Yes, some companies use mathematical models to help guide their pricing decisions.

CHAPTER 3

Section 3.1

1. $p = 7.2 - 0.001x$
2. 3000
3. $4.20
4. The revenue function is different than the profit function since it does not include the effect of the cost function. Since the functions themselves are different, their extrema, in general, will occur at different points. Maximum revenue does NOT imply maximum profit.

Section 3.2

$p = (60{,}000 - x)/20{,}000$; $1.78

Section 3.3

$x = 50$ implies $y = \$62.50$; $x = 70$ implies $y = \$112.70$ so y is increasing. $x = 120$ implies $y = \$259.20$; $x = 150$ implies $y = \$337.50$.

Section 3.4

Answers will vary.

Section 3.5

As the price increases for art by a famous artist, the demand for this art might also increase. Such a demand function would look like the following.

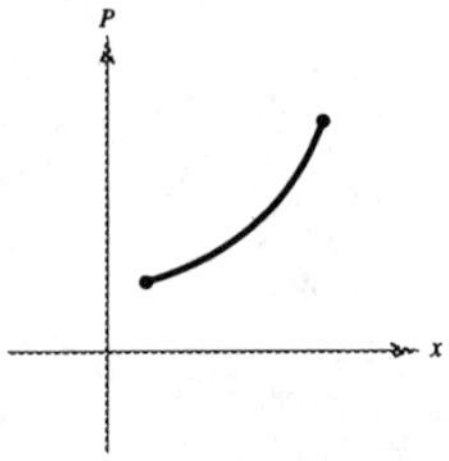

Section 3.7

There are 4 possible types of quartic curves. They may have one extrema (a max or min) or three (max, min, max or min, max, min); they cannot have two, since a cubic must have either one or three real roots.

a. type 1 b. type 2 c. type 1 d. type 2
e. type 2 f. type 1 g. type 1 h. type 2

Section 3.8

a. Percentage error is approximately 2%, which is less than that in the example.
b. Percentage error is approximately 8.57%, greater than that in the example.

CHAPTER 4

Section 4.1

If the interest rates continue to decline after the money is invested, option a would be the best, since the rate is locked in. Option b would be better than c if the rates decline for two years after the investment is made. Option c would be the best if the interest rates increase beyond 7% after two years, or if the money may be needed in a short period of time.

Section 4.2

The maximum of the normal probability density function is 1, which occurs at $x = 0$.

Section 4.3

3%	4%	5%	6%	7%	8%	9%	10%	11%	12%
36.6	27.5	22	18.3	15.7	13.7	12.2	11	10	9.2

Section 4.4

You could approximate the answer to Example 9 by calculating

$$\left(\tfrac{1}{2}\right)^{10,001/5700} - \left(\tfrac{1}{2}\right)^{10,000/5700} \approx -0.000036$$

This approximation corresponds to a secant line approximation, as indicated in the figure.

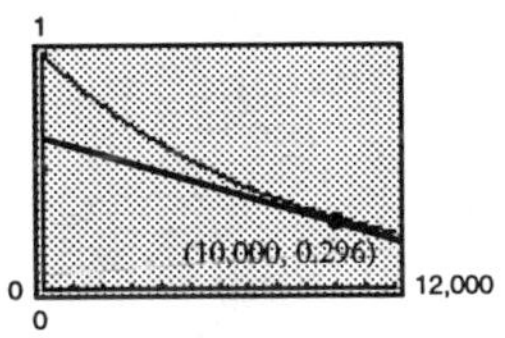

Section 4.5

If the sales followed a linear pattern of decline, after 5 months the sales would be $0. Since this is highly unlikely, the exponential model is probably more realistic. Also the exponential model could never give zero sales or a negative amount of sales but the linear model could.

CHAPTER 5

Section 5.1

This analysis would cease to make sense for any x such that $x > m$, where $(m, C(m))$ is the maximum point on the graph of $y = C(x)$. This is so since, for $x > m$, ΔC would be less than 0, thus implying that making more units of the product actually costs less, which is impossible.

Section 5.2

$35,825.53

Section 5.3

Using a u-substitution technique with $u = bx + c$ we have: $\displaystyle\int \frac{a}{bx + C}\,dx = \frac{a}{b}\ln|bx + c| + C$

a. $\displaystyle\int \frac{1}{2x - 5}\,dx = \frac{1}{2}\ln|2x - 5| + C$

b. $\displaystyle\int \frac{4}{3x + 2}\,dx = \frac{4}{3}\ln|3x + 2| + C$

c. $\displaystyle\int \frac{7}{8x - 3}\,dx = \frac{7}{8}\ln|8x - 3| + C$

Section 5.4

The integral of $\int_{-2}^{2} \sqrt{4 - x^2}\, dx$ represents the area of the semicircle $y = \sqrt{4 - x^2}$ with radius $r = 2$.

Therefore the area is $\frac{\pi r^2}{2} = 2\pi$ without knowing the antiderivative of the function under the integral sign.

Section 5.5

The area under $y = f(x)$ would have units of $x \times y$. Since y is measured in millions of barrels/year and x is measured in years, the area has unit of

$$(\text{billions of barrels/year})(\text{years}) = \text{billions of barrels}$$

Section 5.6

n	10	20	30	40	50	60
Approximation	48.3	97.7	147.0	196.4	245.7	295.1

The midpoint rule is not working for the function $\int_0^1 \frac{1}{x^2}\, dx$ since the function $f(x) = \frac{1}{x^2}$ has an infinite discontinuity at $x = 0$. This integral is an improper integral and does not have a finite value.

Section 5.7

The volume of a cylinder with diameter 7 inches and height 11 inches is about 432 cubic inches. Since this is almost twice the answer obtained in the text and the football could fit inside the cylinder, the answer is reasonable.

CHAPTER 6

Section 6.1

To find b: calculate the integral $\int_0^b f(x)\, dx$ and set it equal to 1. Evaluate the integral using the fundamental theorem of integral calculus and solve for b.

Section 6.2

Yes, it would be advantageous to argue that the inflation rate is lower than 10%, since then the present value would actually be higher than the value given in Example 7.

Section 6.3

Between the first 10 years the rate of growth is increasing; after 10 years the rate of growth is decreasing. The change to decreasing growth would occur later if the deer herd could contain a limiting value of 3000 deer.

Section 6.4

a. Use $\int \frac{du}{u}$. (Formula 2)

b. Use $\int \frac{1}{1 + e^u}\, du$. (Formula 37)

c. Use $\int \frac{1}{1 + e^{nu}}\, du$. (Formula 38)

Section 6.5

$f'''(x) = -4xe^{-x^2}(2x^2 - 3) = 0 \Rightarrow x = 0, \sqrt{\frac{3}{2}}$ (outside interval)

$f''(0) = -2$

$f''(1) = 0.736$

Thus, the maximum of $|f''(x)|$ is 2, as confirmed in the figure.

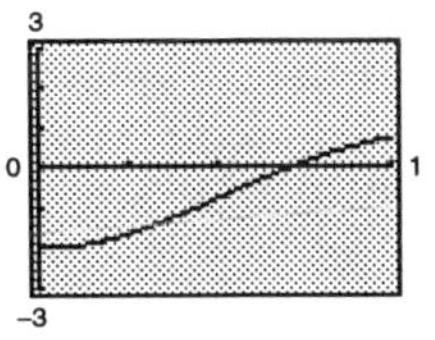

Section 6.6

Both a and b diverge while c converges. The conclusion is that $\int_1^\infty \frac{1}{x^n}\,dx$ converges if $n > 1$ and diverges otherwise.

CHAPTER 7

Section 7.1

In three dimensions, the graph of $ax + by + cz = 0$ is a plane through the origin. The traces are all lines, however.

Section 7.2

a. Elliptic cone　　b. Ellipsoid　　c. Hyperboloid of one sheet

Section 7.3

Option d would have the least amount of interest contained in the payments, if the interest rates do not drastically decline. If there is a drastic decline in interest rates, then option c may have the least amount of interest over the 20 year period. If a minimum payment is desirable, then option b is the one to choose, although the amount of interest here is the highest of all options.

Section 7.4

a. $f_x(x, y) = \dfrac{-x}{\sqrt{1 - x^2 - y^2}}$

$f_x(x, 0) = \dfrac{-x}{\sqrt{1 - x^2}}$

$f_x(x, 0) = 1 \Rightarrow x = \dfrac{-1}{\sqrt{2}}$

At $x = \dfrac{1}{\sqrt{2}}$, the slope on the sphere along $z = \sqrt{1 - x^2}$ is 1.

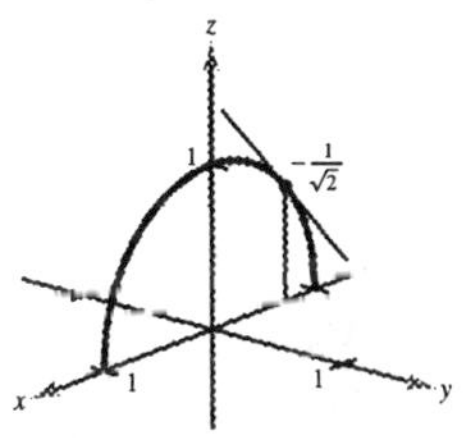

b. $f_y(x, y) = \dfrac{-y}{\sqrt{1 - x^2 - y^2}}$

$f_y(0, y) = \dfrac{-y}{\sqrt{1 - y^2}}$

$f_y(0, y) = 1 \Rightarrow y = \dfrac{-1}{\sqrt{2}}$

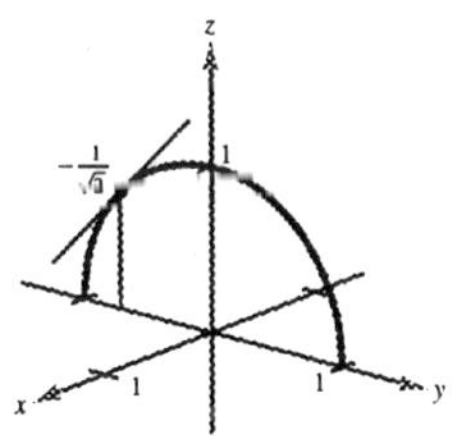

At $y = \dfrac{-1}{\sqrt{2}}$, the slope on the sphere along $z = \sqrt{1 - y^2}$ is 1.

Section 7.5

The first three points (0, 0), (0, 6), (4, 0) have $d < 0$ so there is a saddle point at these points and there cannot be a relative extremum there. But at $\left(\frac{4}{3}, 2\right)$, $d > 0$ and $f_{xx}\left(\frac{4}{3}, 2\right) < 0$ so by the test for relative extrema there is a relative maximum at this point.

Section 7.6

Solve equation 1 for lambda: $\lambda = 2x - \mu$.

In equation 2 replace lambda by the result above: $y = \dfrac{\lambda}{2} = \dfrac{2x - \mu}{2}$.

In equation 4 solve for x in terms of lambda: $x = 3 - y = 3 - \left(\dfrac{2x - \mu}{2}\right) = 3 - x + \dfrac{\mu}{2}$.

Then we have for x: $2x = 3 + \dfrac{\mu}{2} \Rightarrow x = \dfrac{3}{2} + \dfrac{\mu}{4}$.

Using equation 5 we can solve for z in terms of x: $z = 5 - x$.

But from equation 3, $z = \dfrac{mu}{2}$. Putting the last two results together implies:

$$z = 5 - \left(\frac{3}{2} + \frac{mu}{4}\right) = \frac{\mu}{2}$$

so we have mu:

$$\mu = \frac{14}{3}$$

Therefore $z = \dfrac{\mu}{2} = \dfrac{7}{3}$; $x = \dfrac{3}{2} + \dfrac{\mu}{4} = \dfrac{3}{2} + \dfrac{7}{6} = \dfrac{8}{3}$; and $y = 3 - x$ so $y = \dfrac{1}{3}$.

Section 7.7

The exponential and power models have a worse (lower) correlation coefficient.

Section 7.8

a.

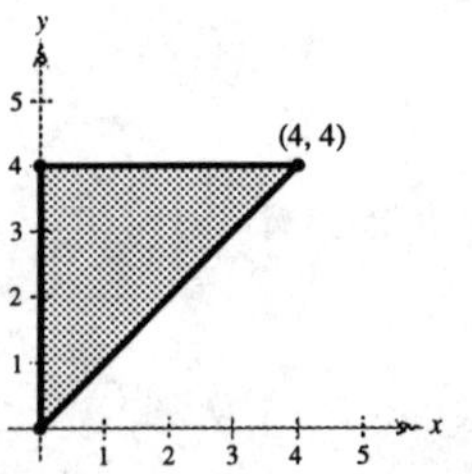

b.

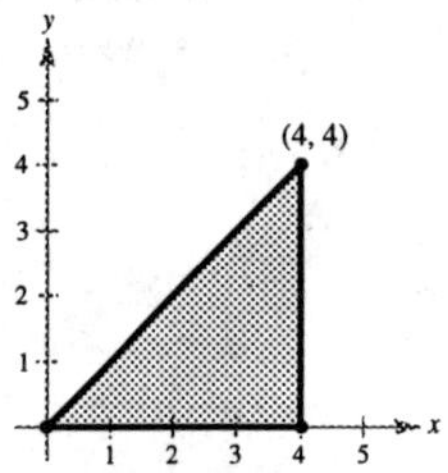

c.

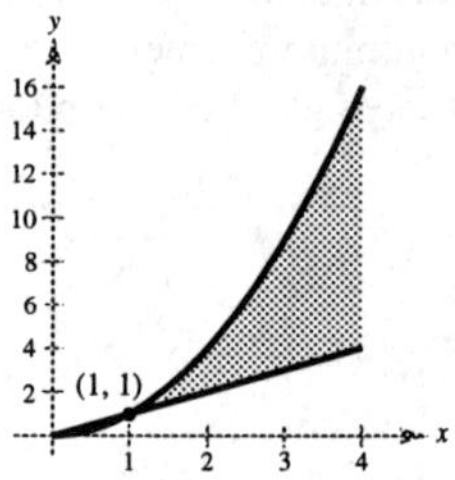

Section 7.9

$$\bar{f} = \frac{1}{A}\iint 100dA = \frac{100}{A}\iint dA = \frac{100}{A}(A) = 100.$$

CHAPTER 8

Section 8.1

The formula for the area of an equilateral triangle is:

$$A = \frac{\sqrt{3}s^2}{4}$$

Section 8.2

a. $\theta = \frac{\pi}{2}$ b. $\theta = \frac{2\pi}{3}, \frac{4\pi}{3}$ c. $\theta = 0, \pi, 2\pi$

d. $\theta = \frac{\pi}{3}, \frac{5\pi}{3}$ e. $\theta = \frac{\pi}{3}, \frac{2\pi}{3}$ f. $\theta = \frac{3\pi}{4}, \frac{7\pi}{4}$

Section 8.3

August 1, 1984 was the person's 7318th day of life.

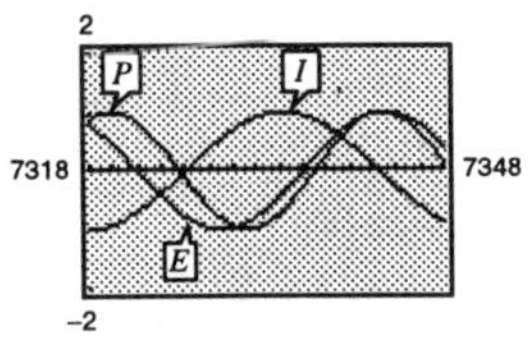

Section 8.4

The minimum temperature occurs at $t = 2$ (2A.M.) and is 55 degrees; the maximum temperature occurs at $t = 14$ (2P.M.) and is 85 degrees. To get these extrema to occur one hour later, replace the $t - 8$ by $t - 9$:

$$y = 70 + 15 \sin \frac{\pi(t-9)}{12}$$

Section 8.5

The average over a 24 hour period is 70 degrees. From midnight to noon the average temperature is 65.2; from noon to midnight the average is 74.8.

Section 8.6

The function $f(x)$ grows faster than $g(x)$ in both a and b.

CHAPTER 9

Section 9.1

a.

x	0	1	2	3	4	5	6	7	8	9
$p(x)$	0.001	0.005	0.99	0	0	0.0028	0	0	0	0

This distribution has a mean of 2 and a standard deviation of 0.187, which is smaller than those in the text. It is possible to have a distribution with standard deviation zero; for example, the distribution below has a standard deviation zero.

x	0	1	2	3	4	5	6	7	8	9
$p(x)$	0	0	0	0	1	0	0	0	0	0

b.

x	0	1	2	3	4	5	6	7	8	9
$p(x)$	$\frac{7}{9}$	0	0	0	1	0	0	0	0	$\frac{2}{9}$

This distribution has a mean of 2, variance is 59 so $\sigma \approx 7.68$, which is larger than those in the text.

Section 9.2

Graph b should be the best graph for the heights of the women, since height should have a normal distribution and graph b is closest to the normal probability density curve. The average height should be approximately 65 inches.

Section 9.3

86.64% of the x values lie within 1.5 standard deviations from the mean; 38.3% lie within 0.5 standard deviations from the mean.

CHAPTER 10

Section 10.1

a. $\left\{\frac{1}{n}\right\}$ converges to 0

b. $\left\{\frac{2n}{3^n - 1}\right\}$ converges to 0

c. $\left\{\frac{2n + 1}{2n - 1}\right\}$ converges to 1

Section 10.2

The ball would have traveled 80 feet, 38 more feet than in the example.

Section 10.3

a. Converges using the ratio rest, since $R = 0 < 1$. Partial sum: 8.5914

b. Converges since it is a geometric series with $r = \frac{1}{2} < 1$. Partial sum: 5.0

c. Converges since it is a constant multiple of a p-series with $p = 2 > 1$. Partial sum: 8.22

Section 10.4

a. Replace x by $-x$ in the series for e^x; take $5 - 3\times$ the series for e^{-x}.

b. Use the series for $1/(x + 1)$:

$$\frac{1}{x+1} = 1 - x + x^2 - x^3 + \cdots. \text{ Since}$$

$$\frac{2}{x-2} = 2\frac{1}{(x+1) - 3} = 2\frac{1}{(x-3)+1} = 2[1 - (x-3) + (x-3)^2 - \cdots]$$

we have

$$\frac{2}{x-2} = 2 - 2(x-3) + 2(x-3)^2 - \cdots$$

c. Using properties of logarithms:

$$\begin{aligned} y &= \ln 4 - \ln x = \ln 4 - [(x+1) - (x-1)^2 + (x-1)^3 - \cdots] \\ &= \ln 4 - (x-1) + (x-1)^2 - (x-1)^3 + \cdots \end{aligned}$$

Section 10.5

If the probability of landing heads is 0.6, the probability of tails is 0.4. The expected value is then given by:

$$\begin{aligned} \sum_{n=0}^{\infty} nP(n) &= \sum_{n=0}^{\infty} n\left(\frac{2}{5}\right)^n \\ &= 0\left(\frac{2}{5}\right) + 1\left(\frac{2}{5}\right) + 2\left(\frac{2}{5}\right)^2 + \cdots \\ \sum_{n=0}^{\infty} nP(n) &= \left(\frac{2}{5}\right)^2\left[1 + 2\left(\frac{2}{5}\right) + 3\left(\frac{2}{5}\right)^2 + \cdots\right] \\ &= \frac{4}{25}\left[\frac{1}{(1-\frac{2}{5})^2}\right] = \frac{4}{9} \end{aligned}$$

The expected value is therefore less than in the example, which agrees with intuition.

Section 10.6

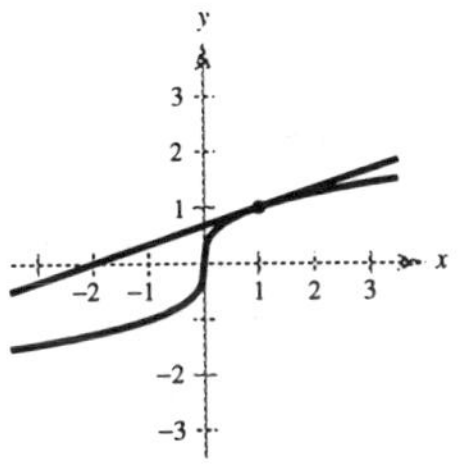

The x-intercept is $x_2 = -2$.

APPENDIX D

Section D.1

$$\frac{d}{dx}(x^2 + y^2) = \frac{d}{dx}(C) \implies 2x + 2yy' = 0 \implies \frac{dy}{dx} = -\frac{x}{y}$$

Section D.2

Variables can be separated in a and c.

Section D.3

$$\frac{d}{dx}\left[ye^{\int p\,dx}\right] = ype^{\int p\,dx} + e^{\int p\,dx}\frac{dy}{dx} = e^{\int p\,dx}\left[\frac{dy}{dx} + py\right]$$

Section D.4

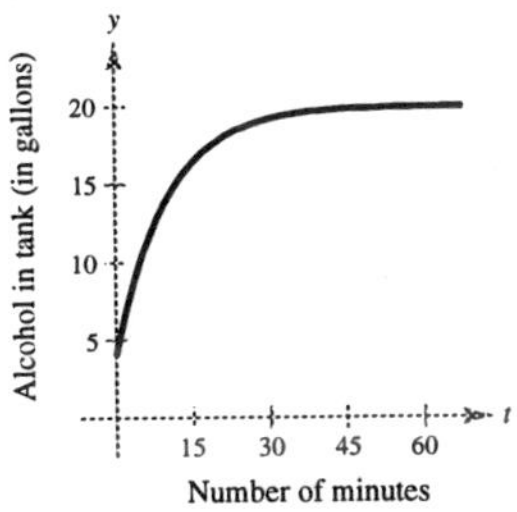

Since $\frac{dy}{dt} = 1.6e^{-t/10} > 0$, the amount of alcohol in the tank increases with time, asymptotically approaching $y = 20$.

Answers to

TECHNOLOGY

CHAPTER 0

Section 0.3

Page 0-14

2055.0632

CHAPTER 1

Section 1.3

Page 31

The lines appear perpendicular in the setting $-9 \le x \le 9$ and $-6 \le y \le 6$.

Section 1.5

Page 51

$$\lim_{x \to 1} \frac{x^3 + 4x - 5}{x - 1} = 7$$

Section 1.6

Page 62

Most calculators set in connected mode will join the two branches of the graph with a nearly vertical line near $x = 2$. This line is not part of the graph.

CHAPTER 3

Section 3.6

Page 225

The horizontal asymptotes are $y = 1$ and $y = -1$.

Section 3.8

Page 242

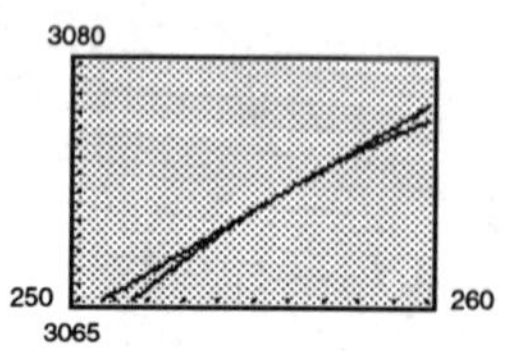

The curves are nearly identical because y is the tangent line approximation to $R(x)$ at their point of intersection (256, 3072).

CHAPTER 4

Section 4.3

Page 280

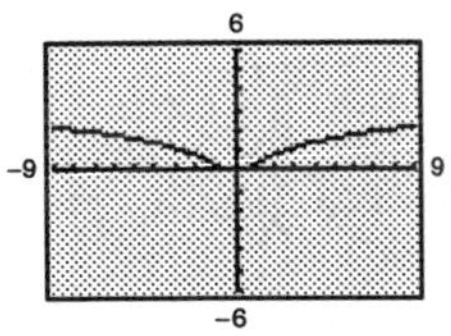

Section 4.4

Page 288

$$\frac{1}{x} + \frac{4x}{x^2 + 1} - \frac{3x^2}{2x^3 + 1} = \frac{1}{x} + \frac{4x(2x^3 + 1) - 3x^2(x^2 + 1)}{(x^2 + 1)(2x^3 + 1)}$$

$$= \frac{1}{x} + \frac{x(5x^3 - 3x + 4)}{(x^2 + 1)(2x^3 + 1)}$$

Page 290

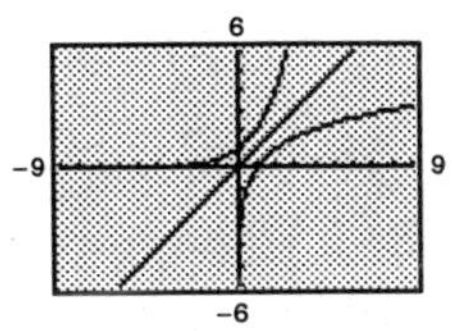

The functions $y_1 = \log_2 x$ and $y_2 = 2^x$ are inverses of each other.

CHAPTER 7

Section 7.5

Page 493

$f(x, y) = x^2 - 2x + y^2 + 4y + 5$: $(1, -2)$ relative minimum

$f(x, y) = \sqrt{x^2 + y^2}$: $(0, 0)$ relative minimum

$f(x, y) = \dfrac{1}{1 + x^2 + y^2}$: $(0, 0)$ relative maximum

$f(x, y) = 8xye^{-x^2-y^2}$: $\left(\frac{1}{\sqrt{2}}, \frac{1}{\sqrt{2}}\right), \left(-\frac{1}{\sqrt{2}}, -\frac{1}{\sqrt{2}}\right)$ relative maxima

$\left(-\frac{1}{\sqrt{2}}, \frac{1}{\sqrt{2}}\right), \left(\frac{1}{\sqrt{2}}, -\frac{1}{\sqrt{2}}\right)$ relative minima

CHAPTER 8

Section 8.4

Page 580

On $(0, 2\pi)$, the relative maxima are $(0.8677, 2.4720)$, $(2.2740, 2.4720)$, $(3.8277, -0.3449)$, and $(5.5971, -0.3449)$.

The relative minima are $(1.5708, 1)$, $(3.2713, -1.1271)$, $(4.7124, -3.0)$, and $(6.1535, -1.1271)$. There are eight relative extrema on $(0, 2\pi)$.

Answers to

D I S C O V E R Y

CHAPTER 1

Section 1.3

Page 24

The line $y = 4x$ rises most quickly.

The line $y = -4x$ falls most quickly.

The slope of a horizontal line is 0.

Section 1.5

Page 52

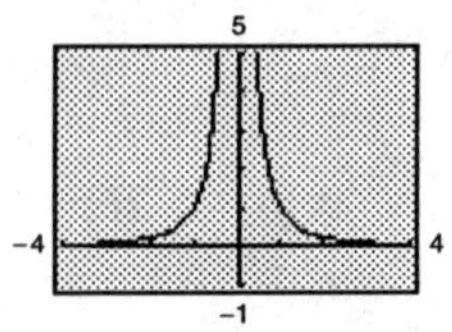

$\lim_{x\to 0} \frac{1}{x^2}$ does not exist.

$\lim_{x\to 1} \frac{1}{x^2} = 1$

Page 54

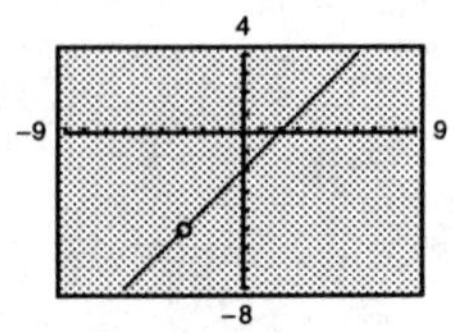

The graph is not a line, since it has a "hole" at $(-3, -5)$.

CHAPTER 2

Section 2.1

Page 86

$y_3 = 4x - 3$ is tangent to $y_1 = x^2 + 1$ at $(2, 5)$.

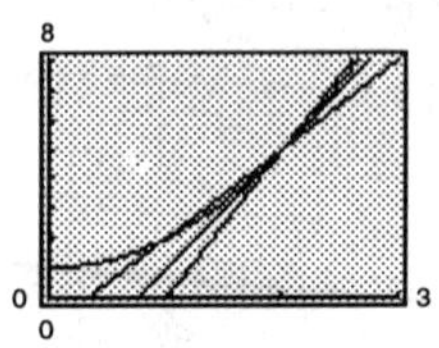

Section 2.3

Page 109

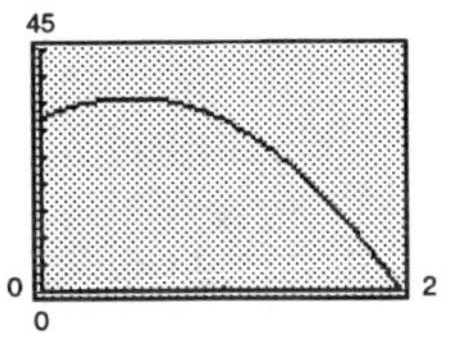

The maximum value of h is 36, which occurs at $t = 0.5$. $h'(0.5) = 0$. In general, maximum and minimum values of differentiable functions occur at points where the derivative is 0 (or at endpoints).

CHAPTER 3

Section 3.1

Page 173

f is increasing on $(-\infty, 0)$.
f' is positive on $(-\infty, 0)$.
If f' is positive, then f is increasing. If f' is negative, then f is decreasing.

g is increasing on $\left(-\infty, -\dfrac{1}{\sqrt{3}}\right)$ and $\left(\dfrac{1}{\sqrt{3}}, \infty\right)$.
g' is positive on the same intervals.

Section 3.2

Page 182

f has a relative minimum at $x = 0$.
f' is negative to the left of $x = 0$, and positive to the right. See the First-Derivative Test.

Section 3.3

Page 192

f is concave upward on $(0, \infty)$.
f'' is positive on $(0, \infty)$. See the Test for Concavity on page 191.

g is concave upward on $(-\infty, -1)$ and $(1, \infty)$.
g'' is positive on $(-\infty, -1)$ and $(1, \infty)$.

Page 194

$f''(x) = 0$ for $x = 2$. The point of inflection occurs at $x = 2$. $g''(x) = 0$ for $x = \pm\sqrt{\frac{5}{6}} \approx \pm 0.9129$. The points of inflection occur at these x-values. In general, if $(c, f(c))$ is a point of inflection, then either $f''(c) = 0$ or $f''(c)$ does not exist.

Section 3.7

Page 234

$$x + \frac{4}{x-2} = \frac{x(x-2)+4}{x-2} = \frac{x^2-2x+4}{x-2} = f(x)$$

You can obtain this alternate form for $f(x)$ by long division. You can see from the graph that $f(x) \to x$ as $x \to \pm\infty$.

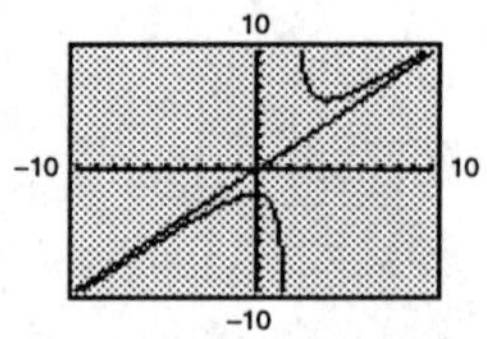

$y = x$ is the slant asymptote for $g(x)$.

CHAPTER 4

Section 4.1

Page 264

n	10	100	1000	10,000	100,000
$\left(1+\frac{1}{n}\right)^n$	2.5937	2.7048	2.7169	2.7181	2.71827

$$\lim_{n\to\infty}\left(1+\frac{1}{n}\right)^n = e \quad (\approx 2.71828)$$

n	−10	−100	−1000	−10,000	−100,000
$\left(1+\frac{1}{n}\right)^n$	2.8680	2.7320	2.7196	2.7184	2.71829

$$\lim_{n\to-\infty}\left(1+\frac{1}{n}\right)^n = e$$

Section 4.2

Page 269

Δx	$e^{\Delta x}$	$1 + \Delta x$
0.1	1.1052	1.1
0.01	1.0101	1.01
0.001	1.0010	1.001

$e^{\Delta x} \approx 1 + \Delta x$. This approximation was used in the derivation of the derivative of $f(x) = e^x$ in the fourth line of the displayed equations.

Section 4.4

Page 286

The slope of the tangent lines tends to zero as you move to the right. The slope is never equal to zero.

$$\frac{d}{dx}(\ln x) = \frac{1}{x}$$

As $x \to \infty, \frac{1}{x} \to 0$.

Page 287

The domain of $f(x) = \ln \sqrt{x + 1}$ is $(-1, \infty)$.
The domain of $f'(x)$ is all $x \neq -1$.
The domain of $y_1 = \ln x^2$ is all $x \neq 0$, whereas the domain of $y_2 = 2 \ln x$ is $x > 0$.

Section 4.5

Page 295

Changing C results in a vertical shift of the graph. Changing the value of k results in a different steepness, or rate of increase. $y = e^x$ grows faster than $y = x^{10}$.

CHAPTER 5

Section 5.1

Page 315

$F_1'(x) = 2x - 2$
$F_2'(x) = 2x - 2$
$F_3'(x) = 2x - 2$
The graphs of F_1, F_2, and F_3 are vertical translations of each other.

Page 329

$$F(x) = (1 - 3x)^{3/2} + C \Rightarrow F'(x) = \frac{3}{2}(1 - 3x)^{1/2}(-3) = \frac{-9}{2}\sqrt{1 - 3x}$$

$$F(x) = \frac{2}{3}(1 - 3x)^{3/2} + C \Rightarrow F'(x) = \frac{2}{3}\left(\frac{3}{2}\right)(1 - 3x)^{1/2}(-3) = -3\sqrt{1 - 3x}$$

$$F(x) = -\frac{2}{9}(1 - 3x)^{3/2} + C \Rightarrow F'(x) = -\frac{2}{9}\left(\frac{3}{2}\right)(1 - 3x)^{1/2}(-3) = \sqrt{1 - 3x}$$

The third function is the correct antiderivative.

CHAPTER 6

Section 6.5

Page 427

Since $h = \dfrac{(b - a)}{n}$, the error formula for the Trapezoidal Rule gives

$$\frac{(b - a)^3}{12n^2}[\max|f''(x)|] = \frac{b - a}{12}[\max|f''(x)|]h^2 \approx ch^2$$

Hence if n is doubled (h is halved), then the error is reduced from ch^2 to $c\left(\dfrac{h}{2}\right)^2 = \dfrac{1}{4}ch^2$. The error in Simpson's rule is reduced by approximately $\dfrac{1}{16}$ if the number of subintervals is doubled.

Section 6.6

Page 439

$$\text{Area} = \int_1^{\infty} \frac{1}{x}\,dx = \lim_{b\to\infty} \ln b = \infty$$

$$\text{Surface area} = 2\pi\int_1^{\infty} \frac{1}{x}\sqrt{1 + \frac{1}{x^4}}\,dx = \infty$$

Paint it by filling it up with paint!

CHAPTER 7

Section 7.1

Page 452

If $x = 0$, (x, y, z) is in the yz-plane.

If $x = 0$ and $y = 0$, (x, y, z) is on the z-axis.

If the point (x, y, z) is on the y-axis, then $x = z = 0$.

If the point (x, y, z) is in the xz-plane, then $y = 0$.

Section 7.4

Page 482

$f_x(x, y) = -x \Rightarrow f_x(0, 0) = 0$

$f_y(x, y) = -2y \Rightarrow f_y(0, 0) = 0$

The slopes are 0 in the x- and y-directions at $(0, 0)$. The graph of f has a maximum at this point.

Section 7.6

Page 504

Minimum would be 0 if no constraint. No maximum value. In Example 4, $\frac{3}{10}$ is the minimum because there is no maximum value.

Section 7.7

Page 513

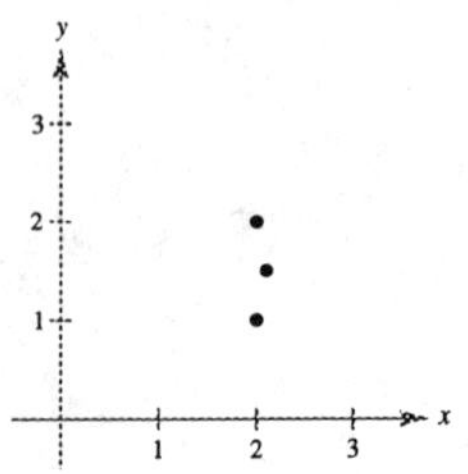

Visually, the line should be something like $x = 2.05$.

CHAPTER 9

Section 9.1

Page 618

Ten students score 100 and ten students score 60. One student scores 99 and 19 score 79. The standard deviation measures how much the scores vary from the average.

CHAPTER 10

Section 10.1

Page 654

$a_1 = 2$ $\quad a_5 = 2.4883$

$a_2 = 2.25$ $\quad a_6 = 2.5216$

$a_3 = 2.3704$ $\quad a_7 = 2.5465$

$a_4 = 2.4414$ $\quad a_8 = 2.5658$

$$\lim_{n\to\infty}\left(1 + \frac{1}{n}\right)^n = e \quad (\approx 2.7183)$$

Section 10.2

Page 661

$S_0 = \frac{1}{0!} = 1$ $\quad S_4 = 2.7083$

$S_1 = \frac{1}{1!} = 1$ $\quad S_5 = 2.7167$

$S_2 = \frac{1}{1!} + \frac{1}{2!} = 1.5$ $\quad S_6 = 2.7181$

$S_3 = 2.6667$ $\quad S_7 = 2.7183$

$$\lim_{n\to\infty} S_n = e \quad (\approx 2.7183)$$

Section 10.3

Page 672

$$\left(1 + \frac{1}{2}\right) + \left(\frac{1}{3} + \frac{1}{4}\right) + \left(\frac{1}{5} + \cdots + \frac{1}{16}\right) + \left(\frac{1}{17} + \cdots + \frac{1}{32}\right) + \cdots > \frac{1}{2} + \frac{1}{2} + \frac{1}{2} + \frac{1}{2} + \cdots$$

which implies that $\sum_{n=1}^{\infty} \frac{1}{n}$ diverges.

APPENDIX D

Section D.3

Page A64

$$y' = \frac{2y + x^2}{x} = x + \frac{2y}{x}$$

At $(1, 0)$, $y' = 1$.

At $\left(e^{-1/2}, \frac{1}{2e}\right)$, $y' = e^{-1/2} + \frac{-2/2e}{e^{-1/2}} = 0$.

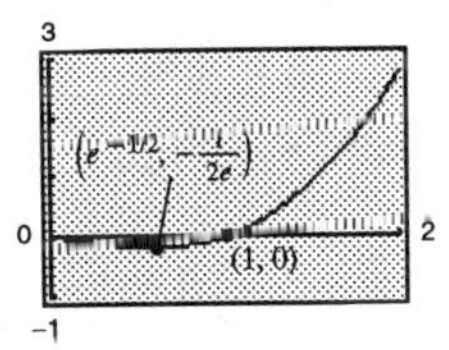

As $x \to 0^+$, $y' \to 0$.

Answers to the

WARM UPS

CHAPTER 1

Section 1.1

1. $3\sqrt{5}$ **2.** $2\sqrt{5}$ **3.** $\frac{1}{2}$ **4.** -2 **5.** $5\sqrt{3}$

6. $-\sqrt{2}$ **7.** $x = -3, x = 9$ **8.** $y = -8, y = 4$

9. $x = 19$ **10.** $y = 1$

Section 1.2

1. $y = \frac{1}{5}(x + 12)$ **2.** $y = x - 15$ **3.** $y = \dfrac{1}{x^3 + 2}$

4. $y = \pm\sqrt{x^2 + x - 6} = \pm\sqrt{(x + 3)(x - 2)}$ **5.** $y = -1 \pm \sqrt{9 - (x - 2)^2}$

6. $y = 5 \pm \sqrt{81 - (x + 6)^2}$ **7.** $(x - 2)(x - 1)$

8. $(x + 3)(x + 2)$ **9.** $\left(y - \frac{3}{2}\right)^2$ **10.** $\left(y - \frac{7}{2}\right)^2$

Section 1.3

1. -1 **2.** $-\frac{7}{3}$ **3.** $\frac{1}{3}$ **4.** $-\frac{7}{6}$

5. $y = 4x + 7$ **6.** $y = 3x - 7$ **7.** $y = 3x - 10$

8. $y = -x - 7$ **9.** $y = 7x - 17$ **10.** $y = \frac{2}{3}x + \frac{5}{3}$

Section 1.4

1. 20 **2.** 10 **3.** $x^2 + x - 6$

4. $x^3 + 9x^2 + 26x + 30$ **5.** $\dfrac{1}{x}$ **6.** $\dfrac{2x - 1}{x}$

7. $y = -2x + 17$ **8.** $y = \frac{6}{5}x^2 + \frac{1}{5}$

9. $y = 3\pm\sqrt{5 + (x + 1)^2}$ **10.** $y = \pm\sqrt{4x^2 + 2}$

Section 1.5

1. **(a)** 7 **(b)** $c^2 - 3c + 3$ **(c)** $x^2 + 2xh + h^2 - 3x - 3h + 3$

2. **(a)** -4 **(b)** 10 **(c)** $3t^2 + 4$

3. h **4.** 4

5. Domain: $(-\infty, 0) \cup (0, \infty)$
Range: $(-\infty, 0) \cup (0, \infty)$

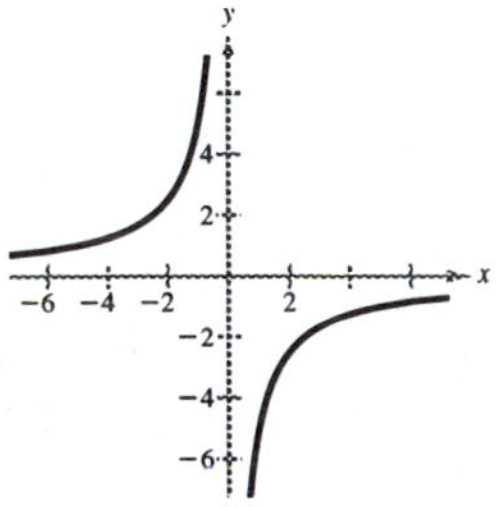

6. Domain: $[-5, 5]$
Range: $[0, 5]$

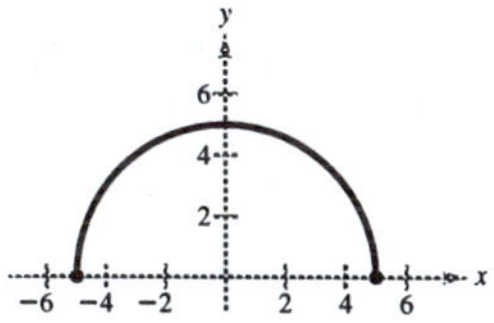

7. Domain: $(-\infty, \infty)$
Range: $[0, \infty)$

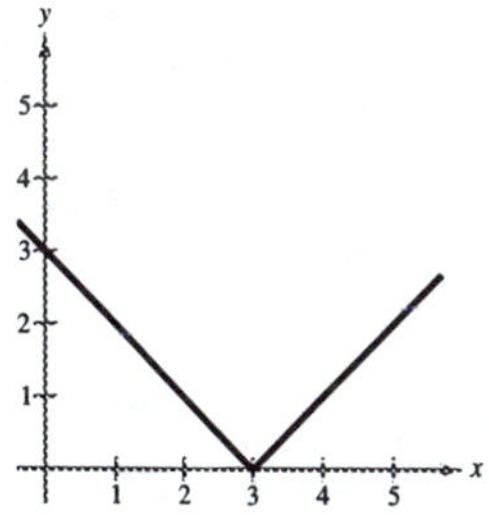

8. Domain: $(-\infty, 0) \cup (0, \infty)$
Range: $-1, 1$

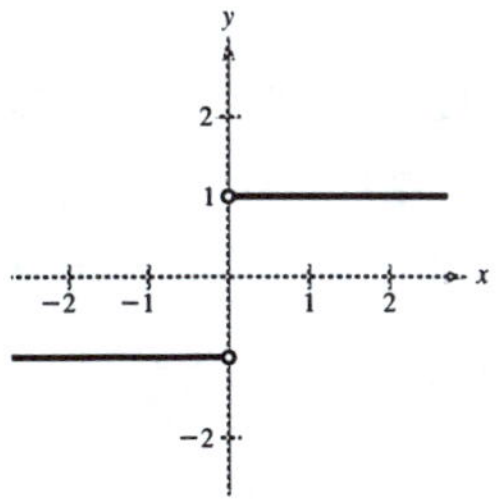

9. y is not a function of x.

10. y is a function of x.

Section 1.6

1. $\dfrac{x+4}{x-8}$ **2.** $\dfrac{x+1}{x-3}$ **3.** $\dfrac{x+2}{2(x-3)}$ **4.** $\dfrac{x-4}{x-2}$

5. $x = 0, -7$ **6.** $x = -5, 1$ **7.** $x = -\frac{2}{3}, -2$ **8.** $x = 0, 3, -8$

9. 13 **10.** -1

CHAPTER 2

Section 2.1

1. $x = 2$ **2.** $y = 2$ **3.** $2x$ **4.** $3x^2$ **5.** $\dfrac{1}{x^2}$ **6.** $2x$

7. $(-\infty, 1) \cup (1, \infty)$ **8.** $(-\infty, \infty)$ **9.** $(-\infty, 0) \cup (0, \infty)$ **10.** $(-\infty, -4) \cup (-4, 3) \cup (3, \infty)$

Section 2.2

1. **(a)** 8 **(b)** 16 **(c)** $\frac{1}{2}$

2. **(a)** $\frac{1}{36}$ **(b)** $\frac{1}{32}$ **(c)** $\frac{1}{64}$

3. $4x(3x^2 + 1)$ **4.** $\dfrac{3}{2}x^{1/2}(x^{3/2} - 1)$ **5.** $\dfrac{1}{4x^{3/4}}$ **6.** $x^2 - \dfrac{1}{x^{1/2}} + \dfrac{1}{3x^{2/3}}$

7. $0, -\frac{2}{3}$ **8.** $0, \pm 1$ **9.** $-10, 2$ **10.** $-2, 12$

Section 2.3

1. 3
2. -7
3. $y' = 8x - 2$
4. $y' = -9t^2 + 4t$
5. $s' = -32t + 24$
6. $y' = -32x + 54$
7. $A' = -\frac{3}{5}r^2 + \frac{3}{5}r + \frac{1}{2}$
8. $y' = 2x^2 - 4x + 7$
9. $y' = 12 - \dfrac{x}{2500}$
10. $y' = 74 - \dfrac{3x^2}{10{,}000}$

Section 2.4

1. $2(3x^2 + 7x + 1)$
2. $4x^2(6 - 5x^2)$
3. $\dfrac{23}{(2x + 7)^2}$
4. $-\dfrac{x^2 + 8x + 4}{(x^2 - 4)^2}$
5. $\dfrac{4x^3 - 3x^2 + 3}{x^2}$
6. $\dfrac{x^2 - 2x + 4}{(x - 1)^2}$
7. 11
8. 0
9. $-\frac{1}{4}$
10. $\frac{17}{4}$

Section 2.5

1. $(1 - 5x)^{2/5}$
2. $(2x - 1)^{3/4}$
3. $(4x^2 + 1)^{-1/2}$
4. $(x - 6)^{-1/3}$
5. $x^{1/2}(1 - 2x)^{-1/3}$
6. $(2x)^{-1}(3 - 7x)^{3/2}$
7. $(x - 2)(3x^2 + 5)$
8. $(x - 1)(5\sqrt{x} - 1)$
9. $(x^2 + 1)^2(4 - x - x^3)$
10. $(3 - x^2)(x - 1)(x^2 + x + 1)$

Section 2.6

1. $t = 0, \dfrac{3}{2}$
2. $t = -2, 7$
3. $t = -2, 10$
4. $t = \dfrac{9 \pm 3\sqrt{10{,}249}}{32}$
5. $\dfrac{dy}{dx} = 6x^2 + 14x$
6. $\dfrac{dy}{dx} = 8x^3 + 18x^2 - 10x - 15$
7. $\dfrac{dy}{dx} = \dfrac{2x(x + 7)}{(2x + 7)^2}$
8. $\dfrac{dy}{dx} = -\dfrac{6x^2 + 10x + 15}{(2x^2 - 5)^2}$
9. Domain: $(-\infty, \infty)$
   Range: $[-4, \infty)$
10. Domain: $[7, \infty)$
    Range: $[0, \infty)$

Section 2.7

1. $y = x^2 - 2x$
2. $y = \dfrac{x - 3}{4}$
3. $y = 1, x \neq -6$
4. $y = -4, x \neq \pm\sqrt{3}$
5. $y = \pm\sqrt{5 - x^2}$
6. $y = \pm\sqrt{6 - x^2}$
7. $\frac{8}{3}$
8. $-\frac{1}{2}$
9. $\frac{5}{7}$
10. 1

Section 2.8

1. $A = \pi r^2$
2. $V = \frac{4}{3}\pi r^3$
3. $S = 6s^2$
4. $V = s^3$
5. $V = \frac{1}{3}\pi r^2 h$
6. $A = \dfrac{1}{2}bh$
7. $-\dfrac{x}{y}$
8. $\dfrac{2x - 3y}{3x}$
9. $-\dfrac{2x + y}{x + 2}$
10. $-\dfrac{y^2 - y + 1}{2xy - 2y - x}$

CHAPTER 3

Section 3.1

1. $x = 0, x = 8$ **2.** $x = 0, x = 24$ **3.** $x = \pm 5$ **4.** $x = 0$

5. $(-\infty, 3) \cup (3, \infty)$ **6.** $(-\infty, 1)$ **7.** $(-\infty, -2) \cup (-2, 5) \cup (5, \infty)$

8. $\left(-\sqrt{3}, \sqrt{3}\right)$

9. $x = -2$: $-\frac{103}{3}$
$x = 0$: $\frac{1}{3}$
$x = 2$: -13

10. $x = -2$: $-\frac{7}{9}$
$x = 0$: 3
$x = 2$: $-\frac{1}{3}$

Section 3.2

1. $0, \pm\frac{1}{2}$ **2.** $-2, 5$ **3.** 1 **4.** $0, 125$ **5.** Negative

6. Positive **7.** Positive **8.** Negative **9.** Increasing **10.** Decreasing

Section 3.3

1. $f''(x) = 48x^2 - 54x$ **2.** $g''(s) = 12s^2 - 18s + 2$ **3.** $g''(x) = 56x^6 + 120x^4 + 72x^2 + 8$

4. $f''(x) = \dfrac{4}{9(x-3)^{2/3}}$ **5.** $h''(x) = \dfrac{15}{2x^5}$ **6.** $f''(x) = -\dfrac{42}{(3x+2)^3}$

7. $x = \pm\dfrac{\sqrt{3}}{3}$ **8.** $x = 0, 3$ **9.** $t = \pm 4$ **10.** $x = 0, \pm 5$

Section 3.4

1. $x + \frac{1}{2}y = 12$ **2.** $2xy - 24$ **3.** $xy = 24$

4. $\sqrt{(x_2 - x_1)^2 + (y_2 - y_1)^2} = 10$ **5.** $x = -3$

6. $x = -\frac{2}{3}, 1$ **7.** $x = \pm 5$ **8.** $x = 4$ **9.** $x = \pm 1$ **10.** $x = \pm 3$

Section 3.5

1. 1 **2.** 1.2 **3.** 2 **4.** $\frac{1}{2}$

5. $\dfrac{dC}{dx} = 1.2 + 0.006x$ **6.** $\dfrac{dP}{dx} = 0.02x + 11$ **7.** $\dfrac{dR}{dx} = 14 - \dfrac{x}{1000}$

8. $\dfrac{dR}{dx} = 3.4 - \dfrac{x}{750}$ **9.** $\dfrac{dP}{dx} = -1.4x + 7$ **10.** $\dfrac{dC}{dx} = 4.2 + 0.003x^2$

Section 3.6

1. 3 **2.** 1 **3.** -11 **4.** 1 **5.** 0 **6.** 1

7. $C = \dfrac{150}{x} + 3$
$\dfrac{dC}{dx} = 3$

8. $\overline{C} = \dfrac{1900}{x} + 1.7 + 0.002x$
$\dfrac{dC}{dx} = 1.7 + 0.004x$

9. $\overline{C} = 0.005x + 0.5 + \dfrac{1375}{x}$
$\dfrac{dC}{dx} = 0.01x + 0.5$

10. $\overline{C} = \dfrac{760}{x} + 0.05$
$\dfrac{dC}{dx} = 0.05$

Section 3.7

1. Vertical: $x = 0$
Horizontal: $y = 0$

2. Vertical: $x = 2$
Horizontal: $y = 0$

3. Vertical: $x = -3$
Horizontal: $y = 40$

4. Vertical: $x = 1, x = 3$
Horizontal: $y = 1$

5. Decreasing on $(-\infty, -2)$
Increasing on $(-2, \infty)$

6. Increasing on $(-\infty, -4)$
Decreasing on $(-4, \infty)$

7. Increasing on $(-\infty, -1)$ and $(1, \infty)$
Decreasing on $(-1, 1)$

8. Decreasing on $(-\infty, 0)$ and $\left(\sqrt[3]{2}, \infty\right)$
Increasing on $\left(0, \sqrt[3]{2}\right)$

9. Increasing on $(-\infty, 1)$ and $(1, \infty)$

10. Decreasing on $(-\infty, -3)$ and $\left(\frac{1}{3}, \infty\right)$
Increasing on $\left(-3, \frac{1}{3}\right)$

Section 3.8

1. $\dfrac{dC}{dx} = 0.18x$ **2.** $\dfrac{dR}{dx} = 1.25 + 0.03\sqrt{x}$ **3.** $\dfrac{dP}{dx} = -\dfrac{0.01}{\sqrt[3]{x^2}} + 1.4$

4. $\dfrac{dA}{dx} = \dfrac{\sqrt{3}}{2}x$ **5.** $\dfrac{dC}{dr} = 2\pi$ **6.** $\dfrac{dS}{dr} = 8\pi r$

7. $A = \pi r^2$ **8.** $A = x^2$ **9.** $V = x^3$ **10.** $V = \frac{4}{3}\pi r^3$

CHAPTER 4

Section 4.1

1. Continuous on $(-\infty, \infty)$ **2.** Removable discontinuity at $x = 4$

3. 0 **4.** 0 **5.** 4 **6.** $\frac{1}{2}$ **7.** $\frac{3}{2}$ **8.** 6 **9.** 0 **10.** 0

Section 4.2

1. $\dfrac{1}{2}e^x(2x^2 - 1)$ **2.** $\dfrac{e^x(x + 1)}{x}$ **3.** $e^x(x - e^x)$ **4.** $e^{-x}(e^{2x} - x)$

5. $-\dfrac{6}{7x^3}$ **6.** $6x - \dfrac{1}{6}$ **7.** $6(2x^2 - x + 6)$ **8.** $\dfrac{t + 2}{2t^{3/2}}$

9. Relative maximum: $\left(-\dfrac{4\sqrt{3}}{3}, \dfrac{16\sqrt{3}}{9}\right)$
Relative minimum: $\left(\dfrac{4\sqrt{3}}{3}, -\dfrac{16\sqrt{3}}{9}\right)$

10. Relative maximum: $(0, 5)$
Relative minima: $(-1, 4), (1, 4)$

Section 4.3

1. $\frac{1}{4}$ **2.** 64 **3.** 1 **4.** $81e^4$ **5.** $x > -4$

6. Any real number x **7.** $x < -1$ and $x > 1$ **8.** $x > 5$

9. \$3462.03 **10.** \$3374.65

Section 4.4

1. $2\ln(x+1)$
2. $\ln x + \ln(x+1)$
3. $\ln x - \ln(x+1)$
4. $3[\ln x - \ln(x-3)]$
5. $\ln 4 + \ln x + \ln(x-7) - 2\ln x$
6. $3\ln x + \ln(x+1) - \frac{1}{2}\ln(x-2)$
7. $-\frac{y(y+y^x \ln y)}{x(y+y^x)}$
8. $-\frac{-3+2xy-y^2}{x(x-2y)}$
9. $-12x+2$
10. $-\frac{6}{x^4}$

Section 4.5

1. $-\frac{1}{4}\ln 2$
2. $\frac{1}{5}\ln\frac{10}{3}$
3. $\frac{1}{3}\ln\frac{25}{16}$
4. $\frac{1}{20}\ln\frac{11}{16}$
5. $7.36e^{0.23t}$
6. $-33.6e^{-1.4t}$
7. $1.296e^{0.072t}$
8. $-0.025e^{-0.001t}$
9. 4
10. 12

CHAPTER 5

Section 5.1

1. $x^{-1/2}$
2. $(2x)^{4/3}$
3. $5^{1/2}x^{3/2} + x^{5/2}$
4. $x^{-1/2} + x^{-2/3}$
5. $(x+1)^{5/2}$
6. $x^{1/6}$
7. -12
8. -10
9. 14
10. 14

Section 5.2

1. $\frac{1}{2}x^4 + x + C$
2. $\frac{3}{2}x^2 + \frac{2}{3}x^{3/2} - 4x + C$
3. $-\frac{1}{x} + C$
4. $-\frac{1}{6t^2} + C$
5. $\frac{4}{7}t^{7/2} + \frac{2}{5}t^{5/2} + C$
6. $\frac{4}{5}x^{5/2} - \frac{2}{3}x^{3/2} + C$
7. $-\frac{5(x-2)^4}{16}$
8. $-\frac{1}{12(x-1)^2}$
9. $9(x^2+3)^{2/3}$
10. $-\frac{5}{(1-x^3)^{1/2}}$

Section 5.3

1. $\left(\frac{5}{2}, \infty\right)$
2. $(-\infty, 2) \cup (3, \infty)$
3. $x + 2 - \frac{2}{x+2}$
4. $x - 2 + \frac{1}{x-4}$
5. $x + 8 + \frac{2x-4}{x^2-4x}$
6. $x^2 - x - 4 + \frac{20x+22}{x^2+5}$
7. $\frac{1}{4}x^4 - \frac{1}{x} + C$
8. $\frac{1}{2}x^2 + 2x + C$
9. $\frac{1}{2}x^2 - \frac{4}{x} + C$
10. $-\frac{1}{x} - \frac{3}{2x^2} + C$

Section 5.4

1. $\frac{3}{2}x^2 + 7x + C$
2. $\frac{2}{5}x^{5/2} + \frac{4}{3}x^{3/2} + C$
3. $\frac{1}{5}\ln|x| + C$
4. $-\frac{1}{6e^{6x}}$
5. $-\frac{8}{5}$
6. $-\frac{62}{3}$
7. $0.008x^{5/2} + 29{,}500x + C$
8. $x^2 + 9000x + C$
9. $25{,}000x - 0.005x^2 + C$
10. $0.01x^3 + 4600x + C$

Section 5.5

1. $-x^2 + 3x + 2$ **2.** $-2x^2 + 4x + 4$ **3.** $-x^3 + 2x^2 + 4x - 5$
4. $x^3 - 6x - 1$ **5.** $(0, 4), (4, 4)$ **6.** $(1, -3), (2, -12)$ **7.** $(-3, 9), (2, 4)$
8. $(-2, -4), (0, 0), (2, 4)$ **9.** $(1, -2), (5, 10)$ **10.** $(1, e)$

Section 5.6

1. $\frac{1}{6}$ **2.** $\frac{3}{20}$ **3.** $\frac{7}{40}$ **4.** $\frac{13}{12}$ **5.** $\frac{61}{30}$
6. $\frac{53}{18}$ **7.** $\frac{2}{3}$ **8.** $\frac{4}{7}$ **9.** 0 **10.** 5

Section 5.7

1. 0, 2 **2.** 0, 2 **3.** 0, 2, −2 **4.** −1, 2 **5.** 2, 4
6. 1, 5 **7.** 53.5982 **8.** 1.9459 **9.** 3.3934 **10.** 1.3896

CHAPTER 6

Section 6.1

1. $5x + C$ **2.** $\frac{2}{5}x^{5/2} + C$ **3.** $\frac{1}{4}(x^2 + 1)^4 + C$ **4.** $e^{6x} + C$
5. $\ln|2x + 1| + C$ **6.** $-e^{-x^2} + C$ **7.** $x(x - 1)(2x - 1)$ **8.** $3x(x + 4)^2(x + 8)$
9. $(x + 21)(x + 7)^{-1/2}$ **10.** $x(x + 5)^{-2/3}$

Section 6.2

1. $12x^3 + 2$ **2.** $\frac{1}{2}x + \frac{1}{8}$ **3.** $\frac{4x^2 + 2}{x^3 + x}$ **4.** $3x^2e^{x^3}$ **5.** $e^x(x^2 + 2x)$
6. $-\frac{4}{(x - 3)^2}$ **7.** $\frac{63}{3}$ **8.** $\frac{4}{3}$ **9.** $\frac{32}{3}$ **10.** 8

Section 6.3

1. $(x - 4)(x + 3)$ **2.** $(x - 5)(x + 5)$ **3.** $x(x - 2)(x + 1)$ **4.** $x(x - 2)^2$
5. $(x - 2)(x - 1)^2$ **6.** $(x - 3)(x - 1)^2$ **7.** $\frac{1}{x - 2} + x$ **8.** $\frac{1}{1 - x} + 2x - 2$
9. $\frac{2}{2 - x} + x^2 - x - 2$ **10.** $\frac{6}{x - 1} + x + 4$

Section 6.4

1. $x^2 + 8x + 16$ **2.** $x^2 + x + \frac{1}{4}$ **3.** $x^2 - 2x + 1$ **4.** $x^2 - \frac{2}{3}x + \frac{1}{9}$
5. $\frac{2}{x} - \frac{2}{x + 2}$ **6.** $-\frac{3}{4x} + \frac{3}{4(x - 4)}$ **7.** $\frac{3}{2(x - 2)} - \frac{2}{x^2} - \frac{3}{2x}$
8. $-\frac{3}{x + 1} + \frac{2}{x - 2} + \frac{4}{x}$ **9.** $2e^x(x - 1) + C$ **10.** $x^3 \ln x - \frac{x^3}{3} + C$

Section 6.5

1. $\frac{2}{x^3}$ **2.** $-\frac{96}{(2x+1)^4}$ **3.** $-\frac{12}{x^4}$ **4.** $6x - 4$

5. $16e^{2x}$ **6.** $e^{x^2}(4x^2 + 2)$ **7.** $(3, 18)$ **8.** $(1, 8)$

9. $n < -5\sqrt{10},\ n > 5\sqrt{10}$ **10.** $n < -5, n > 5$

Section 6.6

1. 9 **2.** 3 **3.** $-\frac{1}{8}$

4. Limit does not exist. **5.** Limit does not exist. **6.** -4

7. (a) $\frac{32}{3}b^3 - 16b^2 + 8b - \frac{4}{3}$ **(b)** $-\frac{4}{3}$ **8. (a)** $\frac{b^2 - b - 11}{(b-2)^2(b-5)}$ **(b)** $\frac{11}{20}$

9. (a) $\ln\left(\frac{5 - 3b^2}{b+1}\right)$ **(b)** $\ln 5 \approx 1.609$ **10. (a)** $e^{-3b^2}(e^{6b^2} + 1)$ **(b)** 2

CHAPTER 7

Section 7.1

1. $2\sqrt{5}$ **2.** 5 **3.** 8 **4.** 8 **5.** $(4, 7)$

6. $(1, 0)$ **7.** $(0, 3)$ **8.** $(-1, 1)$

9. $(x-2)^2 + (y-3)^2 = 4$ **10.** $(x-1)^2 + (y-4)^2 = 25$

Section 7.2

1. $(4, 0), (0, 3)$ **2.** $\left(-\frac{4}{3}, 0\right), (0, -8)$ **3.** $(1, 0), (0, -2)$ **4.** $(-5, 0), (0, -5)$

5. $(x-1)^2 + (y-2)^2 + (z-9)^2 + 1 = 0$ **6.** $(x-4)^2 + (y+2)^2 - (z+3)^2 = 0$

7. $(x+1)^2 + (y-1)^2 - z = 0$ **8.** $(x-3)^2 + (y+5)^2 + (z+13)^2 = 1$

9. $x^2 - y^2 + z^2 = \frac{1}{4}$ **10.** $x^2 - y^2 + z^2 = 4$

Section 7.3

1. 11 **2.** -16 **3.** 7 **4.** 4 **5.** $(-\infty, \infty)$

6. $(-\infty, -3) \cup (-3, 0) \cup (0, \infty)$ **7.** $[5, \infty)$

8. $\left(-\infty, -\sqrt{5}\right] \cup \left[\sqrt{5}, \infty\right)$ **9.** 55.0104 **10.** 6.9165

Section 7.4

1. $\frac{x}{\sqrt{x^2 + 3}}$ **2.** $-6x(3 - x^2)^2$ **3.** $e^{2t+1}(2t + 1)$ **4.** $\frac{e^{2x}(2 - 3e^{2x})}{\sqrt{1 - e^{2x}}}$

5. $\frac{2}{3 - 2x}$ **6.** $\frac{3(t^2 - 2)}{2t(t^2 - 6)}$ **7.** $-\frac{10x}{(4x-1)^3}$

8. $-\frac{(x+2)^2(x^2 + 8x + 27)}{(x^2 - 9)^3}$ **9.** $f'(2) = 8$ **10.** $f'(2) = \frac{7}{2}$

Section 7.5

1. (3, 2) **2.** (11, 6) **3.** (1, 4)

4. (3, −2) **5.** (0, 0), (−1, 0) **6.** (−2, 0), (2, −2)

7. $\frac{\partial z}{\partial x} = 12x^2$ $\frac{\partial^2 z}{\partial y^2} = -6$

$\frac{\partial z}{\partial y} = -6y$ $\frac{\partial^2 z}{\partial x \partial y} = 0$

$\frac{\partial^2 z}{\partial x^2} = 24x$ $\frac{\partial^2 z}{\partial y \partial x} = 0$

8. $\frac{\partial z}{\partial x} = 4x - 3y$ $\frac{\partial^2 z}{\partial y^2} = 2$

$\frac{\partial z}{\partial y} = 2y - 3x$ $\frac{\partial^2 z}{\partial x \partial y} = -3$

$\frac{\partial^2 z}{\partial x^2} = 4$ $\frac{\partial^2 z}{\partial y \partial x} = -3$

9. $\frac{\partial z}{\partial x} = 4x^3 - \frac{\sqrt{xy}}{2x}$ $\frac{\partial^2 z}{\partial y^2} = \frac{\sqrt{xy}}{4y^2}$

$\frac{\partial z}{\partial y} = -\frac{\sqrt{xy}}{2y} + 2$ $\frac{\partial^2 z}{\partial x \partial y} = -\frac{\sqrt{xy}}{4xy}$

$\frac{\partial^2 z}{\partial x^2} = 12x^2 + \frac{\sqrt{xy}}{4x^2}$ $\frac{\partial^2 z}{\partial y \partial x} = -\frac{\sqrt{xy}}{4xy}$

10. $\frac{\partial z}{\partial x} = e^{xy}(xy + 1)$ $\frac{\partial^2 z}{\partial y^2} = x^3 e^{xy}$

$\frac{\partial z}{\partial y} = x^2 e^{xy}$ $\frac{\partial^2 z}{\partial x \partial y} = xe^{xy}(xy + 2)$

$\frac{\partial^2 z}{\partial x^2} = ye^{xy}(xy + 2)$ $\frac{\partial^2 z}{\partial y \partial x} = xe^{xy}(xy + 2)$

Section 7.6

1. $\left(\frac{7}{8}, \frac{1}{12}\right)$ **2.** $\left(-\frac{1}{24}, -\frac{7}{8}\right)$ **3.** $\left(\frac{55}{12}, -\frac{25}{12}\right)$

4. $\left(\frac{22}{23}, -\frac{3}{23}\right)$ **5.** $\left(\frac{5}{3}, \frac{1}{3}, 0\right)$ **6.** $\left(\frac{14}{19}, -\frac{10}{19}, -\frac{32}{57}\right)$

7. $f_x = 2xy + y^2$

$f_y = x^2 + 2xy$

8. $f_x = 50y^2(x + y)$

$f_y = 50y(x + y)(x + 2y)$

9. $f_x = 3x^2 - 4xy + yz$

$f_y = -2x^2 - xz$

$f_z = xy$

10. $f_x = yz + z^2$

$f_y = xz + z^2$

$f_z = xy + 2xz + 2yz$

Section 7.7

1. 5.0225 **2.** 0.0189 **3.** $S_a = 2a - 4 - 4b$, $S_b = 12b - 8 - 4a$ **4.** $S_a = 8a - 6 - 2b$, $S_b = 18b - 4 - 2a$

5. 15 **6.** 42 **7.** $\frac{25}{12}$ **8.** 14 **9.** 31 **10.** 95

Section 7.8

1. 1 **2.** 42 **3.** $\frac{19}{4}$ **4.** $\frac{1}{7}$ **5.** ln 5 **6.** $\frac{e}{2}(e^4 - 1) \approx 72.8474$

7.

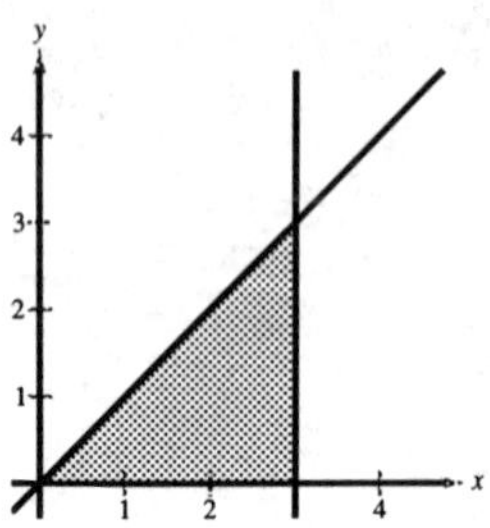

8.

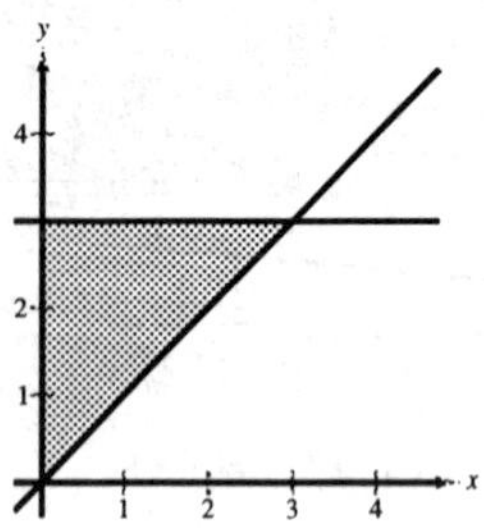

9.

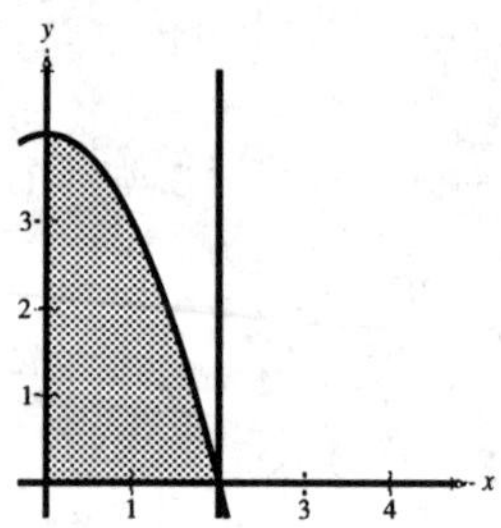

10.

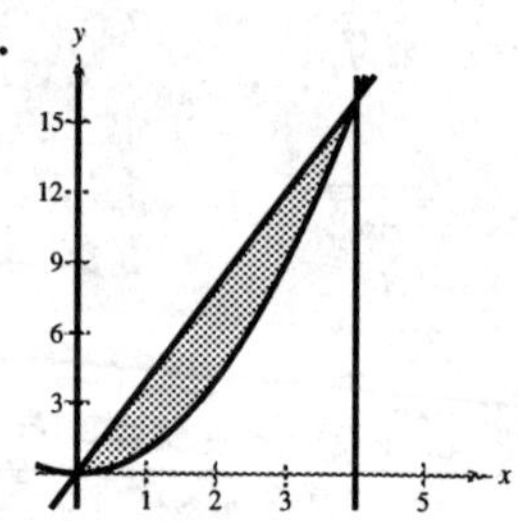

Section 7.9

1. 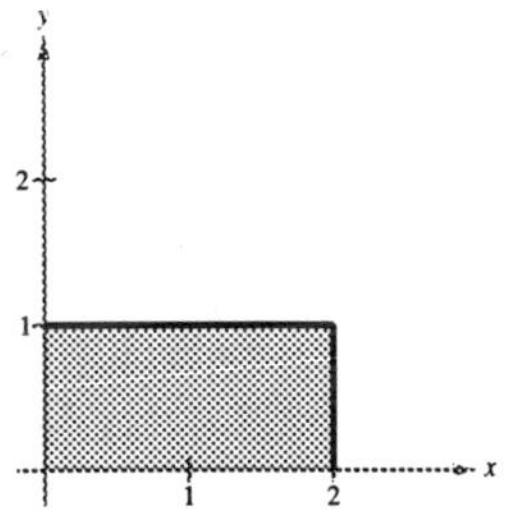

2. 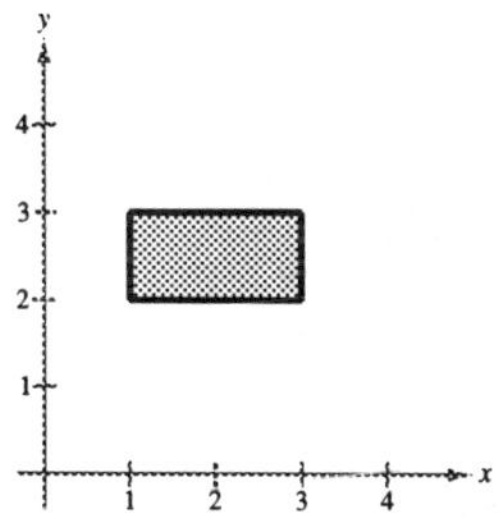

3. 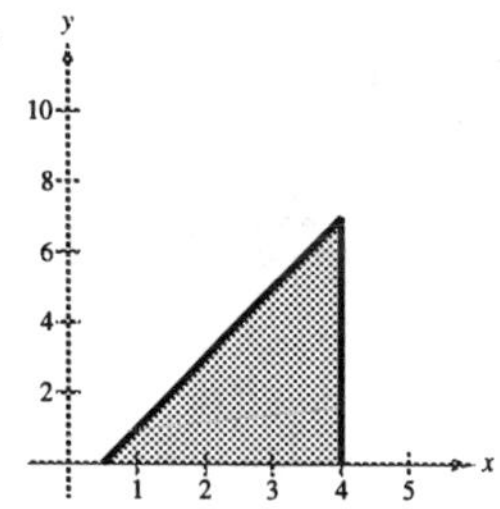

4. 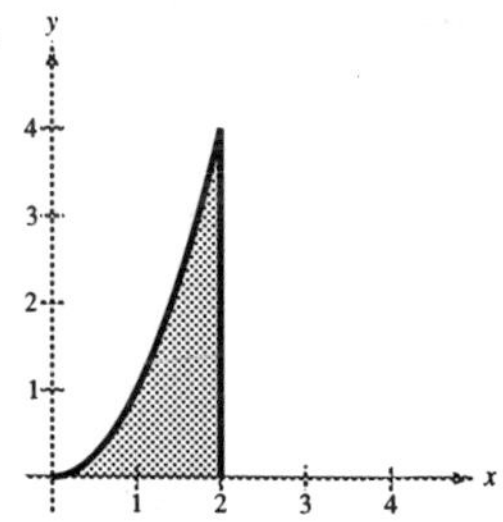

5. 1 **6.** 6 **7.** $\frac{1}{3}$ **8.** $\frac{40}{3}$ **9.** $\frac{28}{3}$ **10.** $\frac{7}{6}$

CHAPTER 8

Section 8.1

1. 35 cm^2 **2.** 12 in.^2 **3.** $c = 13$ **4.** $b = 4$

5. $b = 15$ **6.** $a = 6$ **7.** Equilateral triangle

8. Isosceles triangle **9.** Right triangle **10.** Isosceles triangle and right triangle

Section 8.2

1. $\frac{3\pi}{4}$ **2.** $-\frac{7\pi}{6}$ **3.** $-\frac{5\pi}{9}$ **4.** 3π **5.** $x = 0, x = 1$

6. $x = -\frac{1}{2}, x = 1$ **7.** $x = 1$ **8.** $x = 2, x = 3$ **9.** $t = 10$ **10.** $t = \frac{445}{8}$

Section 8.3

1. 14 **2.** 10 **3.** 0 **4.** 1 **5.** $-\frac{1}{2}$

6. $\frac{1}{2}$ **7.** 0.9659 **8.** -0.9962 **9.** 0.9744 **10.** -0.6494

Section 8.4

1. $f'(x) = 9x^2 - 4x + 4$ **2.** $g'(x) = 12x^2(x^3 + 4)^3$ **3.** $f'(x) = 3x^2 + 2x + 1$

4. $g'(x) = \dfrac{2(5 - x^2)}{(x^2 + 5)^2}$ **5.** Relative minimum: $(-2, -3)$ **6.** Relative maximum: $\left(-2, \frac{22}{3}\right)$

Relative minimum: $\left(2, -\frac{10}{3}\right)$

7. $x = \frac{\pi}{3}, x = \frac{2\pi}{3}$ **8.** $x = \frac{2\pi}{3}, x = \frac{4\pi}{3}$ **9.** $x = \pi$ **10.** No solution

Section 8.5

1. $-\frac{\sqrt{2}}{2}$ **2.** $-\frac{\sqrt{3}}{2}$ **3.** $-\frac{\sqrt{3}}{3}$ **4.** -1 **5.** $\tan x$

6. $\sin^2 x$ **7.** 1 **8.** $\csc x$ **9.** $\frac{88}{3}$ **10.** 4

Section 8.6

1. ∞ **2.** 0 **3.** $\frac{2}{3}$ **4.** ∞ **5.** $-2x \sin x^2$

6. $5 \cos(5x - 1)$ **7.** $4 \sec(4x) \tan(4x)$ **8.** $2x \sec^2(x - 2)$

9. $-4 \sin(2x + 3)$ **10.** $2 \sec^2 x \tan x$

CHAPTER 9

Section 9.1

1. 1
2. 1
3. 2
4. 2
5. $\frac{1}{2}$
6. 1
7. 37.50%
8. $81\frac{9}{11}\%$
9. $54\frac{1}{6}\%$
10. 43.75%

Section 9.2

1. Yes
2. No
3. No
4. Yes
5. 1
6. $\frac{1}{2}$
7. 1
8. $\frac{1}{4}$
9. 1
10. 1

Section 9.3

1. 5
2. 8
3. $3 \ln |2|$
4. $9 \ln |2|$
5. $\frac{4}{3}$
6. $\frac{4}{3}$
7. $\frac{3}{4}$
8. $\frac{2}{9}$
9. (a) $\frac{1}{4}$ (b) $\frac{1}{2}$
10. (a) $\frac{1}{2}$ (b) $\frac{11}{16}$

CHAPTER 10

Section 10.1

1. 0
2. 0
3. 2
4. ∞
5. 0
6. 0
7. $\dfrac{n-2}{n}$
8. $\dfrac{n-3}{n-4}$
9. $\dfrac{3n^2+1}{n^3}$
10. $\dfrac{2n+1}{(n-1)(n+2)}$

Section 10.2

1. $\frac{77}{60}$
2. $\frac{73}{24}$
3. $\frac{31}{16}$
4. $\frac{40}{9}$
5. $\frac{21}{8}$
6. $\frac{31}{32}$
7. $\frac{3}{4}$
8. 0
9. 1
10. $\frac{1}{2}$

Section 10.3

1. $\dfrac{1}{n+1}$
2. $n + 1$
3. $\dfrac{n}{n+1}$
4. $\dfrac{n+1}{n^2}$
5. 1
6. 5
7. 1
8. $\frac{1}{3}$
9. Geometric series
10. Not a geometric series

Section 10.4

1. $f(g(x)) = (x-1)^2$
   $g(f(x)) = x^2 - 1$
2. $f(g(x)) = 6x + 3$
   $g(f(x)) = 6x + 1$
3. $f(g(x)) = \sqrt{x^2+4}$
   $g(f(x)) = x + 4,\ x \ge -4$
4. $f(g(x)) = e^{x^2}$
   $g(f(x)) = e^{2x}$
5. $f'(x) = 5e^x$
   $f''(x) = 5e^x$
   $f'''(x) = 5e^x$
   $f^{(4)}(x) = 5e^x$
6. $f'(x) = \dfrac{1}{x}$
   $f''(x) = -\dfrac{1}{x^2}$
   $f'''(x) = \dfrac{2}{x^3}$
   $f^{(4)}(x) = -\dfrac{6}{x^4}$
7. $f'(x) = 6e^{2x}$
   $f''(x) = 12e^{2x}$
   $f'''(x) = 24e^{2x}$
   $f^{(4)}(x) = 48e^{2x}$
8. $f'(x) = \dfrac{1}{x}$
   $f''(x) = -\dfrac{1}{x^2}$
   $f'''(x) = \dfrac{2}{x^3}$
   $f^{(4)}(x) = -\dfrac{6}{x^4}$
9. $\dfrac{n+1}{3}$
10. $\dfrac{n+3}{n+1}$

Section 10.5

1. $\sum_{n=0}^{\infty} \frac{3^n x^n}{n!}$ 2. $\sum_{n=0}^{\infty} \frac{(-1)^n 3^n x^n}{n!}$ 3. $4\sum_{n=0}^{\infty} (-1)^n (x-1)^n$ 4. $\ln 5 + \sum_{n=1}^{\infty} \frac{(-1)^{n-1}(x-1)^n}{n}$

5. $1 + \frac{x}{4} - \frac{3x^2}{4^2 2!} + \frac{3 \cdot 7x^3}{4^3 3!} - \frac{3 \cdot 7 \cdot 11x^4}{4^4 4!} + \cdots$ 6. $1 + \frac{x}{2} - \frac{x^2}{2^2 2!} + \frac{1 \cdot 3x^3}{2^3 3!} - \frac{1 \cdot 3 \cdot 5x^4}{2^4 4!} + \cdots$

7. $\frac{47}{60}$ 8. $\frac{311}{576}$ 9. $\frac{5}{12}$ 10. $\frac{77}{192}$

Section 10.6

1. $f(2.4) = -0.04$, $f'(2.4) = 2.8$ 2. $f(-0.6) = 0.064$, $f'(-0.6) = 3.48$ 3. $f(0.35) = 0.01$, $f'(0.35) = 4.03$ 4. $f(1.4) = 0.30$, $f'(1.4) = 12.88$

5. $4.9 \le x \le 5.1$ 6. $0.798 \le x \le 0.802$ 7. $5.97 \le x \le 6.03$ 8. $-3.505 \le x \le -3.495$

9. $\left(\frac{\sqrt{13}+3}{2}, \sqrt{13}+2\right), \left(\frac{3-\sqrt{13}}{2}, 2-\sqrt{13}\right)$ 10. $\left(\frac{1-\sqrt{5}}{2}, \frac{3-\sqrt{5}}{2}\right), \left(\frac{1+\sqrt{5}}{2}, \frac{3+\sqrt{5}}{2}\right)$

APPENDIX D

Section D.1

1. $y' = 6x + 2$, $y'' = 6$ 2. $y' = -6x^2 - 8$, $y'' = -12x$ 3. $y' = -6e^{2x}$, $y'' = -12e^{2x}$ 4. $y' = -6xe^{x^2}$, $y'' = -6e^{x^2}(2x^2 + 1)$

5. $\frac{1-x}{y}$ 6. $\frac{2}{3y^2 + 4}$ 7. $-\frac{y}{2x}$ 8. $-\frac{y}{x}$

9. $k = 2\ln 3 - \ln\frac{17}{2} \approx 0.0572$ 10. $k = \ln 10 - \frac{\ln 41}{2} \approx 0.4458$

Section D.2

1. $\frac{2}{5}x^{5/2} + C$ 2. $\frac{1}{4}t^4 - \frac{3}{4}t^{4/3} + C$ 3. $2\ln|x-5| + C$ 4. $\frac{1}{4}\ln|2y^2 + 1| + C$

5. $\frac{1}{2}e^{2y} + C$ 6. $-\frac{1}{2}e^{1-x^2} + C$ 7. $C = -10$ 8. $C = 5$

9. $k = \frac{\ln 5}{2} \approx 0.8047$ 10. $k = -2\ln 3 - \ln 2 \approx -2.8904$

Section D.3

1. $e^x + 1$ 2. $e^{3x} + 1$ 3. $\frac{1}{x^3}$ 4. x^2e^x

5. $2e^x - e^{-x} + C$ 6. $e^{3x}\left(\frac{x}{3} - \frac{1}{9}\right) + \frac{1}{2}e^{2x} + C$ 7. $\frac{1}{2}\ln|2x + 5| + C$

8. $\frac{1}{2}\ln|x^2 + 2x + 3| + C$ 9. $\frac{1}{12}(4x - 3)^3 + C$ 10. $\frac{1}{6}(x^2 - 1)^3 + C$

Section D.4

1. $y = \frac{3}{2}x^2 + C$ 2. $y^2 = 3x + C$ 3. $y = Ce^{x^2}$

4. $y^4 = \frac{1}{2}(x - 4)^2 + C$ 5. $y = 2 + Ce^{-2x}$ 6. $y = xe^{-2x} + Ce^{-2x}$

7. $y = 1 + Ce^{-x^2/2}$ 8. $y = \frac{1}{4}x^2 + Cx^{-2}$ 9. $\frac{dy}{dx} = Cx^2$ 10. $\frac{dx}{dt} = C(x - t)$

Answers to the

POST-GRADUATION EXAM QUESTIONS

CHAPTER 1

1. (d) **2.** (d) **3.** (c) **4.** (a)
5. (e) **6.** (c) **7.** (e) **8.** (b)

CHAPTER 2

1. (c) **2.** (b) **3.** (e)
4. (a) **5.** (c) **6.** (a)

CHAPTER 3

1. (c) **2.** (a) **3.** (c)
4. (d) **5.** (b) **6.** (a)

CHAPTER 4

1. (b) **2.** (c) **3.** (a) **4.** (c)
5. (d) **6.** (a) **7.** (b)

CHAPTER 5

1. (d) **2.** (b) **3.** (c) **4.** (b)
5. (a) **6.** (b) **7.** (d) **8.** (d)

CHAPTER 6

1. (c) **2.** (a) **3.** (c) **4.** (d)
5. (d) **6.** (b) **7.** (a) **8.** (c)

CHAPTER 7

1. (b) **2.** (b) **3.** (c) **4.** (d)
5. (a) **6.** (c) **7.** (b)

CHAPTER 8

1. (b) **2.** (d) **3.** (d) **4.** (c)

CHAPTER 9

1. (c) **2.** (c) **3.** (e) **4.** (a) **5.** (c)

CHAPTER 10

1. (b) **2.** (b) **3.** (b)
4. (b) **5.** (c) **6.** (d)